A Tutorial Guide to Mechanical Desktop 5/ Power Pack

A Tutorial Guide to Mechanical Desktop 5/ Power Pack

An Introduction to Modeling for Engineering Design

Shawna D. Lockhart
Montana State University

Cindy M. Johnson

Prentice Hall

Prentice Hall
Upper Saddle River, NJ 07458

Library of Congress Cataloging-in-Publication Data

Lockhart, Shawna D.,
A tutorial guide to mechanichal desktop 5/ Power Pack
Shawna D. Lockhart, Cindy M. Johnson.
p. cm.
Includes bibliographical references (p.).
ISBN 0-13-029149-8
1. Engineering design—Graphic methods. 2. Engineering graphics.
I. Johnson, Cindy M. II. Title.
TA337.L63 2001
620'.0042—dc21 2001035846
CIP

Vice-President and Editorial Director, ECS: Marcia Horton
Executive Editor: Eric Svendsen
Assistant Vice President of Production and Manufacturing: David W. Riccardi
Marketing Manager: Holly Stark
Marketing Assistant: Karen Moon
Production Editor: Rose Kernan
Executive Managing Editor: Vince O'Brien
Managing Editor: David A. George
Art Director: Carole Anson
Artists: Marita Froimson, Marla Goodman, and Mike Ohar
Page Layout and Composition: Preparé, Inc.
Manufacturing Buyer: Pat Brown and Lisa McDowell

Prentice Hall

Printed in the United States of America

10 9 8 7 6 5 4 3 2 1

ISBN 0-13-029149-8

Pearson Education LTD. *London*
Pearson Education Australia PTY, Limited, *Sidney*
Pearson Education *Singapore*, Pte. Ltd.
Pearson Education North Asia Ltd., *Hong Kong*
Pearson Education Canada, Ltd., *Toronto*
Pearson Educación de Mexico, S.A. de C.V.
Pearson Education—Japan, *Tokyo*
Pearson Education Malaysia, Pte. Ltd.

Contents

PREFACE xiii

GETTING STARTED WITH MECHANICAL DESKTOP 1

Objectives 1
Introduction to the Tutorials 1
Basic Mouse Techniques 1
Context-Sensitive Menus 1
Recognizing Typographical Conventions 2
Configuring Windows for Mechanical Desktop 3
Windows 98/2000 and Windows NT 3
Screen Resolution 3
Customizing Your Shortcuts 4
Creating Multiple Shortcuts 4
Setting the Working Directory 6
Installing Data Files for the Tutorials 7
Mechanical Desktop Configuration for Tutorials 7
User Profiles 9
Changing the Configuration 10
Accepting Configuration Changes 14
Dialog Box Settings 15
Microsoft Windows and Mechanical Desktop 15
Basic Elements of the Mechanical Desktop Screen 16
Minimizing and Restoring Mechanical Desktop for Windows 17
Interacting with Mechanical Desktop 17
Entering Points 17
Entering Command Options 18
Entering Mechanical Desktop Commands 19
Backing Up and Backing Out of Commands 19
Using the Toolbar Icons 20
Using Pull-Down Menus 21
Working with Dialog Boxes 22
Repeating Commands 25
Selecting Objects 25
Removing Objects from a Selection Set 26
Mechanical Desktop Online Help 26
Context-Sensitive Transparent Help 28
Working with Documents 30
Exiting Mechanical Desktop 30

CHAPTER 1 THE MECHANICAL DESKTOP WORKSPACE 32

Objectives 32
Overview of the Tutorials 32
Introduction to Parametric Modeling 32
Launch Mechanical Desktop 33
The Mechanical Desktop Workspace 35
Toolbars 36
The Desktop Browser 36
The Browser and Parametric Modes 36
The Model Tab 36
The Drawing Tab 39
File Management and Save As 41
Using Save As 41
Backing Up Your Files 42
Making Changes in a Parametric Model/Assembly 42
Controlling Your View 44
Realtime Zoom 44
Pan Realtime 45
Preset Views 45
The 3D Orbiter 46
Toolbar Flyouts 47
Setting the Rotation Center 47
3D Continuous Orbit 48
3D Swivel 49
3D Adjust Distance 49
Named Views 49
Clipping 51
3D Adjust Clipping Planes 51
Shading Modes 52
Toggle Shaded/Wireframe 52
Hidden 53
Flat and Gouraud Shaded 53
Printing a Rendered View 54
Command Summary 57
Exercises 58

CHAPTER 2 SIMPLE PART CREATION 59

Objectives 59
Introduction to Part Modeling 59
Starting a New Part File 60
First Decisions for Solid Models 62
Selecting a Base Feature 62
Using 2D Sketching Commands 63
Creating a Profile 65
Over- and Under-constrained Sketches 65
Constraining the Profile: Geometric First 65
Deleting Constraints 66
Adding Constraints 66
Specifying a Fixed Point 67

Re-Solving the Sketch 67
Good Practice for Constraints and Dimensions 67
Adding Dimensions 68
Dimension Placement 69
Creating a Feature from a Sketch 69
Using Extrude 70

Setting Properties in the Browser Window 72
Selecting a New Sketch Plane 73
World versus User Coordinate Systems 73
Creating a Profile for the Second Feature 77
Adding a Colinear Constraint 78
Using List 79
Extruding a Feature with Join 80
Renaming Features 81
Creating the Second Leg Feature 81
Projecting Sketch Geometry 82
Using Profile 83
Adding Placed Features 86
Holes 86
Fillet 88
Modifying Features 89
Saving As a New Part 89
Modifying Dimensions 89
Command Summary 91
Exercises 92

CHAPTER 3 MAKING A DRAWING 96

Objectives 96
Introduction 96
Starting 96
Paper Space 97
Printer Area 98
Adding Multiple Views 98
Moving Dimensions and Drawing Views 101
Improving Drawing Appearance 101
Setting Linetype Scale 102
Layers 103
Changing Layer Color 103
Changing the Text Style 104
Dimension Styles 105
Creating a Named Dimension Style 105
Setting the Dimension Lines 107
Extension Lines 107
Arrow Size and Style 107
Centermark Size and Style 107
Text 108
Fit 109
The Dimension Scaling Factor 110
Primary Units for Dimensions 111

Measurement Scale 111
Zero Suppression for Linear Dimension 111
Angular Dimensions 111
Zero Suppression for Angular Dimensions 112
Alternate Units 112
Tolerances 113
Setting the Current Dimension Style 114
Adjusting Views 115
Pictorial Drawing Views 116
Editing Views 117
Printing Your Drawing 117
Plot Style Table (Pen Assignments) 119
Bidirectional Associativity 120
Changing a Value Using Power Dimension Editing 120
Creating Layout2 122
Adding Text to the Title Block 128
Creating a Template File 129
Command Summary 131
Exercises 132

CHAPTER 4 MORE PART MODELING 134

Objectives 134
Introduction 134
Useful Commands for the Sketcher 134
Using Grid 134
Using Snap 135
Grid and Snap Toggles 135
Using Line 136
Using Relative Coordinates 136
Using Offset 137
Using Undo and Redo 138
Using Extend 138
Using Trim 139
Using Object Snaps 140
Using Mirror 141
Using Erase 142
Using Hot Grips 142
Activating an Object's Grips 143
Using Crossing to Select 143
Making the Profile 145
Extruding the Base Feature 149
Work Features 150
Creating a Parametric Work Plane 150
Sketching the Front View 152
Construction Lines 153
Using Dimension Command Options 154
Intersecting the Feature 155
Removing Material with an Extruded Cut 156
Using Fillet to Trim and Extend 159

Sketching the Center Block with Rectangle 161
Adding the Holes 163
Adding a Work Plane for Assembly 165
Command Summary 166
Exercises 167

CHAPTER 5 MODIFYING PARTS 170

Objectives 170
Introduction 170
Parent–Child Relationships 170
The Base Feature 171
The Model Tree 171
Investigating a Model 171
Renaming features 171
Replaying the Model 172
Suppressing Features 173
Redefining the Model: Editing a Feature 173
Updating a Part 174
Redefining the Model: Editing a Profile 175
Using Undo with Parts 177
Reordering Features 178
Adding a New Feature 179
Creating a Pattern Feature 181
Using Feature Suppression 183
Preparing for Assembly 187
Command Summary 188
Exercises 188

CHAPTER 6 CREATING REVOLVED PARTS AND MORE PART EDITING 190

Objectives 190
Introduction 190
Setting up the Modeling Environment 190
Sketching the Stopper Part 192
Revolving the Profile 194
Creating the Pin Part 196
Controlling the View of Curved Surfaces 198
Editing a Sketch to Append New Geometry 200
Using Arc 200
Object Snap Overrides 203
Starting a New Part from an Existing One 205
Editing a Feature 206
Adding a Work Axis 207
Creating the Bolt 211
Using Polygon 212
Using Dimension Options: Angle, Align, Placement 213
Changing the Default Extrusion Direction 214
Adding the Chamfer 215
Command Summary 216
Exercises 216

CHAPTER 7 CREATING AN ASSEMBLY 219

Objectives 219
Introduction 219
Checking for Data Files 219
Launching Mechanical Desktop 5 Power Pack 219
Starting a New Assembly 220
Assembling the Jack Parts 221
External and Local Parts 221
Organizing Your Parts 221
The Assembly Catalog 221
Adding an External Part 223
Assembly Constraints 226
Mate 226
FlusFlush 226
Insert 226
Angle 227
Using the 3D Manipulator 231
Insert Constraint: Insert and Mate 232
Degrees of Freedom 234
Mating the ThickLeg 234
Constraining the ThinLeg and SupportGuide 239
Using Move to Reposition a Part 241
Constraining the ThinLeg to the JackTop 242
Editing Constraints 245
Constraining the Four Pins 246
Inserting the Bolt 248
Adding Stock Parts (Power Pack) 249
Viewing the Assembly 250
Making a New Scene 250
Explode Factor 250
Using Tweak 250
Activating a Scene 254
Analyzing the Model 254
Checking for Interference 254
Activating and Editing a Part 255
Checking for Interference with the New JackTop 256
Mass Properties 257
Localizing the Assembly's Parts 258
Command Summary 260
Exercises 260

CHAPTER 8 WORKING DRAWINGS FROM MODELS & ASSEMBLIES 263

Objectives 263
Introduction 263
Starting 263
Creating a Layout for the Assembly Drawing 264
Setting Up the Sheet Size and Adding a Title Block 264
Exploded Assembly Drawing 269
Using Scenes in Drawing Views 269

Adding a Parts List 273
Creating the BOM Database 273
Creating the Parts List 274
Balloon Numbers 276
Adding a Shaded Image 279
Printing the Assembly 280
Dimensioned Part Drawing 281
Different Drawing Views 282
Good Dimensioning Technique 283
Adding a Reference Dimension 289
Notes 290
Annotations with Leaders 291
Associating an Annotation 294
Moving a Drawing View 294
General Notes 294
Specifying Tolerance 296
Adding a Section View 299
Command Summary 301
Exercises 302

CHAPTER 9 "SMARTER" PARTS AND BETTER ASSEMBLIES 305

Objectives 305
Introduction 305
Using a Part Template 305
Starting a New Part from a Template 305
Work Planes as Datum Planes 306
Making a Customized Template 308
Sweeping the Side Rail 309
Sketching the Path 309
Making Dimensions Legible 312
Creating the Profile 316
Creating the Swept Feature 318
Using Equations in Dimensions 320
Creating the Cart Assembly 322
Adding a StartPart as the First Part 323
Mating CartRail to StartPart 324
Adding Another Instance of the CartRail Part 326
Mating the Second Rail 327
Modeling a Local Part within the Assembly 328
Assembling the CrossBrace Part 331
Creating a New Instance of a Local Part 335
Creating the Tray Part 336
Assembling the Tray 340
Shelled Features 341
Exporting a Local Part 343
Adding the Caster Subassembly 343
Mating the Bolts 344
Adding the Holes 347
Command Summary 350
Exercises 351

CHAPTER 10 GETTING RESULTS FROM YOUR INTELLIGENT ASSEMBLY 358

Objectives 358
Introduction 358
Table-Driven Parts 358
Creating Active Part Design Variables 359
Creating a Table-Driven Part 361
Changing the Sketch to Use the Design Variables 363
Displaying a Table-Driven Part Configuration 365
Defining Global Design Variables 365
Using Global Variables in a Sketch 367
Using Global Variables in Assembly Constraints 369
Creating a Table-Driven Assembly 371
Editing a Table of Variables 373
Mass Properties 374
Conclusion 376
Command Summary 376
Exercises 377

APPENDIX A COMMAND REFERENCE 382

APPENDIX B GLOSSARY 389

INDEX 395

Preface

This tutorial guide provides a step-by-step introduction to the 3D modeling process and Mechanical Desktop 5, a powerful parametric 3D modeler for desktop computers. Novice modelers, as well as those who have experience with AutoCAD or another 2D CAD program, will experience the power of Mechanical Desktop as they complete parts, assemblies, and drawings for two complete projects. Early tutorials get students started quickly with basic part modeling and drawing creation, then reinforce these basics as students progress with more advanced modeling activities that introduce part modification, assembly modeling, working drawings, intelligent models using design variables, and using the model for visualization and analysis. Optional sections also introduce the use of Mechanical Desktop 5's Power Pack features for standard parts.

Written for independent student use in the lab portion of a beginning engineering graphics course or in an introductory 3D modeling course, this guide's learn-by-doing approach will be useful to anyone who wants to get started using Mechanical Desktop 5 in the context of the modeling tasks encountered in engineering design.

Proven Tutorial Approach

Using a tutorial approach developed and refined for teaching AutoCAD, this manual guides students through the creation and assembly of two complete projects as they follow step-by-step instructions to learn commands and techniques. As students progress, individual steps for common actions are no longer provided, so students can apply what they have learned by completing sequences on their own. Each tutorial activity encourages students to relate the commands and techniques they are learning to the modeling process. Information about Mechanical Desktop is presented in a need-to-know fashion that makes it easy to remember. Tips and shortcuts are included where appropriate to help students become efficient and proficient Mechanical Desktop users. Help Reference boxes identify and locate topics in Mechanical Desktop's on-line Help where students can find more information for future projects. A Command Summary makes it easy to review and use the tutorials as a reference. Each tutorial concludes with exercises that allow students the opportunity to practice their skills with interesting and realistic models.

Key Features

- Step-by-step tutorials written for the novice user
- Introductory chapter on configuring Mechanical Desktop for use with these tutorials
- Procedures follow and reinforce good modeling practice for engineering design
- Tips that anticipate common problems and offer suggestions for efficient practice
- Help References that identify and locate key topics in Mechanical Desktop Help
- Reference boxes refer students to discussion of graphics concepts in optional companion text, *Engineering Design Communication*.
- Challenging end-of-tutorial exercises keyed to the skills presented so far
- Command Summaries at the end of each tutorial and as a reference in the back of the book
- Glossary of modeling and Mechanical Desktop terminology

Overview of the Tutorials

Tutorials 1 through 3 introduce students to the basic components of parametric modeling: parts, assemblies, and drawings. Students become familiar with the 3D environment, tools for visualization, the basic sketching and modeling process, and making and printing a drawing as they complete the first part of the jack assembly.

Tutorials 4 through 8 broaden this introduction with more techniques for modeling parts, assembling parts in an assembly, and creating drawings to document the assembly. Students learn more about sketching techniques and drawing aids, create revolved parts, pattern features, and placed features, and learn how to investigate and modify models created by someone else. By Tutorial 7, all of the parts for the jack have been modeled so students can use them to as they learn to apply assembly constraints. After completing the assembly, student prepare an exploded view, examine mass properties of the assembly, and learn to document their assembly by preparing working drawings of the jack assembly and one of the parts. Students learn to create a parts lists and section view, how to add a raster image to their drawing, and good dimensioning practice.

In Tutorials 9 and 10, students create a second assembly and use techniques for making intelligent parts and assemblies, and for getting information from the model. As they model the stock cart, students learn to make swept and shelled features, work with subassemblies, use equations in dimensions, and use design variables to create configurations of table-driven parts.

Requirements

The instructions in this guide assume the student will be using Mechanical Desktop 5 for Windows® 98/2000 or NT. Mechanical Desktop 5 with Power Pack is used in Tutorials 7 and 8, but these sections are optional and the tutorials can be completed without Power Pack. Any configuration instructions are written for students using Mechanical Desktop in a computer lab setting as well as for those working at home or work.

Data files for the tutorials—and related information about the text—may be found at www.prenhall.com/lockhart.

For students of introductory engineering graphics, reference boxes throughout this guide point to more information about modeling and design practice in *Engineering Design Communication*, a companion text that provides a broad introduction to graphics for engineering design, also available from Prentice Hall. These references are designed to link the modeling exercises in this manual to the graphics concepts covered in the text, but are not required for completion of the tutorials.

Prentice Hall offers discounts when you purchase this book with several other Prentice Hall textbooks. Discounts range from 10–20% off the price of two books separately. To request more specific pricing information, get isbn's for ordering bundles, and learn more about Prentice Hall's offerings in graphics or CAD, either contact your Prentice Hall sales rep, go to http://www.prenhall.com/ and choose AutoCad/ Engineering Graphics/Drafting, or email engineering@prenhall.com. For the name and number of your sales representative, please contact Prentice Hall Faculty Services at 1-800-526-0485.

Acknowledgements

We would like to thank Mark Perkins for working with us to fine-tune and illustrate this guide and for the many excellent exercises he prepared and modeled. We would also like to acknowledge the many hundreds of engineering students at Montana State University, whose enthusiasm for 3D modeling inspired us and whose desire to "get to the modeling" helped us balance the explanations with the hands-on activities in this guide.

We would like to thank the folks at Autodesk, especially Denis Cadu, for assistance with the software. In addition we would like to thank the individuals involved in getting the book in print: Eric Svendsen, Rose Kernan, and many others.

We welcome your comments and encourage you to contact us with your suggestions for future releases of this guide.

Shawna D. Lockhart
lockhart@me.montana.edu
Cindy M. Johnson
cjohnson@seacoast.com

A Tutorial Guide to Mechanical Desktop 5/ Power Pack

Getting Started with Mechanical Desktop

Introduction to the Tutorials

The ten tutorials in this manual will teach you to use *Mechanical Desktop Release 5 Power Pack* through a series of step-by-step exercises. (If you do not have Power Pack, you will still be able to complete the tutorials, except for brief portions of Tutorials 8 and 9.) The tutorials in this manual are based on the assumption that you are using Mechanical Desktop's default settings. If your system has been customized for other uses, some of the settings may not match those used in the step-by-step instruction in the tutorials. Please check your system against the configuration in this chapter so that you can work through the tutorials as instructed.

Basic Mouse Techniques

The tutorials in this manual are based on the assumption that you will be using a two-button mouse or pointing device to work with Mechanical Desktop Release 5. The following terms will be used to streamline the instructions for use of the mouse.

Term	Meaning
Pick or Click	To press quickly and then release the left mouse button
Right-click	To quickly press and release the right mouse button
Double-click	To click the left mouse button twice in rapid succession
Drag	To press and hold down the left mouse button while you move the mouse
Choose	To select an item from a pull-down list of options (generally requires clicking the down arrow on the list, scrolling as necessary to find the item in the list, then picking the item)
Select	To pick a radio button or check box so that the button is dark or a check mark appears
Point	To move the mouse until the mouse pointer on the screen is positioned above the item you want

Context-Sensitive Menus

Many Windows applications, including Mechanical Desktop, offer context menus that appear when the right mouse button is clicked. The commands and options on these menus change depending on the *context* in which the button was clicked. That is, they present the most common tasks you might want to perform at that moment. You will use context menus throughout this guide. When you are instructed to "right click in the graphics area," for example, the next instruction will be the item to pick from the menu that appears. It is important to right-click in the area directed; otherwise you may see a different context menu.

Objectives

This chapter describes how to prepare your computer system and *Mechanical Desktop Release 5* for use with the tutorials in this manual. It also presents an introduction to the Mechanical Desktop environment and its online Help facility.

As you read the chapter, you will

- Understand the instructions for using a mouse to complete these tutorials.
- Recognize the typographical conventions used in this book.
- Create separate Mechanical Desktop shortcuts to use with these tutorials.
- Create and prepare data file and working directories.
- Configure your copy of Mechanical Desktop for use with these tutorials.
- Tour the Mechanical Desktop interface.
- Open and save a drawing.
- Practice using the mouse to pick commands, menu options, and objects.
- Explore Mechanical Desktop's online help facility.

Warning

If you are using Mechanical Desktop on a network at your school, ask your professor or system administrator about how the software is configured. You should not try to reconfigure Mechanical Desktop unless instructed to do so by network personnel.

To complete these tutorials, your system must be running Microsoft Windows 98, 2000, or Windows NT 4.0.

Tip

Your mouse may have more than two buttons. To click (or pick) in Windows and in Mechanical Desktop, use the left mouse button. This button is also referred to as the pick button.

Recognizing Typographical Conventions

When you work with Mechanical Desktop, you will use your keyboard and your pointing device to input information. As you read this manual, you will encounter special type styles that will help you determine the information you must input.

Special symbols illustrate certain computer keys. For example, the Enter or Return key is represented by the symbol [↵].

Tip

The right button on your mouse can perform the same function as the [↵] key. Instead of pressing [↵], right-click and pick Enter from the context menu that appears.

The manual also employs special typefaces when it presents instructions for performing computer operations. Instructions are set off from the main text to indicate a series of actions or Mechanical Desktop command prompts. In general, the item you are to pick or type appears after the colon in the instruction line. In some cases, the instruction line before the colon will mirror the command line prompt:

Command: **LINE**

instructs you to type the Line command.

From point: **pick point A**

instructs you to select point A on your screen by clicking it with the mouse. The words Command and From point represent the Mechanical Desktop prompts that you would see in the command line area on your screen. Items that you are to type at the command prompt appear in UPPERCASE letters so that you can easily identify what to type. You can use uppercase, lowercase, or a combination to type the actual items.

In other cases, instruction words, such as pick, type, select, and choose, appear before the colon. For example, type instructs you to strike or press several keys in sequence. For the following instruction,

Type: **4,4**

you would type 4,4 in that order on your keyboard. Remember to type exactly what you see, including spaces, if any.

Pick tells you to pick an object or an icon or to choose commands from the menu. For example,

Pick: **Line icon**

instructs you to use your pointing device to select the Line icon from the toolbar on your screen. To select an icon, move the pointer until the cursor is over the command and press the pick button.

General directions that you should follow appear stated in bold type. They indicate that you are to follow the steps that are outlined in that instruction on your own before continuing. For example,

Change the color of the feature to blue

directs you to complete the step, using the techniques you have already learned, before you continue.

Reference

Engineering Design Communication

Parametric Modeling Modes, p. 272–274

Reference

Mechanical Desktop On-line Help

The Zoom command has many options for use, which you will learn as you proceed through the tutorials. Use the Command Reference in AutoCAD Help for a complete explanation.

New terms are in italics in the text when they are introduced and are defined in the Glossary at the end of the manual. Cross references to Mechanical Desktop's online Help and to relevant sections of *Engineering Design Communication*, a companion textbook, also appear periodically, as shown in the samples at left, to refer you to additional sources of information about the tasks you are undertaking. These cross references are optional and are not required to complete the tutorials.

Configuring Windows for Mechanical Desktop

Mechanical Desktop Release 5 Power Pack takes advantage of common Windows file operations and allows you to run Mechanical Desktop while running other applications.

Windows 98/2000 and Windows NT

The figures in the tutorials were captured in Windows 98. The Mechanical Desktop screens in Windows NT 4.0 are similar to the screens that you will see in this manual. File operations and some screens may be slightly different running under Windows NT, but the steps are described in sufficient detail that you should be able to follow them on your own.

Screen Resolution

The screens in this book were captured at a screen resolution of 1024 × 768 pixels. If you have a monitor that is capable of higher resolution and want to use a higher setting, your screens will look slightly different from the ones captured for this book, but it should not affect the overall operation of the commands as described in the tutorials. If your screen is set to 800 × 600, you should change it to 1024 × 768 if you have that setting available.

You can change your screen resolution in the Windows Control Panel. From Windows 98/2000,

Pick: **Start, Settings, Control Panel** (from the bottom left of your screen)

Double-click: **Display** (from the Control Panel)

Pick: **Settings** (in the dialog box that appears)

The Display Properties dialog box appears on your screen, as shown in Figure GS.1. You can change the color selection and the pixel selection with the scroll bars. Set your display to True Color (if available) or at least 256 colors and 1024 × 768 pixels (or higher), if available. If you cannot use these settings, use the next closest setting available. Windows

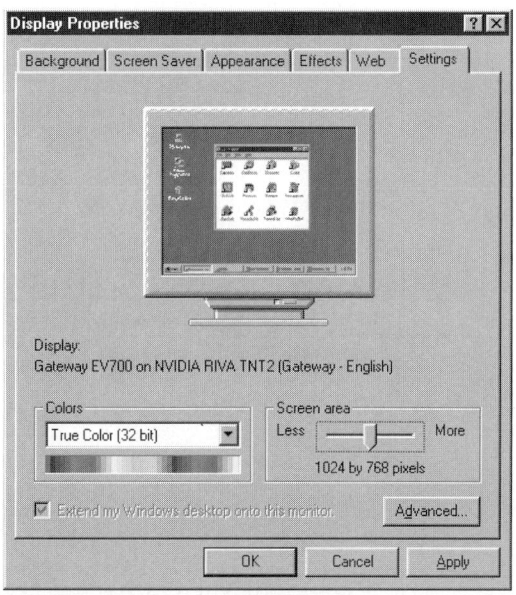

Figure GS.1

will look for and load the appropriate driver for your video display. (Refer to your system documentation if you have trouble with your screen resolution settings.)

Pick: **OK** (to exit the Display Properties dialog box)

Pick: **the close button** on the Control Panel window (the box with an X in the upper right corner of the window)

Customizing Your Shortcuts

To make saving and opening easier as you complete the tutorials, you will create and specify a new working directory for Mechanical Desktop. The working directory is the default directory used by Mechanical Desktop to open and save files. By setting a working directory, you will not have to search through the computer's directories every time you save or open a file.

This is a good practice in general for organizing your files. You should not save your drawing files and other work files in the same directory as the Mechanical Desktop application because doing so makes it likely that they will be overwritten or lost when you install software upgrades. It is also easier to organize and back up your project files and drawings if they are in a separate directory. Making subdirectories within this directory for different types of projects helps you organize your drawings even further.

Tip
This directory name assumes you are using drive C: to store your files. If you are not, your drive letter will be different.

You will create a new shortcut for launching Mechanical Desktop that will use the directory *C:\MDT-Work* as the starting directory. The work directory will be the default choice when you save your Mechanical Desktop files after launching Mechanical Desktop with this shortcut. If you are using a networked computer system, talk to your system administrator before doing this, or make this shortcut on your own network drive space.

Before continuing, use Windows Explorer to create a new directory, C:\MDT-Work. (You can do this by picking File, New Folder and then typing the name MDT-Work as the folder name.)

When you are finished, return to the Windows desktop.

Creating Multiple Shortcuts

Tip
You can add these shortcuts to the Start menu instead of the desktop if you prefer.

You will start Mechanical Desktop from an icon on the desktop. The icons on the Start menu are shortcuts to the executable file in the Mechanical Desktop directory. You can create several shortcuts to the same file, each with different properties. You will add new shortcuts to the desktop using the Settings option on the Start menu.

Pick: **Start, Settings, Taskbar & Start Menu**

Pick: **Start Menu Programs tab**

Pick: **Advanced button**

The list of directories and files on your machine appears similar to Figure GS.2.

Locate the Mechanical Desktop 5 directory in the list of directories on the left. It may be in the Programs directory on your system. When you have located it,

Double-click: **the Mechanical Desktop 5 directory** (so its contents are displayed in the list on the right)

Figure GS.2

You should see shortcuts for Mechanical Desktop 5 and Mechanical Desktop 5 Power Pack in this directory, as shown in Figure GS.3. (You may see a list view instead of the icons shown.) You may use Mechanical Desktop 5 Power Pack for all of the tutorials, but you are required to use it only for Tutorial 6. Make shortcuts for both unless you are certain you will use Power Pack throughout.

Right-click: **the Mechanical Desktop 5 Power Pack shortcut**

Pick: **Send to, Desktop** (create shortcut) (from the context menu that appears)

Right-click: **the Mechanical Desktop 5 shortcut**

Pick: **Send to, Desktop** (create shortcut) (from the context menu that appears)

These actions will create new copies of the Mechanical Desktop shortcuts on the desktop.

Tip

If you do not see copies of the shortcut in the directory, you can change the directory for the command shown on the start menu by browsing to the Windows/Profiles/AllUsers directory and then looking in the Start Menu/Programs folder to find the shortcut.

Figure GS.3

Close: **Windows Explorer** (by clicking its close button)
Pick: **OK** (to close the Taskbar Properties window)

You should see the two new shortcuts on your desktop. Picking once on the shortcut reveals its complete name. Windows allows you to rename a file by right-clicking on its name and picking Rename from the context menu.

Tip
You can also rename the file by picking it twice, but NOT double-clicking. Pick on the shortcut once and then after a few seconds pick on it again; a cursor will start blinking to the right of the name, allowing you to change it.

Right-click: **Mechanical Desktop 5 Power Pack shortcut**
Pick: **Rename**
Rename the shortcut: **Mechanical Desktop 5 Power Pack Tutorials**

Repeat the steps with the Mechanical Desktop 5 shortcut and rename it Mechanical Desktop 5 Tutorials.

You will now set *C:\MDT-Work* as your working directory.

Setting the Working Directory

The Mechanical Desktop session you start with this shortcut will save your drawings to *C:\MDT-Work* by default if you set it as your working directory. To set a working directory, you will modify the settings stored with the new Mechanical Desktop Tutorials shortcuts. (If you created your directory on another drive, substitute that drive letter in the steps that follow.)

Right-click: **the new Mechanical Desktop 5 Power Pack Tutorials shortcut**
Pick: **Properties** (from the context menu)
Pick: **the Shortcut tab** at the top of the Properties dialog box
Pick: **the text box labeled Start in and highlight the existing text**
Type: **C:\MDT-Work**

Your dialog box should look like Figure GS.4. If you have created your work directory on another drive, substitute the correct drive letter.

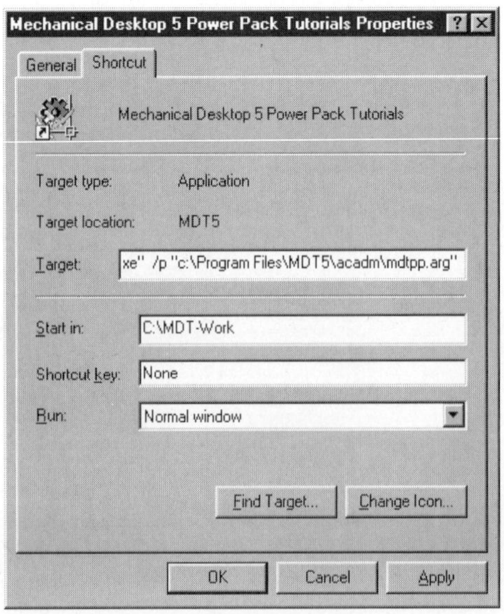

Figure GS.4

Pick: **OK** (to exit the Properties dialog box)

Repeat these steps for the Mechanical Desktop 5 Tutorials shortcut so it, too, starts in the MDT-Work directory.

When you start Mechanical Desktop from these shortcuts, the drawings you save will be saved to *C:\MDT-Work*. The settings necessary for running Mechanical Desktop will not be affected by the change in the working directory. Those drive paths are laid out within Mechanical Desktop's Preferences.

Installing Data Files for the Tutorials

The next step is to create a directory for the data files used in the tutorials and install them. Throughout the tutorials, you will be instructed to work with files already prepared for you. You will not be able to complete the tutorials without these files. You can download these data files with a Web browser such as Netscape Navigator; if you are working in a lab, your professor may have downloaded the files for you and can tell you where they are located. You will create a directory for the data files and then copy them into that directory.

To download the files from a Web browser, use the URL http://www.prenhall.com/lockhart. Choose the icon for this manual to go to its companion site. Enter the site for Chapter 1 and locate the button for the data files. It will be named "Template Files" (or something similar) on the navigation bar at left. Follow the instructions for downloading the file.

If you do not have access to the Internet, you may request a copy of the files from Prentice Hall by calling (201) 236-7364, or by sending e-mail to engineering@prenhall.com.

After you have retrieved the files, use Windows to create a new directory, *C:\MDT-Data*. Copy *MDT-data.exe* into the directory and then run the file to decompress it. You may double-click the file name in the Windows Explorer, or use Start, Run and enter the path and file name in the Run dialog box. When the file is decompressed, you will see an assortment of files in the directory, most of which have the *.dwg* extension.

Now you are ready to customize your Mechanical Desktop menus.

Mechanical Desktop Configuration for Tutorials

The tutorials in this manual are based on the assumption that you are using Mechanical Desktop's default settings. If your system has been customized for other uses, some of the settings may not match those in the step-by-step instructions in the tutorials. You will now set up your system so that you can work through the tutorials as instructed.

To start Mechanical Desktop from Windows,

Pick: **Mechanical Desktop 5 Power Pack Tutorials** icon (that you created on the Windows desktop)

The Mechanical Desktop Today window appears, which you will close.

Pick: **the close box for the Mechanical Desktop today window**

The Mechanical Desktop main screen appears on your screen, as shown in Figure GS.5.

Figure GS.5

You will use the Assist menu on the menu bar at the top of the screen to open the Options dialog box.

Pick: **Assist, Options**

The Options dialog box appears on your screen, as shown in Figure GS.6.

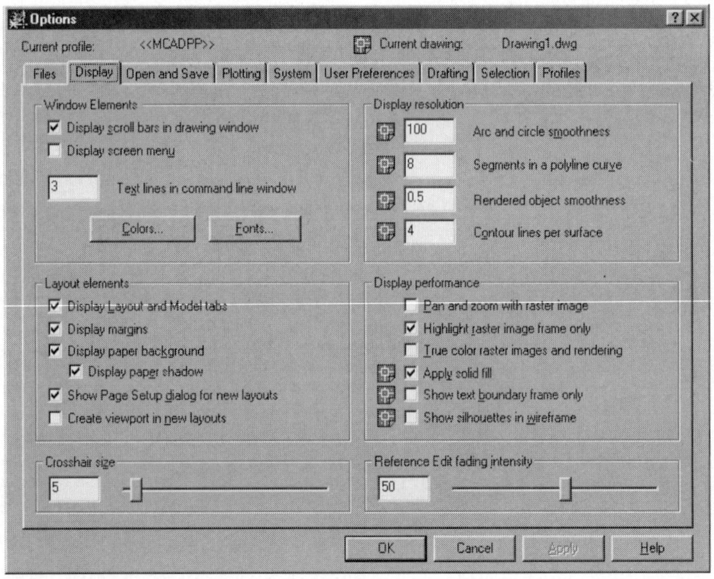

Figure GS.6

The Options dialog box is equivalent to a preferences dialog box. It sets up Mechanical Desktop's search paths, display, pointer, and printer configurations, and other options. The tabs at the top of the dialog box reflect the different groups of options, gathered on "index cards," that can be modified here. To change from one card to the next, you will pick on the name at the top of the card. The card that you select will then appear on top of the stack.

Warning

Changes to Mechanical Desktop's configuration can affect its performance greatly. If you are working on a networked system, you must check with the network administrator before creating your own profile. Whatever changes you make must be made only to your user Profile setting.

The Files card shows the search paths that Mechanical Desktop follows during particular operations. Search paths are what Mechanical Desktop uses to find the configuration, menu, and toolbar files. The Display card allows for changes to the Mechanical Desktop display, including fonts and colors, and allows for changes to the display of an object, including solid, arc, and circle smoothness. The Open and Save card changes the Mechanical Desktop automatic saving times, file extensions for temporary files, and other functions related to opening and saving files. The Plotting card allows you to change the printer and set the default print settings. The System card can be used to change the display driver and pointing device and to set other system variables. User Preferences, Drafting, and Selection all offer options for customizing the software to fit the way you want to work. The Profiles card creates and stores the settings made with the Options dialog box. You will use this card to create your own Profile with the configuration settings used for this manual.

Now you will create the user Profile for these tutorials.

User Profiles

Mechanical Desktop Release 5 allows you to save multiple configuration profiles, making it possible to configure Mechanical Desktop for different uses and peripherals.

Tip

If you are working on a networked computer, your professor or system administrator has probably configured the system already. Ask which profile you should use when in Mechanical Desktop.

If you are working on a shared machine, you can create a special profile to use for these tutorials so that another user doesn't change your settings and defaults between sessions.

You will now create a user-defined profile (or configuration) for the tutorials.

Pick: **Profiles tab**

The Profiles index card appears on top of the other index cards, as shown in Figure GS.7.

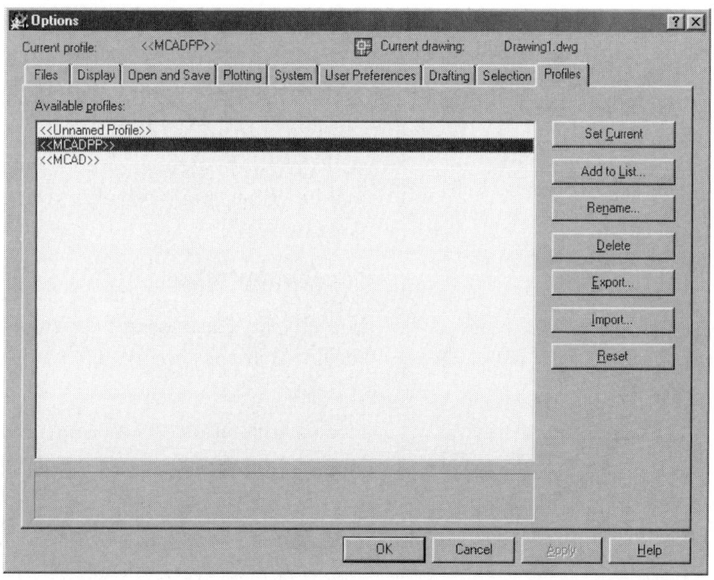

Figure GS.7

Tip

If MCADPP is not the current profile, pick to highlight it in the list and pick the Set Current button. Your new profile will start from the default MCADPP settings and not another user's profile.

The current profile is listed at the top of the dialog box. It should be MCADPP, the default configuration for Mechanical Desktop Power Pack. Any changes made to the other cards will be stored in this Profile because it is selected as current. You will make a new profile for the tutorials, set it as current, and then make the necessary changes to the other cards.

Pick: **Add to List**

The Add Profile dialog box appears, where you will type the name of your new Profile. A description is optional, but not necessary.

Type: **Mechanical Desktop Power Pack Tutorials** (similar to Figure GS.8)

Figure GS.8

Pick: **Apply & Close**

You will see Mechanical Desktop Tutorials appear in the space below Profile name. You will now make it the current profile.

Pick: **Mechanical Desktop Power Pack Tutorials** (to highlight it)

Pick: **Set Current**

Tip

The user profile you create here is for Mechanical Desktop 5 Power Pack. You should repeat the steps to create a user profile starting from the MCAD user profile the first time you use Mechanical Desktop 5 (without Power Pack).

Mechanical Desktop should not change in the background because the configuration settings for this Profile are the same as what was previously current. Now you will make changes to the other cards, as necessary, to follow the tutorials in this manual.

Changing the Configuration

You will start with the Display card. As you review the settings on the cards, note the items contained on each. You may find that you wish to change your profile settings after completing the tutorials to better fit the way you work.

Pick: **Display tab**

The Display card should appear on top of the other cards. On your own, be sure that the settings on your computer are the same as those shown in Figure GS.9.

The background screen color for the screens in this guide is set to white. You do not have to change your background color to white, but you may. If the background on a shared computer is black, changing this setting is an easy way to remember to load your profile each time you start a tutorial session. To change the background screen color,

Pick: **Colors button**

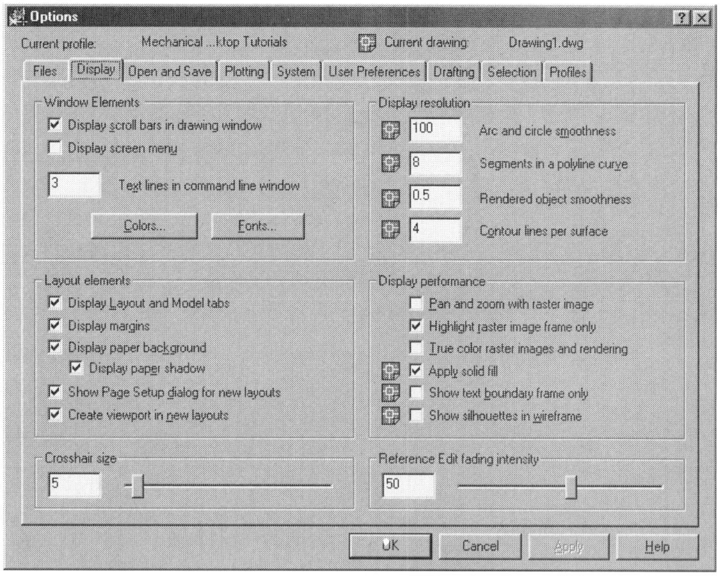

Figure GS.9

Choose Window Element: **Model tab background** (from the drop-down list)

Choose Color: **White** (from the drop-down list)

Pick: **Apply & Close**

You should see the background in the graphics area change as you make this change.

When you have finished,

Pick: **Open and Save tab**

The Open and Save card should appear on top of the other cards. On your own, be sure that the settings on your computer are the same as those shown in Figure GS.10. Mechanical Desktop will automatically save your file if it has not been saved for 120 minutes. In general, you

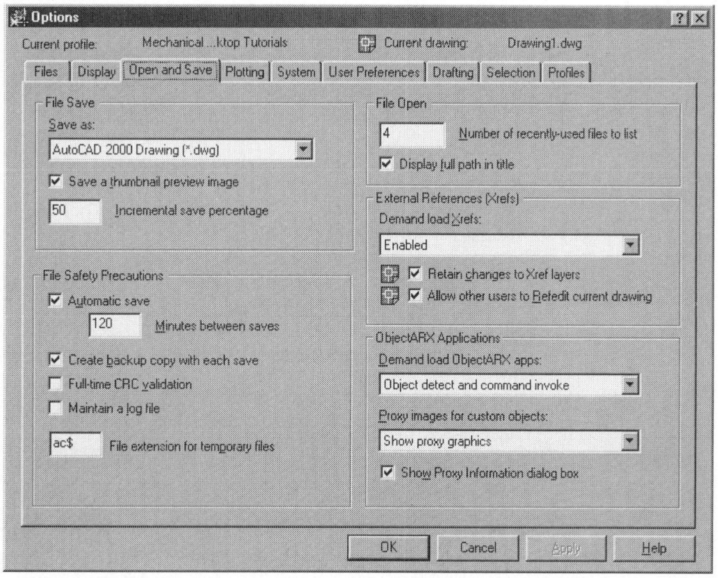

Figure GS.10

will want to save your work periodically instead of relying on this setting. If you are not experienced with using computers, you may want to set this time to 15 minutes. You will notice that the software saves periodically while you are using it. This can be annoying to experienced users.

When you have finished,

Pick: **Plotting**

The Plotting card should appear on top of the other cards, as shown in Figure GS.11.

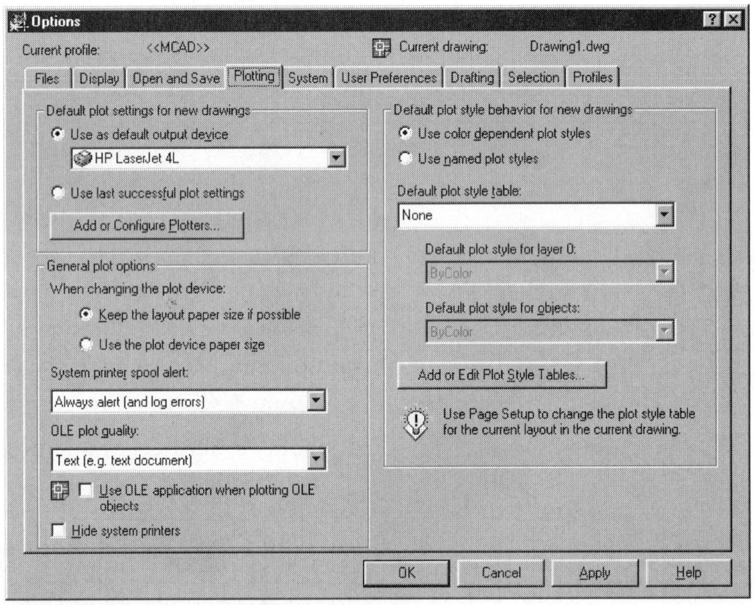

Figure GS.11

You should see your printer configured to work with Mechanical Desktop in the text box labeled default output device. If you did not select a printer when you installed Mechanical Desktop, you can add one here. (You will need it to complete the tutorials.) Available system printers are listed in the list box. Selecting one allows Mechanical Desktop to use the printer that you have already set up to work with Windows. Clicking Add or Configure a Plotter offers the Add a Plotter Wizard, which will walk you through configuring a plotting device. If you want to choose one of the other plotter options, refer to your Mechanical Desktop documentation.

When you have finished,

Pick: **System**

The System card should appear on top of the other cards. The default digitizer is the Current System Pointing Device (your mouse). Mechanical Desktop can also support a tablet. To configure Mechanical Desktop to use a tablet, refer to your Mechanical Desktop documentation. On your own, be sure that the settings on your computer are the same as those shown in Figure GS.12.

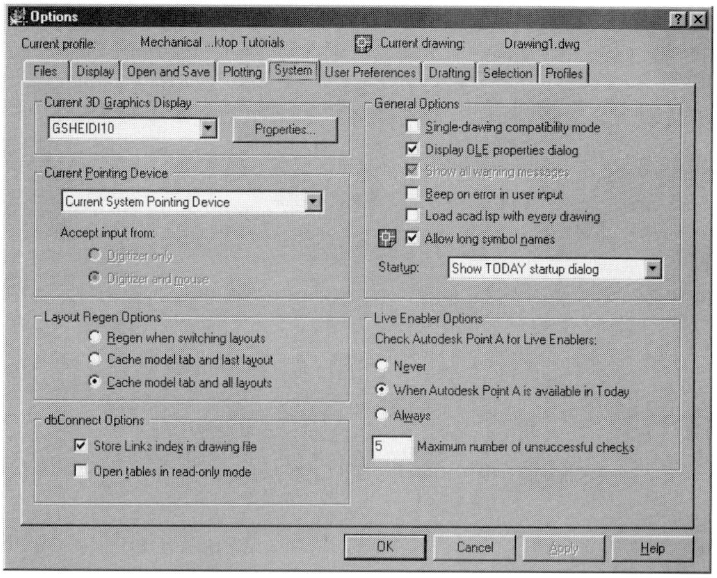

Figure GS.12

When you have finished,

Pick: **User Preferences**

The User Preferences card should appear on top of the other cards. On your own, be sure that the settings on your computer are the same as those shown in Figure GS.13.

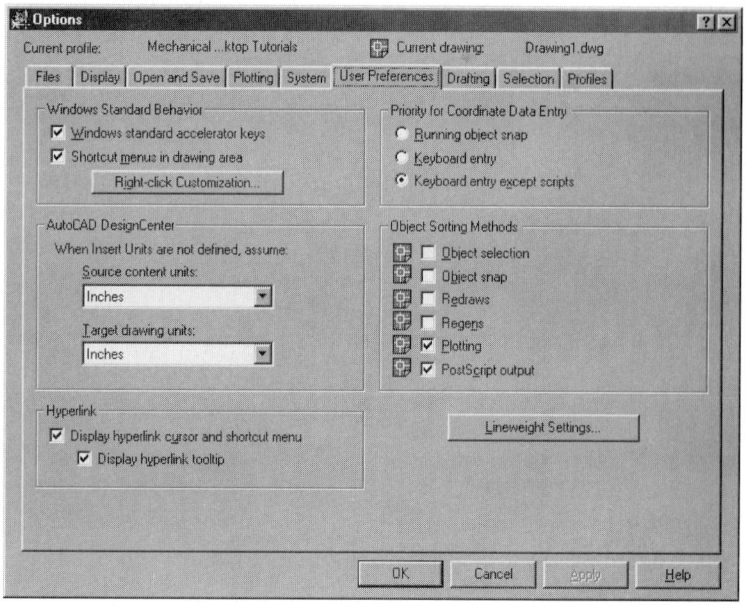

Figure GS.13

Pick: **Drafting**

The Drafting card should appear on top of the other cards. On your own, be sure that the settings on your computer are the same as those shown in Figure GS.14.

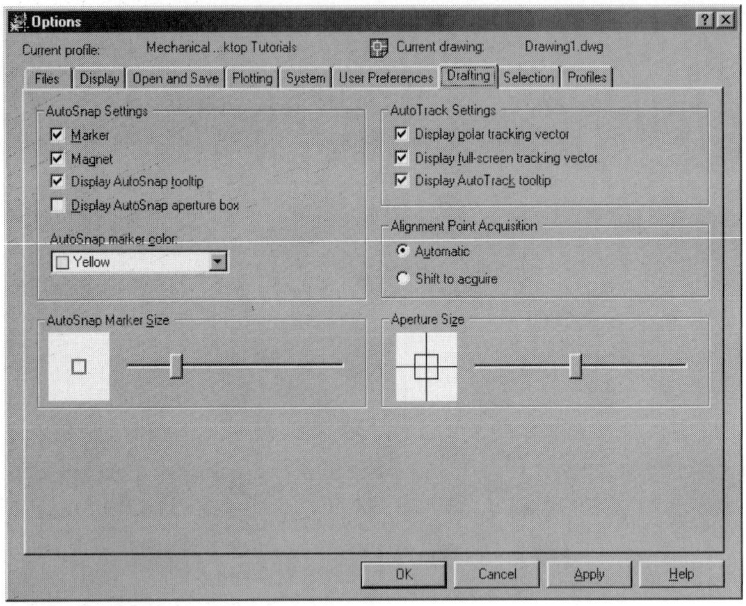

Figure GS.14

Pick: **Selection**

The Selection card should appear on top of the other cards. On your own, be sure that the settings on your computer are the same as those shown in Figure GS.15.

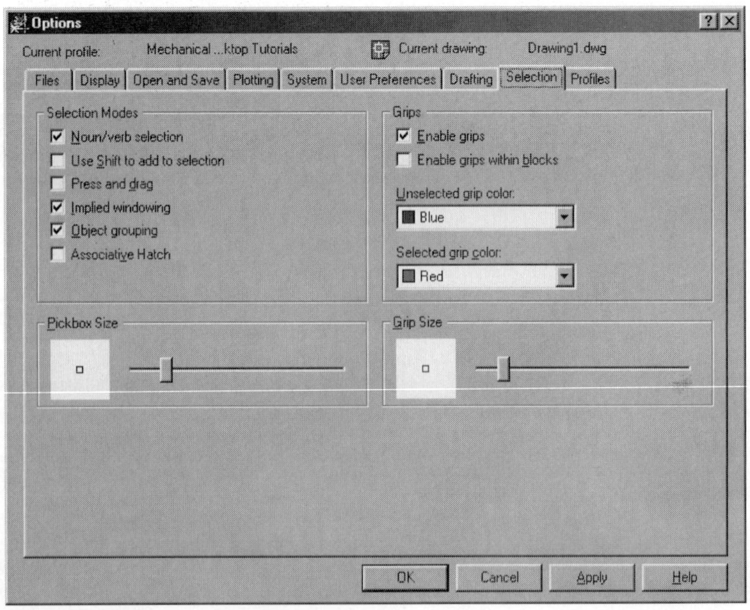

Figure GS.15

Accepting Configuration Changes

Now you will go back to the Profiles card.

Pick: **Profiles**

To accept the changes that you have just made (if any), you will pick the Apply button in the lower part of the dialog box. (If no changes were made, the button will be grayed out.) Be sure that the Current profile is set to Mechanical Desktop Tutorials (the new profile name you created previously). Then

Pick: **Apply**

Pick: **OK**

The Mechanical Desktop screen appears again (with a white background if you changed that setting). Now whenever you begin Mechanical Desktop Power Pack, you can go into the option dialog box and select the Mechanical Desktop Power Pack Tutorials Profile (if it is not already selected) to use this manual. You can also export your profile and save it in a separate directory where it will not be changed by other users. To do this, pick Export from the Profiles tab and use the dialog box to save the file, ending in *.arg*, to a secure folder. When you need to restore your profile, pick Import from the Profiles tab and select your file ending in *.arg* to import.

The first time you start Mechanical Desktop 5 (not the Power Pack), you should repeat these steps to create a profile starting from the MCAD profile so you have a profile for whichever version of the software you launch.

Dialog Box Settings

Three Mechanical Desktop system variables control whether dialog boxes appear when certain commands are issued. Filedia controls the display of dialog boxes for file operations, such as New, Open, and Save. Cmddia controls whether dialog boxes are displayed for Plot and external database commands. Attdia controls whether dialog boxes are displayed with the Insert command. The default value for each of these variables (displayed in the angle brackets) should be 1.

Command: **FILEDIA** [Enter]

New value for FILEDIA <1>: (if the default value is 1, press [Enter] ; if the default value is 0, type 1 [Enter])

Command: **CMDDIA** [Enter]

New value for CMDDIA <1>: (if the default value is 1, press [Enter] ; if the default value is 0, type 1 [Enter])

Command: **ATTDIA** [Enter]

New value for ATTDIA <0>: (if the default value is 1, press [Enter] ; if the default value is 0, type 1 [Enter])

As you follow the instructions in the tutorials, a dialog box should appear on your screen every time one is used in the tutorials. If it does not, recheck the status of these commands.

You have completed the configuration and setup necessary to do the tutorials in this manual. Continue on to learn about the Mechanical Desktop interface and online Help. If you are an experienced AutoCAD user, you may want to read the online Help section (p. 26) of this chapter and then proceed directly to the tutorials.

Microsoft Windows and Mechanical Desktop

Mechanical Desktop Release 5 for Windows uses many of the same conventions as other applications that run in Windows. This section identifies some of the techniques that you will be using to complete the tutorials in this manual.

Basic Elements of the Mechanical Desktop Screen

The Mechanical Desktop screen consists of two main windows and several toolbars, which are also a kind of window. The largest window is the *graphics area*, and it is the main workspace. The Browser Window is on the left; it works like a directory of the assemblies, parts, drawings, and other elements in the file and is useful for navigating within Mechanical Desktop. All windows have elements that are common to all applications written for the Windows environment, although they may not be visible if the window is "docked." These elements are labeled in Figure GS.16 and described as follows.

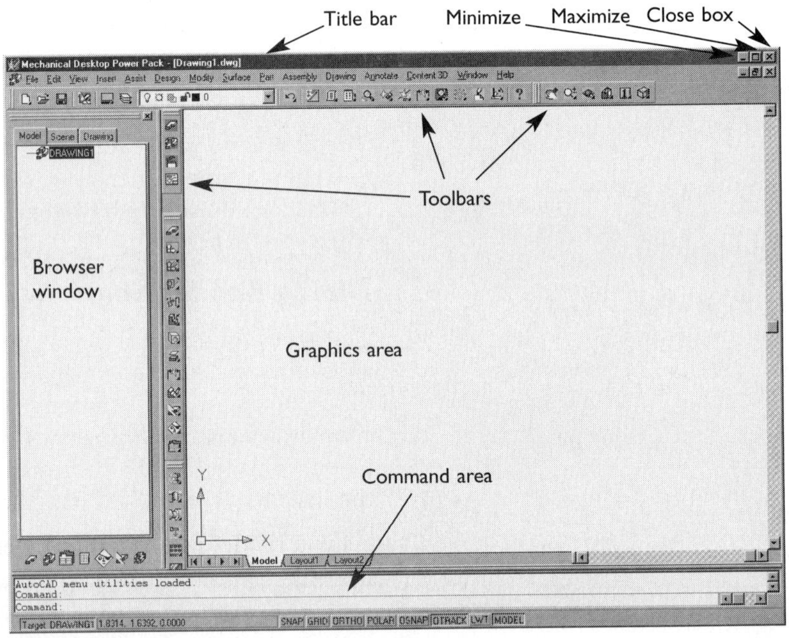

Figure GS.16

- The *close box* is a button in the upper right corner of each window. Picking this button closes the window.
- The *title bar* shows the name of the application (in this case, Mechanical Desktop) followed by the document name and path (in this case, it is only *Drawing1.dwg* because the file has not been saved yet).
- The *window border* is the outside edge of a window. You can change the window size by moving the cursor over the border until it becomes a double-ended arrow. Holding the mouse button down while you move the mouse (dragging) resizes the window.
- The *maximize and minimize buttons* are in the upper right corner of the window. Clicking the maximize button with the mouse enlarges the active window so that it fills the entire screen; this condition is the default for Mechanical Desktop. You will learn to use the minimize button in the next section.

Click and drag on the gray bars above a toolbar or the Browser window (to see it change to a free-floating window.)

Now the Windows elements are visible and can be used to control the window. To return the window to its docked position, you will click and drag on its title bar.

Tip

In a working Mechanical Desktop session, you should always save your work before minimizing Mechanical Desktop.

Click and drag on the Browser window or toolbar again until it snaps into a docked position along the side of the screen.

Minimizing and Restoring Mechanical Desktop for Windows

At times you may want to leave Mechanical Desktop temporarily while you are in the middle of a work session, perhaps to access another application. Minimizing Mechanical Desktop allows you to leave the application and return to it more quickly than you could if you had to exit and start Mechanical Desktop all over again.

Picking the minimize button reduces the Mechanical Desktop window to a title button on the taskbar and makes the Windows desktop accessible. The top set of buttons controls the Mechanical Desktop application window; the lower set controls only the graphics area window. You will pick the top button.

Pick: **the minimize button**

Mechanical Desktop is reduced to a button at the bottom of your screen.

Mechanical Desktop is still running in the background, but other applications are accessible to you. When you are ready to return to Mechanical Desktop, click the minimized Mechanical Desktop program button.

Pick: **the minimized Mechanical Desktop button**

Interacting with Mechanical Desktop

As you model in Mechanical Desktop, you will create geometry by entering commands, choosing command options, and entering points that define geometry. As you work, the command line area at the bottom of your screen will be an important reference point. You will see most commands you enter "echoed" at the command line, and Mechanical Desktop will use this area to prompt you for additional input. You will learn how to interpret the command prompt information as you proceed through this guide. If you have trouble picking a point or completing a step, refer to the command line area to see if Mechanical Desktop is expecting some other input. "Command:" in this window is a signal that Mechanical Desktop is ready for a command.

You will type the Line command to see how the command line works. When you are supposed to type a command, it will appear in all capital letters. You must hit the ↵ key to activate the command.

Type: **LINE** ↵

The command line now reads "Specify first point:", which is a prompt for you to specify the point from which the line should be started.

Entering Points

You can specify points either by typing in the coordinates from the keyboard or by using your mouse to locate the desired points in the graphics window.

When you move the mouse around on the mouse pad or table surface, crosshairs (the small intersecting vertical and horizontal lines on the screen) follow the motion of the mouse. These crosshairs form the Mechanical Desktop cursor in the graphics window. The intersection of the crosshairs will be the selected location when you press the pick button. This is how you pick a point. In some Mechanical Desktop modes, the crosshairs change to arrows or boxes with target areas to facilitate picking an exact point. You will learn about these modes in the tutorials. The cursor on the screen echoes the motion of the mouse.

Specify first point: **pick a location in the graphics area**

Specify next point or [Close/Undo]:

Entering Command Options

The command line now prompts you to enter the next point, but it also offers you options in square brackets. The letters that appear in square brackets, Close/Undo, are command options for the Line command. Subcommands and command options work properly only when entered in response to the appropriate prompts on the command line. When a subcommand or option includes one or more uppercase letters in its name, it is a signal that you can type those letters at the prompt as a shortcut for the option name; for example, X for eXit. If a number appears in the option name—for example, 2P for Circle 2 Points, which creates a circle using two endpoints of the diameter—you need to type both the number and the capital letter(s).

Undo is a common command line option that allows you to "back up" a step without canceling the command.

Type: **U** ↵

The first point is undone and you are prompted again for the first point. You will pick points to make a star, similar to that in Figure GS.17, and then use the Close option to enter the last point.

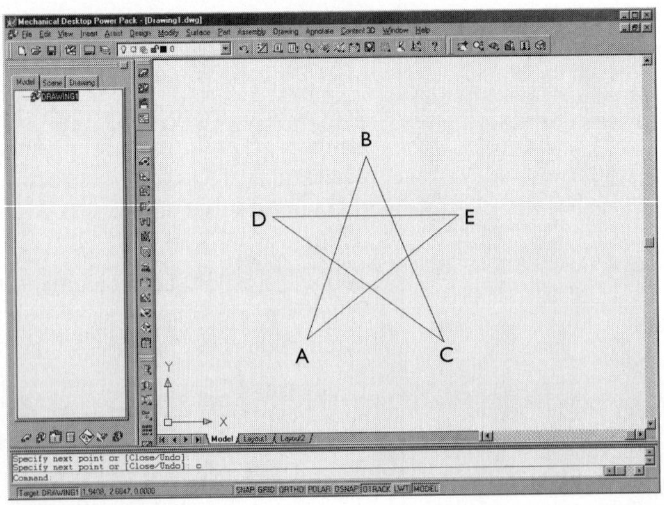

Figure GS.17

Specify first point: **pick points A–E to create the star shape**

Type: **C** ↵

The close option of the Line commands locates the last point at the exact location of the first point. This is a good example of how the Mechanical

Desktop command options augment the use of the mouse in creating exact geometry.

Entering Mechanical Desktop Commands

Mechanical Desktop gives you several options for entering commands. You can type them at the command prompt, as you just did, but it is usually easier to use your pointing device (usually a mouse) to pick them from a context menu, a toolbar, or a pull-down menu from the menu bar. In this guide, you will learn to use all of these methods.

How you invoke a command may affect the order and wording of the prompts displayed on the screen. For the purposes of the tutorials in this manual, you should pick or type commands *exactly as instructed*. As you gain familiarity with commands and are instructed to do more on your own, you may choose commands in the way that seems most convenient for you.

For the most part, you will use context menus that pop up when you right-click with the mouse. The items to be picked from the menu will appear in the next line. For example,

Right-click: in the graphics area

Pick: **2D Sketching, Line**

Many menus have submenus that appear when you pick a menu item, as shown in Figure GS.18. Menu picks separated by commas indicate the menu picks in order as you will pick them.

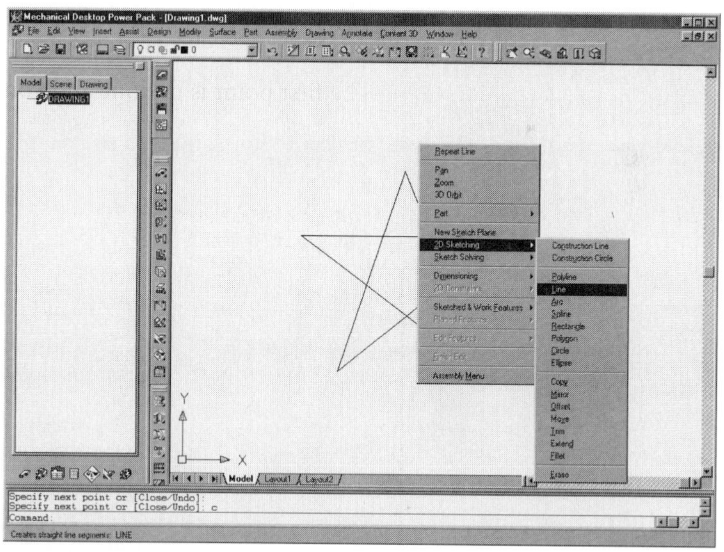

Figure GS.18

Backing Up and Backing Out of Commands

When you type a command name or any data in response to a prompt on the command line, the typed characters wait until you press the spacebar or [↵] to instruct Mechanical Desktop to perform the command or enter and act on the entered data.

If you have not already pressed [↵] or the spacebar, use the [BACKSPACE] key (generally located above and to the right of the [↵] key on standard IBM PC keyboards and represented by a long back arrow) to delete one character at a time from the command/prompt line.

Pressing [ESC] terminates the currently active command (if any) and reissues the *Command*: prompt. You can cancel a command at any time: while typing the command name, during command execution, or during any time-consuming process. A short delay may occur before the cancellation takes effect and the prompt **Cancel** confirms the cancellation.

Pressing [ESC] is also useful if you want to cancel a selection process. If you are in the middle of selecting an object, press [ESC] to cancel the selection process and discard the selection set. Any item that was highlighted because you selected it will return to normal.

If you complete a command and the result is not what you had expected or wanted, use the Undo command or type *U* to reverse the effect.

You will press [ESC] now to cancel the Line command.

Specify first point: [ESC]

Using the Toolbar Icons

Toolbars contain icons that represent commands. Picking the icon once invokes the command. The Content 3D, Desktop Main, Mechanical Main, Mechanical View, and Part Modeling toolbars are visible by default when you launch Mechanical Desktop 5 Power Pack.

Toolbars provide quick access to commands and can be customized to fit the way you work. The toolbars across the top of your screen contain frequently used commands, such as Pan, Zoom, and Undo, and viewing and shading controls. In this guide, the instruction line will include the word "icon" to indicate a selection from a Toolbar; for example, *Pick: Undo icon*.

You can turn toolbars on and off with the Toolbars selection on the View menu. You can also change the contents of toolbars, resize them, and *dock* or *float* them. A docked toolbar attaches to any edge of the graphics window. A floating toolbar, such as the toolbar shown in Figure GS.19, can lie anywhere on the application screen and can be resized. A docked toolbar cannot be resized and doesn't overlap the graphics window. You can move a toolbar around on your screen by picking on the gray bars at its top (or left side) and holding the mouse button down as you drag it to a new location.

When you move your pointing device over an icon, the name of the icon appears below the cursor, as shown in Figure GS.19. This feature is

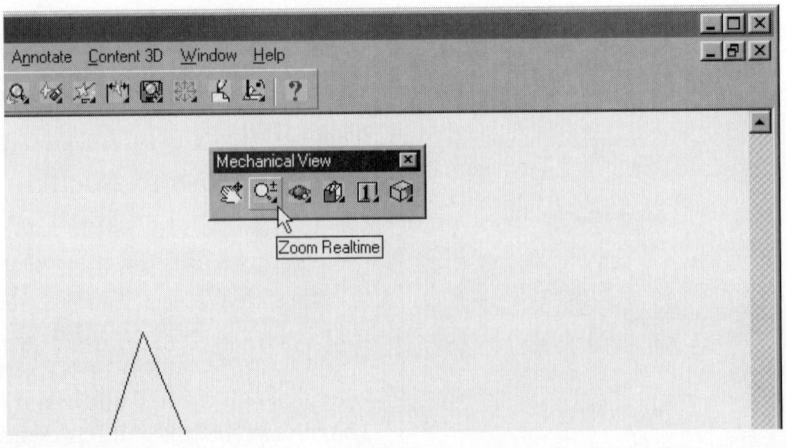

Figure GS.19

called a ***tool tip*** and is an easy way to identify icons. Icons with a small black triangle have ***flyouts*** that contain related commands.

Hold the pick button down with the cursor over the Undo icon (at the top of the screen) **until the flyout appears,** (as shown in Figure GS.20).

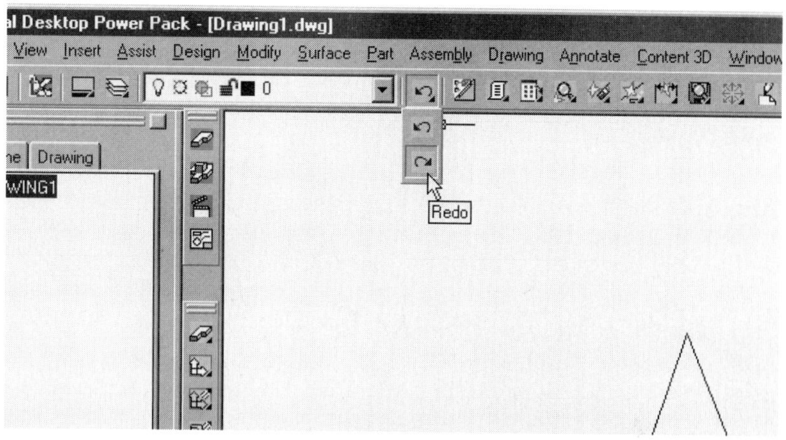

Figure GS.20

The Redo command is available from the flyout; it reverses the last Undo command.

Using Pull-Down Menus

The menu bar and its associated pull-down menus provide another way of executing Mechanical Desktop commands. The menu bar in Mechanical Desktop operates in the same way as menus in other applications for Windows. When in the menu area, the cursor changes to an arrow. You use this arrow to select a menu.

You open a menu in the menu bar by picking its name. To practice, you will open the Annotate menu.

Pick: **Annotate** (from the menu bar)

Picking a menu item executes the commands associated with it, opens a submenu of options, or opens a dialog box of options to be used to control the command. A triangle to the right of a menu item indicates that a submenu is associated with it. You will pick Edit Dimensions to see its submenu.

Pick: **Edit Dimensions** (from the Annotate menu)

The Edit Dimensions submenu shown in Figure GS.21 appears next to the Annotate menu.

A pulled-down menu remains displayed until you

- pick an item from it;
- pick on the menu name again;
- pull down another menu by picking it from the menu bar;
- remove the menu by picking an unused area of the menu bar;
- pick a point on the graphics window;
- type a character on the keyboard;

- pick an item from a tablet or button menu; or
- move your pointing device into the regular screen menu area and click.

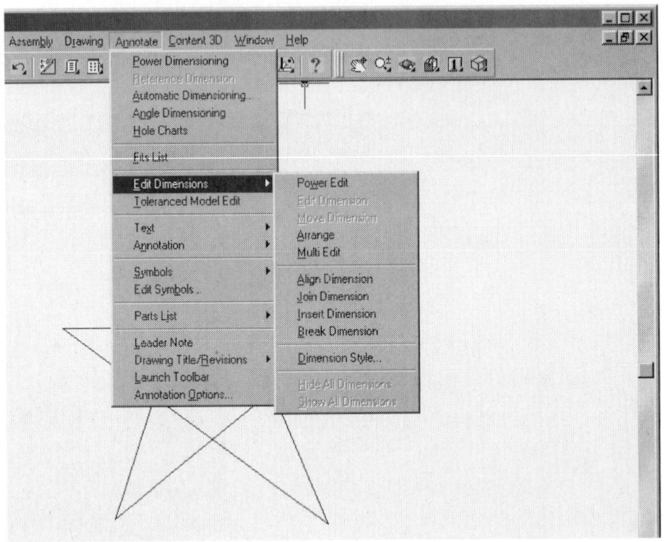

Figure GS.21

To remove the menus from the screen,

Pick: **(anywhere in the blank area of the graphics window)**

Working with Dialog Boxes

Many commands present a dialog box instead of (or in addition to) command line options. These dialog boxes often combine many different options and settings into a single dialog. Because the dialog box groups choices and can include image tiles to illustrate choices, it makes it easier to select the correct option for your task. You will open the Dimension Styles dialog box by picking Dimension Style from the Edit Dimensions submenu.

Pick: **Annotate, Edit Dimensions, Dimension Style**

The dialog box shown in Figure GS.22 appears on your screen.

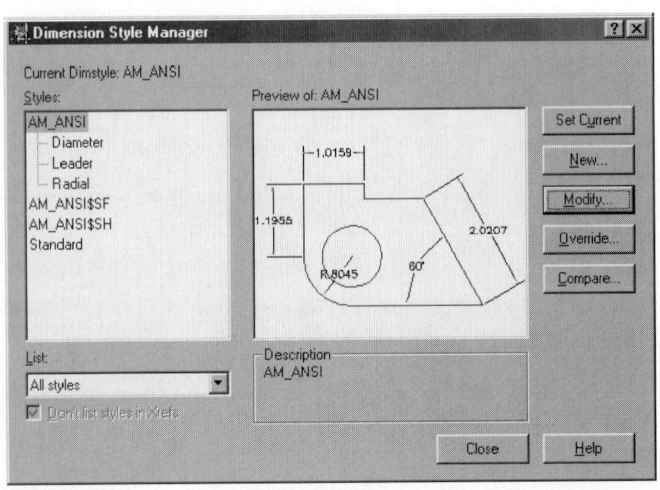

Figure GS.22

Warning

The term "button" used in conjunction with a dialog box refers to these selection options and should not be confused with the buttons on your mouse or digitizer.

The Dimension Style Manager is an example of how many options can be contained in a single dialog box. The buttons at right open additional dialog boxes for creating or changing dimension styles.

Pick: **New button**

Some dialog boxes have subboxes that pop up in front of them. When this occurs, you must respond to the top dialog box and close it before you can continue. The Create New Dimension Style dialog box is an example of such a box.

Most dialog boxes have an OK button (or in this case, Continue) that confirms the settings or options that you have selected in the dialog box. Clicking the OK button is analogous to pressing ↵ to send a command to Mechanical Desktop. Clicking the Cancel button disregards all changes made in the dialog box and closes the dialog box; it has the same effect as pressing ESC. You can use the Help button to get more information about a particular command.

Pick: **Cancel**

Pick: **Modify** (in the Dimension Styles Manager)

Pick: **Primary Units tab**

The Primary Units tab shown in Figure GS.23 appears on your screen. Many dialog boxes contain options grouped by type on index cards, as you saw previously with the Options dialog box. Each of the tabs in this dialog box controls a different aspect of a dimension's appearance. Any item in a dialog box that is grayed cannot be picked, such as the Fraction format option in the Linear Dimensions section.

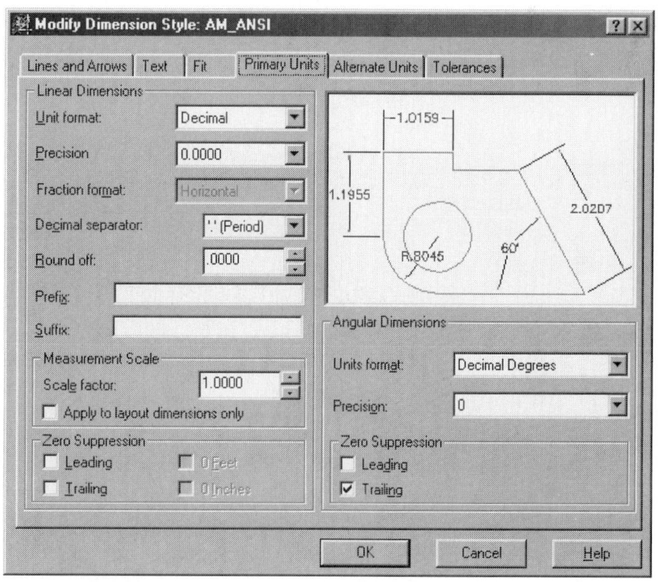

Figure GS.23

Dialog boxes can contain several types of buttons and boxes that control values or commands in Mechanical Desktop.

Input boxes: An input box offers a menu of preset options, such as the Unit format box in Figure GS.23. Input boxes have a small arrow at the right end. Picking the arrow causes the value area to expand into a menu of options that you can use to choose the desired value.

Image tiles: An image tile displays choices as graphic images (image tiles) rather than words. Choosing options from the input boxes causes the image tile to change, illustrating the chosen option. Try this with some of the input boxes on your own.

Text boxes: A text box allows you to specify a value, such as a Prefix for dimensions in the Linear Dimensions area. Picking a text box puts the cursor in it and lets you type values in it or alter values already there. If you enter an invalid value, the OK button has no effect; you must highlight the value, correct it, and select OK again.

Check boxes: A check box is a small rectangle that is either blank or shows a check mark. Check boxes control an on/off switch—for example, turning Leading Zero Suppression on or off—or control a choice from a set of alternatives—for example, determining which modes are on. A blank check button is off.

To see more types,

Pick: **Fit tab**

The Fit section of the dialog box appears, as shown in Figure GS.24.

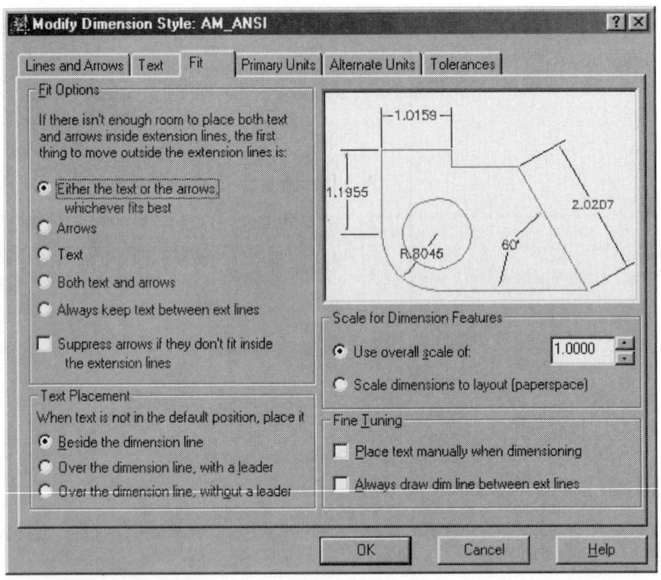

Figure GS.24

Radio buttons: A radio button also turns an option on or off. A filled-in radio button is on. You can select only one radio button at a time; picking one button automatically turns any other button off.

Action buttons: An action button doesn't control a value but causes an action. The OK action button causes the dialog box to close and all the selected options to go into effect. When an action button is highlighted (outlined with a heavy rule), you can press ↵ to activate it.

You can also use the keyboard to move around in dialog boxes. Pressing TAB moves among the options in the dialog box. Try this by pressing TAB several times. Once you have highlighted an input button, you can use the arrow keys to move the cursor in text boxes. The [SPACEBAR] toggles options on and off.

The Cancel button ignores all the selections that you made while this dialog box was open and returns you to the most recent settings. To return to the graphics area without making any changes,

Pick: **Cancel**

Pick: **Close**

You are returned to the main screen.

Repeating Commands

You can press the spacebar or [↵] at the command prompt to repeat the previous command, regardless of the method you used to enter that command. You can also right-click and repeat the command by picking Repeat [last command] at the top of the context menu. (The name of the last command always appears in this location.) Some commands, especially those that prompt you for settings when first invoked, assume default settings when repeated in this manner. You will try this now, and you will use the [ESC] key to cancel the command again.

Press: [SPACEBAR]

Dimension Style Manager is invoked again. You will cancel it with the [ESC] key.

Type: [ESC]

Selecting Objects

A Mechanical Desktop model or drawing is made up of separate elements, called *objects*, that consist of lines, arcs, circles, text, and other elements that you access and draw by using Mechanical Desktop's commands and menu options. Many of Mechanical Desktop's commands ask you to select one or more objects for processing. This collection of objects is called the *selection set*.

In general, when you are prompted to select an object, you pick on an object and Mechanical Desktop highlights it to indicate that it is selected. Most commands allow you to continue to pick objects and add them to the selection set until you signal (by pressing [↵]) that you have finished.

Type: **ERASE**

Select objects: **pick a line on the screen**

Select objects: **pick another line**

Select objects: [↵]

When you press [↵], you signal to Mechanical Desktop that the selection set is complete and the command proceeds to the next step—in this case, erasing the lines.

Another way to select objects is with a selection window, or crossing window. At a prompt to select objects, you can pick two points to define a rectangle that will select a group of objects. A window drawn from left to right selects only those objects completely in the window. A crossing window, drawn from right to left, selects all of the objects it encloses or touches. Try this now after restarting the Erase command.

Right-click: in the graphics area

Pick: **Repeat Erase**

Select objects: **pick a point below and to the left of the objects; then pick a point right of the objects**

Only those objects enclosed by the window are highlighted, if any.

Type: [ESC] (to cancel the selection set)

Select objects: **pick a point below and to the right of the objects; then pick a point left of the objects**

All of the objects crossed by or enclosed in the crossing window are highlighted.

The selection window is a quick way to select a large number of objects. It can also be an aid when selecting hard-to-select objects, as you can remove the one or two unwanted objects after the fact.

Removing Objects from a Selection Set

Pressing R at the select objects prompt allows you to pick objects to remove from the selection set. To resume adding items, press A at the prompt, as follows.

Select objects: **R** [↵]

Remove objects: **pick one or more of the selected lines**

Remove objects: **A** [↵] (to signal you wish to add items to the set)

Select objects: **pick an unselected line**

Select objects: [↵]

All of the selected lines are erased.

Individual Mechanical Desktop commands offer variations on these basic means of selecting objects, which you will learn more about in the tutorials. There are additional methods for selecting, which you can read about using online Help.

Mechanical Desktop Online Help

Mechanical Desktop offers a large amount of information about commands and operations online through its Help feature. Throughout this manual, you will find suggestions for using online Help to learn more about a given topic. Help can tell you all the ways to invoke a command, explain what all the command options are used for, and direct you to related commands and information. Although the tutorials present step-by-step instructions and explanations of common options, Help is a valuable tool for you as you go beyond the tutorials and build your modeling skills.

Under the Help menu, you will find choices for Mechanical Desktop Help and AutoCAD Help. In most cases, you will find Mechanical Help Topics to be most useful for the skills you are learning in this guide. Because AutoCAD commands, such as Line and Erase, are incorporated into Mechanical Desktop, however, AutoCAD Help is included as a reference for those commands. If you are unsure where to look for a command, invoking Help when the command is active will bring up the correct Help topic. You will learn about this context-sensitive feature of Help in the next section.

Pick: **Help, Mechanical Help Topics**

The Mechanical Desktop Help window is shown in Figure GS.25.

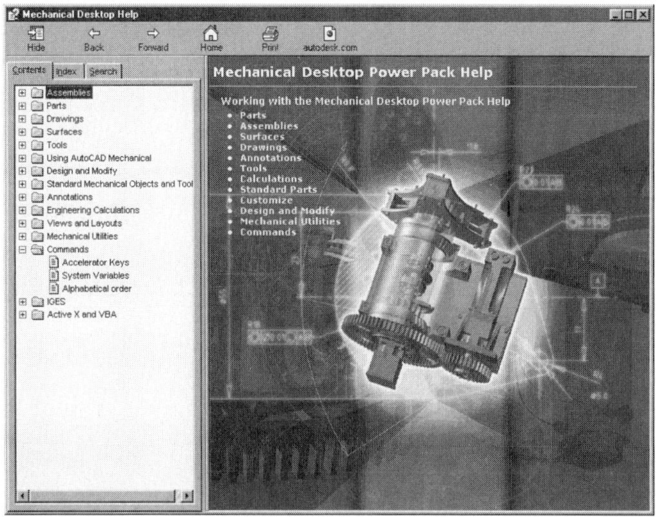

Figure GS.25

Help has two main windows and a set of action buttons at the top that you can use to navigate in Help, print Help topics, or log onto the Autodesk Web site. The window on the left provides three ways of finding Help topics: Contents, Index, and Search. The Contents listing should be uppermost.

Pick: **the Contents tab** (if necessary)

Pick: **the plus sign left of Parts**

Pick: **the plus sign to the left of Commands**

For each topic in Contents, you can read how to accomplish tasks, learn more about the topic, or access reference materials. As a page is displayed, tabs in the right-hand window offer access to additional related topics.

Pick: **Alphabetical order** (in the list under Commands)

The Commands section of the list offers access to all Mechanical Desktop commands by name, as shown in Figure GS.26. You can use the command names in the command reference in this guide to look up the command in this section.

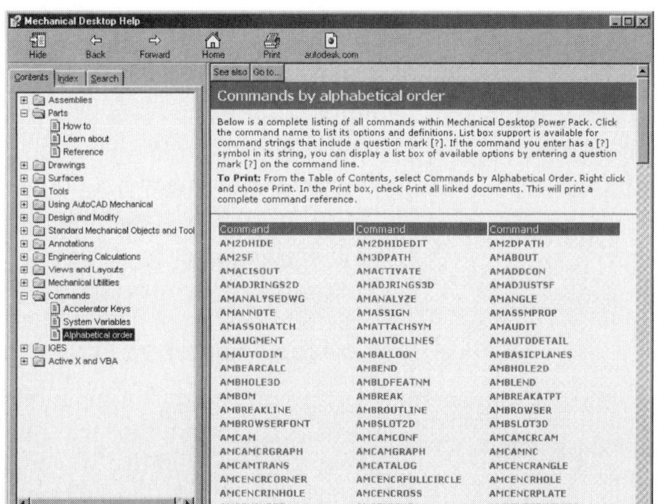

Figure GS.26

Pick: **Index tab**

The Index portion of the dialog box allows you to find information by typing in a keyword in the box at the top of the window. A list of topics related to that key word then appears in the list below. You will try this now.

Type: **equations** (in the box below Index)

Pick: **equations** (in the list below to highlight it)

Pick: **Display** (the button at the bottom of the window)

Information about equations appears in the right-hand window, as shown in Figure GS.27.

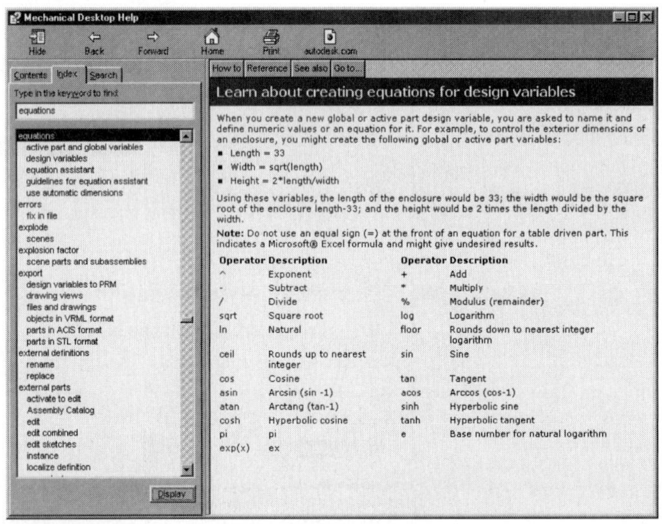

Figure GS.27

Pick: **Search tab**

The Search window works like the Index window, but allows you to search on any combination of words, not just keywords. Mechanical Desktop will list all of the help pages available that contain that set of words. This method is helpful when you do not know the exact function or help topic to search for.

You will use the Windows close box to close the Help window and return to the Mechanical Desktop graphics window.

Pick: **the Windows close box in the top right corner of the Help window**

Context-Sensitive Transparent Help

Mechanical Desktop's Help is *context sensitive*. That is, if you press [F1] during a command, the Help window will appear with information about the command that you are currently using. Accessing Help in this way eliminates the need for searching a topic or navigating through the Help windows.

Help is one of several commands that are transparent. A *transparent* command is one that you can execute during another command. You can pick a transparent command, such as Zoom and Pan, by picking its toolbar icon, or by typing the command with an apostrophe (') in front of it. Typing 'Help or '? is equivalent to pressing F1. You will begin the Line command and then type the transparent Help command to bring up the Help window for the Line command.

Command: **LINE** ↵

From point: **press** F1

The online Help window for the Line command opens, as shown in Figure GS.28.

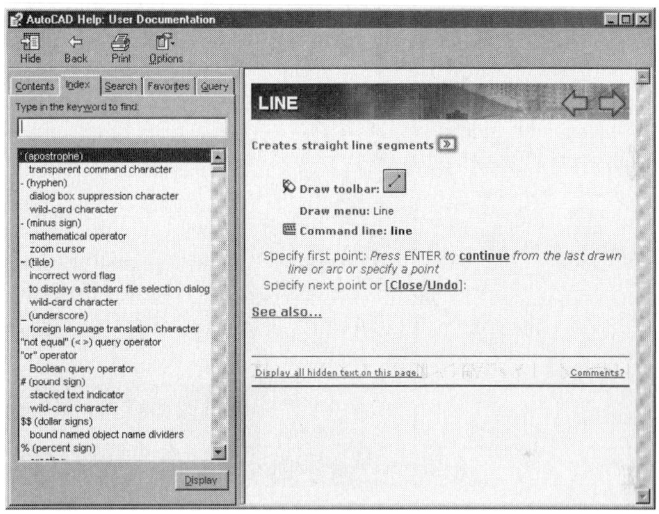

Figure GS.28

Notice that the Help screen is from the AutoCAD Help topics. Because AutoCAD has its own set of menus and toolbars, you may find that some of the information about where the command is located does not match Mechanical Desktop, but the basic information about the operation of the command and its options will be correct.

Pick: **the Windows close box** (to close the Help window.)

Note the message in the Command window: Resuming LINE command. The command is still active. Transparent Help is a great tool when you are unsure how to respond to the command options being presented.

Press: ESC (to cancel the Line command)

Mechanical Desktop's online help is a great resource for learning how to use this powerful software. You should use Help whenever you want more information about commands and options as you work through the tutorials.

Working with Documents

All the documents that you will be instructed to use should now be in the directories *C:\MDT-Data* and *C:\MDT-Work*, which you created earlier in this chapter. The best procedure is to save all files on your hard drive because it has more storage space and is faster than accessing a floppy disk drive. Saving directly from Mechanical Desktop to your floppy disk can result in unrecoverable, corrupted drawing files. After a session is complete, you should back up your files from the hard drive to a removable disk using Windows Explorer or another file management utility.

Exiting Mechanical Desktop

When you have finished a Mechanical Desktop session, you will exit the program by choosing File, Exit.

Pick: **File, Exit** (from the pull-down menu)

Tip

If you did not want to save your drawing, you would have picked No. If you have made some irreversible error and want to discard all changes made in an Mechanical Desktop session, you can choose Exit from the File menu (or type QUIT at the command prompt) and select No so that the changes are not saved to your file.

Because unintentionally quitting Mechanical Desktop could cause the loss of a lengthy editing session, Mechanical Desktop prompts you to save the changes if you want to. A dialog box appears that contains a *Save Changes to Drawing 1*? message and gives you three options: *Yes*, *No*, and *Cancel*.

Pick: **Yes**

The Save Drawing As dialog box appears, as shown in Figure GS.29. (Your dialog box will not list any files because you have not saved anything to your work directory yet.)

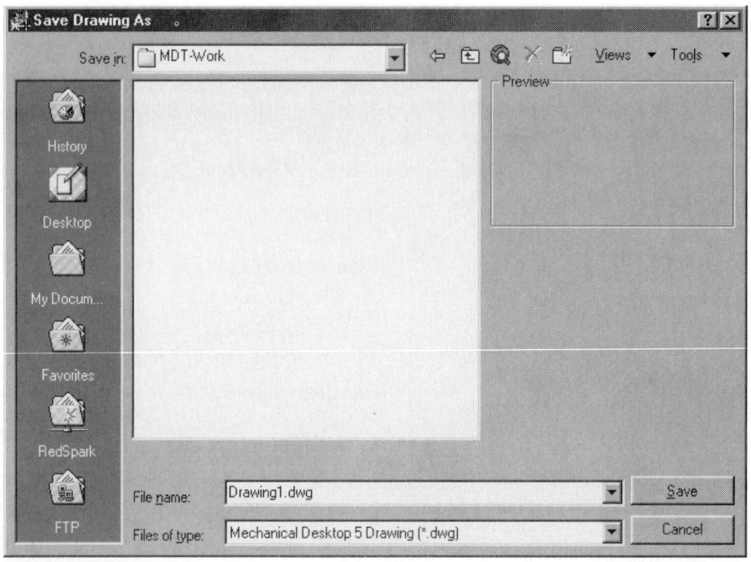

Figure GS.29

This dialog box is similar to other Windows dialog boxes for managing files. The Save in: box contains the directories available on the drives attached to your system; pressing the down arrow will display a list of these drives and directories. By double-clicking on a directory, you can open and select it as the place to store your file.

When you set up Mechanical Desktop, you created a working directory called *C:\MDT-Work* in which to save your files. This directory should be

Warning
You could use this dialog box to save your file to a floppy disk, but this is not a good idea. The best procedure is to create files and save them on your hard drive. When you have finished, use the Windows Explorer to copy the file to a floppy disk as a backup.

open now. If it is not, refer to the Setting the Working Directory section of this chapter (p. 6) to make *C:\MDT-Work* your working directory. This change will eliminate the need to select the correct directory every time you create, open, or save a drawing file.

Now you need to give your file a name. Mechanical Desktop supports long file names; if you do not have long file names enabled on your system, however, your file name cannot be longer than eight characters and cannot contain any spaces. (Mechanical Desktop will automatically add the drawing file suffix *.dwg* to your file name, so you need not type it. All file names in this manual will contain the appropriate suffix for clarity.) In general, it is not a good idea to use punctuation other than underscore (_) and dash (-) in your file names. Likewise it is better not to use spaces in the file name. Though these are legal characters in your file names, AutoCAD and Mechanical Desktop will not let you insert one file into another when it has punctuation or spaces in the name.

To continue practicing Mechanical Desktop basics, name the file *Basics.dwg*. The file name *Drawing1.dwg* should be highlighted in the File Name text box. (If it is not, click in the File Name text box and drag to highlight all the text.)

Type File name: **Basics**

Pick: **Save**

Your file is saved, and Mechanical Desktop automatically returns you to Windows.

You are now ready to start learning Mechanical Desktop by completing the tutorials in this manual.

Tutorial 1

The Mechanical Desktop Workspace

Objectives

When you have finished this tutorial, you will be able to

- Open and save files.
- Navigate and describe in the Mechanical Desktop Workspace.
- Create a directory structure to manage your files.
- Change a feature's dimension within a part.
- View the result of changing parts.
- Zoom, pan, and control the 3D view.
- Create named views.
- Shade and render a view to a file.
- Describe three benefits of parametric modeling.

Reference

Engineering Design Communication

Parametric Modeling Modes pp. 272–274

Overview of the Tutorials

In this guide, you will create parametric models and drawings for two complete assemblies. Tutorials 1–3 introduce you to the basic components of parametric modeling: parts, assemblies, and drawings. Tutorials 4–8 broaden your skills with Mechanical Desktop as you learn more techniques for modeling parts, assembling parts in an assembly, and creating drawings to document the assembly. In Tutorials 9 and 10, you will create a second assembly and incorporate labor-saving Mechanical Desktop techniques for making intelligent parts and assemblies, and for getting information from the model. You will begin in this tutorial by exploring the Mechanical Desktop 3D environment and learning about the commands you will use throughout this guide to make your modeling effective and efficient.

Introduction to Parametric Modeling

Most parametric modeling software, including Autodesk's Mechanical Desktop, has three or more aspects to the process of modeling: parts, which are models of objects that are typically manufactured from a single material in a single piece; assemblies, in which multiple parts are combined together into a functioning device; and drawings, which show 2D projected views suitable for printed documentation.

Separating these different modeling aspects in the software parallels the process of designing an assembly. Each individual part must be designed or specified so that it can be manufactured. Parts must fit together correctly so they can be assembled into a device or system. Drawings must be produced to document the parts and assemblies, to pass information along to the manufacturer and assembly workers, and for record keeping. Dividing software functions into these three functional areas allows specialization that improves performance and ease of use in doing each of those tasks. However, each of these things relies on the other. If you hand draft a part, you must draw it again to show it in the assembly. Using parametric modeling software, you can create an assembly by telling the assembly modeling portion of the software where to find the part and how to display it with respect to the other parts in that assembly. Likewise, to make a drawing, you only need to select a part or assembly and tell the software how you want the views arranged on the drawing sheet. The interaction between the part, assembly, and drawing modes is a key time saver in parametric modeling.

Tip

If you have not reviewed the system setup sections of the Getting Started chapter of this book, you should do so now.

Launch Mechanical Desktop

If you have completed the Getting Started chapter of this book, you are ready to launch Mechanical Desktop Release 5 and begin. For the early tutorials in this book, you do not need to launch Mechanical Desktop 5 Power Pack but may start from the Mechanical Desktop 5 shortcut on your desktop that you created in Getting Started. (You may choose to launch the Power Pack if you wish; your toolbars and menus will not match the screens in this book exactly, but you will be able to complete the tutorials.)

Pick: **Mechanical Desktop 5 shortcut** (on your desktop)

The Mechanical Desktop software appears on your screen. The Mechanical Desktop Today window appears in a central position on the screen, as shown in Figure 1.1

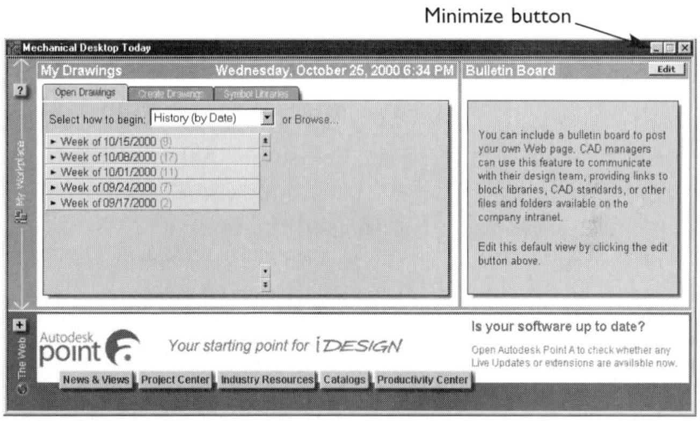

Figure 1.1

This window allows you to personalize your workspace and easily access resources on the Internet. The three tabs under My Drawings allow you to quickly access files you have used recently, start a new drawing, or load libraries of symbols to be used in your file. You will not use this window at this time. To remove it from your screen,

Pick: the minimize button in the upper right corner of the window

The Mechanical Desktop software is visible on your screen, as shown in Figure 1.2.

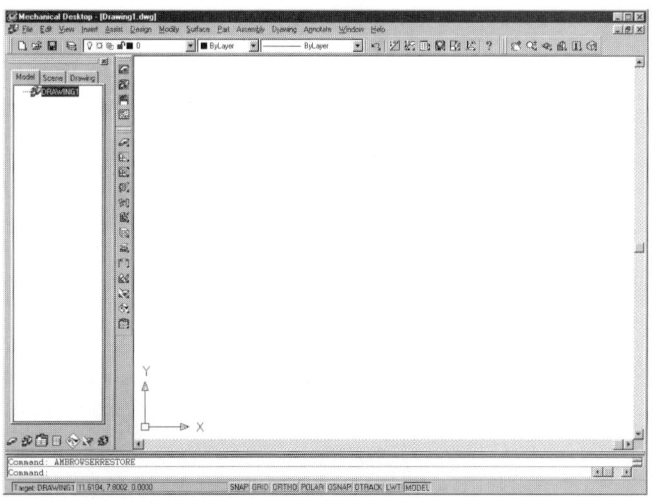

Figure 1.2

If you are using Mechanical Desktop in a lab setting, or one where you share it with other users, take a moment now to load the profile you created in Getting Started.

Pick: **Assist, Options** (from the menu bar)

Pick: **Profiles** tab

Pick: **Mechanical Desktop Tutorials** (to highlight it)

Pick: **Set Current**

Pick: **OK**

You should repeat these steps at the start of each tutorial after you launch Mechanical Desktop to set the preferences to those saved with your profile.

Now you will use the menu bar to open a finished version of the jack project that you will create during these tutorials.

Pick: **File, Open** (from the menu bar)

The Select File dialog box appears, as shown in Figure 1.3. The left side of the dialog box offers quick access to different locations on your hard drive and the Internet. The drop-down "Look in" menu offers access to the drives and directories on your hard drive. You will use the dialog box to select a data file that was provided with this text.

Warning

You should have previously copied these data files to your hard drive in a directory named C:\MDT-Data. You should also have made C:\MDT-Work the default Start in directory for Mechanical Desktop 5. If you have not done this, return to the Getting Started chapter and follow the directions for creating the directory and copying the data files to it.

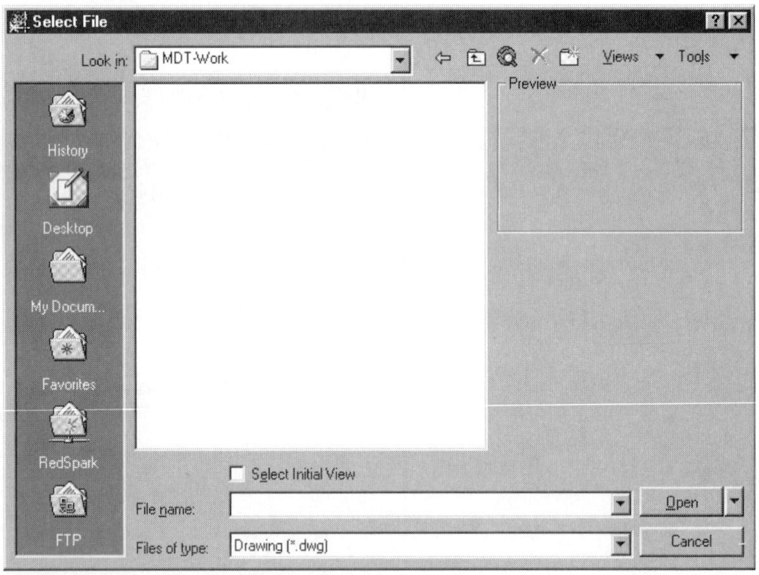

Figure 1.3

Pick: **the down arrow to the right of the Look In box**

The list of drives and directories for your computer will expand from the selection. On your own, select the drive and directory where your data files reside. Once you have this directory selected, the list should appear similar to Figure 1.4.

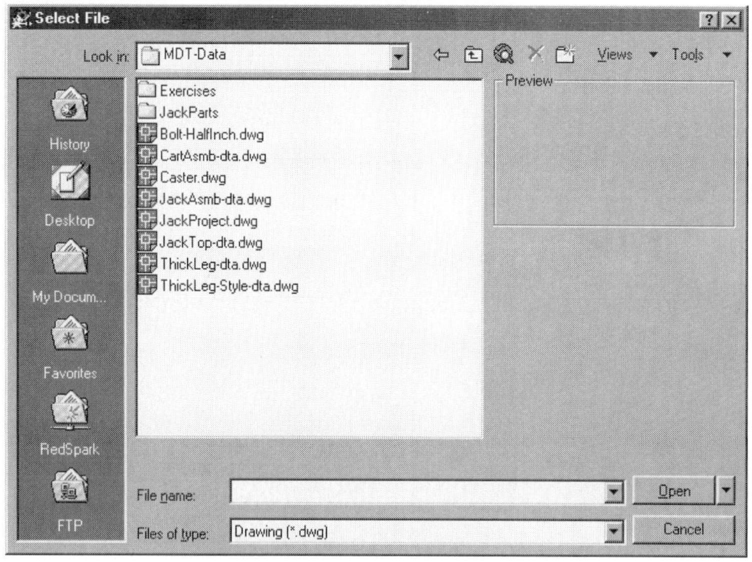

Figure 1.4

Tip

If you pick the file once, you will see a preview of its contents. You can then open it by picking Open.

Double-click: **JackProject.dwg**

The Mechanical Desktop Workspace

The jack project appears in the graphics area of your screen. Figure 1.5 shows the areas of the Mechanical Desktop screen.

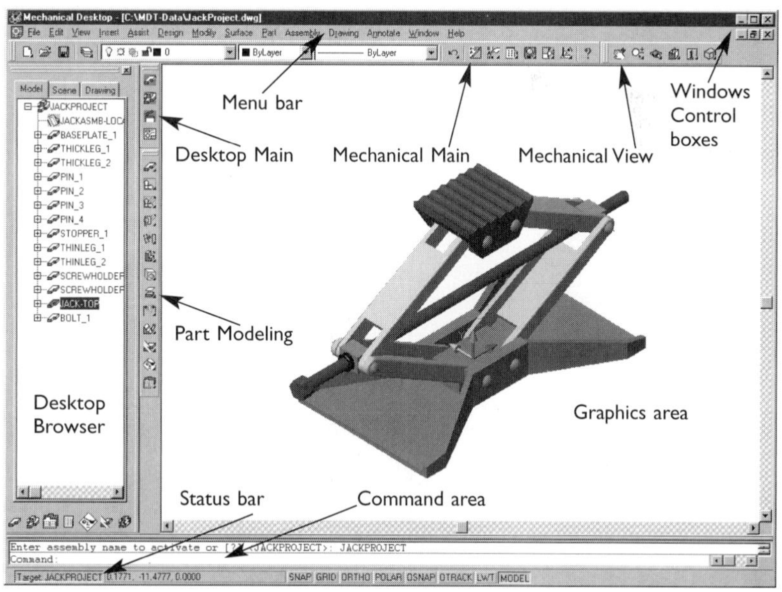

Figure 1.5

The screen is organized into windows according to the conventions used in most Windows applications. The two main areas of the screen are the Desktop Browser and the graphics area. The menu bar across the top offers access to all of the Mechanical Desktop commands. As you work, the status bar at the bottom of the screen displays information about active settings, and the command line area reports actions and prompts you for selections. You will refer to the command line area frequently to confirm the active command and to respond to prompts for user input.

Toolbars

Below the menu bar and arranged along the graphics area are command icons, which are organized into toolbars. These icons provide quick access to Mechanical Desktop commands. On your screen, you should see the Mechanical Main, Mechanical View, Desktop Main, and Part Modeling toolbars. You can turn toolbars on and off with the Toolbars selection on the View menu. You can also use it to display additional toolbars for commands you use frequently. For this guide, you will not need more than the default toolbars.

The Desktop Browser

The Desktop Browser and toolbars are "docked" so they fit snugly against the edges of the active windows. You can move them around on your screen by picking on the gray bars at their top (or left side) and holding the mouse button down as you drag to a new location. Figure 1.6 shows the Desktop Browser being moved on the screen. Try this now on your own.

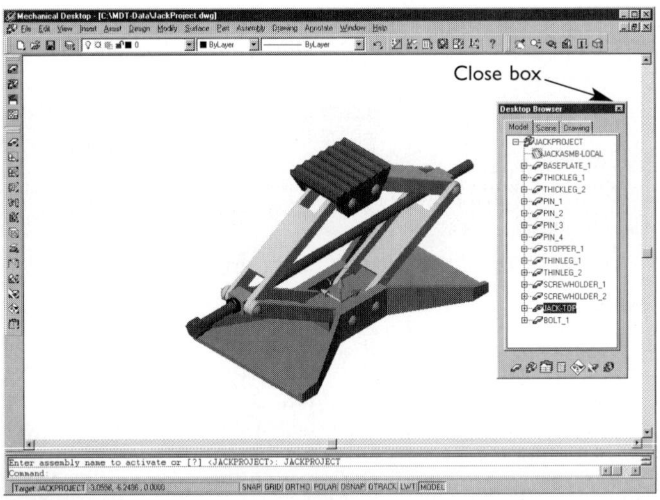

Figure 1.6

Drag the Desktop Browser so it becomes a floating window.

You can also turn the Desktop Browser on and off to make more working space on your screen. Next you will turn off the Desktop Browser, using the Windows close box (which looks like an X) located in its upper right corner. This is shown in Figure 1.6.

Pick: **the Browser's close box** (to close the Desktop Browser)

The Desktop Browser should now be removed from your screen. To turn it back on, use the View menu on the menu bar.

Pick: **View, Display, Desktop Browser**

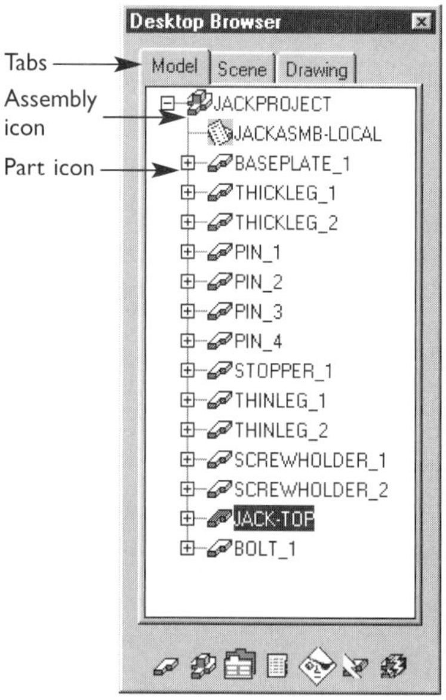

Figure 1.7

The Desktop Browser should reappear. When you have finished, locate the browser where you can use it conveniently or return it so that it is docked to the left side of the screen.

Dock the Desktop Browser to the left side of the screen.

The Browser and Parametric Modes

Each Mechanical Desktop project can contain different types of information in a single drawing file. Unlike some modelers, a single Mechanical Desktop drawing file may include information that often is stored as separate part files, assembly files, and drawing files in other parametric modeling software. The tabs on the Desktop Browser let you view different information stored in the file. It has three main tabs, identified in Figure 1.7: the Model tab, the Scene tab, and the Drawing tab.

The Model Tab

The Model tab shows the assemblies, parts, and features that make up the 3D parametric model.

Pick: **the Model tab** (so it is uppermost)

The Model tab lists the items making up the model in a "model tree," as shown in Figure 1.7. The list is organized to show the hierarchy of the items in relation to each other. At the upper end is the JackProject assembly. It is represented by an assembly icon—a pile of blocks that represent the multiple parts in an assembly. Beneath this level are individual part models, such as the Baseplate_1 and Thickleg_1 parts of the jack. Their icons look like a single L-shaped block to represent a single part.

Reference

Engineering Design Communication

Feature-Based Modeling, pp. 251–254

You can use the model tree to identify parts and features. You will use the tree to select the Jack-Top part.

Pick: **the Jack-Top part in the tree** (so it is highlighted)

Notice that this part is now highlighted on the actual model shown in the graphics window. You will expand the items below the Jack-Top part by selecting on the Plus sign (+) to its left.

Pick: **the + sign left of the Jack-Top part in the model tree**

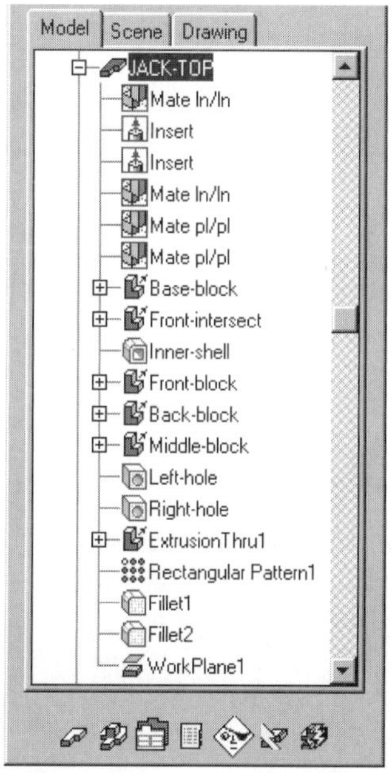

Figure 1.8

Each part model is made up of features, and the model tree displays a part's features in the order in which they were created. It also shows how the part is constrained to other parts in the assembly using mates and inserts. The constraints and the features making up the Jack-Top part appear listed below it, as shown in Figure 1.8. (You may need to use the scroll bar on the side of the Browser window to see the entire list.)

You can select these features in the tree and they will appear highlighted in the model shown in the graphics window.

Pick: **Left-hole**

The hole on the left side of the Jack-Top part becomes highlighted. Using Mechanical Desktop, you can easily access features to change their dimensions, location, or even type—even after they have been added to the model!

Next you will take a look at the Scene tab.

Pick: **Scene tab**

It should now appear uppermost in the stack, as shown in Figure 1.9.

Figure 1.9

The Scene Tab

The Scene tab allows you to show an assembly in different views to make it easier to visualize. You can quickly create exploded views like the one shown in Figure 1.10. A tree similar to the one used in the model tree shows the assembled and exploded scenes listed for the jack assembly. If you select the plus (+) sign, the parts will appear listed below each. You can use the Scene tab to change to the exploded scene.

Double-click: **Exploded** (from the Scene tree)

The exploded view of the assembly appears in the graphics window as shown in Figure 1.10.

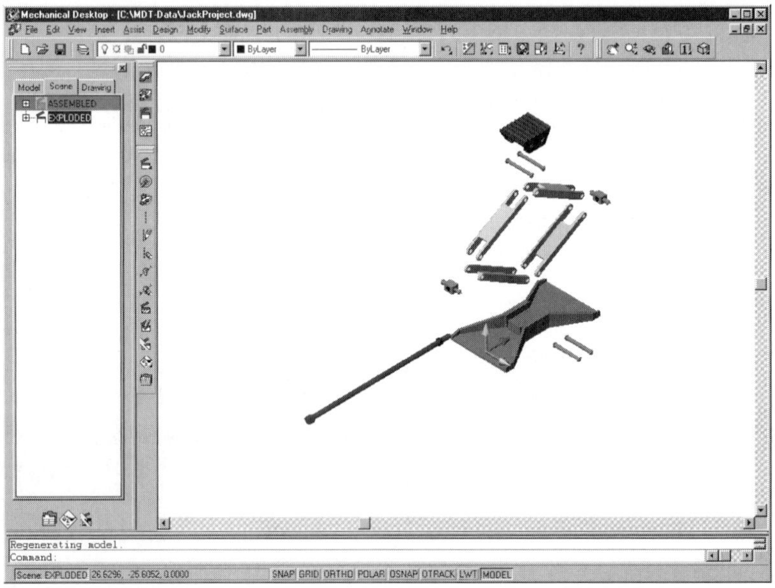

Figure 1.10

You can use the Scene tab to set the explode factor for the assembly or to tweak the location, rotation, or translation of individual parts to improve the appearance of the scene. Context menus associated with items in the Browser Window change to reflect those commands that relate to the active mode. To see the options available for items in the Browser window when the Scene tab is uppermost, you click on an item with the right mouse button. You will do this now to see how the Explode factor works.

Right-click: **EXPLODED** (in the Scene window)

Pick: **Explode factor** (from the menu that appears)

The Explode factor dialog box appears on your screen. The Explode factor is a numerical value used to set the magnitude with which parts are moved away from each other in an exploded view. To change it, you will highlight the value contained in the data entry box and enter a new value, as shown in Figure 1.11.

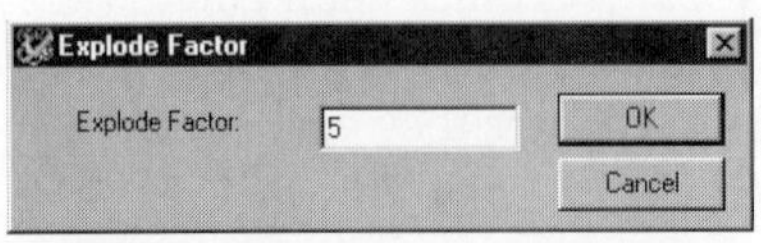

Figure 1.11

Type Explode Factor: **5**

Pick: **OK**

The scene should now appear as shown in Figure 1.12 Each element in the assembly was moved farther apart than before. Since the exploded view was originally created by moving each part individually by tweaking (which you will learn about in Tutorial 8), the change is not an improvement. You will reset the explode factor to zero on your own.

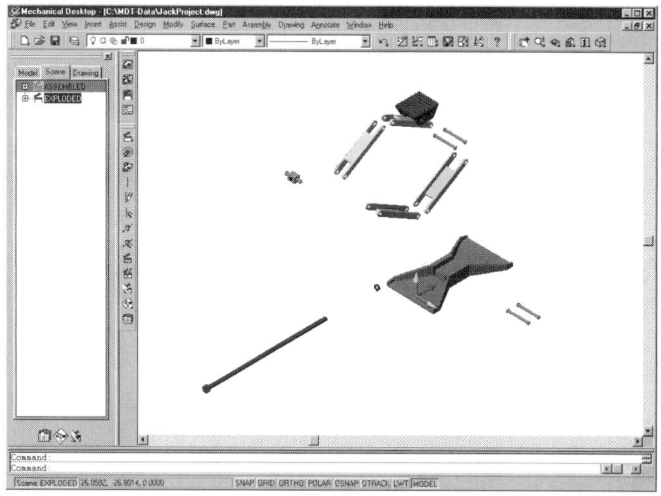

Figure 1.12

Repeat the steps above to the reset the explode factor to zero.

To switch to the Drawing tab,

Pick: **Drawing tab**

The Drawing Tab

The Drawing tab should now appear on top, as shown in Figure 1.13. The Drawing tab is used to create and organize layouts. Each layout is equivalent to a drawing sheet that shows printable views of the modeled parts or assemblies. Layout1 for the JackProject shows a set of orthographic views of the Top part. These views were generated directly from the model of the part. They were not redrawn as two-dimensional views.

Tip

Refer to the Getting Started chapter of this guide for more information about context menus and the many ways in which you may activate commands in Mechanical Desktop.

Tip

If you do not see the New View selection, you may have to select Drawing Menu from the bottom of the context menu.

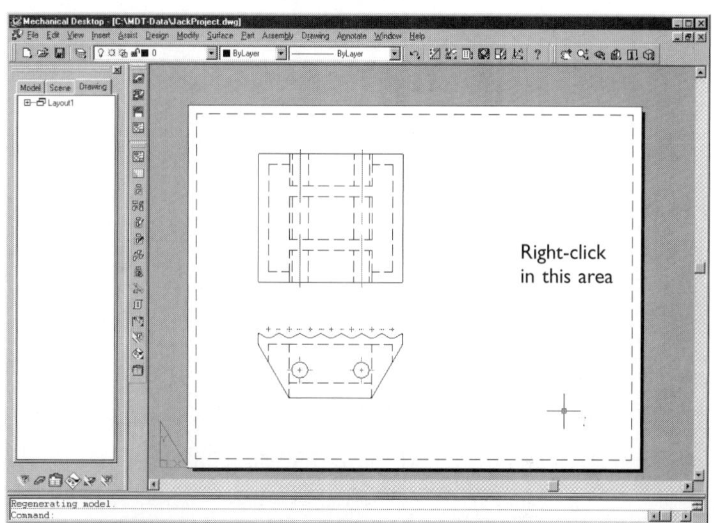

Figure 1.13

In the following steps, you will add the right side view to the drawing. To do so, you will right-click in the graphics area and select from the context menu for layouts.

Right-click: in the graphics area

Pick: **New View** (from the context menu that appears)

The Create Drawing View dialog box appears on your screen. You will use the drop-down menu next to View Type to set it to the selection shown in Figure 1.14.

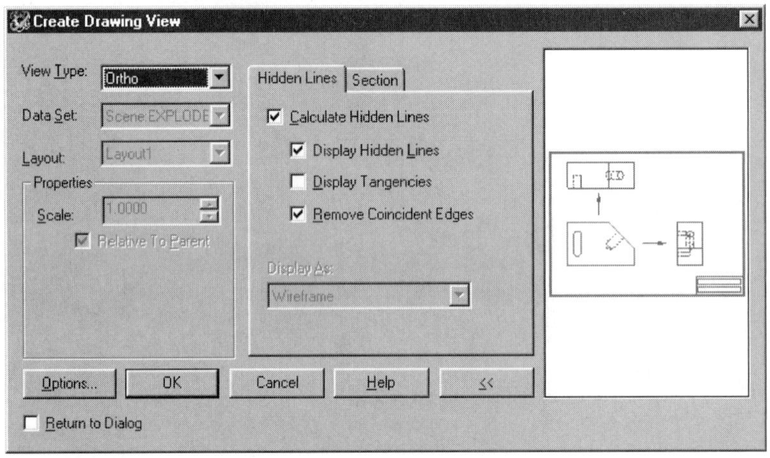

Figure 1.14

Choose View Type: **Ortho**

Pick: **OK**

You will return to the graphics screen for the final selections.

Select parent view: **pick the lower view as shown in Figure 1.15.**

Specify location for orthogonal view: **pick to the right of the previous selection** (referring to Figure 1.15.)

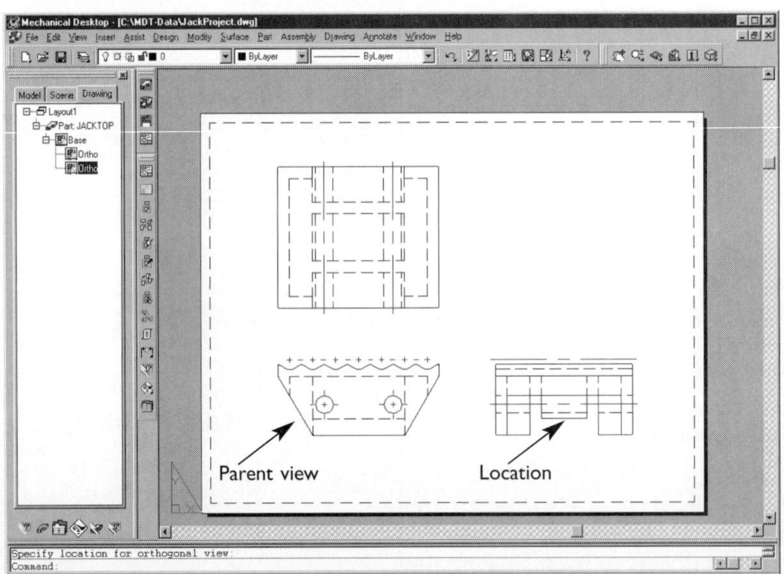

Figure 1.15

Tip

To end commands you will often press ↵. Many commands continue so you can repeat an action, such as selecting a view location, without restarting the command.

Tip

The tutorials assume that you have set MDT-Work to be your default directory. If you did not, refer to Getting Started, pp. 6, for instructions.

Reference

Engineering Design Communication

Storing Electronic Files, pp. 344-347

Specify location for orthogonal view: ↵

The new view should appear on your screen as shown in Figure 1.15. The Drawing tree in the Browser Window expands to show the new view in relation to the part and to the other views of the part model.

Notice that the correct orthographic view and properly shown hidden and center lines were generated automatically. If you have ever created orthographic drawings by hand, this should have been extremely satisfying. But it gets even better, as you will see after you have set up the file for further changes!

File Management and Save As

Before you make additional changes to the JackProject file, you will save it in your MDT-Work directory so the original data file is not changed.

Managing your files is an important part of learning to use Mechanical Desktop effectively. Throughout these tutorials, you will be encouraged to name files descriptively and save them in an organized manner. Generally, you should keep all of the files for a project together, with all of the files for an assembly in a directory, and with subdirectories as needed for subassemblies or other file types. For these tutorials, you will save all of the files you create in the MDT-Work directory, but you will create a separate subdirectory for the files you make for the jack assembly and the cart assembly you will make later. If you did not create the MDT-Work directory in the Getting Started chapter of this book, do so now before continuing.

Using Save As

When you work with a data file (such as the JackProject that you opened from MDT-Data), you should use Save As to save the file to your MDT-Work directory. The Save As command opens the Save As dialog box, where you will specify a different name and/or location for the file. (You don't need to enter the *.dwg* suffix; it is added automatically.)

Pick: **File, Save As**

Choose Save in: **MDT-Work** (or the directory you are using for your files)

Type File Name: **JackProject-2**

Pick: **Save**

The file is saved to the MDT-Work directory and becomes the currently active file. The new name, *JackProject-2.dwg*, appears at the top of your screen and as the name of the assembly on the Model tab. The original file is unchanged in the MDT-Data directory.

Backing Up Your Files

Saving your files in an organized way also makes it easy to make back-up copies on a regular basis. Each time you save your file using the Save command, Mechanical Desktop writes a new copy of the file to the current location and file name and creates a copy of the file with a *.bak* extension. If the *.dwg* file is corrupted or lost, you can use the *.bak* file to return to the last saved version. Save your file now, then use the Explorer to see the files written to your MDT-Work directory.

Pick: **File, Save**

Pick: **Start, Windows Explorer** (and navigate to the MDT-Work directory)

You should see two files in your MDT-Work directory: one *.bak* and one *.dwg*. The *.bak* files saved are no protection against hard disk failure. You should use the operating system (Windows Explorer) to copy new and changed files to a different back-up disk after each session so none of your data is lost.

Close Windows Explorer on your own.

Making Changes in a Parametric Model/Assembly

The advantage of using parametric modeling (sometimes called feature- or constraint-based modeling) is that the model can easily be changed by changing the sizes and relationships that define it. You can change the size of one feature and that change will be made throughout the assembly, scenes, and drawings for that project. You will change the length of the ThickLeg part and see the effect of that change on the assembly and drawings.

Pick: **the Model Tab**

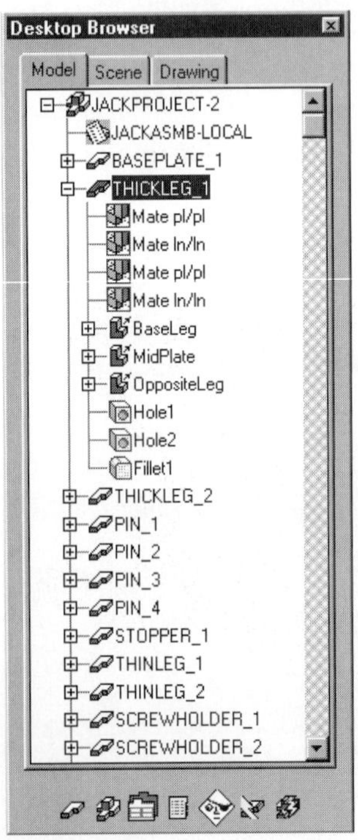

Figure 1.16

You see the tree structure showing the jack assembly and its parts. Pick to expand the tree below the Thickleg_1 part, as shown in Figure 1.16.

Select the ThickLeg_1 part of the jack. It becomes highlighted in the graphics window. Next you will change the overall length of the ThickLeg part by modifying its base feature. You will see how easy it is to make changes to a part and assembly.

Right-click: **ThickLeg_1** (in the model tree)

Pick: **Activate Part**

Right-click: **BaseLeg** (under ThickLeg_1 in the model tree)

Pick: **Edit**

The Extrusion dialog box appears on your screen, as shown in Figure 1.17. You can use it to change the distance that this feature is extruded, its draft angle, and the termination for the feature.

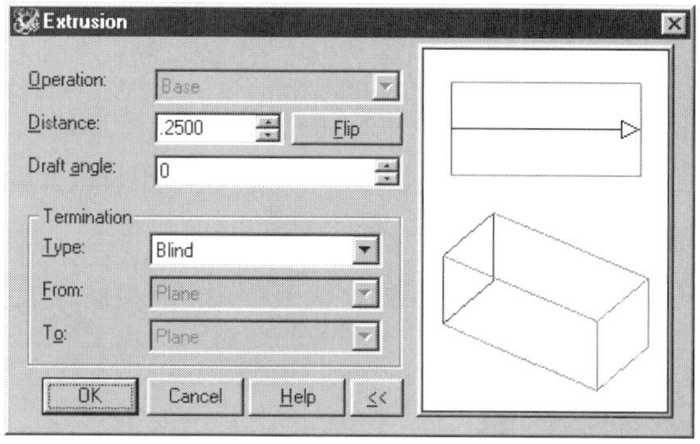

Figure 1.17

You will learn more about making features using extrusions and its options in the next tutorial. For now,

Pick: **OK**

The extrusion dialog box is removed from the screen, but the dimensions for the BaseLeg feature still remain on the screen. At the command prompt,

Select object: **pick on the 6.50 dimension** for the BaseLeg feature

Enter dimension value <6.5>: **7.50** ↵

Solved fully constrained sketch.

Select object: ↵

Right-click: in the graphics area

Pick: **Update Part**

The ThickLeg part updates to show the 7.50 length, as shown in Figure 1.18.

To update the assembly to reflect this change,

Right-click: in the graphics area

Pick: **Assembly Menu**

Pick: **Update Assembly**

The assembly updates to fit with the new length of the ThickLeg. Figure 1.19 shows the result.

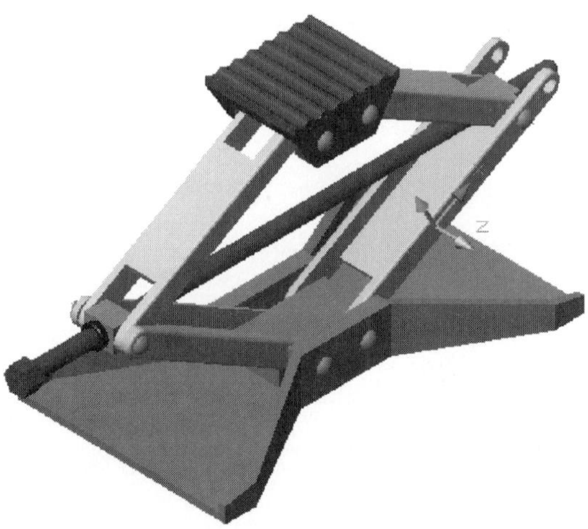

Figure 1.18

Figure 1.19

Reference

Engineering Design Communication

Transformations, pp. 187–191

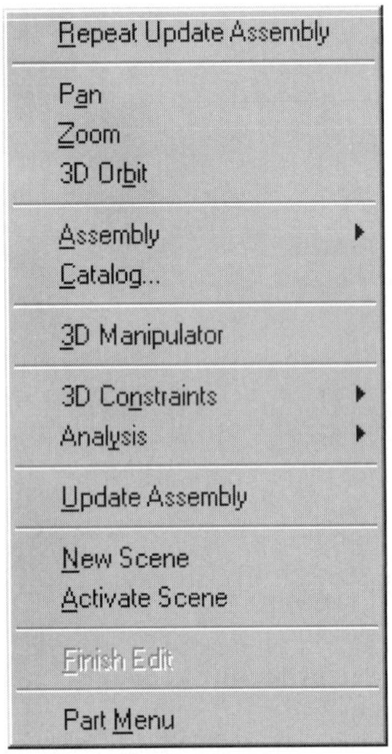

Figure 1.20

Reference

Mechanical Desktop On-line Help

The Zoom command has many options for use, which you will learn as you proceed through the tutorials. Use the Command Reference in AutoCAD Help for a complete explanation.

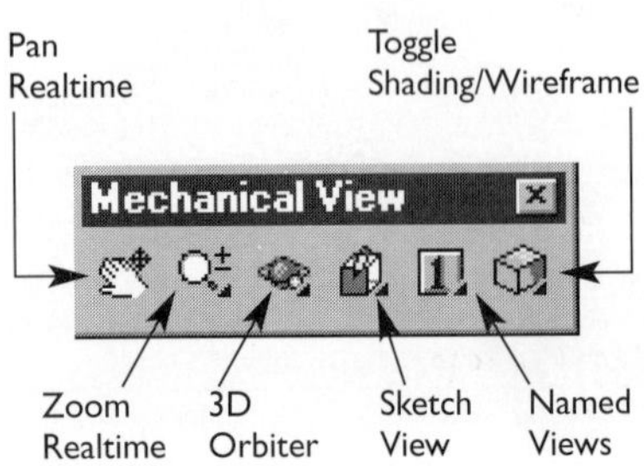

Figure 1.21

Now you have seen some of the power of parametric modeling as a design tool. Next you will learn to control your view of parts and assemblies so that you can work with them better.

Controlling Your View

You may already be familiar with the types of viewing controls provided in Mechanical Desktop. Many of them are standard AutoCAD viewing tools, and some are common to other graphics programs as well. Knowing how to use the viewing tools is essential to effective modeling. You need to be able to orient objects on your screen to make it easier to select the portions you are working on, and also to help you (and others) visualize features. There are several ways to access the viewing controls that will be presented in these tutorials. As you become familiar with them, you can use the means that suits you best. First, you will use the Context menu to explore the viewing controls.

Right-click: in the graphics area

The Context menu shown in Figure 1.20 appears on your screen. You can use it to quickly activate Realtime Pan and Zoom. You will use the Realtime Zoom command to enlarge the view of the object.

Pick: **Zoom** (from the context menu)

Realtime Zoom

Realtime Zoom invokes the Zoom command and keeps it active until you exit. The cursor changes to a magnifying glass with a plus (+) and minus (−) sign next to it. To enlarge the view by zooming in, hold down the mouse button and roll the mouse away from you. To make the view smaller, or zoom out, hold down the mouse button and roll the mouse toward you.

On your own, enlarge the view of the jack so that it fills the drawing area.

While you are in Realtime Zoom (or Pan), the context menu changes to offer more commands for controlling the view, so you can change from parallel to perspective projection, change the shading method used, select default views, and reset the view among other things. To exit Realtime Zoom, you can press [↵] or [ESC] or use the Context menu, as you will here.

Right-click: in the graphics area

Pick: **Exit** (from the context menu)

To explore the Realtime Pan command, instead of using the context menu, you will use the Mechanical View toolbar. The Mechanical View toolbar is shown in Figure 1.21. (If it is not on, you can turn it on by picking View, Toolbars, Desktop Toolbars, Mechanical View.) You can use the band at the left end of the toolbar to drag it around your screen, as you did with the Desktop Browser earlier in this tutorial. Most of the common viewing commands on the View menu are also located on this toolbar for easy access.

Next you will use the Mechanical View toolbar to pick the Pan Realtime command.

Tip

If you keep the cursor over a toolbar button for a length of time, its tooltip will appear, listing the name of the button. A help line describing its function and command name will appear in the status bar.

Pan Realtime

The Pan Realtime command allows you to position the view of the object on your screen.

Pick: **Pan Realtime icon**

When Pan is active, the cursor changes to the shape of a hand. Place the hand where you would like to grab the object and hold the mouse's pick button down while dragging the object view around on the screen. The Pan command moves the view of the object on the screen only, and does not move the objects themselves on the coordinate system. You must exit Pan Realtime as you did Realtime Zoom; you will do so by pressing [ESC].

On your own, position the view near the center of the screen.

Press: [ESC]

Preset Views

Preset views let you orient your view of the 3D space on your screen. All of the standard orthographic views (top, front, right side, etc.) and several isometric views are available as preset views. All of the viewing controls are available from the View menu, which you will use to explore the preset views. Make sure the Model tab is uppermost in the Desktop Browser. The jack assembly should be visible in your graphics area.

Pick: **View, 3D Views, Back Left Isometric** (from the menu bar)

The view of the jack assembly changes to an isometric view seen from the back and left, as shown in Figure 1.22.

Figure 1.22

Next you will view the jack assembly from the top.

Pick: **View, 3D Views, Top**

Your view changes to a top view of the jack, as shown in Figure 1.23.

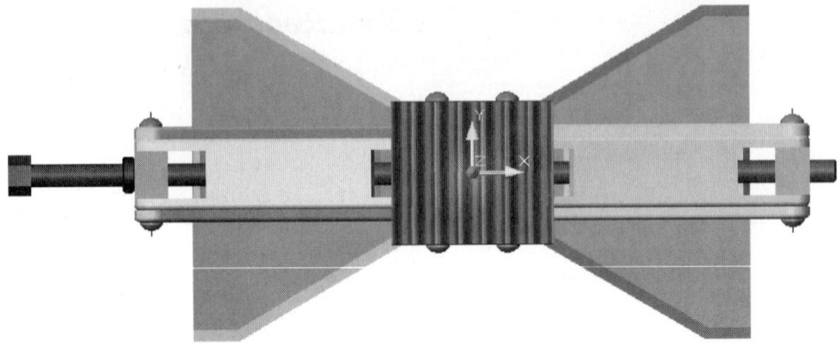

Figure 1.23

The front, back, left, right, and bottom selections from the menu work similarly, as do the other isometric selections. Experiment with them on your own. Next, return your view to an isometric selection.

Pick: **View, 3D Views, Back Right Isometric**

The back right isometric view of the jack should show in your graphics window.

The 3D Orbiter

In addition to the preset views, you can use the 3D Orbiter to spin your view so that it is oriented in whatever direction you want. The 3D Orbiter, as well as the other viewing controls, is accessible from the Mechanical View toolbar.

Pick: **3D Orbit icon** (from the toolbar)

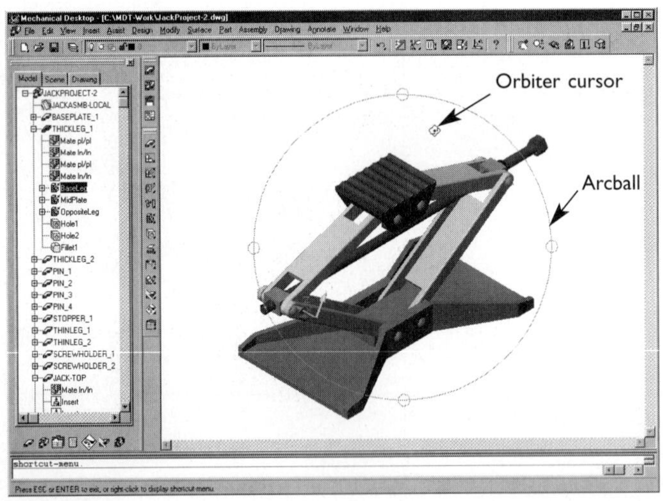

Figure 1.24

The cursor changes to the orbiter cursor and the ***arcball*** appears on your screen, as shown in Figure 1.24. The arcball is a circle divided by four small circles into quadrants. In the 3DORBIT command, the center of the arcball (the target) stays stationary. The point from which you are viewing (the camera) moves around the target.

Now, move your cursor over different parts of the arcball. Notice the change in the orbiter cursor. When you move your cursor inside the arcball circle, the cursor looks like a sphere with circular orbits. Hold down the pick button and drag the sphere around inside the arcball. You

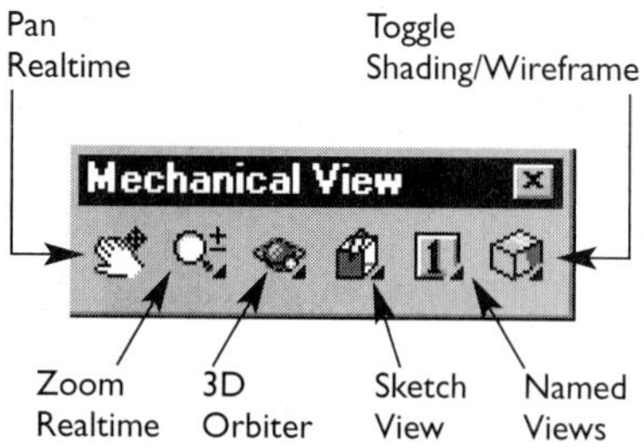

can rotate the view freely. Your view might appear something like that shown in Figure 1.25. In this mode, it is as if your cursor were grabbing a sphere surrounding the objects and dragging it around the target point. Try rolling your mouse horizontally, vertically, and diagonally.

Figure 1.25

Now move the cursor outside the arcball. The cursor changes to a sphere circled by an arrow. Click and drag the cursor around, keeping it outside the arcball. Notice that the view "rolls" around an axis that extends through the center of the arcball, perpendicular to the screen. Practice on your own rolling the view using this mode.

Move your cursor over the small circle at the left of the arcball. The cursor now looks like a sphere surrounded by a horizontal ellipse with a horizontal line representing the *Y*-axis. Click and drag your mouse from this point. Notice that the view rotates around the *Y*-axis through the center of the arcball. The same thing results if you select the circle on the on the right side of the arcball.

Next move the cursor to the small circle on the top of the arcball. Click and drag your mouse here to see how the view rotates about the *X*-axis. The *X*-axis is shown as a horizontal line on the cursor icon.

Toolbar Flyouts

Some toolbar icons, such as the 3D Orbit, Zoom, and Sketch View icons, are also *flyouts*. Flyouts are indicated by a small triangle in the lower right corner of the button. A flyout button expands so that you can select additional tools when you pick on it and keep the mouse button held down.

Pick and hold: **the 3D Orbit flyout** (from the Mechanical View toolbar)

The flyout expands, as shown in Figure 1.26. The flyout contains tools that allow for more advanced choices in viewing the object in 3D. The buttons available on this flyout are labeled in Figure 1.26.

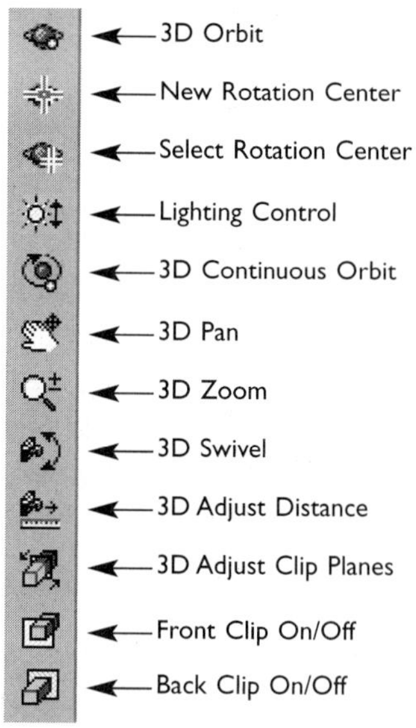

Figure 1.26

Tip

The Zoom button is a special kind of flyout that makes it easy to keep your favorite Zoom option available. After you choose one of the Zoom options, such as Zoom All, it becomes the topmost button so you can select it without opening the flyout.

Setting the Rotation Center

This option allows you to set the center about which the objects rotate when you are using the 3D Orbiter.

Pick: **New Rotation Center icon** (from the flyout)

Reference

Mechanical Desktop On-line Help

You can review AutoCAD selection methods by selecting AutoCAD Help from the Help menu and then expanding the heading under User's Guide in the Contents list. Expand Editing Drawings in the list and choose Selecting Objects.

The objects may reposition on your screen, near the center of the 3D Orbiter. Use it now on your own, to spin the jack assembly.

You may not always want to rotate a part or assembly about its calculated center point. If so, you can select a center of rotation by using Select Rotation Center. This command allows you to type in coordinates of the rotation center or select a point using any AutoCAD selection methods.

Pick: **Select Rotation Center icon** (from the flyout)

Select rotation center: **0.9,–6.9,2.75** [Enter]

The jack assembly now is positioned so that the lower right corner of the base is the rotation center in the 3D Orbiter, as shown in Figure 1.27.

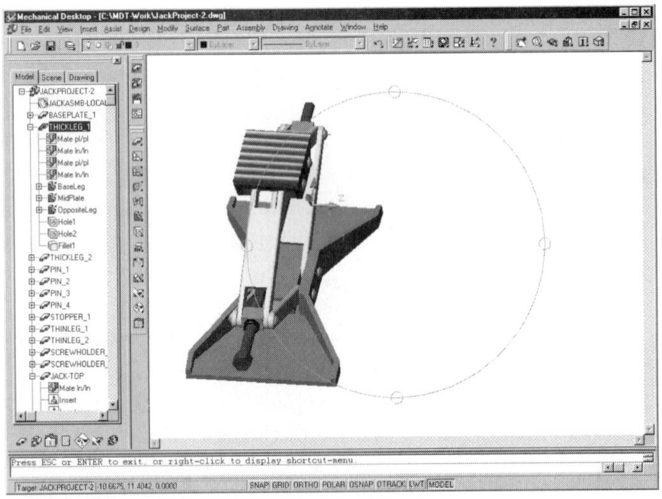

Figure 1.27

Use the orbiter on your own to spin the jack around this center.

To restore the assembly center for spinning the part

Pick: **New Rotation Center icon**

The center of the assembly should now be central in the orbiter and the parts should spin around that center.

3D Continuous Orbit

The 3D Continuous Orbit command allows you to start the part spinning in a direction set by the motion of your mouse and have it continuously spin in that direction until you press [Enter]. You can use this as a general way to check the part. It is also useful when showing the part to someone else, so that they can inspect all sides of it.

Pick: **3D Continuous Orbit icon**

Drag your mouse in the direction you want to start the part spinning.

Dragging the mouse slowly spins the part more slowly on the screen. Dragging the mouse more quickly spins the part quickly. To stop the parts' spinning,

Press: [ESC]

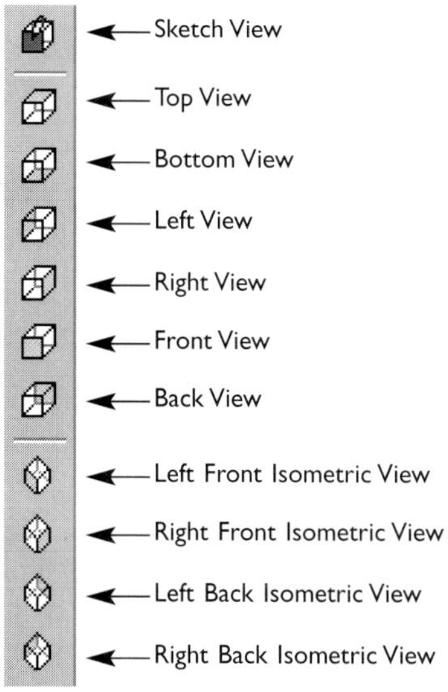

Figure 1.28

The parts will stop in the position that you stopped spinning them. Use the Sketch View flyout to return to a better view. The Sketch View flyout contains all the preset views listed on the View menu, as shown in Figure 1.28.

Pick: **Left Front Isometric View icon** (from the Sketch View flyout)

3D Swivel

The 3D Swivel changes the view as if you were turning a camera on a tripod. For example, turning the camera to the right would move the view of the object to the left. If you point the camera up, the view of the objects would move down. You can think of the location of the cursor as the camera's viewfinder.

Pick: **3D Swivel** (from the 3D Orbiter flyout)

The cursor changes to the shape of a camera.

On your own, experiment with the 3D Swivel command to change the 3D view.

3D Adjust Distance

The 3D Distance command controls view magnification similar to the way objects look larger when they are closer to the camera. When it is selected, the cursor changes to a line with one upward and one downward arrow. Dragging the cursor toward the top of the screen moves the camera closer to the objects, making them appear larger. Dragging the cursor toward the bottom of the screen moves the camera away from the objects, making them appear smaller.

Experiment with the 3D Distance command on your own.

When you are finished, exit the 3D Orbit command.

Right-click: in the graphics area

Pick: **Exit**

Named Views

When you are working with a model, you will switch between views to help you select and view the model. In addition to the preset views, Mechanical Desktop allows you to assign a name to any view and save it so it can be activated as preset views are. Named views can save the center point, viewing direction, zoom factor, and perspective (distance from the model). To create a named view,

Pick: **View, Named Views** (from the menu bar)

The View Dialog box opens, as shown in Figure 1.29.

The Named View tab displays the list of named views; the preset orthographic and isometric views are listed on the other tab.

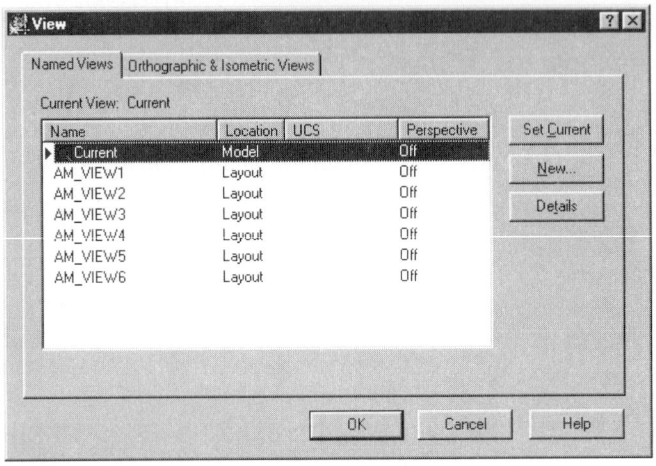

Figure 1.29

Pick: **New**

The New View dialog box appears, as shown in Figure 1.30.

The dialog box allows you to assign a name to the view as it currently appears in the graphics area. It also allows you to specify whether you want to save a coordinate system with the view. (You will learn how to align the user coordinate system with your model in the next tutorial.) To save the view settings as they are now on your screen, you will assign a name to the current display.

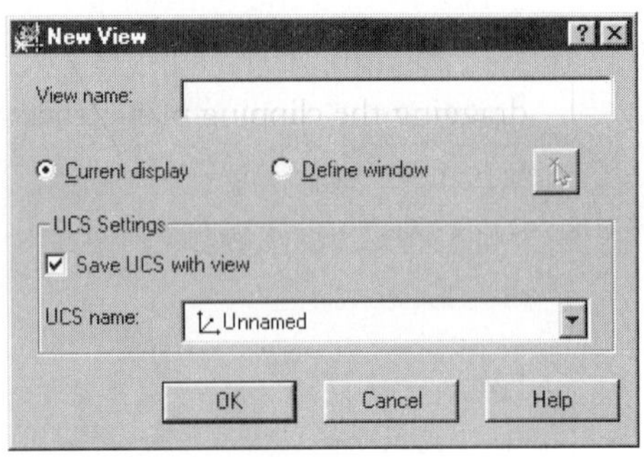

Figure 1.30

Type View name: **Sample View**

Pick: **OK**

The Sample View appears in the list in the View dialog box. You will change to an orthographic view, then return to your saved view.

Pick: **Orthographic and Isometric Views tab**

Pick: **Top**

Pick: **Set Current**

Pick: **OK**

The display changes to show the top view. To make your saved view current,

Pick: **View, Named Views**

Pick: **Named Views tab**

Pick: **Sample View** (in the list on the Named View tab)

Pick: **Set Current**

Pick: **OK**

Your saved view is restored in the display. Saved views are saved with your *.dwg* file.

Clipping

Clipping is the general technique of displaying only that portion of the object on one side of an invisible plane, called the *clipping plane*. You can think of the clipping plane as slicing through the object. You can use clipping to create section views of the objects you are modeling so you can see how interior parts fit together. You can also use it dynamically as a 3D viewing control to create section views of the part on your screen.

3D Adjust Clipping Planes

The 3D Adjust Clipping Planes command allows you to position a front clipping plane, a back clipping plane, or position them both to create a slice through your object(s). To make it easier to see the position of the planes, you will change to a Top view before invoking the command.

Pick: **View, 3D Views, Top View** (from the menu bar)

Pick: **3D Adjust Clip Planes icon** (from the 3DOrbit flyout)

The Adjust Clipping Planes window appears on your screen, as shown in Figure 1.31.

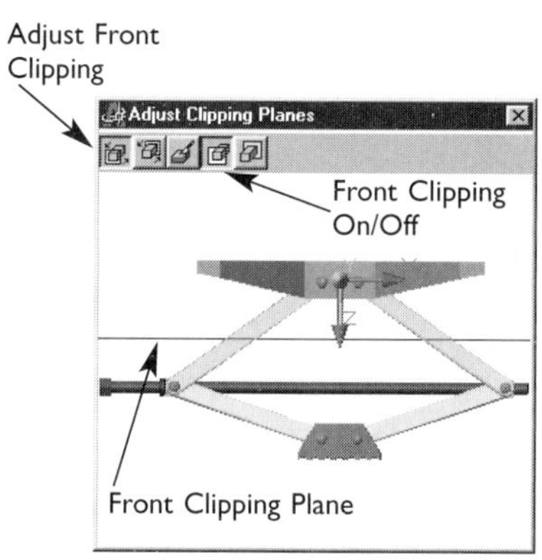

Figure 1.31

The window shows the objects in your current view rotated at 90 degrees. The Adjust Clipping Planes window has its owns icons. Using them, you can adjust the front clipping plane and back clipping plane, create a slice by dragging the clipping planes together as a group, or toggle the view so that it is either clipped or not by the front or black clipping planes.

Pick: **Adjust Front Clipping icon**

First you will adjust the clipping planes by dragging the line that represents the edge view of the clipping plane. While you do this, Mechanical Desktop will update the view displayed in the 3D Orbiter and clip the view of the object. The Front Clipping On/Off icon should appear highlighted (depressed) as shown in Figure 1.31, in order for you to see the effect of clipping in the 3D Orbiter.

Drag the clipping plane line in the Adjust Clipping Planes window.

Your view will be clipped similarly to Figure 1.32.

Next, you will turn Back Clipping On and set the location of the back clipping plane.

Pick: **Back Clipping On** (so it is highlighted)

Pick: **Adjust Back Clipping Plane**

Drag the green/red line in the Adjust Clipping Planes window
(to see the back of the object clipped off).

With both clipping planes positioned, you can use your mouse to drag both clipping plane lines as a group to create a slice though your object. Front and Back Clipping should be on in order to view this.

Pick: **Create Slice icon**

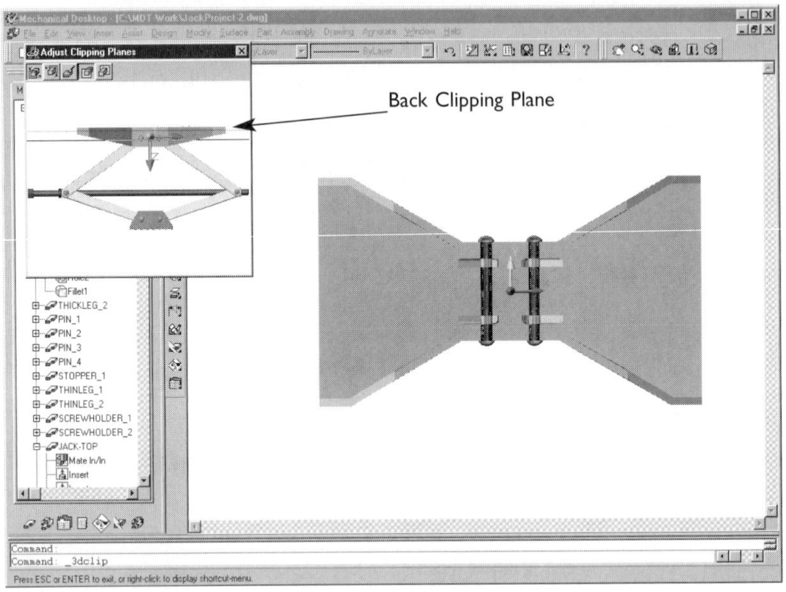

Figure 1.32

Drag both clipping planes in the Adjust Clipping Planes window (to move the slice).

When you have finished experimenting with this feature,

Turn Front and Back Clipping off on your own.

Pick: **the close box on the Adjust Clipping Planes window** (which looks like an X in the upper right corner)

Press: ESC (to exit the 3D Clipping command)

Shading Modes

You can also control the 3D appearance of the object by selecting a shading method. Shaded views add realism and aid visualization, but the more realistic the shading, the more computing power must be used to generate the view on your screen. This can make it take longer to spin in the orbiter or complete other viewing operations.

Fully shaded objects can also be more difficult to select for modeling operations. When you are working on an object, you will often want to be able to see vertices, edges, and surfaces on the back of the part to make selecting easier. You can do this by changing from a shaded to a wireframe view. A wireframe view depicts the object as if were a sculpture made of wires.

Toggle Shaded/Wireframe

Buttons for controlling the shading of the model are easily accessible on the Mechanical View toolbar (and the View menu). You will change back to an isometric view before exploring these.

Pick: **Left Front Isometric icon** (from the Sketch View flyout)

Pick: **Toggle Shading/Wireframe icon** (from the toolbar)

The view of the jack assembly changes from the shaded view you have been seeing to the wireframe view shown in Figure 1.33.

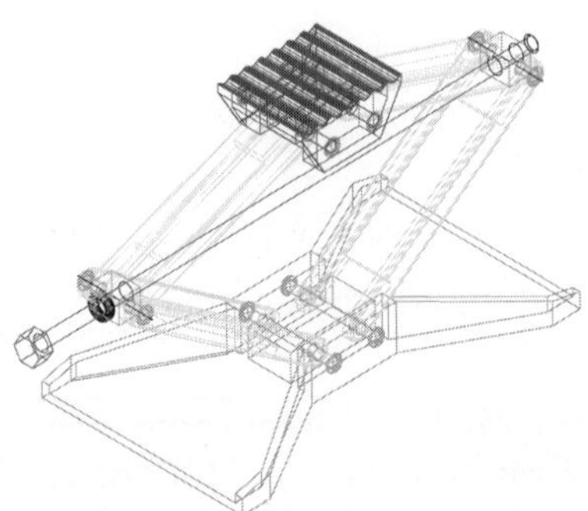

Figure 1.33

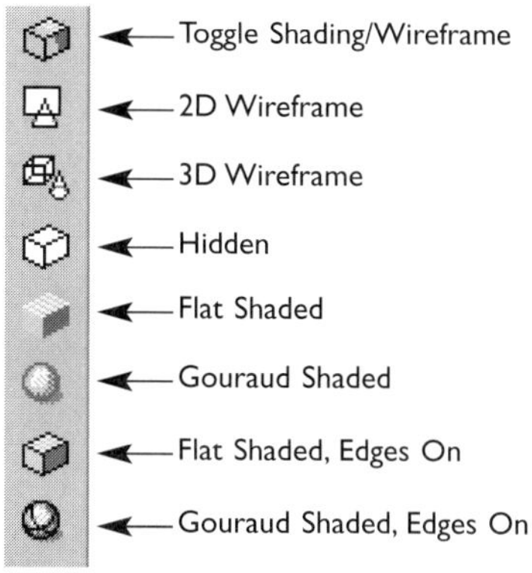

Figure 1.34

To change back to a shaded view,

Pick: **Toggle Shading/Wireframe**

Your view appears shaded once again.

Other shading commands are on the flyout under the Toggle Shaded/Wireframe icon. They are labeled in Figure 1.34.

Hidden

The Hidden option displays the model as wireframe but hides the back edges of the model for a more realistic view.

Pick: **Hidden**

The view changes to show the model with the back edges hidden, as shown in Figure 1.35.

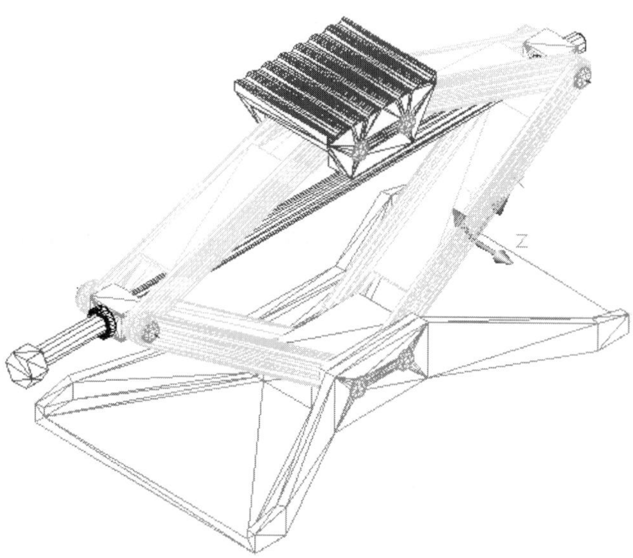

Figure 1.35

Flat and Gouraud Shaded

Two shading options are available, both with the option of showing the model edges over the top of the shading or not. You will first enlarge your view of the Jack-Top and pins so you can see the difference between the options more clearly.

Pick: **Zoom Realtime icon**

Drag to enlarge the view so you see the Jack-Top and pins in it.

Pick: **Flat Shaded**

Figure 1.36

Figure 1.37

Flat shading is now used to display the model, as shown in Figure 1.36. Flat shading uses a continuous tone for the color of each surface that is determined by how much light is striking it. For planar objects flat shading appears reasonably realistic. Flat shading takes less computer processing time than does Gouraud shading.

Pick: **Gouraud Shaded**

Now the display of the model changes to use Gouraud shading, as shown in Figure 1.37. Gouraud shading calculates a color for each vertex on the object depending on how much light is striking it, and then blends each set of vertices to determine the color for the pixels making up the flat surface. Rounded objects look more realistic with this type of shading, as you can see in the display of the rounded ends of the pins.

On your own, experiment with the selections Flat Shaded, Edges On and Gouraud Shaded, Edges On.

Return to a Left Front Isometric view when finished.

Printing a Rendered View

You can print the shaded display of your model using the Print command, but Mechanical Desktop offers more options for printing shaded views with its Render command. This command offers additional shading algorithms, allows you to control the colors and resolution of the printed image, and lets you assign materials and control the lighting and background for your image. You can render an image to the screen or to a file that you can import into another application. You will render the Jack assembly to a file that you can print from your word processor to see how Render works.

Reference

Engineering Design Communication

Rendering Options and Algorithms, pp. 580–585

Reference

Mechanical Desktop On-line Help

For a complete explanation of rendering options, choose AutoCAD Help from the Help menu and select Render from the list of Commands.

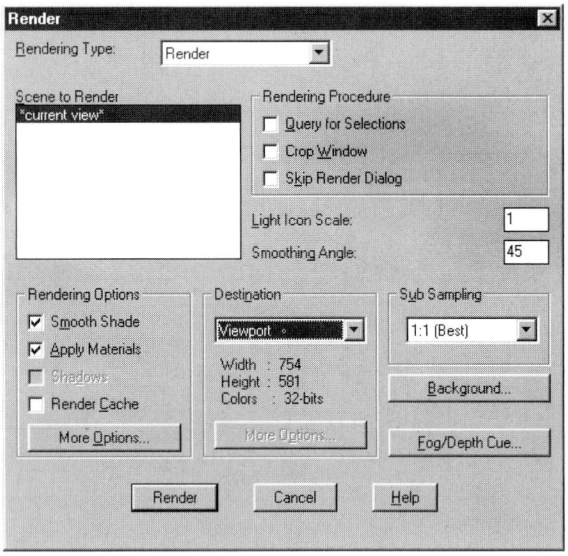

Figure 1.38

Pick: **View, Render, Render**

The Render dialog box appears, as shown in Figure 1.38.

In the bottom left, you can choose the rendering options to apply.

Pick: **More Options**

The dialog box that appears allows you to choose a rendering method. In addition to Gouraud shading, you can select a higher-quality option, Phong shading. Leave it set to Gouraud, the default setting.

Pick: **Cancel**

In the Destination area, you may choose to render to a file, a viewport in Mechanical Desktop, or the Render window. When you invoked the Render command, a Render window opened. It is now minimized on your taskbar. The Render window is useful for quickly previewing, saving, and printing files in *.bmp* format. Rendering to a file, however, offers more file formats and resolutions to choose from and will generally produce better results. You will render the current view of the Jack assembly to a file, so you will choose File in the Destination area of the dialog box and then choose more options to adjust the file type and resolution.

Choose Destination: **File** (from the drop-down menu)

Pick: **More Options** (in the Destination area)

Reference

Engineering Design Communication

Graphics Formats for Reports and Presentations, pp. 572–574

The File Output Configuration dialog box appears. You will set the options as shown in Figure 1.39 to render a file that is 1024 pixels wide (so it will be approximately 3 inches wide on a 300-dpi [dot per inch] printer) and with 256 colors. The higher the resolution and the greater

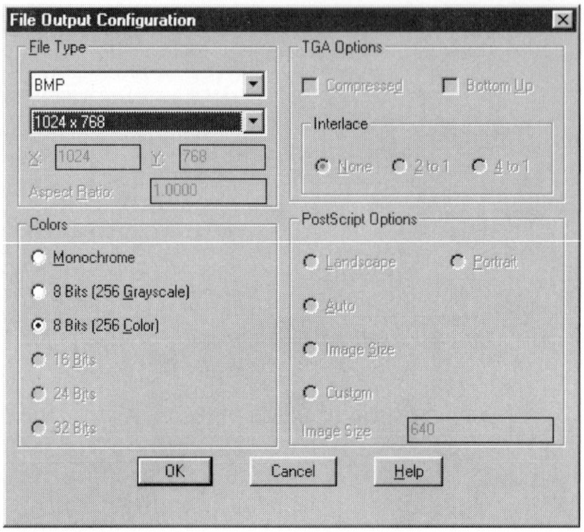

Figure 1.39

the number of colors, the more realistic the image will be. However, the time required to render the file will also increase as quality increases. You should calculate the resolution and number of colors appropriate for the final use—if your printer cannot print color, a grayscale rendering is the right choice. If you are rendering a file for the Web, 96 pixels will make up one inch of the image on a typical monitor. Notice that some of the options are grayed out for *.bmp* files; each file type has different limitations as to the number of colors and resolution it will capture.

Choose File Type: **BMP** (a common Windows graphics format)

Choose resolution: **1024 × 768**

Select Colors: **8 bits (256 Color)**

Pick: **OK**

Pick: **Render** (in the Render dialog box)

The Rendering File window opens, where you can select the location for the file.

On your own, set the destination to your MDT-Work directory and name the file JackASMB.bmp.

The file is rendered to the MDT-Work directory (or the location you specified), where you can access it for use with other applications

On your own, import the .bmp file into your word processor and print it.

You have completed this tutorial and your introduction to Mechanical Desktop's parametric modeling power. Save your work and exit, or continue on to the exercises.

Save your file.

Exit Mechanical Desktop.

Command Summary

Action	Command Name	Description
3D Adjust Clip Plane	3DCLIP	Enables and sets the location of dynamic clipping planes that create section views through a part or assembly
3D Adjust Distance	3DDISTANCE	Switches to 3D interactive view so that cursor movement simulates the effect of moving the camera closer or farther away
3D Orbiter	3DORBIT	Controls the interactive viewing of objects in 3D
3D Swivel	3DSWIVEL	Switches to 3D interactive view so that cursor movement simulates the effect of turning the camera
Activate Part	AMACTIVATE	Activates a local or external part in an assembly so it may be edited
Add Drawing View	AMDWGVIEW	Creates and adds 2D views to a drawing layout
Change active mode between Drawing and Model	AMMODE	Controls whether Model Mode (Part, Assembly, Scene) or Drawing Mode (Layout) is active
Change the 3D View	AMVIEW	Controls the view in Model mode
Continuous 3D orbit	3DCORBIT	Enables the user to set objects in the 3D view into continuous orbit
Edit Feature	AMEDITFEAT	Edits a feature by changing the operation and settings used to create it
Help	AMHELP	Opens online help: Mechanical Help Topics
New Rotation Center	AMROTCENTER	Calculates a center of rotation at the 3D center of all visible objects
Open a File	OPEN	Opens an existing *.dwg* file
Pan Realtime	3DPAN	Switches to the interactive 3D view so you can click and drag the view of the object in the graphics window
Render	RENDER	Creates a photorealistic image of the 3D model
Save As	SAVEAS	Saves the current file under a new file name and makes the new file name current
Set Explode Factor	AMXFACTOR	Controls the explosion factor for a scene
Set Rotation Center	AMSELROT	Lets the user set a point about which the model will rotate
Toggle Desktop Browser	AMBROWSER	Controls the display of the Desktop Browser
Toggle Shading/Wireframe	SHADEMODE	Controls whether an object appears shaded or wireframe in the current view
Update	AMUPDATE	Recalculates the model to show the effect of changes made to features, constraints, or dimensions on parts, assemblies, scenes or drawing views
Zoom	ZOOM	Controls the magnification at which objects are viewed on screen
Zoom Realtime	3DZOOM	Switches to the interactive 3D view so you can click and drag to change the magnification of the view

Exercises

1. **Bearing** Open the Mechanical Desktop part named *Bearing.dwg* from the *MDT-Data\Exercises* folder. Use your knowledge of the software to answer the following questions about the part.

   How many through and blind holes are there in the Bearing part? Describe their location.

2. **Dice** Open the Mechanical Desktop part named *Dice.dwg* from the *MDT-Data\Exercises* folder. Use your knowledge of the software to answer the following questions.

   List the number on the die face associated with each of the six orthographic views.

3. **Sphere** Open the Mechanical Desktop part named *Sphere.dwg* from the *MDT-Data\Exercises* folder. Use your knowledge of the software to answer the following questions about the sphere.

   List the objects found inside the Sphere. (*Hint*: It may be helpful to use the 3D Adjust Clip Plane command to view the features.)

4. **Cube** Open the Mechanical Desktop part named *Cube.dwg* from the *MDT-Data\Exercises* folder. Use your knowledge of the software to answer the following questions about the shape.

   Describe, in order, the objects that are found as one slices through the Cube. (*Hint*: It may be helpful to use the 3D Adjust Clip Plane command to find the features.)

5. **Modifications *(challenge)*** Open the Mechanical Desktop part named *Mod.dwg* from the *MDT-Data\Exercises* folder. Next you will make some changes to the model and answer questions about the results.

   a) Right-click on the Center Hole feature in the model tree and pick edit. Use the dialog box to change the distance for the feature to 10. Don't forget to update the part after you make the change. What happens when the Center Hole feature is changed to cut only a distance of 10?

   b) Right-click on the Profile 1 below the Base feature and pick Edit. Change the dimension for the rounded end from R1.00 to R1.50. What happens when the ends of the Base sketch are changed from a radius of 1 to 1.5?

   c) Right-click on 1st Legs and pick Edit. What happens when the Join distance of the 1st Legs is changed from 10 to 20?

   d) What happens when the 2nd Legs operation is changed from a join to a cut?

Tutorial

2 Simple Part Creation

Objectives

When you have finished this tutorial, you will be able to

- Describe the basic part modeling process.
- Create a sketch with Polyline and Rectangle.
- Convert a sketch to a profile.
- Show, add, and delete geometric constraints.
- Add dimensions.
- Use extrusion to create sketched features.
- Select and orient a new sketch plane.
- Project geometry to a sketch plane.
- Add holes and fillets as placed features.
- Modify a feature to create a new part.

Introduction to Part Modeling

Parametric (or feature-based or constraint-based) modeling software, including Mechanical Desktop, has a common approach to creating many basic 3D objects: sketching a planar shape and then using that shape to create the 3D feature.

Figure 2.1 shows a planar sketch of a feature used as a *profile*, or 2D shape, that will be used to generate 3D geometry. Figure 2.2 shows that sketched profile *extruded* to create a feature. Figure 2.3 shows the same profile *revolved* to create a feature. *Sweeping* is another method for creating shapes, by extruding the profile along a 3D path. Figure 2.4 shows the same profile swept along a curved path. The shapes of many objects that you use in daily life, or would design as an engineer, can be modeled using the basic methods of extrusion, revolution, or sweeping.

In this tutorial, you will learn the basics of part modeling as you create some of the simpler parts for the jack assembly. You will gain an overview of the part modeling process by using extrusion to create features. First you will create features by starting with a sketch; then you will add built-in placed features to your part. In later tutorials, you will build on these basics

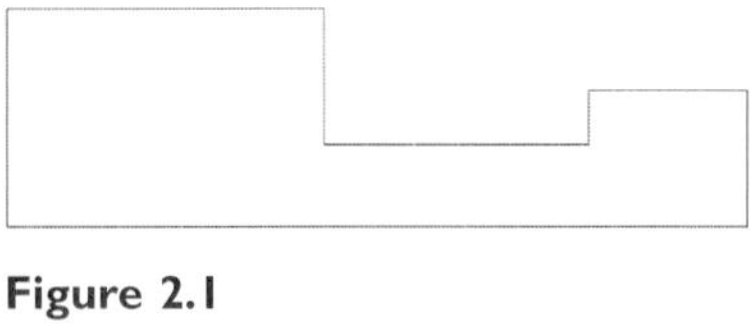

Figure 2.1

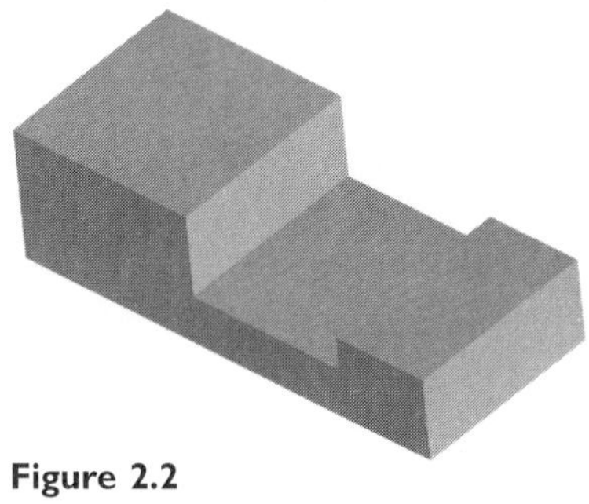

Figure 2.2

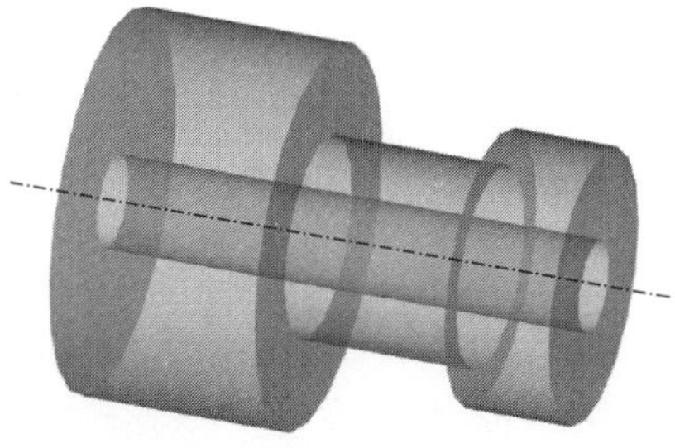

Figure 2.3

Figure 2.4

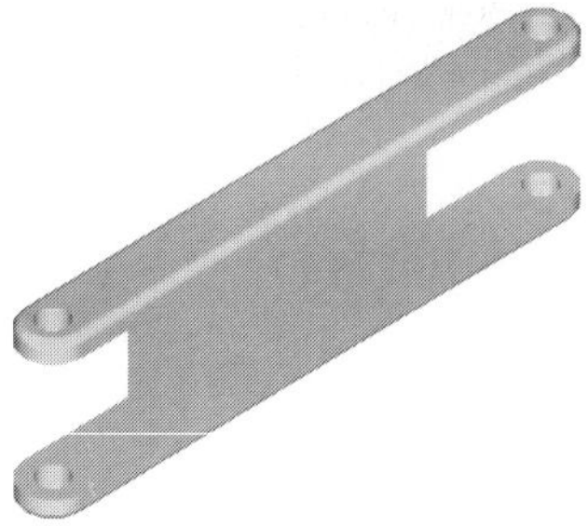

Figure 2.5

to learn more advanced sketching and modeling techniques. The first part you will model is the thicker leg of the jack, as shown in Figure 2.5.

Starting a New Part File

You will start by launching Mechanical Desktop 5 as you did in the last tutorial.

Pick: **Mechanical Desktop 5 icon**

Load your profile using the Options dialog box (if necessary)

The Mechanical Desktop Today window appears on your screen. When you start a new drawing in Mechanical Desktop, the default is the assembly modeling environment you saw in the last tutorial. When creating individual parts, however, Mechanical Desktop allows you to work in a part modeling environment that contains menus and toolbars for the part modeling tools you will need. Using this option can result in smaller file sizes and allows you to create standard parts that can be used in a number of assemblies. You will use the File menu to specify that you want to open a new part file.

Pick: **File, New Part File**

The Mechanical Desktop Today window reappears with the Create Drawings tab on top, as shown in Figure 2.6. When you start a new file, you can use the drop-down menu on this tab to start from scratch, start from a template, or start by using a wizard. You must also specify the units you wish to use in the new file; you will select English units for this tutorial.

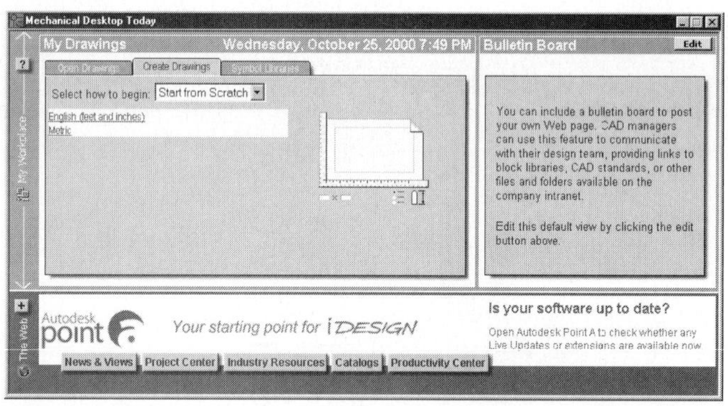

Figure 2.6

Choose Select how to begin: **Start From Scratch** (from the drop-down menu)

Pick: **English** (feet and inches)

The Mechanical Desktop part modeling environment appears on your screen, as shown in Figure 2.7. Notice that the Desktop Browser now has only two tabs: Model and Drawing. The Scene tab used with assemblies is not visible.

You will save your new part file immediately so it has a name and location.

Pick: **File, Save**

The Save Drawing As dialog box appears, as shown in Figure 2.8. The first time you save a file, this dialog box appears because no name or

Warning

The default display for the graphics area in these tutorials is black lines on a white background. If your screen is not displaying black lines on a white background, refer to the Getting Started chapter for information about changing your screen display and creating a profile that you can load each session.

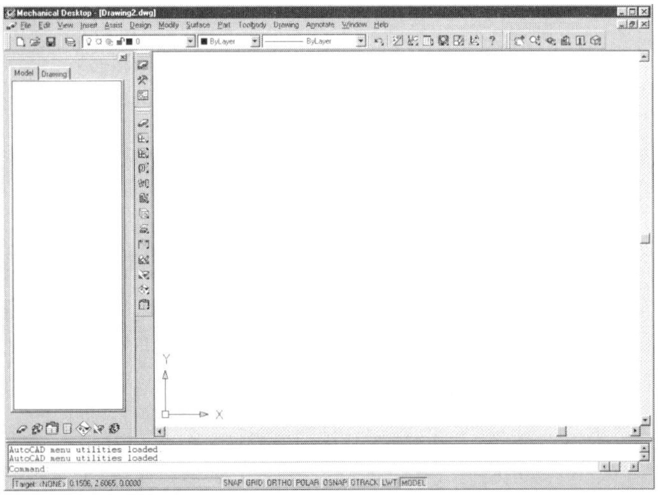

Figure 2.7

location has been assigned to the file yet. You will select the drive and directory where you will store your parts. For these tutorials, it is assumed that you are saving your parts on drive C: in the MDT-Work directory. You will create a folder in the Work directory called Jack, where you will store all the files you make for the Jack assembly.

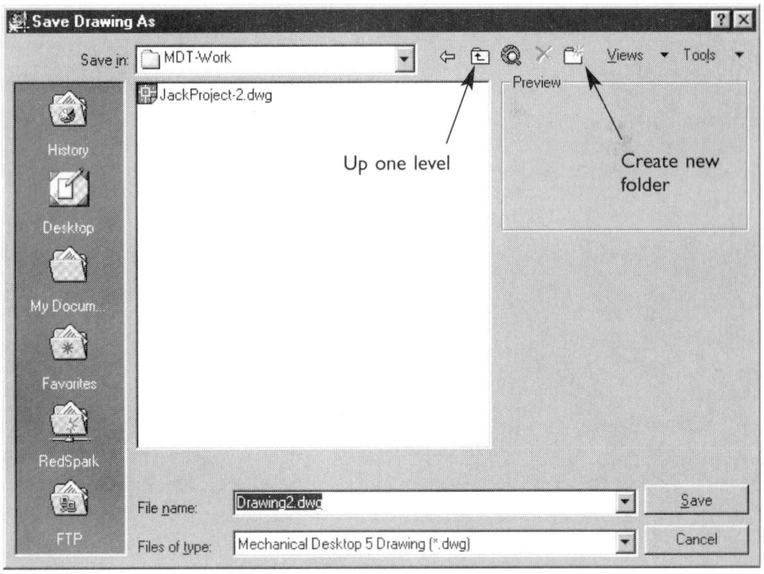

Figure 2.8

Tip

You can use the Up One Level icon, which is labeled in Figure 2.8 to move up out of the folder you are currently viewing.

Choose Save In: **MDT-Work** (if it is not already displayed)

Pick: **Create New Folder icon**

Type: **Jack** (to replace the highlighted New Folder name)

Double-click: **the Jack folder**

The Jack folder should now be in the Save In area. You will store all of the parts for the jack assembly in this folder. This will make them easy to find and manage when you create the assembly.

In the File Name area of the dialog box,

Type File name: **ThickLeg**

Pick: **Save**

The title bar at the top of the Mechanical Desktop window should now display the current drawing name, *ThickLeg.dwg*. The Desktop Browser should be open, with the Model tab appearing uppermost. Your screen should appear similar to Figure 2.9.

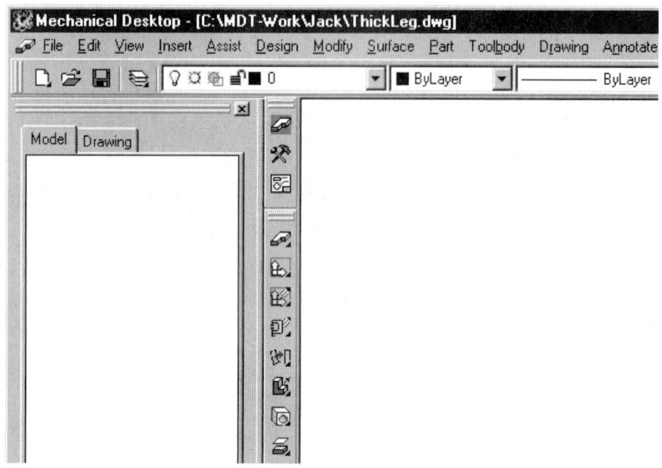

Figure 2.9

Reference

Engineering Design Communication

The Base Feature, pp. 260–262

Now you are ready to begin to model. The first feature you create is called the *base feature*. It is special because other features depend on it for their location in the model. Once the first feature has been created, other features are added to or subtracted from it to form more complex objects.

First Decisions for Solid Models

One of the first steps in modeling is to think about what would make a good base feature for the part. Since other features are added or subtracted from the base feature, it is important to start with a feature that is significant and is not likely to be deleted from the design or changed radically, and to which the other features can be added. You may also want to consider how to orient your initial sketch with respect to the coordinate system or to a set of default planes that will help when you begin to assemble parts (which you will learn more about in Tutorial 8). Some thought up front about these issues saves you time later and makes changing the model easier.

Reference

Engineering Design Communication

Datum Planes and Surfaces, pp. 263–265

Selecting a Base Feature

Begin by mentally breaking the part down into simpler shapes. Think about the intended function of the part as you are deciding on the base feature. Does this feature play a significant role in the design of the part? How is it likely to change as the design changes? Will you ever want to eliminate that feature from the design? Will you be able to add other features to it? Does it connect to other features or are they connected to it? While it is easy to make changes to parts created using feature-based modeling, this ease is dependent on planning the model carefully. You

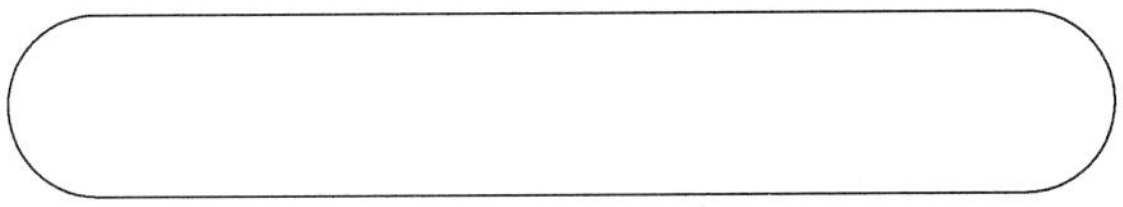

Figure 2.10

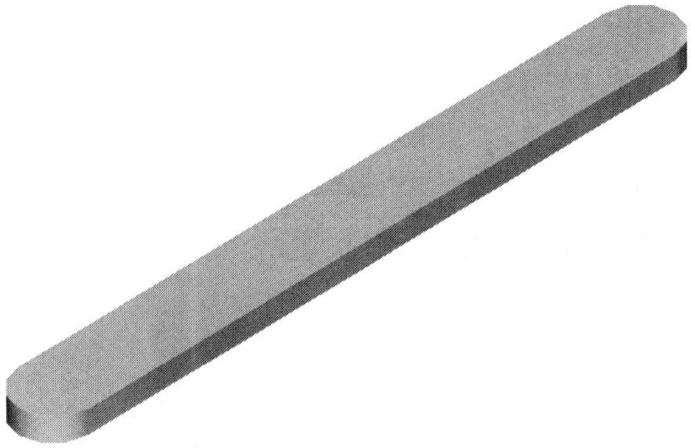

Figure 2.11

will want to begin your model with a base feature that is relatively simple and usually a physically large feature of the part.

You will be modeling the thicker leg of the Jack assembly shown in Figure 2.5. For the thick leg, you will sketch the flat shape of the leg, shown in Figure 2.10, and then extrude it to create the base feature shown in Figure 2.11.

Reference

Mechanical Desktop On-line Help

To learn more about the AutoCAD drawing commands mentioned here, choose AutoCAD Help from the Help menu. You may look up the commands by name in the Command Reference, or use the User's Guide to read about Creating Objects.

Using 2D Sketching Commands

If you have used AutoCAD in the past, you will find it easy to sketch using commands you already know. If you have not used AutoCAD, you only need to learn a few basic commands to create most 2D objects that you will use in sketching, such as Lines, Arcs, Circles, Polylines (lines and arcs that are connected), Rectangles, Polygons (regular polygons with equal length sides), Splines, and Ellipses. 2D objects are created with these commands by picking points on the screen or by entering the coordinates of those points by typing them at the command prompt at the bottom of your screen.

You will use the Polyline command to sketch the basic shape for the thick leg part. A ***Polyline*** is a connected series of lines and arcs that behave as a single object.

AutoCAD commands often have ***options*** that appear at the command prompt. You can select an option by typing the capital letter(s) that appear at the command prompt and pressing [↵]. You will do this to change between drawing straight line segments and arc segments in the Polyline command.

Tip

You should always monitor the command prompt area. As you work, the software reports in the command prompt area which command is active and what kind of user input it expects. If you are unable to select a toolbar button or object on the screen, it may be that you have inadvertently started a command and the software is waiting for a different response. You can press [ESC] at any time to cancel any unwanted command and remove the prompt from the command line.

You will use the context menu to start sketching:

Right-click: in the graphics area

Pick: **2D Sketching** (from the context menu)

The menu selections shown in Figure 2.12 should be on your screen. This context menu contains all of the drawing and editing commands you will use to create sketches.

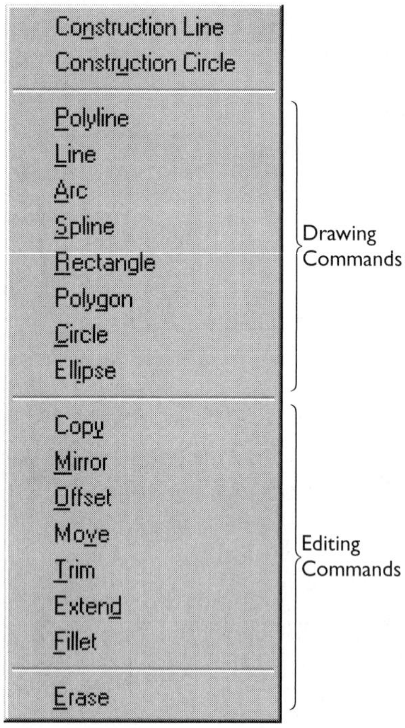

Figure 2.12

Pick: **Polyline**

You will draw a polyline by picking points on the screen. Refer to Figure 2.13 for the points to pick, so you achieve approximately the same shape shown. Sketches for parametric drawings do not have to be exact. As you will see, when you create a profile from the sketch, the software uses rules to neaten up what you have drawn, including things like assuming that a line that is close to horizontal or vertical is intended as such. However, it is often just as easy to create them accurately as not, as you will do in this sketch.

To make straight lines,

Press **F8** or pick the Ortho button on the status bar (so it is depressed, or on).

Now you are ready to begin the sketch. Refer to Figure 2.13 for where to pick.

Specify start point: **pick the upper left endpoint of top line labeled A**

Current line-width is 0.0000

Specify next point or [Arc/Close/Halfwidth/Length/Undo/Width]: **pick the right end of top line labeled B**

Specify next point or [Arc/Close/Halfwidth/Length/Undo/Width]: **A** ↵

Specify endpoint of arc or

[Angle/CEnter/CLose/Direction/Halfwidth/Line/Radius/Second pt/Undo/Width]: **pick straight below the previous point at point C**

Specify endpoint of arc or

[Angle/CEnter/CLose/Direction/Halfwidth/Line/Radius/Second pt/Undo/Width]: **L** ↵

Specify next point or [Arc/Close/Halfwidth/Length/Undo/Width]: **pick left endpoint of lower line labeled D**

Specify next point or [Arc/Close/Halfwidth/Length/Undo/Width]: **A** ↵

Specify endpoint of arc or

[Angle/CEnter/CLose/Direction/Halfwidth/Line/Radius/Second pt/Undo/Width]: **CL** ↵

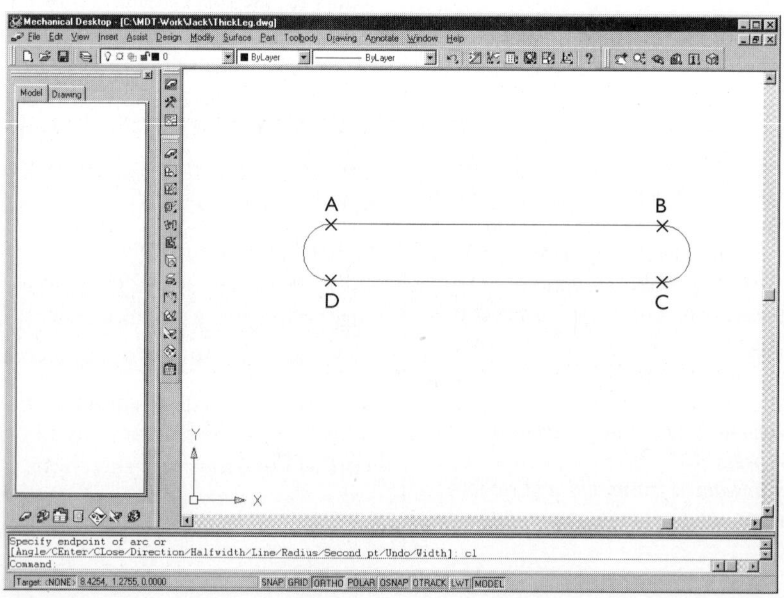

Figure 2.13

The arc should connect back to the first point to close the polyline. Using a closed polyline is an easy way to make 2D sketch geometry, as you can be sure that the lines and arcs used to create it meet to create a closed outline. Your drawing should appear similar to Figure 2.13.

Once you have created a sketch, you are ready to tell Mechanical Desktop that you want to use that sketch shape as a profile.

Creating a Profile

A *profile* is a constrained sketch that is the cross section or outline of a shape (such as an extrusion, revolution, or a swept feature). In the process of creating the 2D shape as a profile, the software analyzes what you have drawn and applies constraints. *Constraints* are conditions that determine the geometric and size properties of the sketch. The constraints applied to a sketch determine the ways in which it will change when it is modified later. Geometric constraints control properties such as tangency, perpendicularity, parallelism, concentricity, and others. Dimensions can be thought of as another type of constraint that control the size of features.

Reference

Engineering Design Communication

Sketch Constraints, pp. 256–259

Next you will turn your sketch into a profile, a process that will apply constraints to the shape you have drawn. All of the lines you have drawn are a part of a single object (a polyline), so you can use the Single Profile option.

Right-click: in the graphics area

Pick: **Sketch Solving, Single Profile** (from the context menu)

You see the message, "Computing… Solved under constrained sketch requiring 2 dimensions or constraints."

Over- and Under-constrained Sketches

The message tells you that the software has analyzed the shape and applied some constraints but still needs at least two dimensions or additional geometric constraints before the geometry is fully defined. (Depending on the way your sketch was drawn, you may see a message that you need three or more constraints.) Your sketch is "underconstrained," as most sketches will be until you add dimensions to them. As you add constraints and dimensions, Mechanical Desktop will not allow you to overconstrain the sketch. In the next section, you will review the constraints and add those that are needed.

Constraining the Profile: Geometric First

The first step is to check the geometric constraints that have been added to your sketch.

Right-click: in the graphics area

Pick: **2D Constraints, Show Constraints**

Symbols appear on your screen identifying the constraints that have been applied to the sketch, as shown in Figure 2.14. The circled numbers

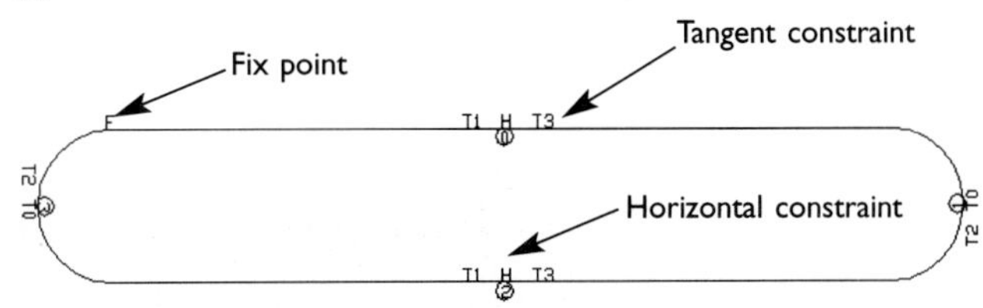

Figure 2.14

identify each 2D element making up the sketch. You should have two lines and two arcs making a total of four elements in the sketch. (The elements may be numbered differently if you created the elements in a different order.)

Mechanical Desktop provides for 13 different geometric constraints. The following table identifies the symbols and provides an explanation for the sketch constraint. The constraint symbol appears next to the sketch entity and includes the number of the other entity the constraint relates to. For example, the arc at the right end of the sketch is tangent to both the upper and lower horizontal lines. The right arc displays T0 and T2, meaning that it is tangent to the upper line (labeled 0) and the lower line (labeled 2). Each line should display T3 and T1, meaning that they are tangent to arc 1 and arc 3.

Symbol	Name	Expanded Description
H	Horizontal	Entity is horizontal.
V	Vertical	Entity is vertical.
L	Perpendicular	Entity is perpendicular to numbered entity.
P	Parallel	Entity is parallel to numbered entity.
T	Tangent	Entity is tangent to numbered entity.
C	Co-linear	Entity is colinear with numbered entity.
N	Concentric	Entity is concentric to numbered entity (has the same center).
J	Project	The vertex joins the infinite projection of a line, a point, an arc, or a circle.
—	Join	Used to automatically close gaps in sketch.
X	*X*-Value	Entity has the same *X*-coordinate value for its placement. Can be used with profile lines, circles, arcs, and part edges.
Y	*Y*-Value	Entity has the same *Y*-coordinate value for its placement. Can be used with profile lines, circles, arcs, and part edges.
R	Radius	Entity is equal radius with numbered entity.
E	Length	Entity is equal in length with numbered entity.
M	Mirror	Entity is mirrored from numbered sketch geometry.
F	Fix	The indicated vertex is the fixed point for the sketch.

Deleting Constraints

You can delete unwanted constraints by selecting 2D Constraints, Delete Constraint from the context menu. Select the constraint symbol you want to delete and press ↵. The selected constraint will be deleted.

Adding Constraints

To add constraints, select 2D Constraints from the context menu. From the submenu that appears, select the name of the constraint you want to add. You will be prompted to select the entity to be constrained ("reoriented") and then to select the entity it will get its properties from. To add a horizontal or vertical constraint, you only need to select the entity, as these constraints are not relationships between entities. Press [⏎] twice to exit the command when you are finished adding constraints.

Specifying a Fixed Point

The fixed point is the location on the sketch that stays fixed on the coordinate system. When you change the size of a dimension, the fixed point remains where it is on the coordinate system and the other vertices move to accommodate the change. The fixed point is typically located by default as the first point you start drawing in the sketch. You can move the fixed point for the sketch using the context menu selection 2D Constraints, Fix. Then follow the prompt to select an endpoint or sketch segment to lock.

Press: [ESC] (to exit the Show Constraints command)

If necessary, add or delete constraints so that your sketch is constrained like the one in Figure 2.14.

Re-Solving the Sketch

If you make changes to the sketch, you may need to select 2D Sketch Solving, Re-Solve to apply the changes to the geometry. Re-Solving the sketch reapplies the constraints using the sizes that you sketched. This is an important step before adding dimensions, which will change the size of features and may make it difficult to select or apply the geometric constraints that you have overlooked.

Now you are ready to add the necessary dimensions to define the size of the features.

Good Practice for Constraints and Dimensions

In general, it is good practice to control the shape with geometric constraints first and add dimensions second. You can add or delete constraints as needed to stabilize the shape of the profile before you add dimensions to control the size. Try to use geometric constraints that reflect your design intent and do not add any unnecessary constraints. For example, if one line is horizontal and the other one vertical, do not add the unneeded perpendicular constraint between them.

It is also a good idea to draw the features somewhat near the size you want them, rather than having the software drastically change the size. As each individual dimension is changed and the sketch is solved, segments that are much smaller than the others may become eliminated or difficult to see. Starting with the approximate size for the feature is a good practice. You can pay attention to the coordinates located in the status bar to help you with a general indication of the size, or use a grid and snap to approximate sizes (as you will in Tutorial 4).

When the software is analyzing the shape and applying constraints, line segments or features that are small compared with others may cause the geometric constraints to be applied oddly. Those small items may also be difficult to dimension later. For this reason it is better not to change the dimensions until the shape is constrained geometrically. For the same reason, when you do add the dimensions, it is better to set the sizes for large features first and then for the smaller ones.

Try to keep in mind these general rules:

- Consider design intent for the feature.
- Use geometric constraints to stabilize shape before adding any dimensions.
- Re-solve the sketch as needed before adding dimensions.
- Dimension larger elements before smaller ones.
- Keep working to create a combination of constraints and dimensions that make your sketch function in a way that reflects the design intent for the part and can update accordingly.

Adding Dimensions

When adding dimensions, it is important to consider how you will want the part to update. Have the dimensions reflect the critical dimensions on the part, and organize them according to their relationships in the part. Also keep in mind where the fixed point is located. Locating dimensions from the fixed point will help prevent unwanted results when you update the part. The dimensions you add will later control the part when it updates.

Reference

Engineering Design Communication

Setting the Base Point, p. 260

Mechanical Desktop's Dimensioning command uses what it knows about the sketch entities to make assumptions about the dimensions you are adding. That is, if you select a line, it will assume you are placing a linear dimension and will offer the current size of the entity as the default value for the dimension. This semiautomatic dimensioning is convenient but also easy to override to achieve the dimensions you want if you watch the prompts at the command line.

You will now add the two dimensions you need to fully define the size of this feature: a radius for the rounded end and a length for one of the horizontal segments. You will start with the right-hand arc.

Right-click: in the graphics area

Pick: **Dimensioning, New Dimension**

Select first object: **pick on the right arc**

Select second object or place dimension: **pick to place dimension, as shown in Figure 2.15**

Enter dimension value or [Undo/Diameter/Ordinate/Placement point] <1.2603>: **.38** ↵

Solved under constrained sketch requiring 1 dimensions or constraints.

The sketch will update to show the feature at the new size. Use Zoom on your own as necessary to enlarge your view. Next you will add the dimension for the length of the feature.

Select first object: **pick on the upper horizontal line**

Select second object or place dimension: **pick above the line to place the dimension**

Enter dimension value or [Undo/Hor/Ver/Align/Par/aNgle/Ord/Diameter/pLace] <2.4892>: **6.5** ↵

Solved fully constrained sketch.

Select first object: ↵ **(to exit the command)**

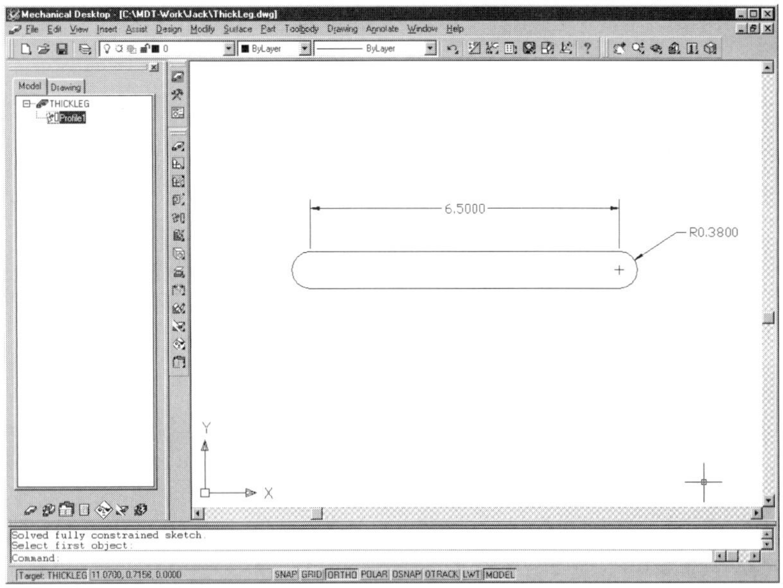

Figure 2.15

Your drawing should now look similar to Figure 2.15. Notice the message, *Solved fully constrained sketch*. This tells you that you have a combination of geometric constraints and dimensions that fully define the sketch.

Dimension Placement

It is not critical to place dimensions strictly according to good drawing practices for manual or plotted drawings while you are modeling. You will make proper engineering drawings from the model using the Drawing mode of the software. In Drawing mode, it is important for you to clean up the appearance of the dimensions to meet standard engineering practices. However, using some of the same general rules for good dimensioning when modeling will help you maintain legible dimensions that are easy to select when updating the model:

- Place the dimensions outside the drawing outline when possible.
- Place the dimensions a visible distance away from the feature.
- Locate the dimension near the feature being dimensioned.
- Keep shorter dimensions closer to the object, longer dimensions farther away.

Following these practices will help you to easily read the dimensions associated with the model.

Now you are ready to select what type of feature you want to create from the fully defined profile. To save your work thus far,

Pick: **File, Save**

Creating a Feature from a Sketch

Once you have created a profile, you can use it to create a sketched feature. *Sketched features* include extrusions, revolutions, sweeps, lofts, and face splits. You will use extrusion to create the base feature for the ThickLeg part.

Using Extrude

Extrusion is the process of forcing material through a shaped opening. The Extrude command uses the profile in the same way that the extrusion process uses the shaped opening. The profile determines the shape of the resulting feature. You will use the dialog box for the Extrude command to explore its options before you create your feature.

Right-click: in the graphics area
Pick: **Sketched & Work Features, Extrude**

Operation

The Extrusion dialog box appears. Notice that Operation is set to Base and is grayed out. You are creating the base feature for the model. It is the first feature you create, so you are not presented with other options, such as subtracting or adding material. When you create subsequent features you can select whether the material will be cut from, joined to, or intersected with the existing part.

Draft Angle

You can use the Draft angle option to produce a taper as the model is extruded. Figure 2.16 shows a part extruded with a draft angle of 0 (no taper). Figure 2.17 shows the same profile extruded with 10 degrees of draft.

Termination Options

You can select from the following options for terminating the extrusion operation: Blind, Mid-Plane, To Face/Plane, and From-To.

Blind	Feature is created to one side of the profile a specified distance.
Mid-Plane	Feature is created to each side of the profile by the specified distance.
To Face/Plane	Feature is created from one side of the profile to a selected face or plane.
From-To	Feature is created between two specified planes using the profile.

The first option, Blind, is illustrated in the image tile on the right side of the dialog box.

Figure 2.16

Figure 2.17

Use the Type pull-down menu to choose each option to see its illustration in the image tile.

The first feature you create is usually extruded using the Blind option. Blind gets its name from the term used for the depth of holes. A blind hole is drilled to a specified depth and does not go entirely through the part. (A through hole goes entirely through a part.) A blind extrusion requires a distance setting.

Flip Direction

When you create extruded features, the flat sketch is in the current X-Y plane. The feature is typically extruded in the positive Z-direction, using a right-hand rule. You can pick the Flip button to change the direction for the extrusion. If the extruded feature is going in the wrong direction, you can return to the Extrude dialog box to flip it at a later time. You will learn more about editing features in Tutorial 6.

Extruding the ThickLeg Base Feature

Now you are ready to extrude the profile to create a 3D feature. You will set the distance for the extrusion to .25 so your settings are the same as those in Figure 2.18.

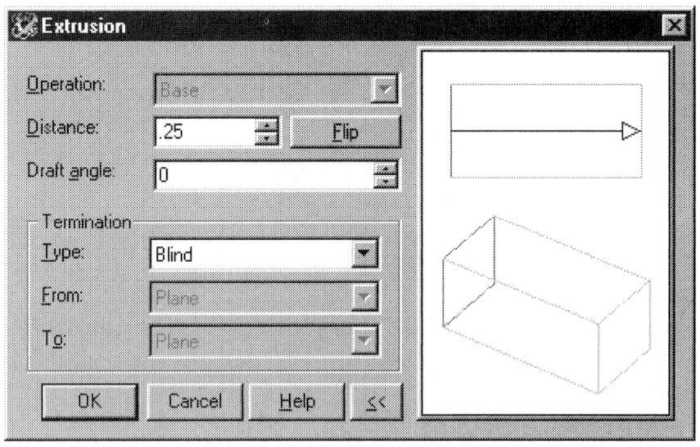

Figure 2.18

Set Distance: **.25**

Set Draft Angle: **0**

Choose Termination: **Blind**

Pick: **OK**

Notice that the model tree now shows one feature, ExtrusionBlind1. Features have icons that reflect the operation used to create them; in this case, the extrusion icon appears.

Pick: **the + sign to expand the model tree under ExtrusionBlind1.**

Profile1 appears nested below ExtrusionBlind1 in the model tree, as shown in Figure 2.19. Profiles for each sketched feature remain associated with the feature.

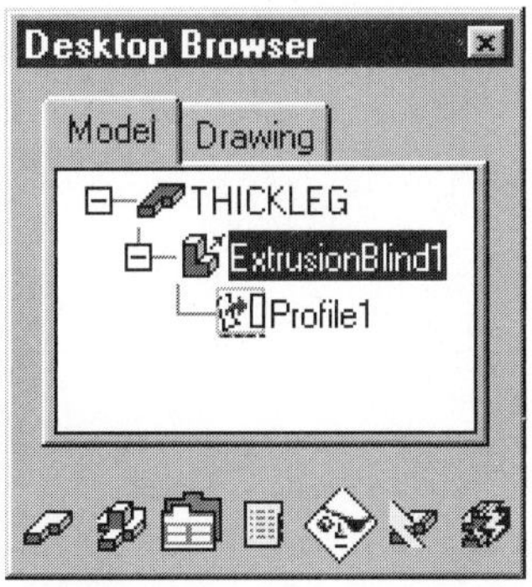

Figure 2.19

Pick: **Right Front Isometric View icon**

Pick: **Hidden icon** (from the Toggle Shading/Wireframe flyout)

Your view changes to a hidden isometric view, as shown in Figure 2.20, where you can see the extruded 3D feature.

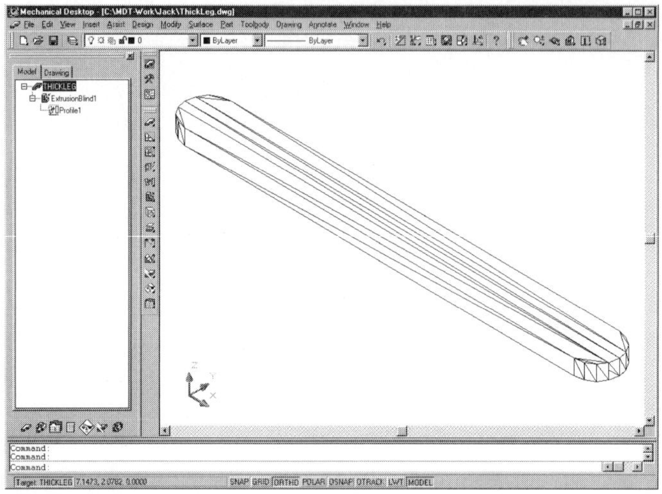

Figure 2.20

Setting Properties in the Browser Window

You can use the Browser Window to activate editing commands related to the items that appear there. Only those commands that relate to that part or feature will be available. You will use the context menu to change the color property of the part.

Right-click: **ThickLeg** (in the model tree)

Pick: **Properties**

Pick: **Color**

Tip

It is advisable to select a lighter-toned color. This will allow for easy viewing of the part in shaded mode.

The select color dialog box appears on your screen, as shown in Figure 2.21. You can select one of the standard colors or use the full 256-color palette to select a custom color.

Pick: **Green** (from the Standard colors at top)

Pick: **OK**

Pick: **Gouraud Shaded icon**

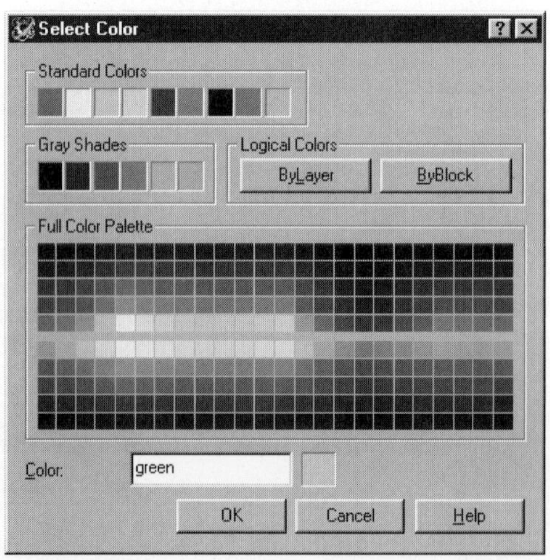

Figure 2.21

Figure 2.22

Reference

Engineering Design Communication

User Coordinate Systems, pp. 186–187

The feature appears shaded in the color you selected, as shown in Figure 2.22.

On your own, spin the part using the commands you learned in Tutorial 1. When you are finished,

Save your part.

Return to a Front Right Isometric View.

Switch to a 3D Wireframe display.

Selecting a New Sketch Plane

Next you will select a new plane to sketch on as you create the profile for the center portion of the thick leg part. For the base feature, you sketched on the default *X-Y* plane. As you add features, you will change the *X-Y* plane that your profile will be sketched on so that it will be oriented correctly when it is extruded (or revolved or swept) into a feature.

Any planar surface of the part can be used as a sketch plane. In Mechanical Desktop, a *sketch plane* refers to a temporary, infinite plane with *X*- and *Y*-axes that correspond to a planar surface on a feature. You can also create planes to sketch on that are not associated with an existing planar surface. These are called *work planes* and can be defined anywhere on the coordinate system. You will learn more about using work planes later. For this feature, you will use the upper surface of the base feature to define the sketch plane.

World versus User Coordinate Systems

When you select (or define) a sketch plane, you are telling Mechanical Desktop that you want to use a different set of coordinate axes to specify the location of points in 3D space. This user-defined set is called a User Coordinate System (or UCS). So far, you have sketched on the default set of axes, called the World coordinate system. In this system, the origin is in the lower left of your screen, the *X*-axis is horizontal, the *Y*-axis is vertical, and the *Z*-axis is perpendicular to the *X-Y* plane pointing out of your screen. The UCS icon in the lower left of your screen indicates the orientation of these axes. You will change to the Sketch view to see their default orientation. (Sketch view always looks straight down onto the current *X-Y* plane.)

Pick: **Sketch view icon** (from the Mechanical View toolbar)

Pick: **Zoom All icon** (from the Zoom flyout)

The UCS icon changes to indicate that you are looking straight onto the *X-Y* plane. The *Z*-axis is pointed directly out of the screen toward you, as shown in Figure 2.23.

Figure 2.23

When you create a User Coordinate System, you are prompted to provide the information that defines the new *X-Y-Z* axes. Generally, you must pick points that define the origin, the positive portion of the *X*-axis, and the positive portion of the *Y*-axis. However, Mechanical Desktop includes many commands that simplify choosing a new UCS. Because UCSs can be saved and named, a set of preset UCSs is built into Mechanical Desktop. You will change the UCS to align with a right-side orthographic view.

Pick: **Assist, Orthographic UCS, Right**

The UCS icon changes to indicate that you are looking onto the *X-Z*, with the *Y*-axis pointed toward you, as shown in Figure 2.24.

Figure 2.24

Pick: **Sketch View icon**

Your view of the base feature changes to a right-side view, as in Figure 2.25, and you are now looking down on the current *X-Y* plane.

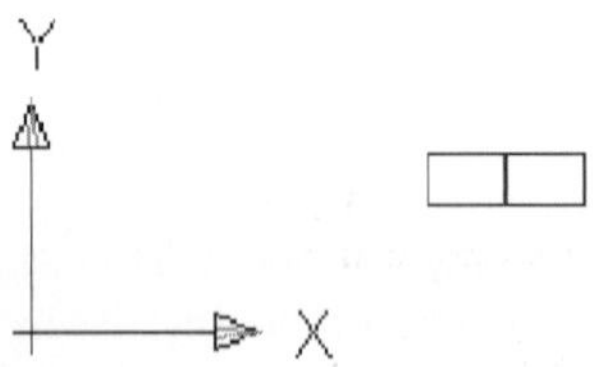

Figure 2.25

The UCS does not always have to be displayed at the origin of the axes. You can change it so that that it is located in the lower left of your screen. Setting the UCS icon to display at the origin is another aid to navigation in 3D space. At other times, it is more convenient to have it in the lower left of the screen. To change it,

Pick: **View, Display, UCS Icon, Origin** (so the check disappears)

Your UCS icon moves to the lower left of the screen; it is no longer at the origin.

Pick: **Assist, Named UCS**

Pick: **Orthographic UCS tab**

The UCS dialog box appears, as shown in Figure 2.26. Like Named Views, UCSs can be named and stored, and you can use this dialog box to select a preset UCS or name and save UCSs you create. As you can see, the dialog box reports that the current UCS is Right. You will use this dialog box to restore the World coordinate system and set the UCS icon to appear at the origin.

Pick: **Named UCSs tab**

Pick: **World**

Pick: **Set Current**

Pick: **Settings tab**

Pick: **Display at UCS origin point**

Pick: **OK**

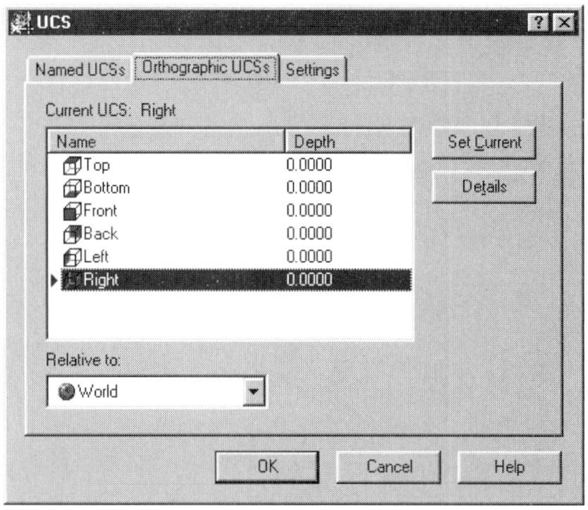

Figure 2.26

Pick: **Front Right Isometric View**

The World UCS is restored and your drawing shows a front-right isometric view, as shown in Figure 2.27.

Reference

Mechanical Desktop On-line Help

For a complete explanation of the UCS icon, use the Command Reference in AutoCAD Help to look up UCSICON. In addition to the 3D icon that is displayed as the default, you can also show a 2D icon that has variations to show whether you are at the origin, in the WCS, and viewing from above or below the Z-axis.

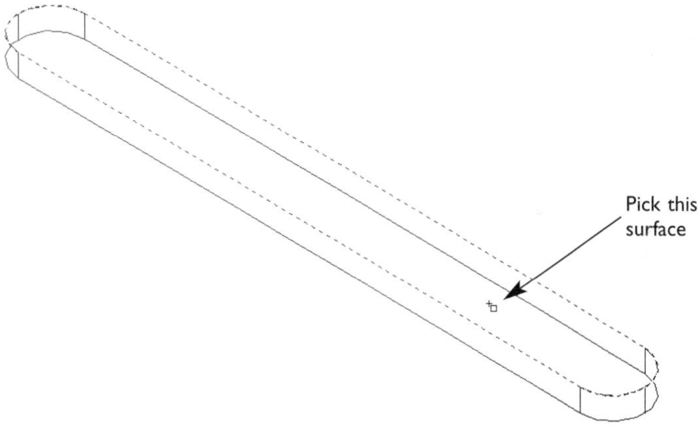

Figure 2.27

The simplest way to define the UCS for 3D modeling, however, is to use the built-in commands for choosing a new sketch or work plane and use an existing plane or planar face to define it. This is the method you will use as you select a new sketch plane for the next feature. Notice how the prompts ask you to provide the information needed to orient the sketch plane: You will first highlight a planar surface, confirm your choice, then select an edge to align the X-axis, and confirm that choice.

Right-click: in the graphics area

Pick: **New Sketch Plane**

Select work plane, planar face or [worldXy/worldYz/worldZx/Ucs]: **pick on the top surface of the object, as shown in Figure 2.27.**

Enter an option [Next/Accept] <Accept>: **if the top surface is highlighted, right-click; if not, pick again until it becomes highlighted and then right-click to accept.**

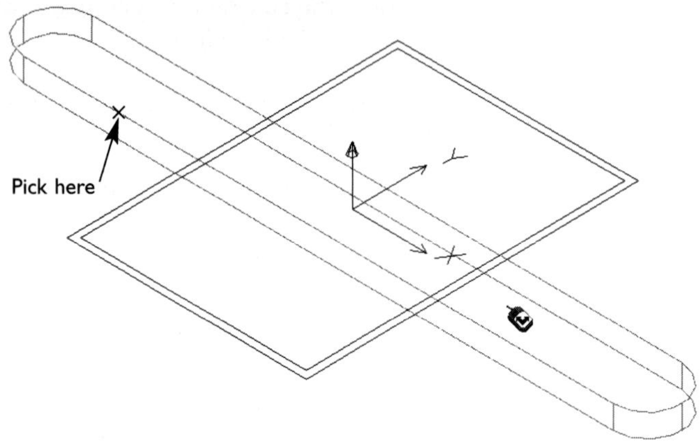

Figure 2.28

A representation of the sketch plane and a UCS icon will appear along with an animated illustration of a mouse as shown in Figure 2.28. You have selected the plane and now need to specify the orientation of the *X*-*Y*-*Z* axes. You will pick an edge to align the *X*-axis. You can use the right mouse button to accept the displayed orientation or press R to rotate the axes. If a red arrow appears, move the mouse until it is aligned with the positive *X*-axis, as shown in the figure.

Select edge to align X axis or [Z-flip/Rotate] <Accept>: **pick the straight lower edge of the top surface indicated in Figure 2.28.**
Plane=Parametric
Select edge to align X axis or [Z-flip/Rotate] <Accept>: **if the XYZ axis is oriented correctly, right-click to accept. If not, press R to cycle through the orientations.**

Tip

You can type the option letter F to flip the positive direction of the Z-axis.

The UCS icon appears in the center of the part, as shown in Figure 2.29, showing you the orientation for the user coordinate system you have created using the New Sketch Plane command. Verify that the *X*-*Y* axis is aligned with the plane you want to sketch on—in this case, the top surface of the existing part. The actual *X*- and *Y*-directions do not matter greatly as this new feature will be related to the previous one by dimensions and constraints, not by the orientation of the coordinates.

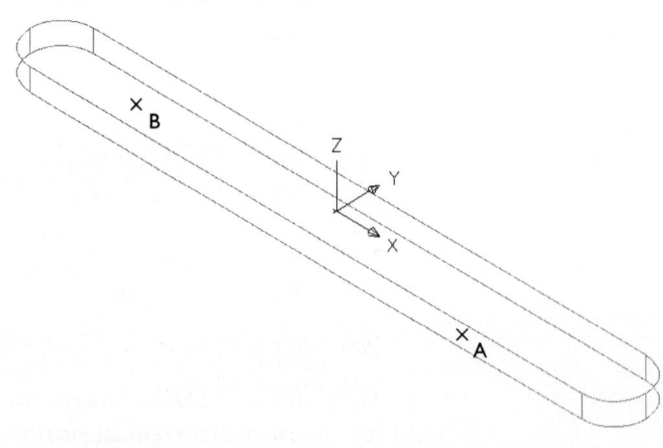

Figure 2.29

Now the coordinate system is aligned so you can sketch on the top surface of the base feature.

Creating a Profile for the Second Feature

Now you are ready to start sketching the new profile. This time you will sketch in the isometric view. Sometimes this is helpful so that you can see how your sketch fits in relation to the rest of the part. The cross member for the ThickLeg part should be aligned with the bottom edge of the leg. It is not symmetrical, so you will provide dimensions that can be used to move it on the part.

You will use the Rectangle command, which creates a polyline with vertical and horizontal edges. You will draw the rectangle so that it is not directly over the existing edges, which will make it easier to see and select. You will add contraints to align it to the edge.

Right-click: in the graphics area

Pick: **2D Sketching, Rectangle**

Specify first corner point or [Chamfer/Elevation/Fillet/Thickness/Width]:
pick near point A in Figure 2.29

Specify other corner point: **pick near point B**

The rectangle should appear similar to Figure 2.30.

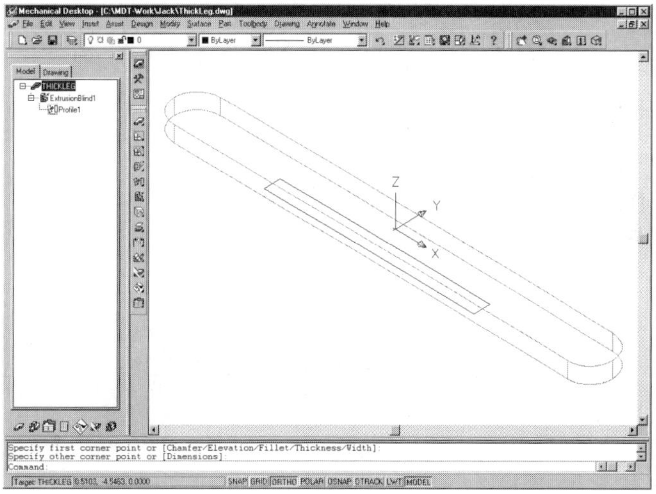

Figure 2.30

Next you will solve this sketch to create a profile.

Right-click: in the graphics area

Pick: **Sketch Solving, Single Profile**

The sketch is solved to use as a profile. Next you will display the constraints and add any necessary ones. To see them more clearly, you will pick the sketch view.

Pick: **Sketch View icon**

Pick: **Zoom All icon**

Your view changes so that you are looking directly perpendicular to the sketch plane.

Right-click: in the graphics area

Pick: **2D Constraints, Show Constraints**

The constraints are shown on the screen, as in Figure 2.31.

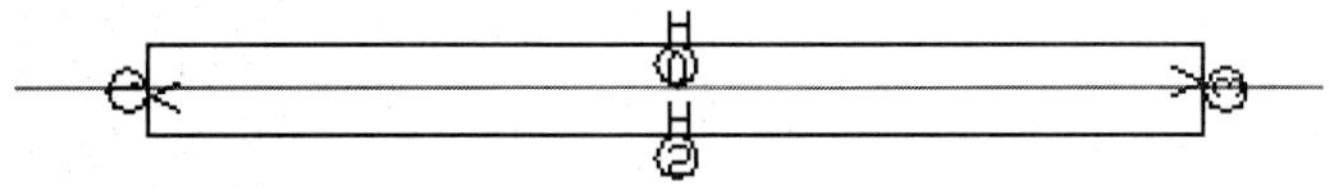

Figure 2.31

Examine the constraints that were applied to create the profile. You should have two horizontal and two vertical constraints. Because you used the Rectangle command, you should not have to add or delete these constraints, as the command creates a polyline with edges aligned to the *X*- and *Y*-axes.

Enter an option [All/Select/Next/eXit] <eXit>: ↵

Adding a Colinear Constraint

Next you will add a colinear constraint so that the bottom edge of the sketch stays aligned with the bottom edge of the leg.

Press: ↵

Right-click: in the graphics area

Pick: **2D Constraints, Collinear**

Valid selections: line or spline segment

The message at the command line tells you what type of entities can be used with this constraint.

Select object to be reoriented: **pick on the bottom horizontal line of the sketch**

Select object to be made colinear to: **pick on the bottom horizontal line of the existing part**

Solved under constrained sketch requiring 3 dimensions or constraints.

Valid selections: line or spline segment

Select object to be reoriented: ↵ ↵ (to exit the command)

The sketch geometry changes to meet the colinear requirement for the bottom lines. Your drawing should look like Figure 2.32.

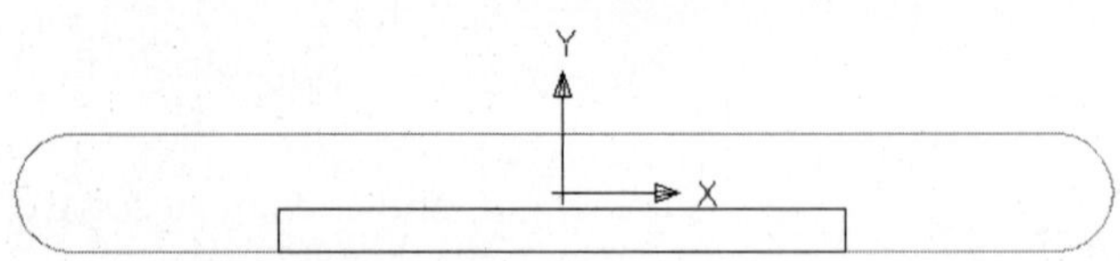

Figure 2.32

Now you are ready to add the necessary three dimensions to control the size and location of this feature, shown in Figure 2.33. You will add the dimensions from the short vertical lines to the centerpoints of the arcs first.

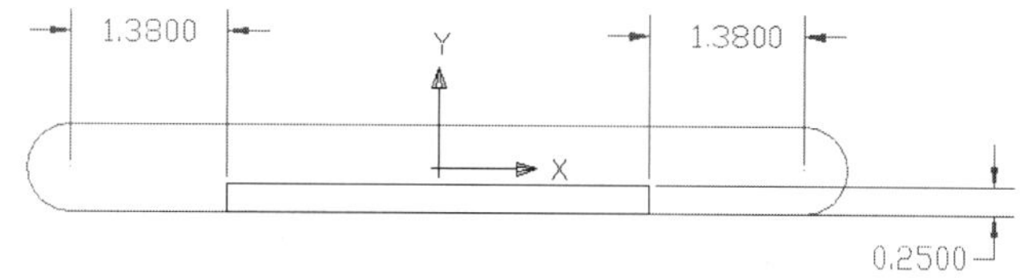

Figure 2.33

The Dimensioning command will automatically select the centerpoint of the arc to dimension to when you select the arc element.

Right-click: in the graphics area

Pick: **Dimensioning, New Dimension**

Select first object: **pick on the right arc**

Select second object or place dimension: **pick on the right vertical line of the new sketch**

Specify dimension placement: **pick above the part**

Enter dimension value or [Undo/Diameter/Ordinate/Placement point] <1.2603>: **1.38**

Repeat these steps for the left end of the part

On your own, add the .25 dimension for the short vertical line

Press: ↵ (to exit the command)

Tip

If you make a mistake, you can pick the Undo icon on the Mechanical toolbar to undo the last command. If you have exited the Dimensioning command, you can pick on a dimension and press Delete to erase it.

When you have finished these steps, you should see the message, *Solved fully constrained sketch*.

Using List

You can use AutoCAD's List command to find information about your drawing. The List command lets you select drawing objects and reports back the object type; its layer; *X*-, *Y*-, *Z*-coordinates; color; extrusion direction; and other useful information. You will use the List command to pick on one of the horizontal lines of the sketch lines to find out its length.

Command: **LIST** ↵

Select objects: **pick on the upper horizontal line of the sketch** ↵

An AutoCAD text window appears and reports:

ACAD_LINE Layer: "0"

Space: Model space

Handle = 213 from point, X= <value> Y= <value> Z= 0.0000

to point, X= <value> Y= <value> Z= 0.0000

Length = 3.7400, Angle in XY Plane = 0

Delta X = 3.7400, Delta Y = 0.0000, Delta Z = 0.0000

This information also appears in the command line area at the bottom of your screen.

Close the Text window by clicking its close box.

Extruding a Feature with Join

Next you will extrude the fully constrained sketch to a blind distance of 1.5, with no draft angle to create the feature. You will use the Join operation to extrude the part and combine it to the base feature in a single step. Join is equivalent to the Boolean operation union.

Right-click: in the graphics area

Pick: **Sketched & Work Features, Extrude**

The Extrusion dialog box appears on your screen. You will choose the settings shown in Figure 2.34.

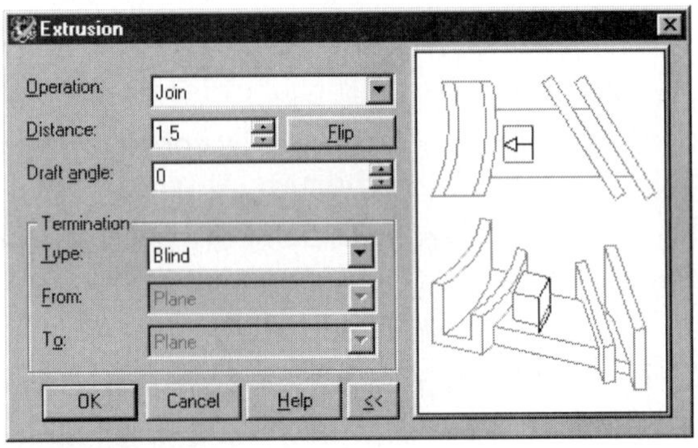

Figure 2.34

Choose Operation: **Join**

Set Distance: **1.5**

Set Draft angle: **0**

Choose Termination: **Blind**

Pick: **OK**

Pick: **View, 3D Views, Back Right Isometric**

Figure 2.35 shows the new feature added to the part.

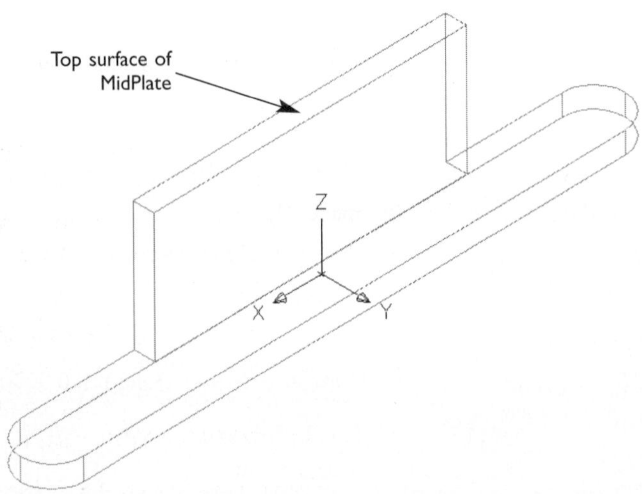

Figure 2.35

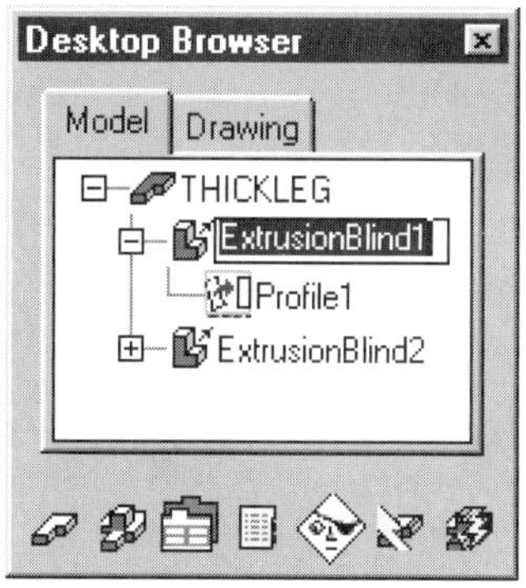

Figure 2.36

Tip

Double-clicking a feature will automatically open the dialog box that was used to create it so that you can edit the feature. Make sure to click once; then wait and click again to rename features. If you have trouble, right-click and pick Rename from the context menu.

Renaming Features

Notice that the Model Tree now shows a second extruded feaure, ExtrusionBlind2. You can use the Model Tree in the Desktop Browser to rename parts, features, and profiles to make it easier to keep track of them as the model gets more complex. To rename a feature, you can right-click it and select Rename from the context menu to highlight it. Or, you can click twice to highlight the name and type over it, as you will here. Figure 2.36 shows the feature name highlighted so it can be renamed.

Pick: **ExtrusionBlind1** (so that it becomes highlighted)

Wait a moment.

Pick: **ExtrusionBlind1** (so that the typing cursor appears)

Select: **the text in the box** (if necessary)

Type: **BaseLeg** ↵

The name ExtrusionBlind1 should be replaced with the new name BaseLeg.

On your own, rename ExtrusionBlind2 to MidPlate.

Save your part file.

The Desktop Browser should now look similar to Figure 2.37.

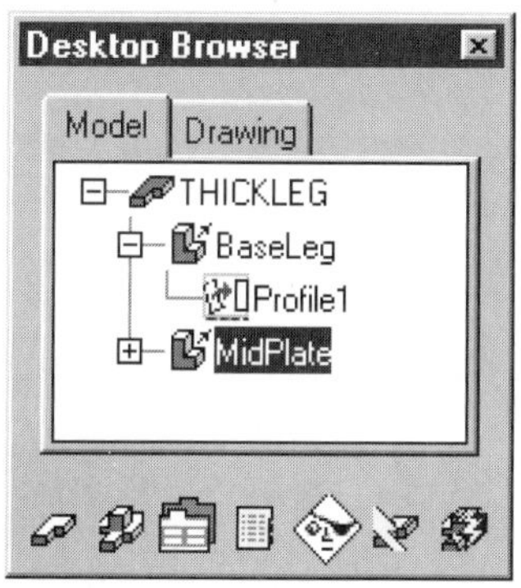

Figure 2.37

Tip

Make sure to press ↵ *after typing the name. If you do not, it will return to the old name.*

Reference

Engineering Design Communication

Recognizing Symmetry, pp. 182–184

Creating the Second Leg Feature

Next you will create the second leg of the ThickLeg part. It is the mirror image of the first feature that you created. Mechanical Desktop allows you to copy a feature and flip it to create a mirrored copy to use as a new feature, but unfortunately, you cannot copy the base feature. If you could, you could make the other portion of the leg by mirroring the leg that is already created. For the ThickLeg part and other base features, it is often just as easy to create the other leg from a sketch. Mechanical Desktop lets you project the sketch geometry from the first leg feature to create the new sketch quickly.

Next you will create the other rounded leg feature as a sketched feature. First you will create a new sketch plane on the other side of the middle plate.

Right-click: in the graphics area

Pick: **New Sketch Plane**

Select work plane, planar face or [worldXy/worldYz/worldZx/Ucs]: **pick the top surface of the MidPlate identified in Figure 2.35**

Enter an option [Next/Accept] <Accept>: **pick with the left mouse button as needed until the correct surface is highlighted and then press the right mouse button to accept**

Select edge to align X axis or [Z-flip/Rotate] <Accept>: **pick on the right end of the lower long edge**

Select edge to align X axis or [Z-flip/Rotate] <Accept>: **press right mouse button to accept**

The coordinate system should be aligned with the part as shown in Figure 2.38. If it is not, repeat the steps to create a new sketch plane.

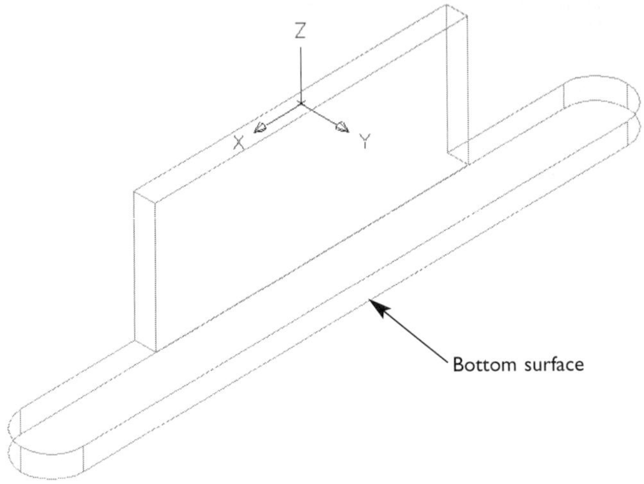

Figure 2.38

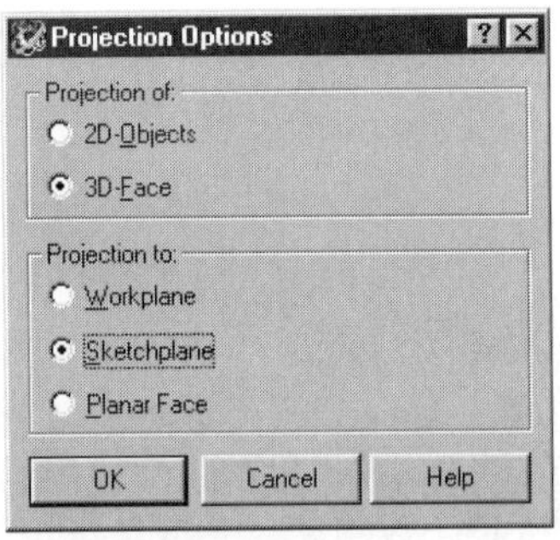

Figure 2.39

Projecting Sketch Geometry

The Project Objects to Plane command allows you to "copy" existing geometry for use in making a new 2D sketch. You can project 2D objects or the edges of 3D faces to a sketch or work plane to create an exact copy. You will project the bottom surface of the BaseLeg feature to sketch the second leg feature.

Right-click: in the graphics area

Pick: **Sketch Solving, Project Objects to Plane**

Using the dialog box that appears similar to Figure 2.39, make the following selections.

Select Projection of: **3D-Face**

Select Projection to: **Sketchplane**

Pick: **OK**

You are returned to the graphics area for the remaining selections.

Select face: **pick the bottom surface of the object, as shown in Figure 2.38**

Enter an option [Next/Accept] <Accept>: **if the bottom surface is not highlighted, cycle to the next surface by pressing the pick button** (left mouse button)

Enter an option [Next/Accept] <Accept>: **when the correct surface is highlighted press** ↵ (or the right mouse button)

The surface is projected onto the sketch plane, as shown in Figure 2.40, where you can use it as the basis for a profile.

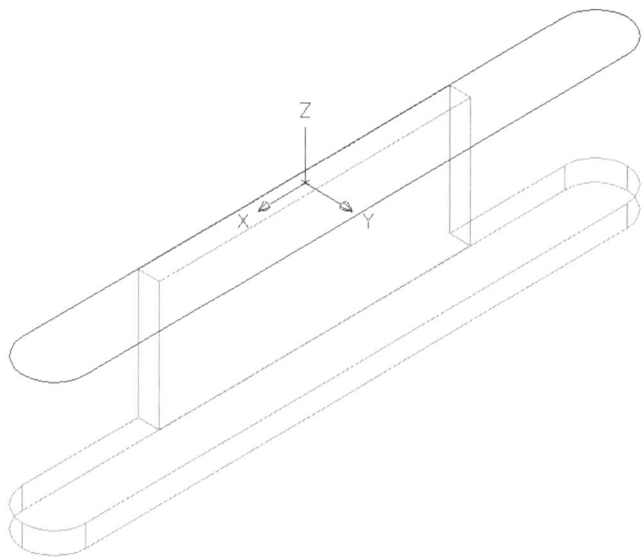

Figure 2.40

Using Profile

You will use Profile to create the profile, as your projected geometry consists of two lines and two arcs, not a single polyline entity.

Right-click: in the graphics area

Pick: **Sketch Solving, Profile**

Select objects for sketch: **pick the two lines and two arcs of the projected sketch** ↵

Solved under constrained sketch requiring 4 dimensions or constraints.

Now view the constraints added when solving the sketch.

Right-click: in the graphics area

Pick: **2D Constraints, Show Constraints**

You should see horizontal and tangency constraints. Sometimes the constraints are shown too small to be clearly visible. To change their size,

Pick: **Part, Part Options**

The Mechanical Options dialog box appears on your screen. The Part tab should be uppermost, as shown in Figure 2.41. If it is not, pick the Part tab now. You will use the dialog box to set the constraint size larger.

Pick: **Constraint Size**

Use the slider bar in the Constraint Size dialog box, as shown in Figure 2.42, to increase the size of the constraint labels.

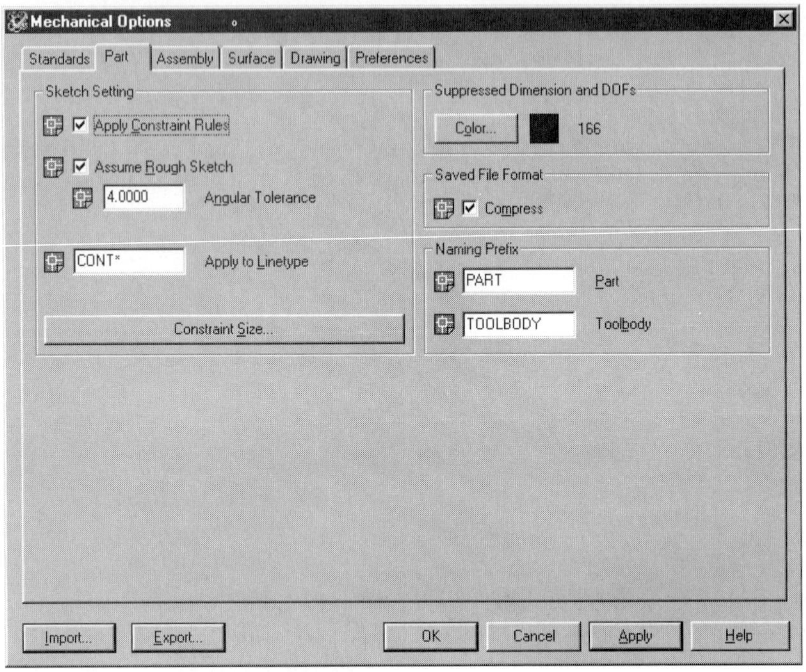

Figure 2.41

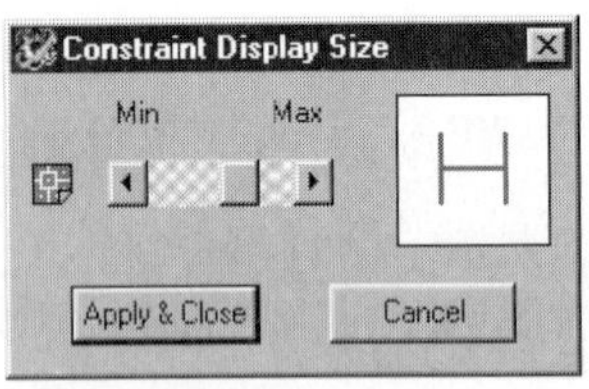

Figure 2.42

Pick: **Apply & Close**

Pick: **OK**

Right-click: in the graphics area

Pick: **2D Constraints, Show Constraints**

The constraints should appear as shown in Figure 2.43. These are similar to the constraints for the sketch for the original leg. The message displayed when you solved the sketch stated that four additional dimensions or constraints were required to fully define the profile. You will add colinear, concentric, and equal radius constraints to maintain the same sizes as the original leg. You will start the command and then type options at the command prompt to specify the concentric constraints.

Tip

Some modelers keep projected sketches always aligned with the original feature, but Mechanical Desktop does not, so you must add desired constraints.

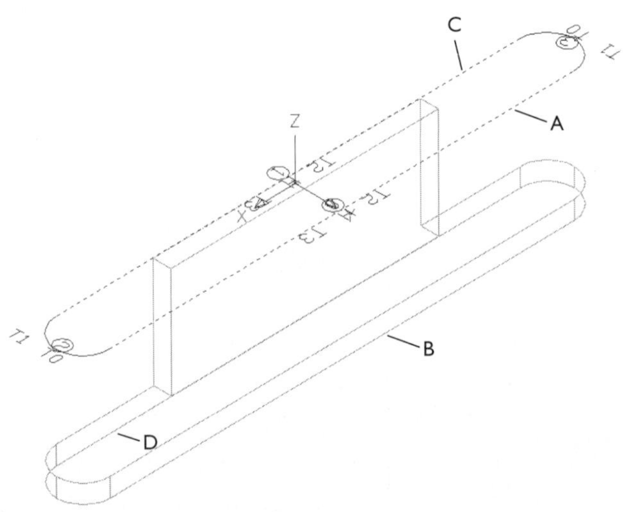

Figure 2.43

Press: ↵

Right-click: in the graphics area

Pick: **2D Constraints, Collinear**

Valid selections: **line or spline segment**

Select object to be reoriented: **pick on the lower straight line of the sketch labeled A in Figure 2.43.**

Select object to be made colinear to: **pick on the corresponding line on the bottom surface of the part, labeled B**

Select object to be reoriented: **pick on the upper straight line of the sketch labeled C**

Select object to be made colinear to: **pick on the corresponding line of the part labeled D**

Select object to be reoriented: ↵

[Hor/Ver/PErp/PAr/Tan/CL/CN/PRoj/Join/XValue/YValue/Radius/Length/Mir/Fix/eXit] <eXit>: **CN** ↵

Valid selections: arc, circle, or ellipse

Select object to be reoriented: **pick on the left arc in the sketch**

Select object to be made concentric to: **pick on the left rounded end of the base feature**

Select object to be reoriented: **pick on the right arc in the sketch**

Select object to be made concentric to: **pick on the right rounded end of the base feature**

Select object to be reoriented: ↵

Enter an option

[Hor/Ver/PErp/PAr/Tan/CL/CN/PRoj/Join/XValue/YValue/Radius/Length/Mir/Fix/eXit] <eXit>: ↵

Next you will resolve the sketch.

Right-click: in the graphics area

Pick: **Sketch Solving, Re-Solve**

You should see the message, *Solved fully constrained sketch*. If you do not, show the constraints again and check to see that you have added the two colinear and two concentric constraints. Add and delete constraints on your own if necessary so that the sketch can be fully solved.

Now you are ready to extrude the profile into a feature.

Right-click: in the graphics area

Pick: **Sketched & Work Features, Extrude**

Choose Operation: **Join**

Set Distance: **.25**

Set Draft angle: **0**

Choose Termination: **Blind**

Pick: **OK**

Pick: **Gouraud Shaded icon**

Now your part should look like Figure 2.44.

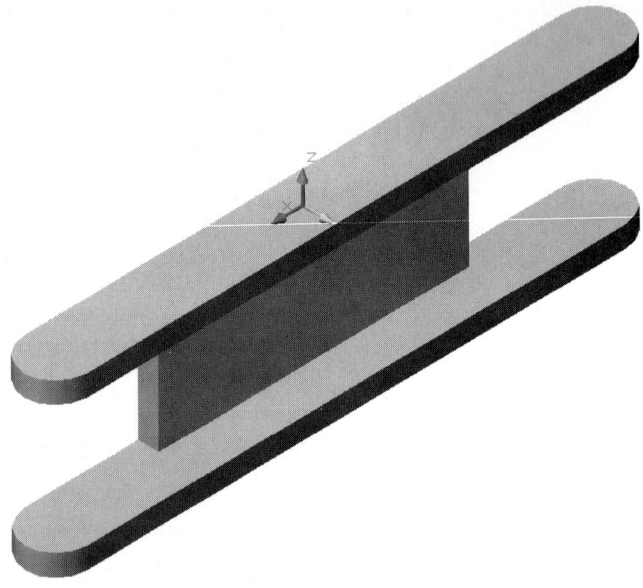

Figure 2.44

On your own, rename ExtrusionBlind1 to OppositeLeg and save your part.

Adding Placed Features

Next you will add a placed feature for the holes. Placed features are those that you do not need to sketch. You control their characteristics by specifying values for their parameters. Placed features make it easy for you to add holes, fillets, and chamfers to your parts. You can also do more advanced operations, such as adding face draft, hollowing out a shell, or creating a surface cut. You can also use this part of the menu to array features, to combine two parts into a single part, and to split a part.

Holes

When you add a placed feature, the settings you select automatically create and constrain the geometry. Notice which selections reflect the sketch geometry, geometric constraints, and dimensions for the feature as you create the holes.

Right-click: in the graphics area

Pick: **Placed Features, Hole**

The Hole dialog box, shown in Figure 2.45, appears on your screen. You can make drilled, countersunk, or counterbored holes using this dialog box and specify different options for terminating the hole.

Use the drop-down menus for Operation and Termination to see the different options illustrated in the image tile.

You can choose to locate the hole by specifying a distance from two nonparallel edges on the part, a concentric center on an existing feature, a point, or a distance from another hole. You can also specify a tap drill size using this dialog box. You will specify the options shown in Figure 2.45.

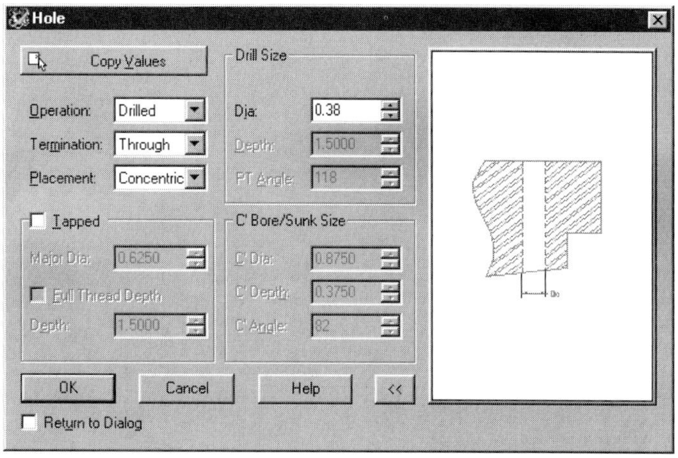

Figure 2.45

Choose Operation: **Drilled**

Choose Termination: **Through**

Choose Placement: **Concentric**

Set Dia(meter): **.38**

Pick: **OK**

You are returned to the drawing for the remaining selections.

Select work plane or planar face [worldXy/worldYz/worldZx/Ucs]: **pick the top surface of the part**

Enter an option [Next/Accept] <Accept>: **if the correct surface is not highlighted, press the left mouse button until the correct surface becomes highlighted and then press the right mouse button to accept**

Select concentric edge: **pick on one of the rounded ends**

Computing ...

Select work plane or planar face [worldXy/worldYz/worldZx/Ucs]: **pick the top surface again**

Enter an option [Next/Accept] <Accept>: **if the correct surface is not highlighted, press the left mouse button until the correct surface becomes highlighted and then press the right mouse button to accept**

Select concentric edge: **pick the other rounded end**

Computing ...

Select work plane or planar face [worldXy/worldYz/worldZx/Ucs]: ↵

The holes should appear through both sides of the part, as shown in Figure 2.46.

On your own, spin the part.

Save your file.

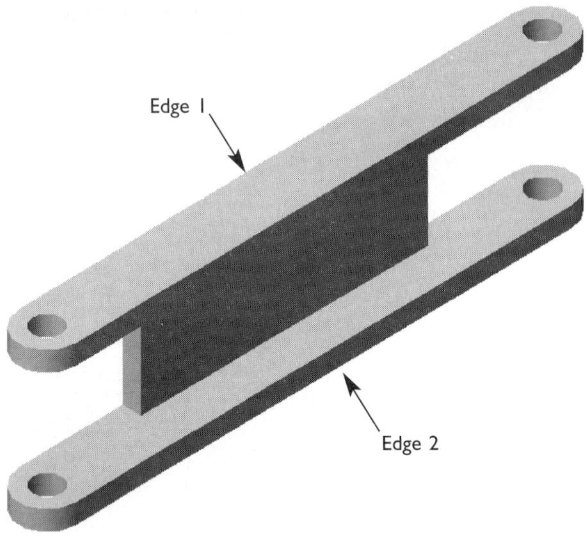

Figure 2.46

Fillet

The final feature for the ThickLeg part is another placed feature, a fillet, to round the outside edges of the leg features.

Right-click: in the graphics area

Pick: **Placed Features, Fillet**

The Fillet dialog box shown in Figure 2.47 appears on your screen. You can make fillets with constant, fixed width, cubic, or linear radii using this dialog box.

Pick the radio buttons to see the kinds of fillets illustrated in the image tile in the dialog box.

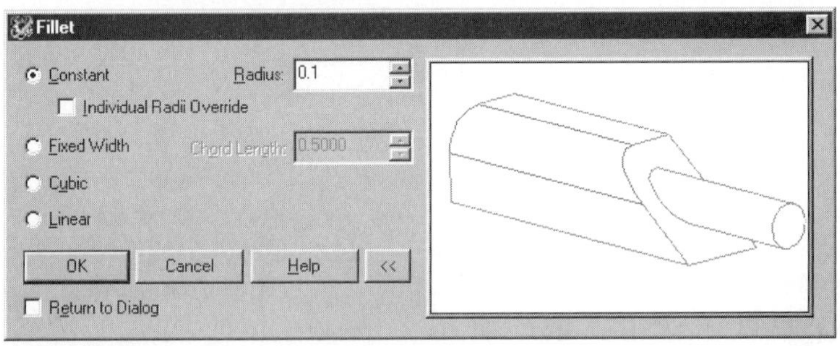

Figure 2.47

Select: **Constant**

Set Radius: **0.1**

Pick: **OK**

You are returned to the drawing for the remaining selections.

Select edges or faces to fillet <continue>: **pick edges 1 and 2, as shown in Figure 2.46.**

Select edges or faces to fillet<continue>: ↵

Computing ...

The two edges you selected are now filleted and should appear as shown in Figure 2.48.

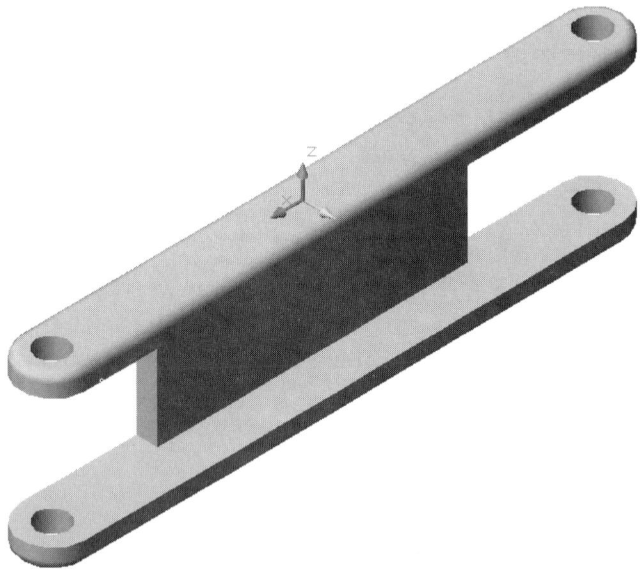

Figure 2.48

On your own, spin the part.

Save your file.

Modifying Features

The advantage to parametric modeling is that it is easy to change the model. When modeling very similar parts, it can be easier to modify a copy of an existing part than to start from scratch. You will save this model and make changes to the new copy to use as the ThinLeg part since they are both very similar in shape. The key difference between the two is size: The middle plate for the thin leg is much thinner.

Saving As a New Part

You can use the Save As command to save your project to a new file name. The original file will still be on the disk.

Pick: **File, Save As**

Use the dialog box that appears to be sure you are in the *MDT-Work\Jack* directory and

Type File name: ***ThinLeg.dwg***

Pick: **Save**

Notice that now the title bar displays the name *ThinLeg.dwg*. Any changes you make will now be to the file *ThinLeg.dwg*. The ThickLeg part that you created previously will remain the same.

Modifying Dimensions

Now you are ready to modify the dimensions for the MidPlate feature. Because the width of the MidPlate was not specified in the sketch, but rather in the extrusion operation, you will modify the feature.

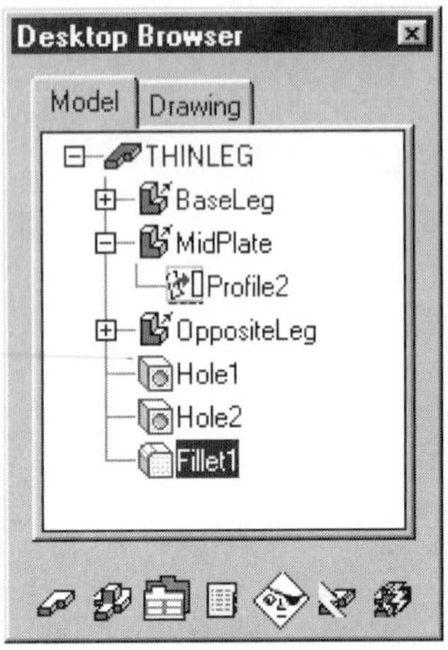

Figure 2.49

Pick: **to expand the model tree below the MidPlate feature** (as shown in Figure 2.49).

Right-click: **MidPlate** (in the Browser Window)

Pick: **Edit** (from the context menu)

The dialog box used to define the extrusion of the MidPlate feature appears. You will change the distance value for the extrusion from 1.5 to 1.0.

Set Distance: **1.0**

Pick: **OK**

Select object: ↵

You will notice that the MidPlate feature in the feature tree has been highlighted with a yellow bar. This indicates that the feature has been edited and is waiting to be updated. Now you are ready to update the part to see the change.

Right-click: in the graphics area

Pick: **Update Part**

The part updates to show the change. Figure 2.50 shows the updated part. Wasn't that fun and easy?

Now save your changes to this part and exit the software.

Good job; you created two parts for the jack assembly!

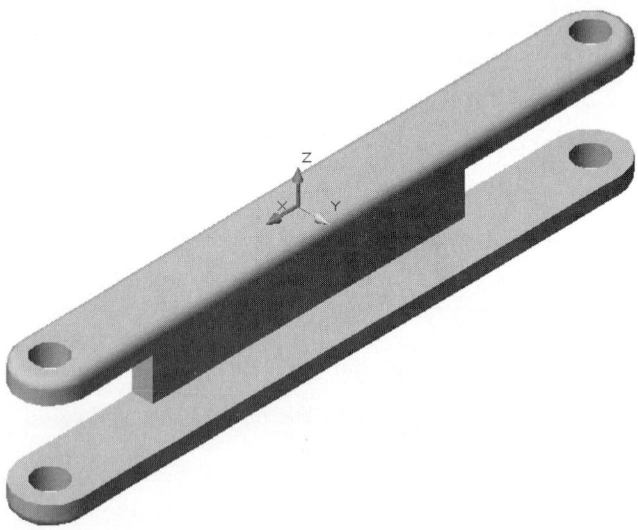

Figure 2.50

Command Summary

Action	Command Name	Description
3D Fillet	AMFILLET	Creates a fillet or round (placed feature) between faces of a 3D part
Add Constraints	AMADDCON	Adds constraints to a 2D sketch
Adjust Constraint Size	AMOPTIONS	Manages part, assembly, surface, scene, drawing, annotation, and desktop preferences; can set size of constraint display
Change Color	PROPERTIES	Changes the color of a 3D object
Extrude	AMEXTRUDE	Creates an extruded solid feature from a 2D profile
Extrude, Join	AMEXTRUDE, J	Extrudes a feature and adds it to the model (equivalent to Boolean union)
Hole	AMHOLE	Creates parametric hole features
List	LIST	Displays information such as length, layer, and location about any object
Make a Profile	AMPROFILE	Solves 2D geometry to create a constrained sketch
Named UCS	UCSMAN	Displays and modifies defined user coordinate systems, and activates named and orthographic UCSs
New Dimension	AMPARDIM	Adds a parametric dimension to a sketch
New Part	AMNEW	Create a new part or part instance
New Sketch Plane	AMSKPLN	Sets the sketch plane location and the orientation of its coordinate axes
Ortho	ORTHO	Constrains cursor movement to horizontal and vertical on the current UCS (or with the Grid rotation angle)
Polyline	PLINE	Creates a 2D polyline (of lines and arcs) that may also have a width and linetype pattern
Project Objects to Plane	AMPROJECT2PLN	Projects 2D objects or edges of planar faces to a work plane, sketch plane, or 3D face so they can be used in new sketch
Rectangle	RECTANGLE	Creates a rectangular polyline object
Rename (in Browser)	AMRENAME	Renames an element (part, feature, assembly, scene) in the Browser window
Re-solve Sketch	AMRSOLVESK	Displays the number of constraints needed to fully constrain a sketch, or appends new geometry to a sketch
Show Constraints	AMSHOWCON	Displays 2D constraint symbols on the selected sketch
Single Profile	AMPROFILE	Solves a single 2D entity to create a constrained sketch
Sketch View	AMVIEW, Sketch	Controls the view in Model mode; the Sketch option sets the view to a plan view of the current sketch
Zoom, All	ZOOM, A	Zooms to the maximum magnification possible so that all of the objects are visible

Exercises

1. **Plywood Puzzle** A 4 × 8′ sheet of plywood is shown in Figure 2.51 below. Use your knowledge of 2D sketching and constraints to determine the largest diameter circular shape that can be cut from this sheet using only two pieces.
   Hint: The layout in Figure 2.52 suggests a solution method.

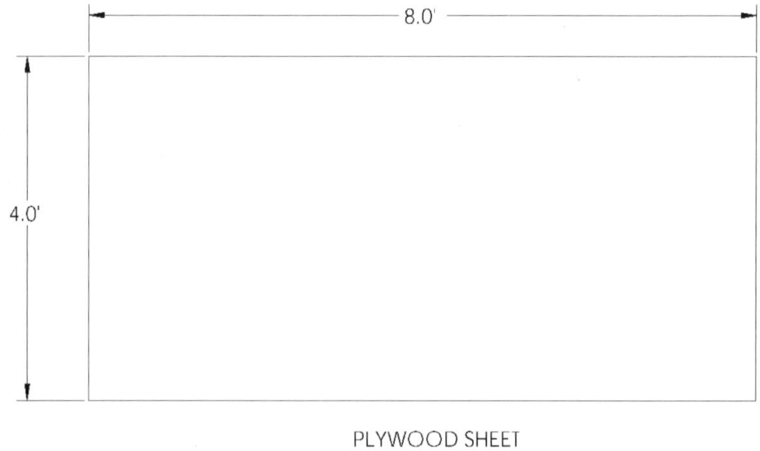

Figure 2.51

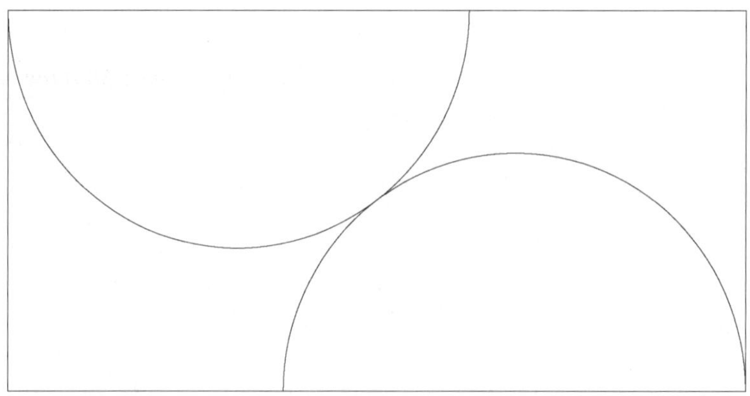

Figure 2.52

2. **Triangle Puzzle** Construct an equilateral triangle like the one shown in Figure 2.53. Use your knowledge of sketching and 2D constraints to fit three equal circles inside the triangle so that they don't overlap. What is the largest possible diameter?

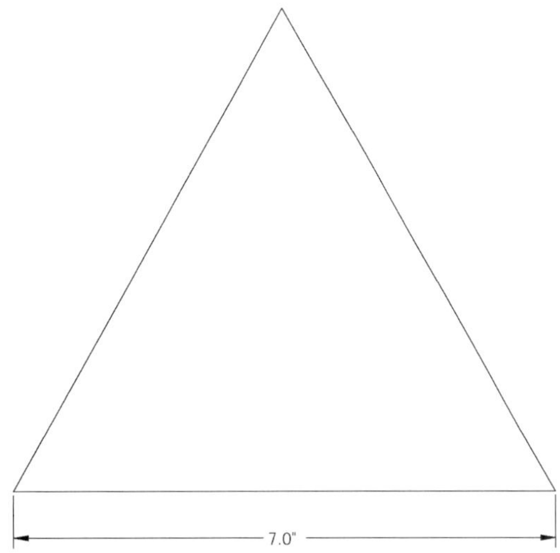

Figure 2.53

Model the parts shown in Figures below. (Additional parts for modeling may be found in the assembly exercises at the end of Tutorials 7, 9, and 10.)

3. **Hex Cap** *Note*: All dimensions are in millimeters.

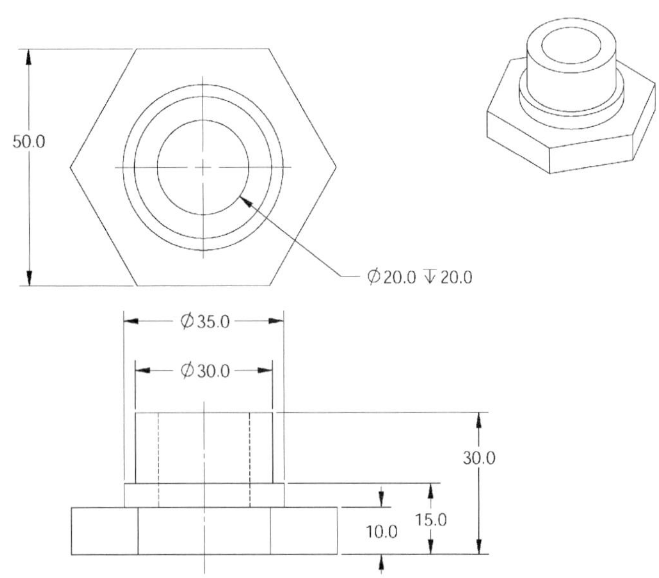

Figure 2.54

4. **L-Bracket** *Note*: All dimensions are in inches.

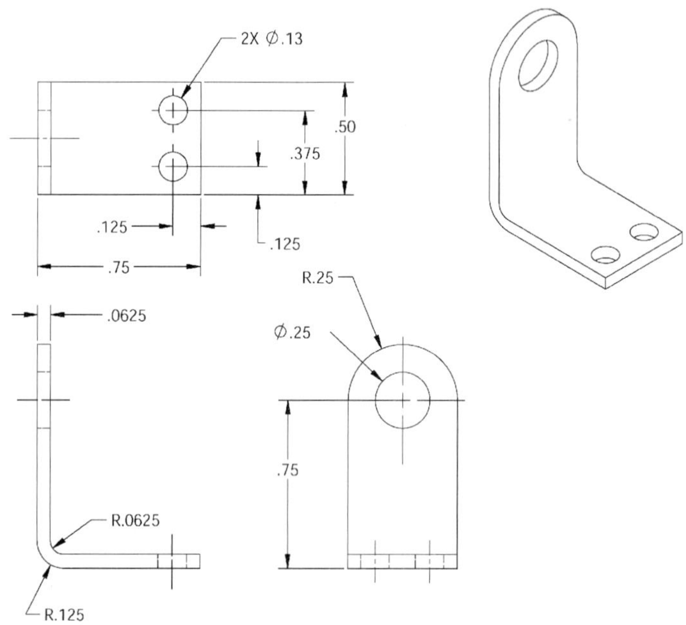

Figure 2.55

5. **Shaft Clamp** After the part is modeled, add 1-mm fillets and rounds. *Note*: All dimensions are in millimeters.

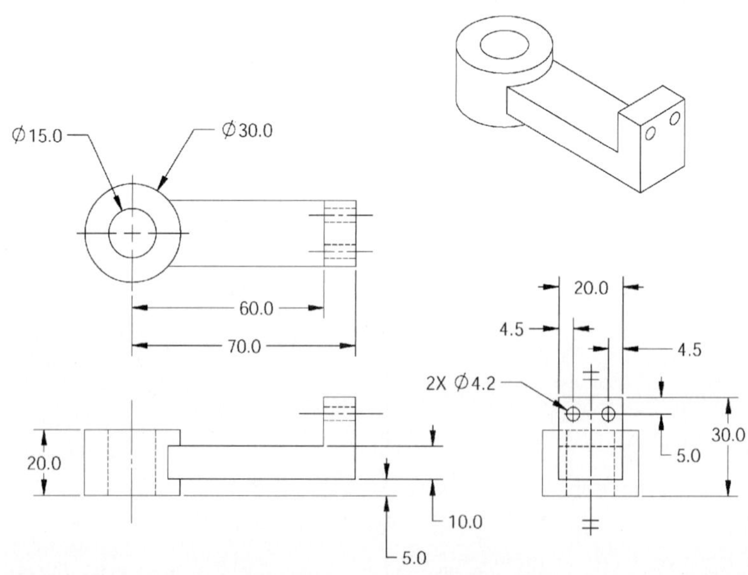

Figure 2.56

6. **U-Bracket** *Note*: All dimensions are in inches.

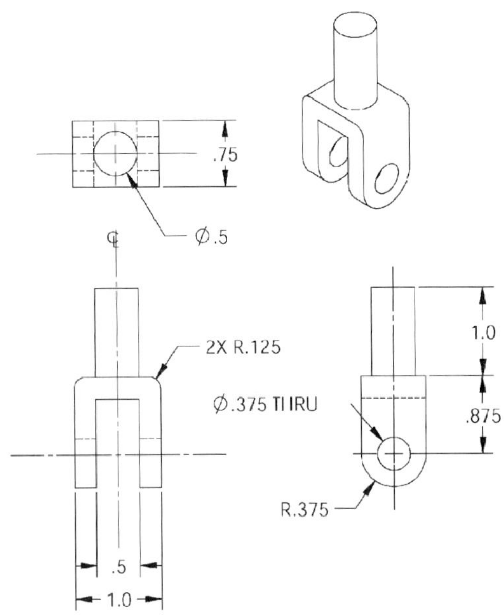

Figure 2.57

Tutorial

3 Making a Drawing

Objectives

When you have finished this tutorial, you will be able to

- Make a multiview drawing from a part model.
- Change Views and Linetype scale.
- Modify Layer properties.
- Create and save a Text Style.
- Create and save a Dimension Style.
- Add a pictorial view to your drawing.
- Move and Edit views in your drawing.
- Plot a drawing layout.
- Use bidirectional associativity to modify your model and drawing.

Reference

Engineering Design Communication

Documentation Drawings, pp. 468–469

Introduction

In the last two tutorials, you explored Model mode for assemblies and for part files. In this tutorial, you will use Drawing mode to create an engineering drawing that shows multiple views of the ThickLeg part.

Engineering drawings are often used in combination with the part model file to communicate important information about how the part is to be manufactured. Some manufacturers rely solely on the information in the printed drawing to manufacture the part. Others can create the actual part directly using the information in the 3D model. Even when parts are manufactured directly from the model, the printable drawing serves as a way to provide tolerances, surface finish information, and other specifications that cannot be conveyed easily using only the model. If you have ever created multiview drawings by hand or using 2D CAD software, you will be amazed at how easily you can create multiview drawings using Mechanical Desktop.

Starting

You will create a drawing of the ThickLeg part that you created in the last tutorial. To ensure that you get the same results as shown here, you will work with the data file named *ThickLeg-dta.dwg* that you copied to the MDT-Data directory. (You may use your file from Tutorial 2 if you wish.)

Launch Mechanical Desktop 5 (as you did in previous tutorials).

Pick: **Browse** (in the Open Drawings tab of the Mechanical Desktop Today window)

Use the methods that you have learned to

Open: ***ThickLeg-dta.dwg*** (from the MDT-Data directory)

The ThickLeg-dta file drawing name appears in the title bar area of your screen. You see the ThickLeg part in the graphics window. You will save this file with a new name in your MDT-Work directory so that the data file remains unchanged.

Pick: **File, Save As**

Choose Save in: **MDT-Work**

Type File name: **ThickLeg-drawing**

Pick: **Save**

The title bar should now display the current name ThickLeg-drawing as shown in Figure 3.1. Changes that you save will occur in this file.

If you are working in a shared computer lab, use the Options dialog box to set your user profile to current.

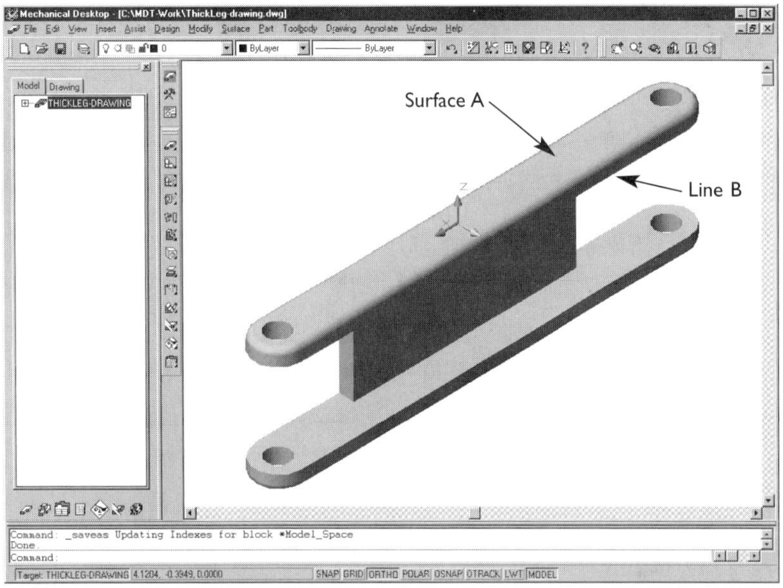

Figure 3.1

Now you are ready to create a drawing. In the Desktop Browser,

Pick: **Drawing tab**

Mechanical Desktop uses AutoCAD's *paper space* mode to layout the drawing. When you pick the drawing tab, you switch from model space into paper space. Your screen looks like Figure 3.2, showing what looks like a blank sheet of drawing paper. You will use this mode to lay out your 2D multiview engineering drawing.

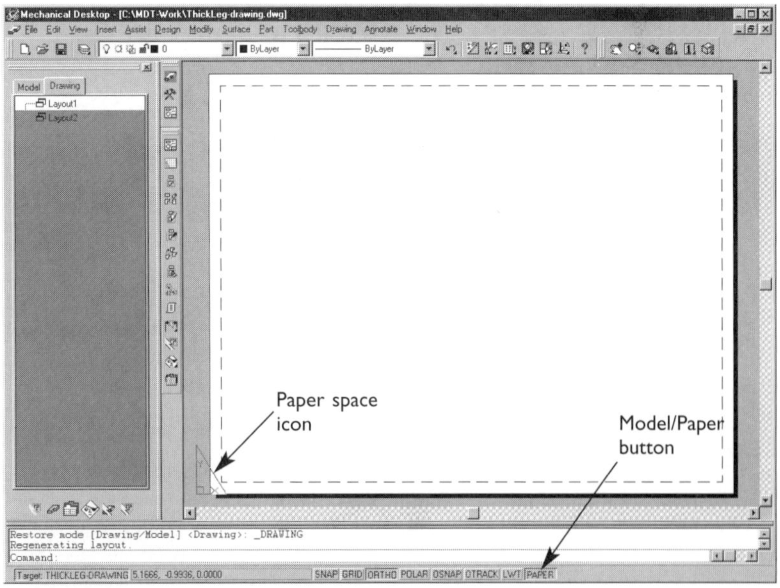

Figure 3.2

Paper Space

Mechanical Desktop allows you to set up drawings that are similar to those drawn on a sheet of paper. You arrange 2D views of the model on a paper space "sheet" to create arrangements of 2D views that you can plot to any scale. Up to this point you have been working in AutoCAD's *model space* to model part geometry. Note that the UCS icon is no longer in lower left corner of the graphics window on your screen; it has been replaced by the *paper space icon*, which looks like a triangle. Notice also that PAPER appears where MODEL used to be on the status bar. The area around the paper space sheet is shown with a gray background. These things all indicate that you are in paper space.

Paper space allows you to arrange views of your model on the sheet of paper shown in the graphics window. Your model geometry is created in model space, whereas paper space is used to create and add drawing elements such as borders, title blocks, text notes, and views of the model.

Notice that the Desktop Browser displays Layout1 and Layout2. A *layout* is like a single piece of paper. You can have as many drawing layouts as necessary for a model space part (or assembly). Each layout can be a different size and have its own settings, border, text, and drawing views.

Printer Area

The white area in the graphics window represents the sheet of paper. The default paper size is the one used by the printer or plotter you have configured. For this tutorial, we are assuming an 8.5 × 11″ sheet used by most printers. If your printer uses a different paper size, your screen may appear slightly different due to the proportions of your sheet size.

The dashed line represents the limits of the printer. The distance from the edge that a printer can reach is different for each printer. The distance from the left edge isn't necessarily the same as it is from the right edge or the top and bottom edges. (When you preview a plot in the Print/Plot Configuration dialog box, the outer line in the preview shows the printer limits.) The limits of a printer may also be different for each sheet size. For this drawing you will use the default 8.5 × 11″ sheet.

Adding Multiple Views

You will begin by adding multiple views that show the thick leg of the jack.

Right-click: in the graphics area

Pick: **Multiple Views**

The Create Drawing View dialog box appears on your screen, as shown in Figure 3.3. Notice that the View type is set to Multiple, which will allow you to continue to add orthographic views until you press ↵ to indicate that you are done.

Use the drop down menu for View type and pick through the selections to see an illustration of the other options shown in the image tile.

The Data Set is set to the Active Part. In this case you are working from a model that only contains the ThickLeg part and not an assembly of multiple parts, so ThickLeg is the active part. If there were other parts in the assembly, you could choose one here.

The Layout area is set to Layout1. This is the first Layout for this part, so that will be a good choice. Later in this tutorial you will learn how to use different sheet sizes for your drawing and to make a template that will use different sheet sizes for Layout1 and Layout2.

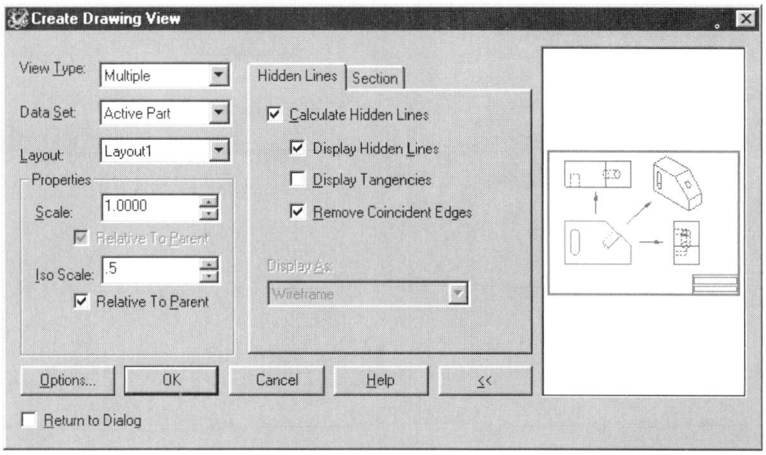

Figure 3.3

The Scale area lets you select the scale that will be used for the drawing. For this part, you will select the default, 1.0000, which is a scale of 1 = 1. You will plot the sheet at full scale on your printer. This will produce a drawing where the part is shown actual size. For drawings where the full-size part will not fit on the sheet or is too small to see, you can use this area to change the scale. If you include an isometric view, it can use a different scale from the orthographic views. This is a common practice. You will choose .5 for the isometric scale. Checking the Relative to Parent box shows the isometric view scaled relative to the scale of the parent (orthographic) views. In this case, because the scale of the parent view is 1.0000, the scale for the isometric view is .5 (half-size or at a scale of 1 = 2). If the parent view were at .5 (1 = 2), selecting Relative to Parent for the isometric view would make its scale .25 or a scale of 1 = 4.

You can also use this dialog box to control the calculation of and appearance of hidden lines. Typically in a multiview drawing you want to display hidden lines. Often in assembly drawings you will not display hidden lines. Most drawings do not display tangencies (lines indicating where a contoured surface is tangent to a plane surface). Coincident edges are where there is more than one line lying behind another in a view. Typically these are removed.

Choose View Type: **Multiple**

Choose Data Set: **Active Part**

Choose Layout: **Layout1**

Set Scale: **1.0000**

Set Iso(metric) Scale: **.5**

Select: **Relative to Parent** (so it is checked)

Select: **Display Hidden Lines** (so it is checked)

Deselect: **Display Tangencies** (so it is unchecked)

Select: **Remove Coincident Hidden Lines** (so it is checked)

Pick: **OK**

You are switched back to model space, where you will select the direction of sight for the central view, usually the front view of the part. You see the message, Regenerating model. You will orient the viewing plane with command options similar to those you use to orient a sketch plane (or a new UCS). First, you will specify a face to use for the *X-Y* plane, then select an edge to align the *X*-axis to, and finally confirm the line of sight. At the prompt,

Select planar face, work plane, or [Ucs/View/worldXy/worldYz/worldZx]: **pick surface A in Figure 3.1**

Enter an option [Next/Accept] <Accept>: **if the correct surface is not highlighted, press the left-mouse button to cycle through the surface; when the correct surface is highlighted, right-click to accept**

Define X-axis direction:

Select work axis, straight edge, or [worldX/worldY/worldZ]: **pick the bottom horizontal line of the top leg labeled B in Figure 3.1**

Adjust orientation [Flip/Rotate] <Accept>: **if the direction arrow is pointing upward, right-click to accept**

You now see the message, Regenerating layout. You are switched back to paper space to continue laying out the views.

Specify location of base view: **pick near the center of the sheet.**

The view appears on the sheet. If it is not placed in a manner that will allow you to fit a view above it and one to the right of it, at the prompt,

Specify location of base view or [Done] <next view>: **pick a new location for the view, or right-click and pick Enter to continue**

Specify location of projected view or [New parent view]: **pick above the base view**

Specify location of projected view or [Done] <next view>: ↵

Specify location of projected view or [New parent view]: **pick to the right of the base view**

Specify location of projected view or [Done] <next view>: ↵

Specify location of projected view or [New parent view]: ↵

You should now have a multiview drawing on your sheet that looks similar to Figure 3.4.

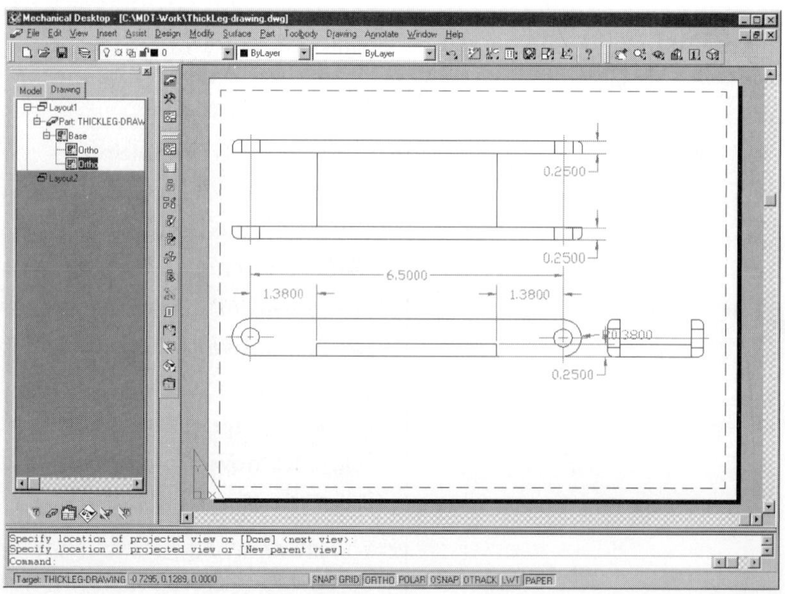

Figure 3.4

Moving Dimensions and Drawing Views

If, after placing the views, you aren't happy with the location of some dimensions, you can adjust their placement with the Move Dimensions command, as follows. You will use these steps and the Move Views command on your own as needed, watching the prompts at the command line as you work, so your drawing looks like the one in Figure 3.5.

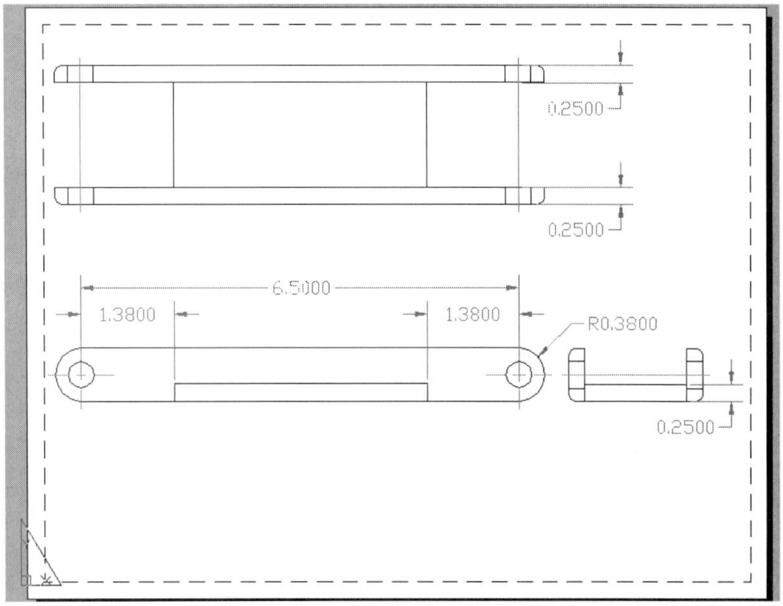

Figure 3.5

Pick: **Annotate, Edit Dimensions, Move Dimension** (from the menu bar)

Enter an option [Flip, Move, move mUltiple, Reattach]<Move>: ↵

Select dimension to move: **pick on the dimension you want to move**

Select destination view: **pick the view where you would like the dimension shown**

Select location: **pick the new location**

Select location: **pick again if necessary, or ↵ to exit**

If the views are too close, you can use the Move Views command to move them, as follows.

Pick: **Drawing, Move View** (from the menu bar)

Select view to move: **pick a view**

Specify new view location: **pick a new location for the view**

Specify new view location: **pick again or press ↵ to exit the command**

To make the layout fill the screen better,

Pick: **View, Zoom, All**

Improving Drawing Appearance

The rounded ends of the thickleg may appear as straight lines as shown in Figure 3.6. You can use AutoCAD's Viewres command to increase the number of segments that are used to approximate circles on the screen,

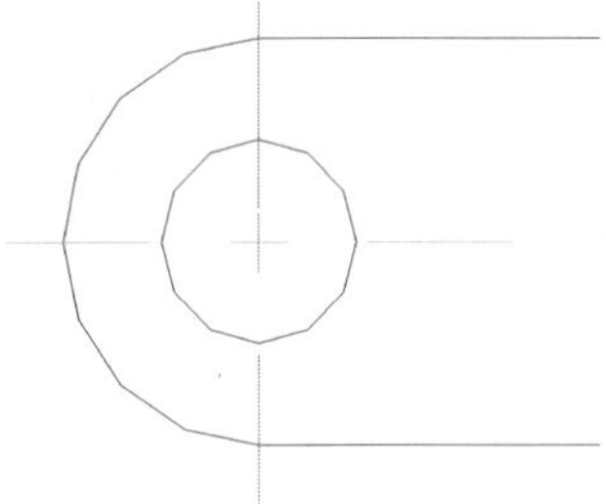

Figure 3.6

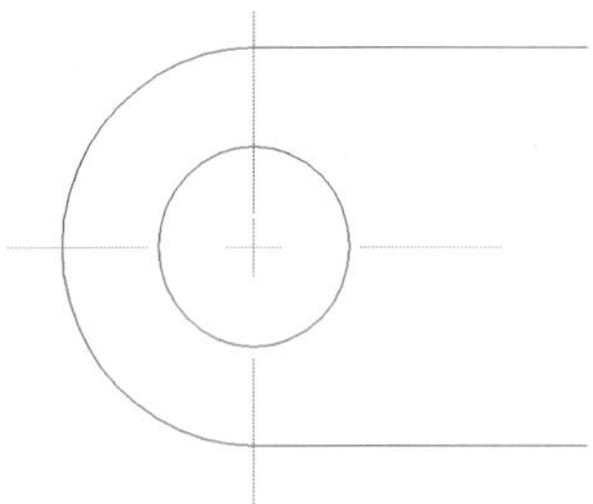

Figure 3.7

Tip

If your hidden lines do not regenerate with shorter dashes, it may be due to the linetype being used to generate the pattern. Experiment with different settings for LTSCALE until you find one that works for your line pattern. (Smaller values generate shorter dashes.) You can also pick the Layers icon and set the linetype for layer AM_HID, the layer for the hidden lines, to the linepattern HIDDEN and then use an LTSCALE setting near .5. You will learn more about managing layers next.

making the ends look more rounded. After making a change, you usually will need to update the drawing. To see the effect of this change better, first use Zoom, Window to enlarge the left end of the front view.

Pick: **Zoom Window icon**

Pick: **above and to the left of the rounded end in the front view, then below and right to form a window around the rounded end.**

Your screen should be zoomed up like Figure 3.6.

Command: **VIEWRES** ↵

Do you want fast zooms? [Yes/No] <Y>: ↵

Enter circle zoom percent (1-20000) <100>: **2000** ↵

Right-click: in the graphics area

Pick: **Drawing menu**

Pick: **Update Layout**

The option to update the current layout will update all views that are on it.

3 drawing views being updated.

The layout updates so that the rounded end of the part appears nicely rounded as shown in Figure 3.7.

Pick: **Zoom All icon**

Your view zooms to show the full sheet again.

Setting Linetype Scale

The AutoCAD command for linetype scaling can also affect the appearance of your drawing. The hidden lines in your drawing should appear as dashed lines but do not. Depending on the size of the part, the pattern for the line may not appear the correct size in the drawing. You will set LTSCALE to a smaller value, so that hidden lines, centerlines, and other patterned lines use a smaller scale.

Command: **LTSCALE** ↵

Enter new linetype scale factor <1.0000>: **.03** ↵

The layout is regenerated with the patterns shown with shorter dashes, as shown in Figure 3.8.

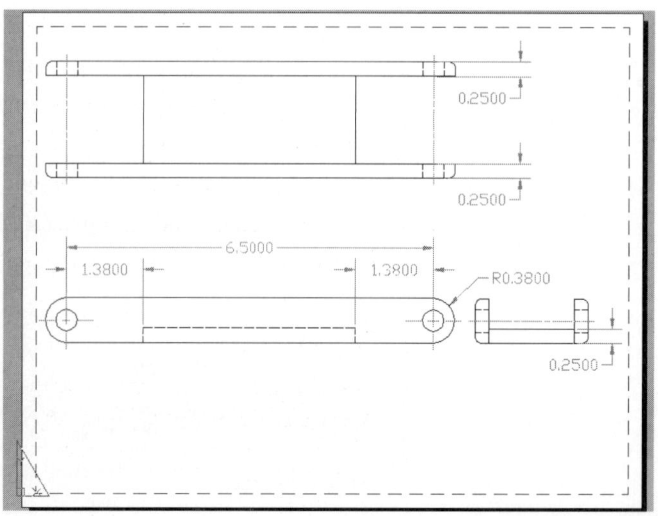

Figure 3.8

Reference

Engineering Design Communication

Interpreting Linetypes, pp. 143–145

Layers

Mechanical Desktop organizes drawing information into logical groups using *layers*. You can think of layers as a set of clear overlay sheets in your drawing. When Mechanical Desktop generates dimensions, hidden lines, centerlines, and other information for drawing views, it automatically creates layers as needed. All of the hidden lines in your drawing, for example, are stored on the AM_HID layer of your drawing. You should not delete or rename these automatically generated layers or your file may be corrupted.

Layers can be turned on and off so that they are visible or not in your drawing. For complex drawings, you can create layers to organize groups of objects in your drawing. For example, all of the parts associated with a subassembly could be stored in a single layer group so they can be turned on and off as desired.

Layers have properties that can be used to control the appearance of objects on the layer, such as color and linetype. When objects are created, the default is to set color and linetype by layer. That is, the color and linetype of the object are controlled by the color and linetype property of the layer the object is on. Using layer properties in this way helps to standardize types of objects in the drawing, which in turn helps standardize their plotted appearance.

Warning

Mechanical Desktop automatically creates drawing layers for dimensions and other elements of your drawing file. Do not delete or rename these layers. Doing so can destroy or corrupt your drawing or model.

When you plot your drawing, layer properties can be used to control how kinds of objects are printed. Object color is one way to control the pen selection when you are printing to a printer or plotter. You can also use the line thickness setting to achieve the same thing on printers that are capable of printing different line weights. By changing the line thickness for the hidden line layer, for example, you can change the thickness of the printed line for all hidden lines in your drawing. In general, the default settings work well and you should not make changes to the layer properties unless you encounter a problem. However, you can change the layer color to improve the appearance of your drawing and the visibility on your screen. You will do so to familiarize yourself with the Layer Properties Manager.

Reference

Mechanical Desktop On-line Help

Read more about layers by choosing Mechanical Utilities in the Contents listing in Mechanical Desktop Help.

Changing Layer Color

All layers and layer properties can be managed from a single dialog box, the Layer Properties manager, shown in Figure 3.9. You will use it to change the color of the dimension layer, AM_PARDIM, to blue (color #5). You will use the Layers icon (which looks like a stack of papers) on the Mechanical Main toolbar to open it.

Pick: **Layers icon**

Pick: **the color box corresponding to layer AM_PARDIM**

Pick: **blue** (from the top row of standard colors, number 5 from the left)

Pick: **OK**

Pick: **OK**

You should now see the dimensions on the screen have changed to the color blue. This is very similar to the centerline color, so you will use the dialog box on your own to change the color of the centerline layer, AM_7, to red.

On your own, change the color of AM_7 to red.

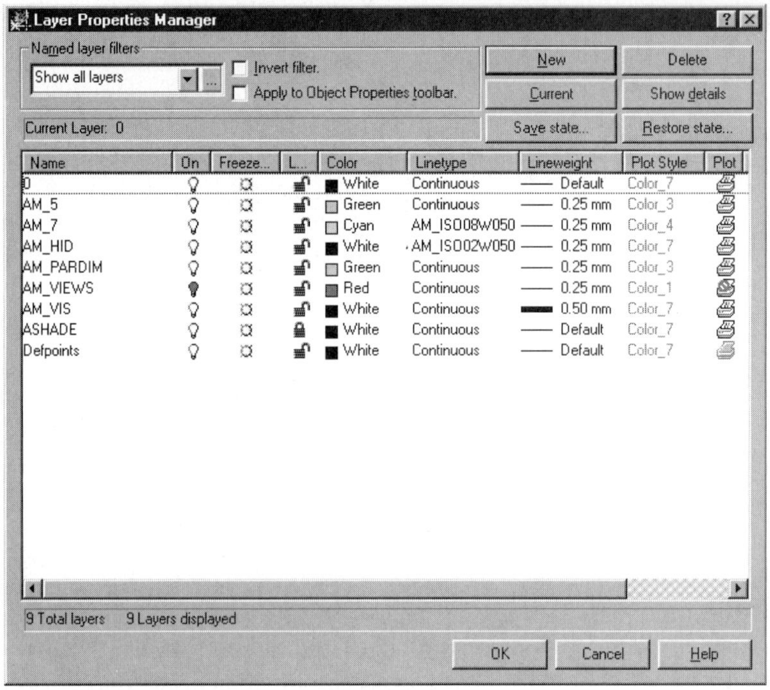

Figure 3.9

Changing the Text Style

The style of the text on your drawing can also be customized. The default text style used by Mechanical Desktop is AutoCAD's Standard style. This uses the *txt.shx* font. A *font* is a set of characters with a particular shape. The best font for lettering engineering drawings is called *romans.shx*, for Roman Simplex. The *.shx* fonts are provided with AutoCAD. Anyone using AutoCAD will have this font. True type fonts and other system fonts may look great in your drawing but may not be available on someone else's system. This will require that they substitute another font, which may make the lettering not fit correctly in the drawing.

Mechanical Desktop lets you create and save a text style as an easy way to save settings for the font and size of text you want to use. When you create a style, you can assign it any name you want, but you must specify the name of an existing font for the style to use.

Pick: **Annotate, Text, Text Style** (from the menu bar)

The Text Style dialog box appears, showing AutoCAD's choices of Style name, Fonts, and Effects.

Pick: **New**

Type Style Name: **MyText**

Pick: **OK** (or press ↵)

Choose Font Name: **romans.shx** (from the drop-down menu)

You will use the defaults in all other areas of the dialog box. Your dialog box should look like Figure 3.10.

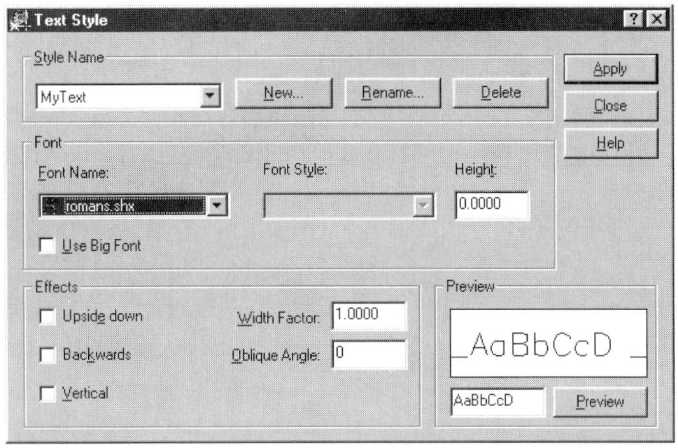

Figure 3.10

Pick: **Apply**

Pick: **Close**

The lettering in your drawing does not automatically update to use this style. Any lettering created after you change the font will use the new style. You will change the text style for the existing text after you change some other aspects of the dimension appearance.

Dimension Styles

In addition to a Text style, you can create styles for the appearance of dimensions. Many features controlling the appearance of the dimensions are set using dimension variables (dim vars), which you can set in the Dimension Styles dialog box.

Pick: **Annotate, Edit Dimensions, Dimension Style** (from the menu bar)

The Dimension Style Manager appears on your screen, as shown in Figure 3.11.

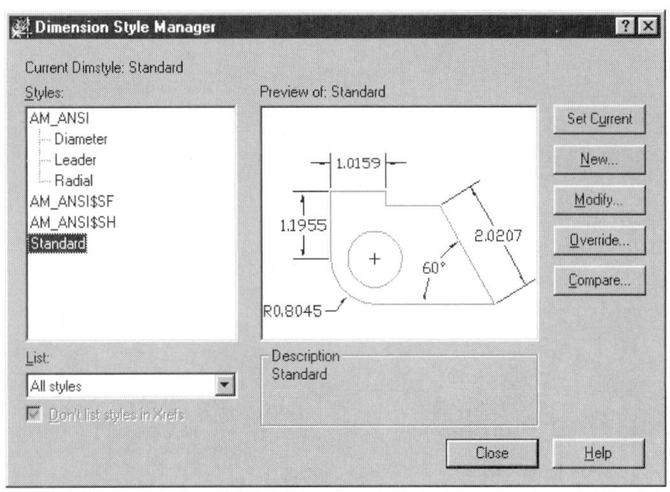

Figure 3.11

Creating a Named Dimension Style

Note that the current style name is Standard. This is the default appearance for dimensions in Mechanical Desktop. Other dimension styles, including the ANSI styles available from Mechanical Desktop, may be listed in the dialog box. You can see how they change the appearance of the dimensions by choosing them and seeing the illustration in the image tile change.

You can create your own dimension style (or styles) with a name that you specify. This way you can save your own dimension appearances. You will pick the New button to create a new dimension style named Mechanical.

Pick: **New**

The Create New Dimension Style dialog box appears on your screen. In the New Style Name area, highlight the existing text, Copy of Standard, and

Type New Style Name: **Mechanical**

In the Start With area, Standard is listed as the style to use to begin the settings for the new Mechanical style. Near the bottom of the dialog box you will see the Use for area. This style will be used for all dimension types: linear, angular, radial, and so forth, but you may create dimension styles that only apply to certain types of dimensions. When you have entered the new style name the dialog box should look like Figure 3.12.

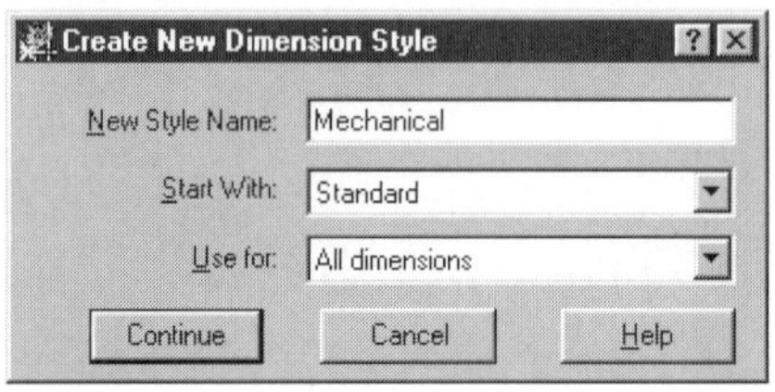

Figure 3.12

Pick: **Continue**

The New Dimension Style dialog box now appears. It has tabs for groups of dimension characteristics: Lines and Arrows, Text, Fit, Primary Units, Alternate Units, and Tolerances. Each tab allows you to control these aspects of dimensions that use the style Mechanical.

Reference

Engineering Design Communication

Standard Dimension Appearances, pp. 501–504

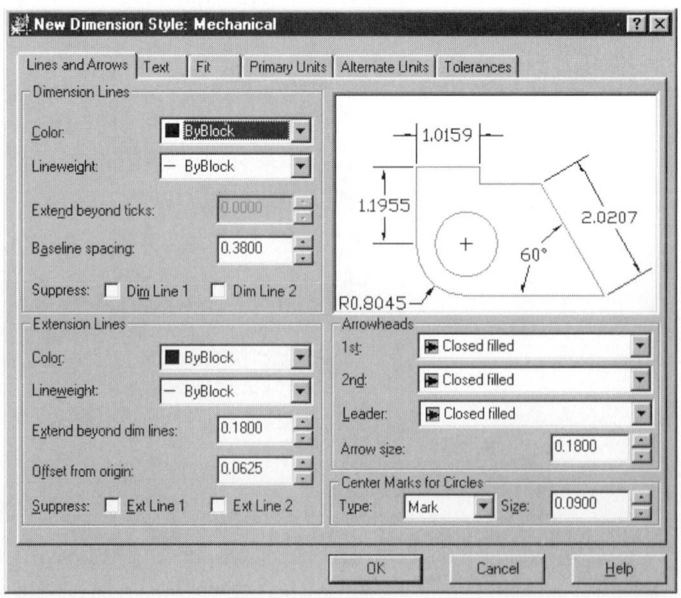

Figure 3.13

The Lines and Arrows tab should be uppermost, as shown in Figure 3.13. This tab allows you to control how the dimension lines, extension lines, arrowheads, and centermarks are drawn.

The settings that you will make are based on an 8.5 × 11" sheet size. Larger drawing sheets often use larger sizes.

Setting the Dimension Lines

In the Dimension Lines area at the upper left of the dialog box, you can control the appearance of the dimension lines. You can set the color, lineweight, and baseline spacing. Baseline spacing controls the distance between successive dimensions added using baseline dimensioning. The default value, 0.38, will work for now. (The ANSI standard states that successive dimensions should be at least 1/4" or 0.25" apart, so 0.38 will meet this criterion.) You can choose to suppress the first or second dimension line when the dimension value divides the dimension line into two parts by checking the boxes near the bottom of this area. On your own make sure that none of the suppress dimension line boxes are checked. You will set the color for dimension lines so that they are controlled by the dimension's layer.

Choose Color: **ByLayer** (from the Color drop-down menu)

Extension Lines

The Extension Lines area lets you control the color and lineweight for extension lines. In addition, you can set the distance that extension lines extend beyond the dimension line, and the distance from the selected dimension placement that the extension line will begin (offset from origin). You can also suppress either extension line. Make sure that neither box is checked to suppress the extension lines. You will use this area to control the color by layer, then set the distance to extend beyond dimension lines to .125 (the ANSI standard for this size drawing).

Choose Color: **ByLayer** (from the drop-down menu)

Type Extend beyond dim lines: **.125** (or use the downward arrow to set the value to .125)

Arrow Size and Style

To set the arrow size, use the right side of the dialog box. A list of the available arrow styles pulls-down adjacent to the first or second arrowhead boxes.

Scroll through the available arrow styles to see their illustration in the image tile.

Choose 1st: **Closed filled** (when you are finished).

Notice that when you select an arrow style for the first arrowhead, the second arrowhead box automatically changes to reflect that. If you want to select a different style for the second arrowhead, choose it last. Next set the size for the arrowhead.

Type Arrow size: **.125** (to replace the value)

Centermark Size and Style

When you create radius or diameter dimensions, you can use the Center Marks for Circles area to control the size and style of the centermark or to suppress it so that no centermark is created. The Line option draws a full set of centermarks. The Mark option draws just a tick at the center of the circle. Choosing None will cause no centermarks to be drawn. You can see these illustrated in the image tile as you choose them.

Tip

If the value for the centermark size is too large, AutoCAD will not draw full centermarks when set to the Line style; instead, it will draw just the Mark style in the center of the circle.

Choose Type: **Line**

You can set the size of the centermark by using the Size box. The value for Size sets the distance from the center to the end of one side of the center cross of the centermark. To make a smaller center cross, set it to a smaller value; for a larger center cross, set the size larger. Leave the value for your centermarks set to the default value, 0.09, for now. When you have finished making the selections described, the Lines and Arrows tab should appear as shown in Figure 3.14.

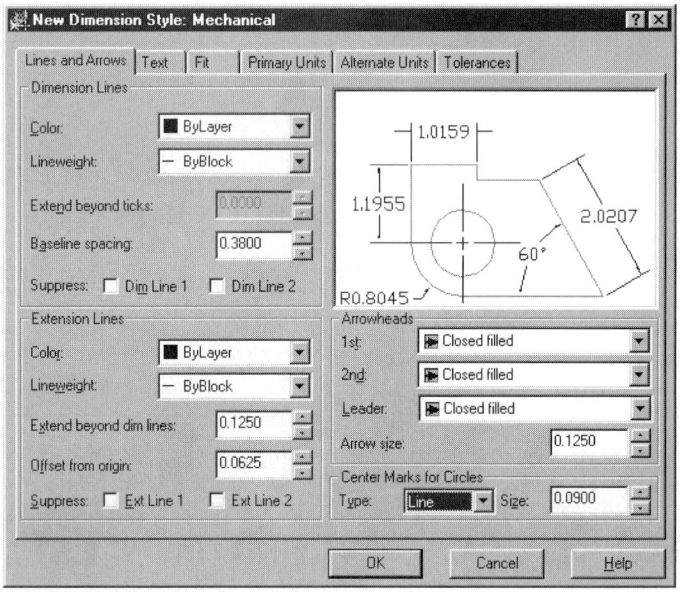

Figure 3.14

Text

The Text tab allows you to control the text style, height, color, placement, alignment, and the gap between the end of the dimension line and the start of the dimension text.

Pick: **the Text tab** (so it is uppermost)

In the Text Appearance area, you will choose your new text style, MyText. In order for a style to be listed here, you must have previously created it in the drawing, as you did with MyText. You will set the text height to 0.125, the standard height for 8.5 × 11 drawings, and set color ByLayer.

Choose Text style: **MyText** (from the drop-down menu)

Choose Text color: **ByLayer**

Type Text height: **.125**

The Text Placement area has a control for horizontal and vertical text placement. You can use the options for Horizontal to center text, place it near the first or second extension line, or place it over the first or second extension line. The default choice, Centered, is usually used for mechanical drawings. On your own try each setting and notice the effect of the on the image tile. When you are finished, to return the setting to the default,

Choose Horizontal: **Centered**

The Vertical placement control sets the placement of text relative to the dimension line. The options for vertical justification are Centered, Above, Outside, and JIS (Japanese Industrial Standard). Choose them to see their illustration in the image tile. For ANSI mechanical drawings, select Centered to create dimension values that are centered vertically, breaking the dimension line. The ISO (International Standards Organization) standard usually uses the options for text above or outside the dimension line. The JIS option orients the text parallel to the dimension line and to its left. When you are finished examining this portion of the dialog box,

Choose Vertical: **Centered**

The Text Alignment area controls whether text always reads horizontally (i.e., is unidirectional text) or whether it will be aligned with the dimension line. Choose each setting to see its illustration; then

Select: **Horizontal** (so the button is darkened)

The Text tab should now appear as shown in Figure 3.15.

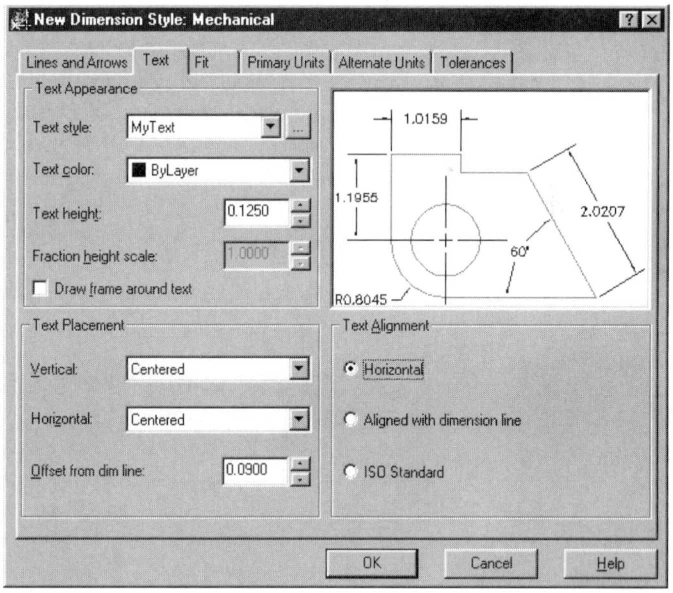

Figure 3.15

Pick: **Fit tab**

Fit

The Fit tab lets you set the options for how the dimensions will be placed in the drawing. These settings are especially helpful when there is not enough room to place both the dimension value, dimension lines, and arrowheads inside the extension lines.

The setting *Either the text or the arrows, whichever fits best*, places both the text and arrows inside the extension lines, if possible. If that is not possible, either the text or the arrows will be placed inside the extension lines. If neither will fit, both the arrows and text are placed outside the extension lines. Generally, these are all accepted practices, as long as the dimension can be correctly interpreted. Otherwise you can select to have just the text, just the arrow, text and arrows, to always place the text between the extension lines, or to suppress the arrows if they don't fit inside the extension lines.

Select: **Either the text or the arrows, whichever fits best**

The Text placement area of this dialog box allows you to control how text is placed when it is not in the default location (between the extension lines). You can select to place it beside the dimension line, or over the dimension line, either with or without a leader. Leave it set at *Beside the dimension line*.

The Fine Tuning area in the lower right of the dialog box lets you place text manually or to force a line to be drawn between the extension lines, even when text and arrows are placed outside the extension lines. If you select Place text manually when dimensioning, when you pick or type the location for the dimension line, the dimension value will appear at that position. Leave these boxes unchecked.

The Dimension Scaling Factor

There are essentially two ways to scale the dimension features: *Use an overall scale factor* (usually relative to model space) or *Scale dimensions to layout (paper space)*. Overall scale sets a scaling factor to use for all of the dimension variables that control the sizes of dimension elements. Its value appears in the box to the right of Overall Scale. Setting the overall scale to 2 doubles the size of all the dimension elements in the drawing.

Picking *Scale dimensions to layout (paperspace)* sets the scaling for dimension elements so that the values you enter in the dialog box (such as .125 for text height) are the sizes on the drawing when the layout is plotted at a scale of 1 = 1.

For this tutorial,

Select: **Scale dimensions to layout (paper space)**.

Your Fit tab should appear as shown in Figure 3.16.

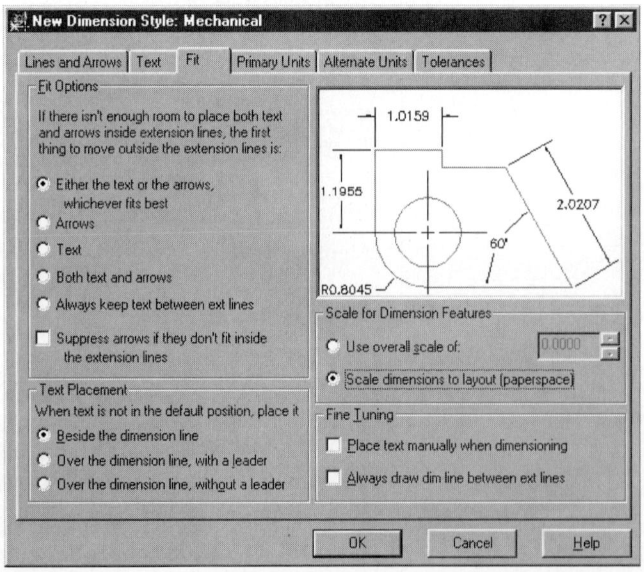

Figure 3.16

Next, you will set up the primary units for the dimensioning.

Pick: **Primary Units tab**

Primary Units for Dimensions

You can use this tab of the dialog box to select the type of units, set the number of decimal places for dimension values, and assign a prefix or suffix for dimensions. There are two main areas, one for Linear dimensions and one for Angular dimensions, as they use different types of units.

When dimensioning a drawing, you must consider the precision of the values to be used in the dimensions. Specifying a dimension to four decimal places, which is the default, implies that accuracies of 1/10,000th of an inch are appropriate tolerances for this part. Standard practice is to specify decimal inch dimensions to two decimal places (an accuracy of 1/100th of an inch) unless the function of the part makes a tighter tolerance desirable.

You may also suppress leading or trailing zeros in the dimension value, or suppress dimensions of 0 feet or 0 inches. In decimal dimensions of less than one unit, common practice is to show leading zeros in metric dimensions and to suppress them in inch dimensions.

You will now set up the display for Linear dimensions. You will leave the type of units set to Decimal and set precision to two decimal places.

Choose Unit format: **Decimal**

Choose Precision: **0.00**

Measurement Scale

The Scale area of the Primary Units tab is used to specify a scaling factor by which all the actual dimension values are multiplied. This factor is different from the overall scaling factor because it affects the dimension *value*, not the size of the arrowheads and other dimension elements. You should always model your parts to the real size and not use this setting to change the values for dimensions.

Leave the Scale factor set to 1.0000.

Zero Suppression for Linear Dimension

Zero Suppression is available for both linear dimensions and angular dimensions, in four categories: leading 0s, trailing 0s, 0 feet, and 0 inches. Selecting the box to the left of any of these words suppresses those zeros; you will suppress leading zeros only,

Select Zero Suppression: **Leading** (so it is checked)

Angular Dimensions

The type of units used to dimension angles can be set using the Angular Dimensions area. Choose the various options to see them illustrated in the angular dimension in the image tile. When finished,

Choose (Angular) Units format: **Decimal Degrees**

Zero Suppression for Angular Dimensions

You will suppress the leading zeros for angular dimensions as you did for linear dimensions.

Select (Angular) Zero Suppression: **Leading** (so it is checked)

When you have finished setting this tab of the dialog box, it should appear as shown in Figure 3.17.

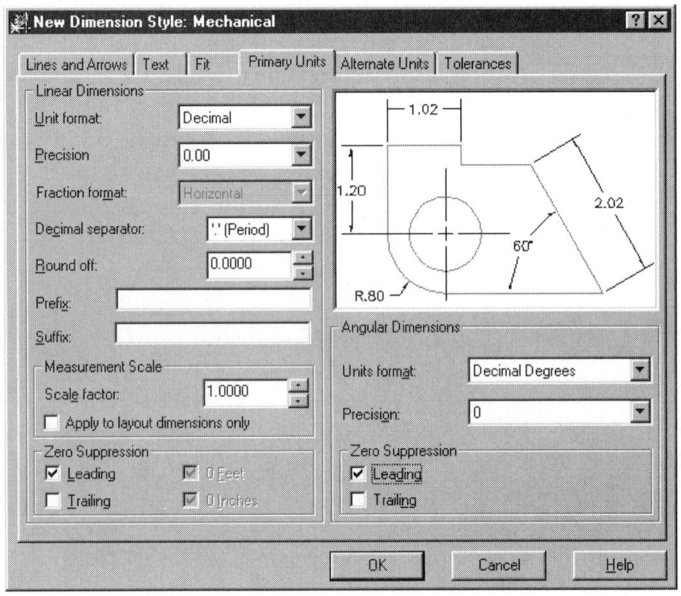

Figure 3.17

Next you will look at the settings that can be controlled using the Alternate Units tab.

Pick: **Alternate Units tab**

Alternate Units

Alternate units are useful when you want to dimension a drawing with more than one system of measurement. For example, dimensioning drawings with both metric and inch values is fairly common. When you use alternate units, Mechanical Desktop automatically converts your drawing units to the alternate units, using the scaling factor you provide. You can enable alternate units by picking the box to the left of *Display alternate units* in the Alternate Units tab, shown in Figure 3.18. Until you do, the remaining selections are grayed out.

Select: **Display alternate units** (so it is checked)

Use the image tile to explore the options for alternate units by choosing different Unit formats. The default settings are for metric units, which use Decimal units and would require a multiplier of 25.4 to convert inches to millimeters. If you enter text to be used a prefix, it will appear in front of the dimension value; a suffix, after the dimension value.

Alternate units can be placed after the primary dimension value or below primary value. The default places the alternate units inside square brackets following the normal units. For this drawing you will not use alternate units, so

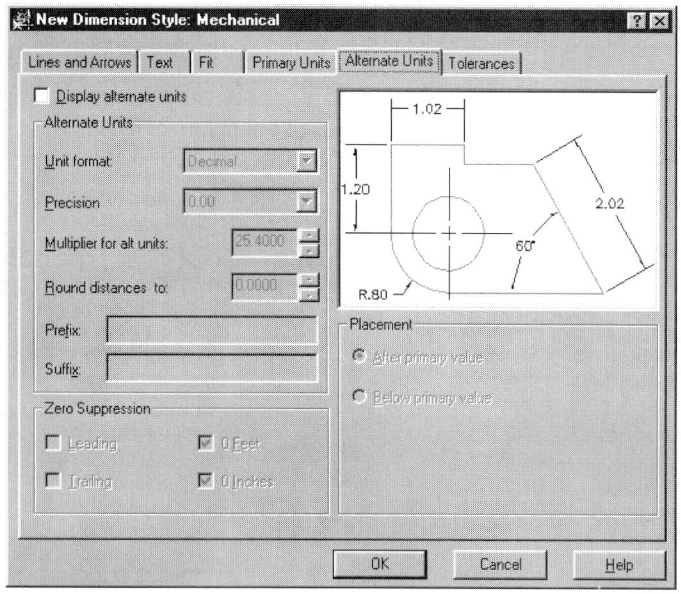

Figure 3.18

Deselect: **Display alternate units** (so it is not checked)

Next you will select the Tolerances Tab to see this area of the dialog box. You will learn more about setting tolerances in Tutorial 8.

Pick: **Tolerances tab**

The Tolerances tab appears on your screen.

Tolerances

You can use the Tolerances tab to specify limit or variance tolerances with the dimension value. The various built-in methods for displaying tolerances are illustrated in Figure 3.19.

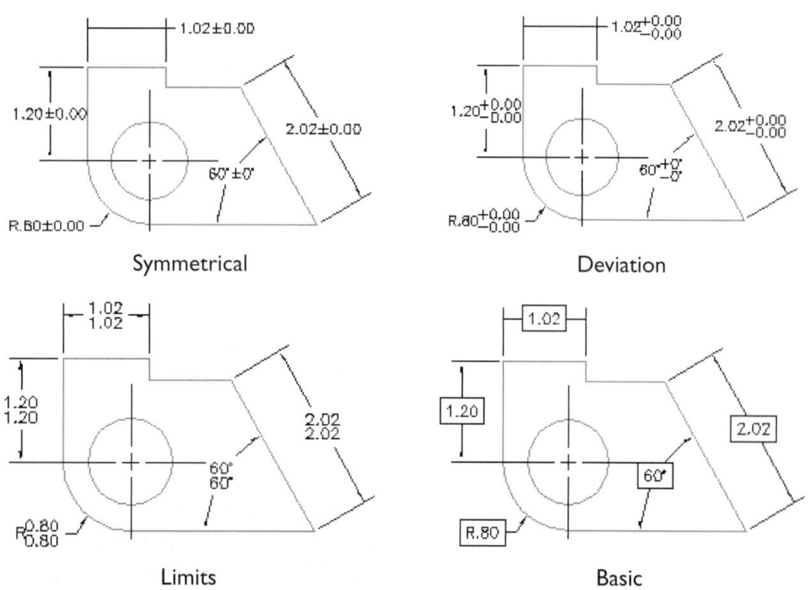

Figure 3.19

Use the Method drop-down menu to see the options, Symmetrical, Deviation, Limits, Basic, or None, in the image tile.

The image tile reflects all the choices you have made thus far in setting up the Dimension style. You will not use tolerances on the drawings for this tutorial, so set the Method to None so your Tolerances tab looks like Figure 3.20.

Choose Method: **None**

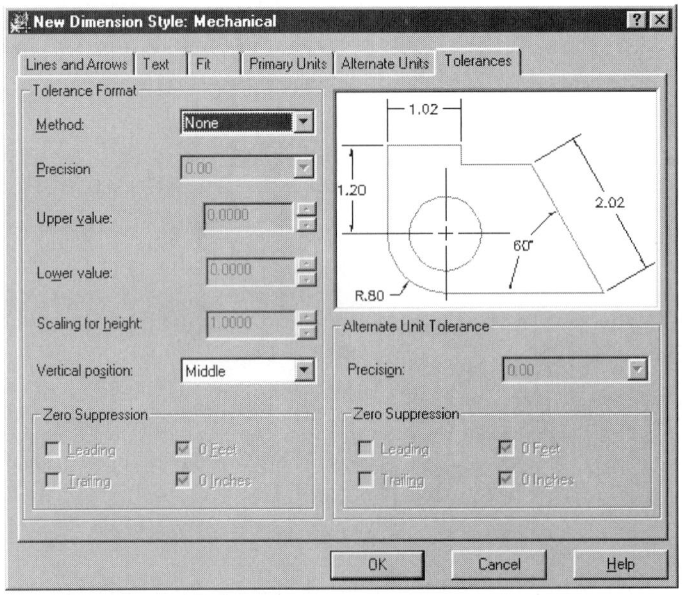

Figure 3.20

Now you have set up your basic style for the dimensions.

Pick: **OK**

You are returned to the Dimension Style Manager dialog box.

Setting the Current Dimension Style

The selections you have made apply to dimension style Mechanical only. In order to apply your new dimension style so that new dimensions will use this style, pick the Set Current button near the top of the dialog box.

Pick: **Set Current**

Pick: **Close** (to exit the Dimension Style Manager)

Next you will update all of the dimensions in your drawing so that they use the current style (the new style you created). You will type DIM to enter Dimensioning mode so you can use the Dimension Update command.

Command: **DIM** [↵]

DIM: **UPDATE** [↵]

Select objects: **ALL** [↵]

Select objects: [↵]

DIM: [ESC]

The dimensions in your drawing update to the new dimension style, as shown in Figure 3.21.

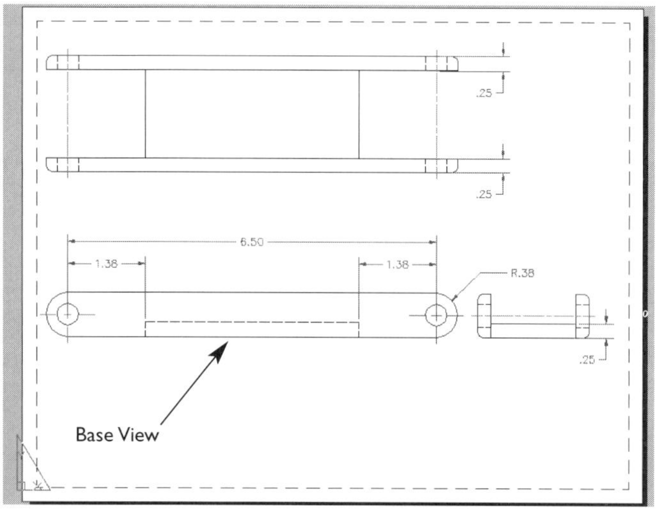

Figure 3.21

Next you will adjust the locations of the views and add a pictorial view to the drawing.

Adjusting Views

You can easily move, edit, and delete views in the layout. You will practice by moving the views as needed to make space for a pictorial view.

Right-click: **in the graphics area**

Pick: **Move View**

Select view to move: **pick the base view as indicated in Figure 3.21**

2 descendant views will also be moved.

Specify new view location or [Layout]: **pick slightly to the left and down**

The views all move relative to the new location selected for the base view, as it is the parent for the other two views. On your own, select until you have located the views to make space in the upper right of the sheet for a half-scale isometric view, as shown in Figure 3.22. When you have finished selecting,

Press: [Enter] (to exit the command)

Tip

You can always move the views again later if it doesn't fit the first time.

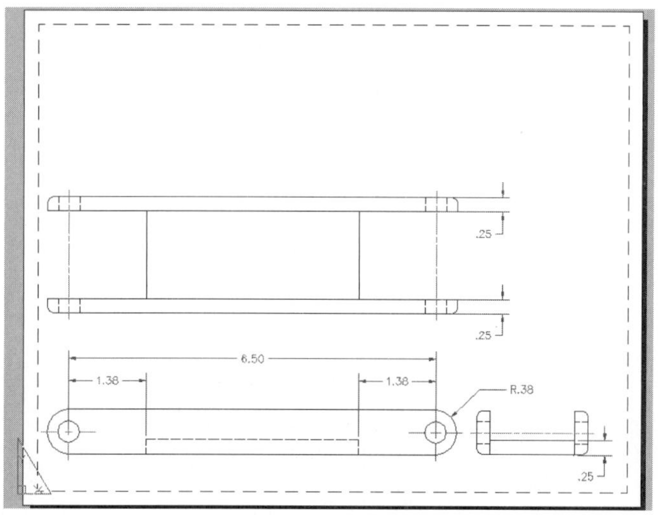

Figure 3.22

Pictorial Drawing Views

Isometric views are easy to generate directly from the model. They make a nice addition to multiview drawings, making them easy to interpret. Typically, pictorial views are not dimensioned and they are often shown to a smaller scale to fit them on the sheet easily. You will add an isometric pictorial view to the drawing.

Right-click: in the graphics area

Pick: **New View**

The Create Drawing View dialog box appears. You will make the following selections.

Choose View Type: **Iso**

Choose Scale: **.5000**

Select: **Relative to Parent** (so it is checked)

Select: **Display Tangencies** (so it is checked)

Select: **Remove Coincident Edges** (so it is checked)

Deselect: **Display Hidden lines** (so it is unchecked)

Pick: **OK**

Select parent view: **pick the base view**

Specify location for isometric view: **pick near the middle of the blank area in the upper right; pick again if you do not like the location when the view appears**

Specify location for isometric view: [↵]

The isometric view should be added to your drawing as shown in Figure 3.23. The view you chose as the parent view was used as the front view in the isometric drawing. Because you clicked above the parent view and to the right, you generated a front-right isometric view. Had you clicked the left side of the drawing, the view would be front-left. Clicking below generates back-right and back-left views. Try the command again to see how this works. Press [ESC] instead of [↵] when finished to avoid placing another view.

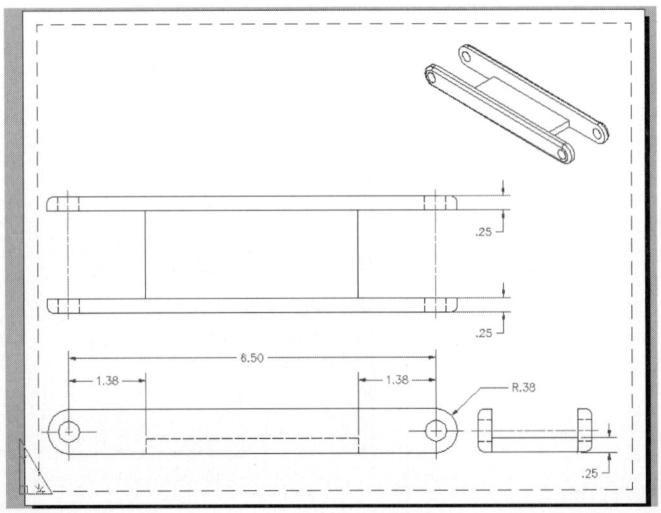

Figure 3.23

Reference

Engineering Design Communication

Intersections and Tangencies, pp. 134–137

Editing Views

The properties that you set when you create a view can be edited at any time. Next you will change the isometric view so that tangencies are not displayed. Notice how the tangencies appear in the isometric view in Figure 3.23. Typically when a curved surface is tangent to a plane surface, no line is displayed in the drawing.

Right-click: **in the graphics area**

Pick: **Edit View**

Select view to edit: **pick the isometric view**

The Edit Drawing View dialog box appears on your screen. It displays the name of the view and which part was active when the view was created in its title bar (if more than one part is present). You will turn off the display of the tangent edges so your view will appear as in Figure 3.24.

Deselect: **Display Tangencies** (so it is unchecked)

Pick: **OK**

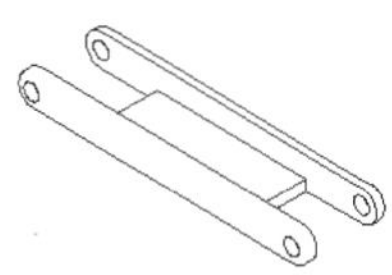

Figure 3.24

Printing Your Drawing

You have set up your drawing layout to produce a good looking plot. You will use the print command and your Windows printer (or whatever printer you configured Mechanical Desktop to use) to print it. Before you do, be sure that you are still in Drawing mode. If the Model mode is active, you will be plotting the model and not the multiview drawing.

Pick: **File, Plot**

The Plot dialog box appears on your screen with the Plot Settings tab uppermost, as shown Figure 3.25. You will use the settings shown there for your layout. Because you have the views set up correctly on an 8.5 × 11″ sheet, you will use 1 = 1 for the Plot Scale. This just means that you want to plot your layout full size on the actual sheet of paper, regardless of the scale of the views to the model.

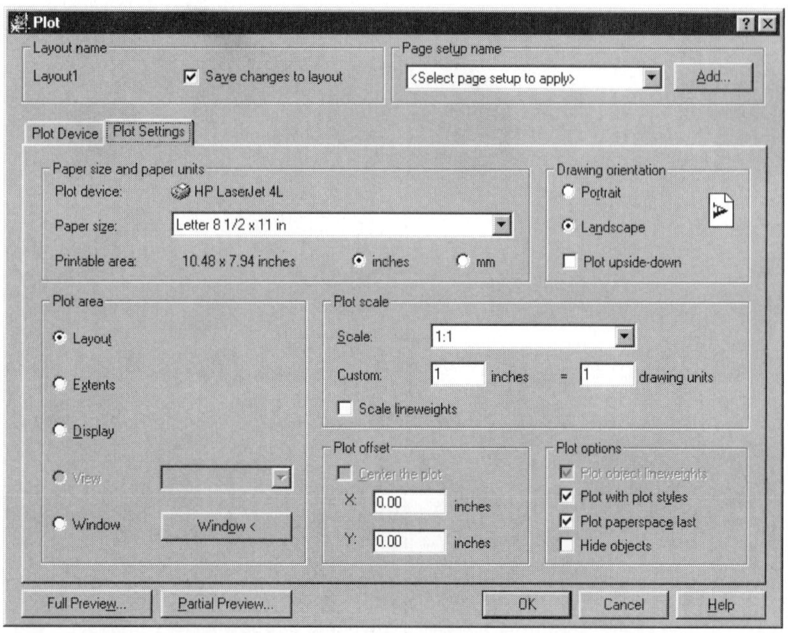

Figure 3.25

Choose Paper size: **Letter**

Select Plot area: **Layout**

Select Drawing orientation: **Landscape**

Choose Scale: **1:1**

Next you will use the Plot Device tab to select the printer and set up the plot style.

Pick: **Plot Device tab**

The Plot Device Tab appears uppermost in the dialog box, as shown in Figure 3.26. You can use it to select the printer (or plotter) that will be used, to plot all layouts or only the current tab, and to set the plot style or pen assignments. AutoCAD and Mechanical Desktop are generally configured to use the Windows System Driver to set up your plotter or printer. If you do not have a printer installed or need to change the printer, you may need to use the Windows operating system to install and set a default printer. In most cases, you will be able to select any of your Windows printers from this tab.

Be sure your printer is selected in the Plotter configuration area.

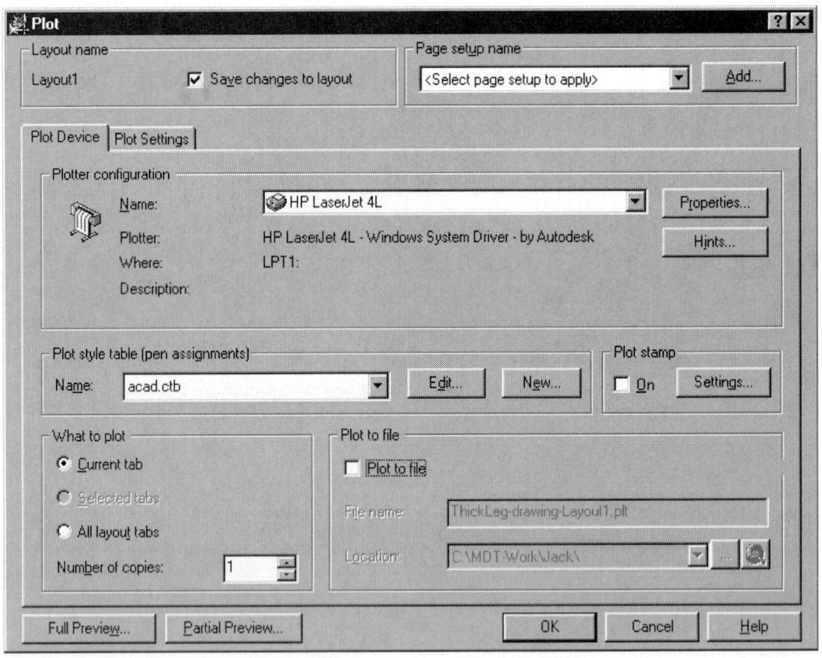

Figure 3.26

Plot Style Table (Pen Assignments)

A plot style table lets you assign properties to objects that will determine how they appear when printed. Properties such as color, line weight, line fill, screening, and dithering can be assigned to objects based on their name or on the color property they have in Mechanical Desktop. For example, all red objects can be set to print to a different color or with a different line weight. The plot style settings can override the properties for the object for printing purposes.

Plot style tables are independent of the output device selected. If a particular device does not have a property called for in the plot style (e.g., screening), that property is just skipped. In this way, a single plot style table can be used with a variety of printers. These plot style tables are saved separately from the drawing. Tables ending in *.ctb* set properties according to object color. Those ending in *.stb* set properties by object name. Changing the plot style changes how any drawing that uses that style will be output. This is an important new feature in Mechanical Desktop (and AutoCAD 2000). For this example, you will use the default style named *acad.ctb*.

Choose (Plot style table) Name: **acad.ctb**

Pick: **Edit** (to the right of the menu)

The Plot Style Table Editor appears as shown in Figure 3.27. As you can see, *acad.ctb* sets plot styles by color. Most of the plot properties are set to use the existing object properties (which are generally controlled by the layer settings you saw earlier in this tutorial), so this plot style will reflect the settings in your file. If you use a plotter that changes pens, you can use this table to set the pen number for each color in the drawing. Conversely, to plot an object in a certain way, you create a plot style for one of the 256 colors and set the object's color to that color. If necessary, set the

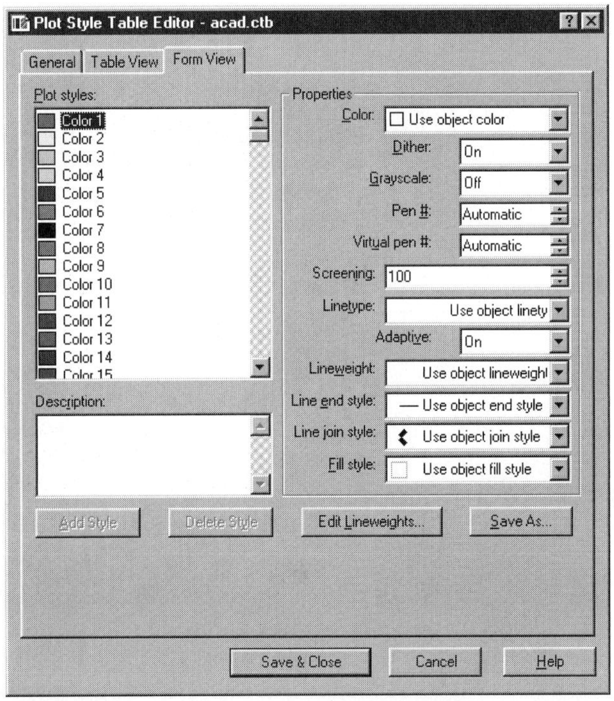

Figure 3.27

parameters for your plotter. For most Windows system printers, the default settings will work well.

Pick: **Save & Close** (to save changes) **or Cancel** (to leave the file unchanged)

Pick: **Full Preview button**

A preview of your plot appears on your screen, as shown in Figure 3.28.

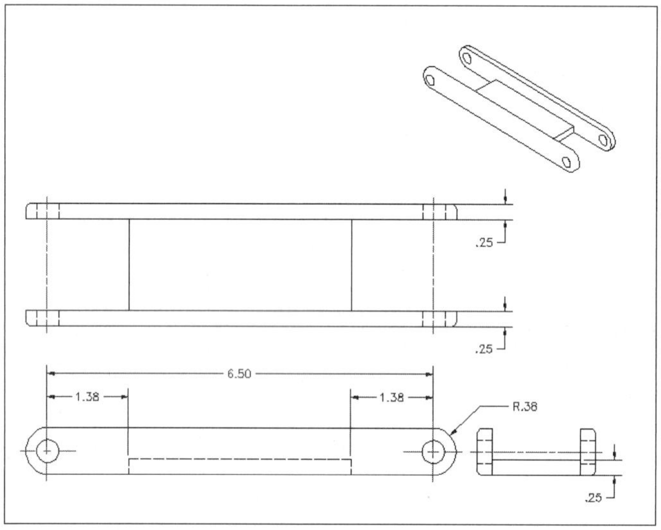

Figure 3.28

If the drawing does not appear correctly in the preview, you may need to use the Windows operating system to select the correct printer or to change its default paper orientation or other selections. You might try using Portrait orientation for the drawing instead of landscape. Make changes as necessary so that the plot fits well on the sheet. Preview your drawing again as needed until you like the way it looks. Once you have the drawing previewing correctly,

Right-click and pick: **Plot** (or press [ESC] to cancel printing)

Now you are well on your way to mastering Mechanical Desktop. You know how to model simple parts and how to create multiview drawings and print them. But what if you need to change a dimension? Now you have two places where you can do it: in the model or in the drawing.

Bidirectional Associativity

Bidirectional associativity means that you can change the dimension value in the drawing and the change will propagate to the model, or you can change the value in the model and the change will propagate to the drawing. This is a powerful feature that makes it easy to change your drawings and models easily as the design evolves.

Changing a Value Using Power Dimension Editing

You will change the value for the distance between the holes in the thick leg from 6.50 to 6.00. To do this you will use Noun/Verb selection. This means that you will first select the object in the drawing and then you will select the commands that will happen to that object (the noun being the object and the verb being the command). Make sure that you are still in the Drawing mode of the Desktop Browser. To change the 6.50 value,

Pick: **the 6.50 dimension**

The 6.50 dimension becomes highlighted and its grips appear as shown in Figure 3.29. *Grips* are small boxes that appear on control locations for an object when it is selected (with no command active).

Tip

If the grips do not appear, you may have this capability turned off in the software. To turn it on, pick Assist, Options, Selection tab, and make sure that Noun/verb selection is checked.

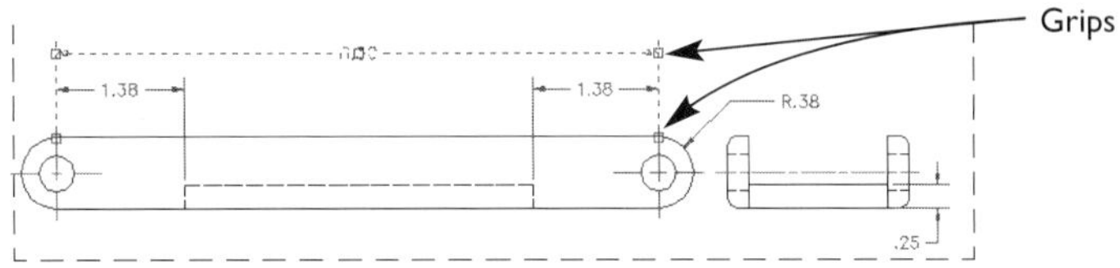

Figure 3.29

Right-click: to pop-up context menu

Pick: **Power Edit**

The Power Dimensioning dialog box appears as shown in Figure 3.30.

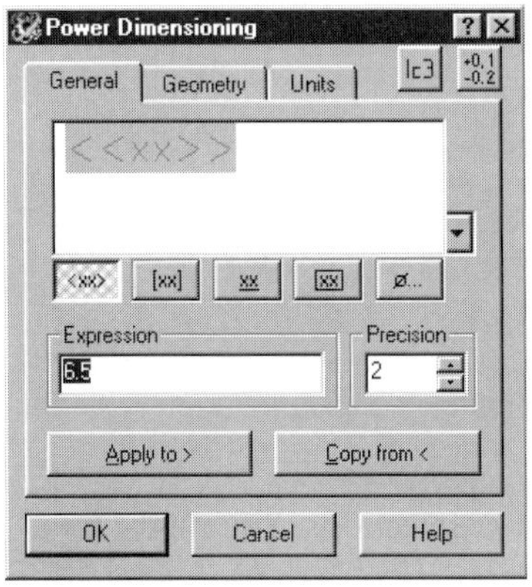

Figure 3.30

You can use this dialog box to control the number of decimal places for displaying dimension values, to add a tolerance, to add fit symbols to your drawings, and to control other dimension settings that you saw in the Dimension Style dialog box. You will learn more about these features in Tutorial 8. For now you will change the value of the dimension.

Type Expression: **6.00**

Pick: **OK**

The dimension value changes to display 6.00 instead of 6.50, but the drawing does not update automatically. Next you will update the model and the drawing.

Right-click: in the graphics area

Pick: **Update Part**

You are automatically switched to the model, where you see it update, and then back to the drawing, where you see it update. The drawing should now look similar to Figure 3.31.

Tip

If you have inadvertently enabled tolerances or fits, use Power Edit again and make sure the Enable box under the Fit and Tolerance tabs is not checked.

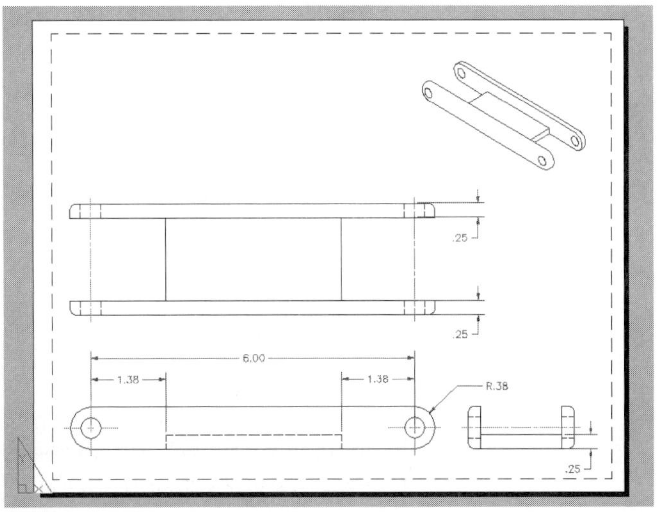

Figure 3.31

Figure 3.32

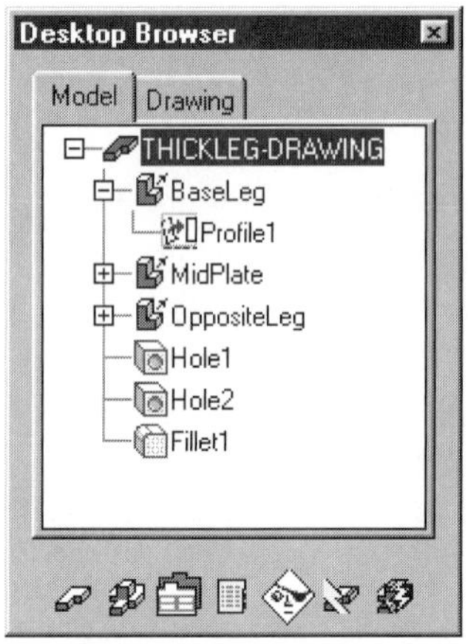

Figure 3.33

Pick: **Model tab** (in the Desktop Browser)

You switch to the view of the part. It has updated in size to 6.00 between holes. It should look similar to Figure 3.32.

Next you will change the distance between holes back to 6.50 in the model and see how the change updates in the drawings.

Pick: **the plus sign** (to expand the model tree)

Pick: **to expand the tree under the BaseLeg feature**

The model tree should appear similar to Figure 3.33.

Right-click: **Profile1** (under Baseleg to pop up the context menu)

Pick: **Edit**

Select object: **pick the 6.00 dimension**

Enter dimension value (<6.0): **6.5** ↵

Solved fully constrained sketch.

Select object: ↵

Right-click: in the graphics area

Pick: **Update Part**

The part updates to show the 6.50 size.

Pick: **Drawing tab** (in the Desktop Browser)

You see the drawing update to show the feature at size 6.50, as shown in Figure 3.34.

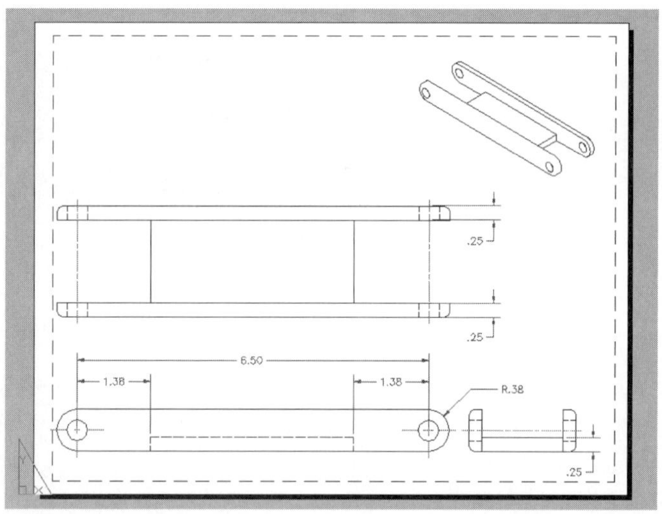

Figure 3.34

Save your work.

Creating Layout2

Next you will add a larger format sheet and create a second layout of the part using a standard title block.

Double-pick: **Layout2** (in the model tree)

The blank sheet for Layout2 shows on your screen as in Figure 3.35.

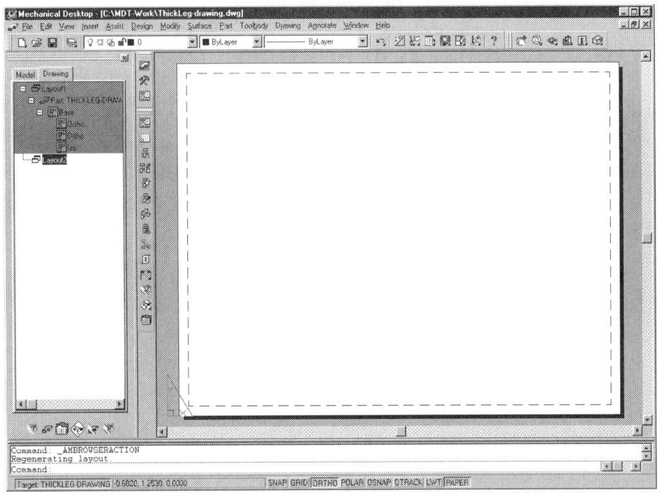

Figure 3.35

Right-click: **Layout2** (in the model tree)

Pick: **Page Setup**

The Page Setup dialog box appears on your screen. You will use it to select the DWF Classic.pc3 printer, so that you can select larger sheet sizes. If your default printer only allows 8.5 × 11″ paper, you cannot set up a larger sheet size without changing to a printer that is capable of larger sheets or using this Classic.pc3 printer setting. In the Plotter Configuration area,

Set Name: **DWF Classic.pc3**

The plotter name should be set as shown in Figure 3.36.

Pick: **Properties**

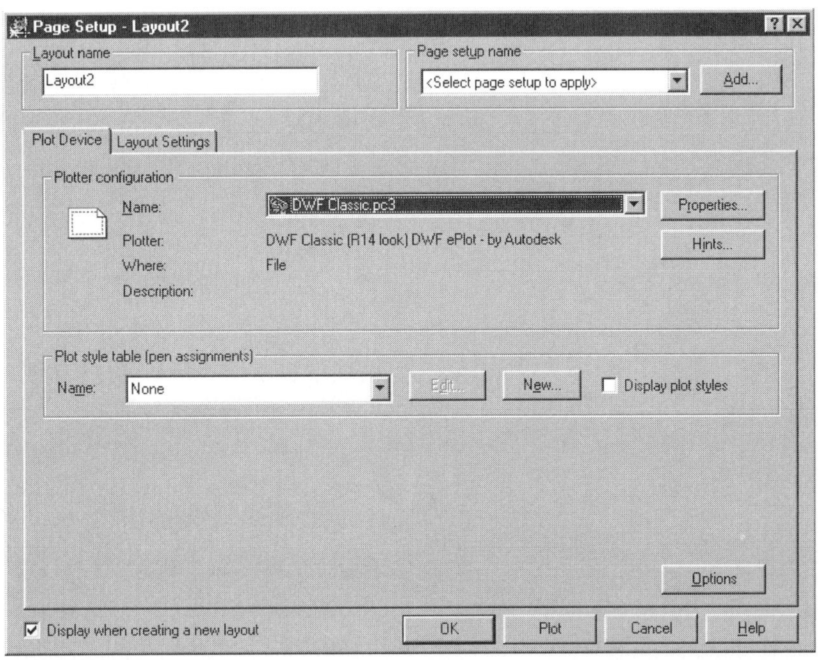

Figure 3.36

The Plotter Configuration Editor appears on your screen. When you have finished making your selections, it will appear as shown in Figure 3.37. You will use it to set the media size to ANSI C (22.00 × 17.00 inches).

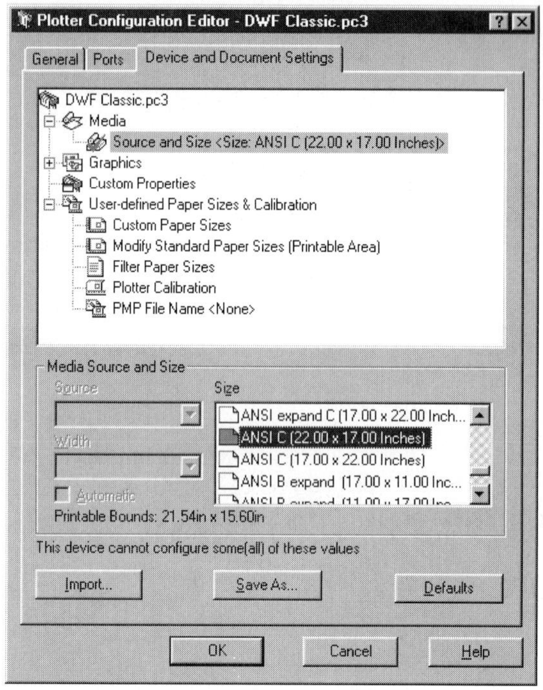

Figure 3.37

Pick: **ANSI C (22.00 × 17.00 inches)**

Pick: **OK**

You may see a warning, *You have made changes to a PC3 Printer Configuration File. Save changes to the following file?*

Pick: **OK**

Pick: **OK** (to close the Page Setup dialog box)

Now you return to Layout2 using the larger sheet size. Verify this on your own by moving your cursor around on the screen and noticing the coordinate values that are reported on the status line.

Next you will insert a title block.

Right-pick: **Layout2** (in the model tree)

Pick: **Insert Title Block**

A list of the available title blocks shows in a text window (and at the command line). You will type the number of the one you select. In this case, number 9, ANSI-C.

Available title blocks: ...

0: None

1: ISO A4 Size(mm)

2: ISO A3 Size(mm)

3: ISO A2 Size(mm)
4: ISO A1 Size(mm)
5: ISO A0 Size(mm)
6: ANSI-V Size(in)
7: ANSI-A Size(in)
8: ANSI-B Size(in)
9: ANSI-C Size(in)
10: ANSI-D Size(in)
11: ANSI-E Size(in)
12: Arch/Engineering (24 × 36in)
13: Generic D size Sheet (24 × 36in)

Enter number of title block to load or [Add/Delete/Redisplay]: **9** ↵

Create a drawing named ansi_c.dwg? <Y>: **N** ↵

Enter an option [Align/Create/Scale viewports/Options/Title block/Undo]: ↵

The title block is added to the layout, as shown in Figure 3.38. It is slightly too large for the sheet size. You will scale it down slightly, then move it to center it horizontally on the page.

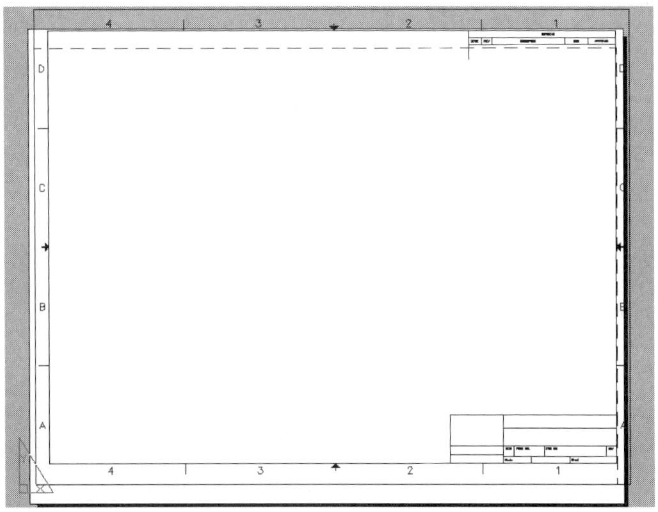

Figure 3.38

Pick: **Modify, Scale** (from the menu bar)

Select objects: **ALL** ↵ ↵

Specify base point: **0,0,0** ↵

Specify scale factor or [Reference]: **.92** ↵

Type: **MOVE** ↵

Select objects: **ALL** ↵ ↵

Specify base point or first point of displacement: **pick a point on screen**

Specify second point of displacement: **pick a point to the right** (to center the border and title block)

The title block now fits on the sheet, as shown in Figure 3.39. Most companies have their own standard title block that they use. It is generally worth the effort to make or modify a standard title block so that it fits the sheet perfectly and contains all of the typical information that your company wants provided on the sheet.

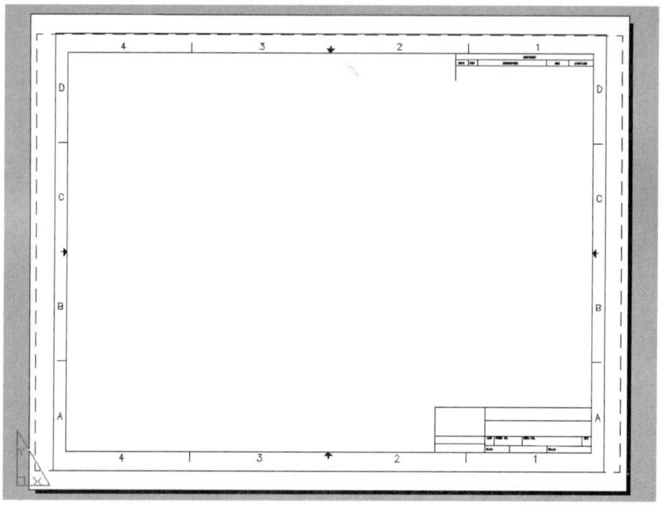

Figure 3.39

Now you are ready to use this layout to add multiple views.

Right-click in the graphics window

Pick: **Multiple Views**

The Create Drawing Views dialog box appears on your screen. Make the following settings.

Tip

If you did not see a dialog box when you selected the Multiple Views selection, at the Command prompt, type CMDDIA ⏎. You will see the prompt, "Enter new value for CMDDIA <0>:" Type 1 and press ⏎ to restore the use of command dialog boxes.

Choose View Type: **Multiple**

Choose Data Set: **Active Part**

Choose Layout: **Layout2**

Set Scale: **1.0000**

Set Iso (metric) Scale: **.5**

Select: **Relative to Parent** (so it is checked)

Select: **Display Hidden Lines** (so it is checked)

Deselect: **Display Tangencies** (so it is unchecked)

Select: **Remove Coincident Hidden Lines** (so it is checked)

Pick: **OK**

You return to the drawing for the following selections. You will select the front face of the lower leg as the plane.

Select planar face, work plane, or [Ucs/View/worldXy/worldYz/worldZx]: **pick surface A in Figure 3.40**

Enter an option [Next/Accept] <Accept>: **if the correct surface is not highlighted, press the left-mouse button to cycle through the surface, when the correct surface is highlighted, right-click to accept**

Define X axis direction:

Select work axis, straight edge, or [worldX/worldY/worldZ]: **pick the bottom horizontal line of the top leg labeled B in Figure 3.40**

Adjust orientation [Flip/Rotate] <Accept>: **if the direction arrow is pointing forward, right-click and pick Enter to accept**

You now see the message, *Regenerating layout*. You are switched back to paper space to continue laying out the views.

Specify location of base view: **pick near the center of the sheet**

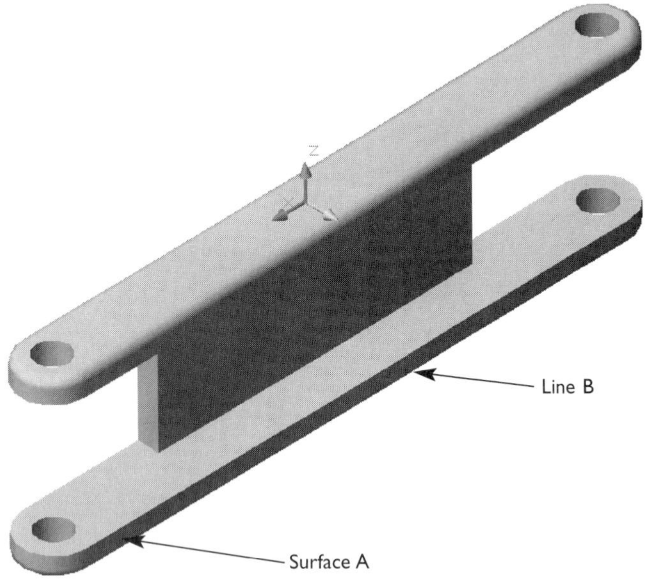

Figure 3.40

The view appears on the sheet. If it is not placed in a manner that will allow you to fit a view above it and one to the right of it, at the prompt,

Specify location of base view or [Done] <next view>: **pick a new location for the view, or right-click and pick Enter to continue**

Specify location of projected view or [New parent view]: **pick above the base view**

Specify location of projected view or [Done] <next view>: ↵

Specify location of projected view or [New parent view]: **pick to the right of the base view**

Specify location of projected view or [Done] <next view>: ↵

Specify location of projected view or [New parent view]: ↵

You should now have a multiview drawing on your sheet that looks similar to Figure 3.41.

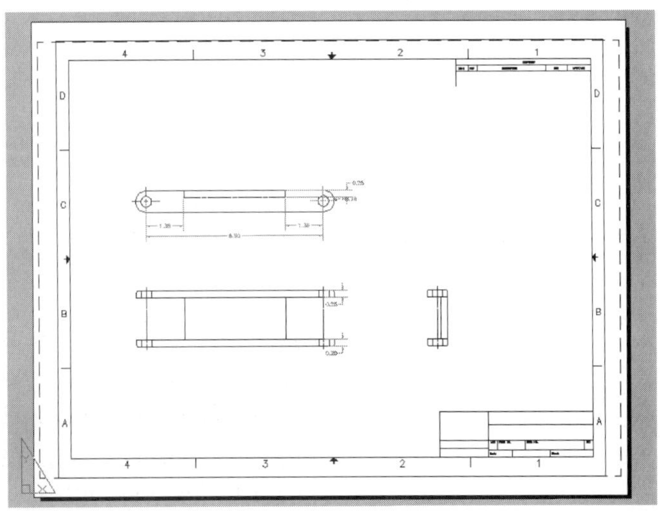

Figure 3.41

Adding Text to the Title Block

You will use the Text command to add text to the title block, as shown in Figure 3.42. The Text command allows you to add single lines of text to a drawing (in contrast to MTEXT, which adds multiline text). (Text is the same as the Dtext command in previous versions of AutoCAD.) Formatting options for each line of text are set at the command line, such as the height of the text, its rotation, and how it should be aligned left to right, or top to bottom. The Text command starts from a predefined text style for the font to be used. You will change to your new Text Style, MyText, then respond to prompts at the command line to format the text.

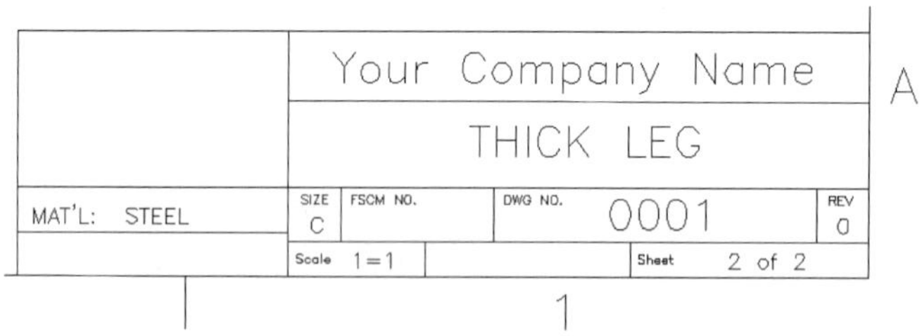

Figure 3.42

Pick: **Zoom Window icon**

Enlarge the title block in the lower right of the drawing on your own.

Command: **TEXT** ↵

Current text style: "ADESK1" Text height: **0.2500**

Specify start point of text or [Justify/Style]: **S** ↵

Enter style name or [?]<ADESK1>: **MYTEXT** ↵

Current text style: "MyText" Text height: **0.2000**

Specify start point of text or [Justify/Style]: **J** ↵

Enter an option [Align/Fit/Center/Middle/Right/
    TL/TC/TR/ML/MC/MR/BL/BC/BR]: **C** ↵

Specify center point of text: **pick near the lower middle of the top block**

Specify height <0.2000>: ↵

Specify rotation angle of text <0>: ↵

Enter text: **Your Company Name** ↵

Enter text: ↵

The text appears centered in the box as shown in Figure 3.42. To add the rest of the text shown in the figure, you will respond to the command prompts to center each line and change the height for the smaller items to 0.1.

Press: **the** SPACEBAR **to repeat the command** (or right-click and pick Repeat TEXT from the context menu)

On your own, add the remaining text, as shown in Figure 3.42.

Pick: **Zoom All icon**

On your own, clean up the dimensions and add a full-size isometric view, as shown in Figure 3.43.

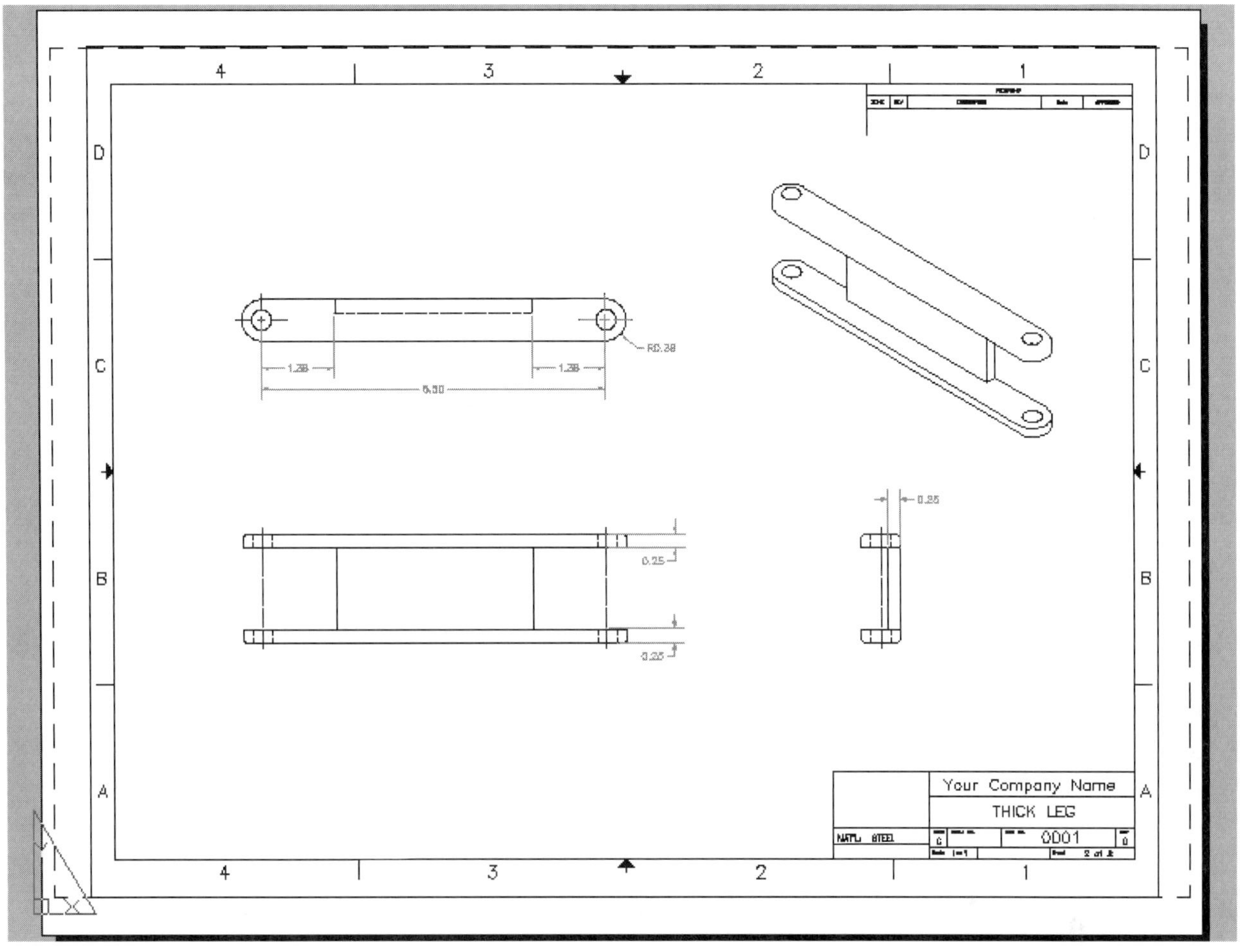

Figure 3.43

Creating a Template File

The text and dimension styles and layouts you have created in this tutorial are stored with the *ThickLeg-drawing.dwg* file. To use them in other files, you can create a template file that you can use when you start a new part drawing.

To create a template file, you will first delete the part features in this file so they will not be saved with the template. Be sure you have saved your ThickLeg-Drawing file first.

Pick: **File, Save**

Pick: **Model tab**

Right-click: **BaseLeg feature**

Pick: **Delete**

Highlighted features will be deleted. Continue? [Yes/No] <Yes>: ↵

Computing ...

Pick: **Drawing Tab**

You see the drawing views removed from the sheet. Next you will update so that the deletions to the part are up to date in the model.

Command: **AMUPDATE** ↵

Enter an option [active Part/Assembly/aLl parts/linKs] <active Part>: **L** ↵

All local parts have been recomputed.

Pick: **Drawing, Update View**

The option to update the current layout will update all views that are on it.

Select view to update or [All layouts/current Layout]: **A** ↵

3 drawing views being updated

Now you are ready to save your file as a template. Template files have the *.dwt* extension and can be used again and again to start new files. (Don't be concerned about the title block text; you will learn how to edit this in Tutorial 8.)

Pick: **File, Save As**

Choose Files of Type: **Mechanical Desktop 5 Drawing Template (*.dwt)***

Choose Save in: **navigate to the MDT-Work directory**

Type File name: ***MyDrawing.dwt***

Pick: **Save**

The Template Description dialog box appears.

Type description: **Template for MDT R5 Tutorials**

Pick: **OK**

Your file is now saved as a template file in the MDT-Work directory. When you start a new part file, you can select Template in the Mechanical Desktop today window and choose your template as a base file for the new part.

You have completed your first drawing and this tutorial!

Command Summary

Action	Command Name	Description
Add Multiple Drawing Views	AMDWGVIEW, Multiple	Creates and adds 2D views to a drawing layout
Add Text	TEXT (or DTEXT)	Adds single line of text; text is displayed on screen as it is typed and formatting options are presented at the command line
Add Title Block	MVSETUP	Sets up the specifications of a drawing, including the insertion of a title block
Change active mode between Drawing and Model	AMMODE	Controls whether Model Mode (Part, Assembly, Scene) or Drawing Mode (Layout) is active
Delete Feature	AMDELFEAT	Deletes a feature and all of its dependent features
Dimension Style	DIMSTYLE	Creates and modifies dimension styles; displays the Dimension Style Manager dialog box
Edit View	AMEDITVIEW	Edit scale, visibility, and other aspects of drawing views
Layer Control	AMLAYER	Opens the Layer Control dialog box for defining and managing layer definitions
Move Dimension	AMMOVEDIM	Moves dimensions between views in a drawing without losing their association with the geometry
Move View	AMMOVEVIEW	Moves a drawing view within a layout or to a new layout
Plot	PLOT	Plots to a plotting device or file; opens the Plot dialog box
Power Edit Dimensions	AMPOWEREDIT	Opens the appropriate dialog box for editing a power object; when a dimension is selected, it opens the Power Dimensioning dialog box
Scale	SCALE	Enlarges or reduces selected drawing objects
Scale linetype	LTSCALE	Sets the linetype scale factor so line patterns can be adjusted for different size drawings
Set screen resolution	VIEWRES	Sets the resolution at which objects are displayed on screen; can be used to change the number of line segments used to display arcs and circles
Text Style	STYLE	Creates and modifies named styles and sets the current style; opens the Text Style dialog box
Update Layout	AMUPDATE	Recalculates the model to show the effect of changes made to features, constraints, or dimensions on parts, assemblies, scenes, or drawing views
Update Dimensions	DIM, UPDATE	Updates dimension objects to reflect the current dimension style and units settings; this is a special dimensions command available only at the DIM prompt

Exercises

Create multiview drawings for the parts shown in Figures 3.44 through 3.47. You may start with the models you created for these parts in Tutorial 2.

1. **Hex Cap** *Note*: All dimensions are in millimeters.

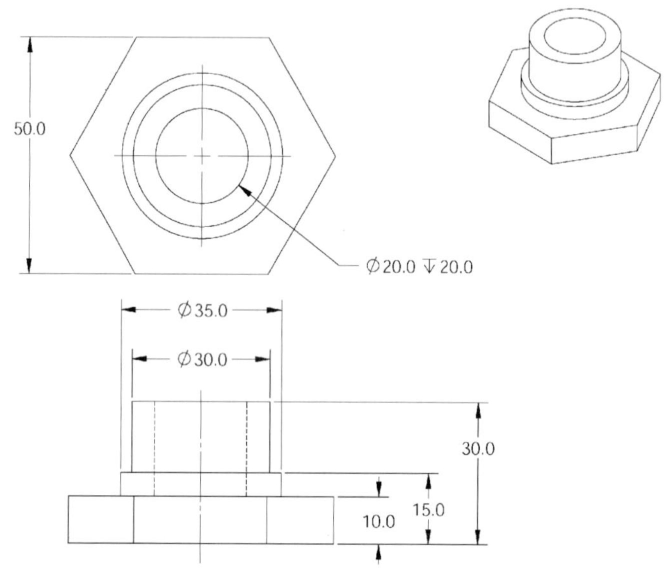

Figure 3.44

2. **L-Bracket** *Note*: All dimensions are in inches.

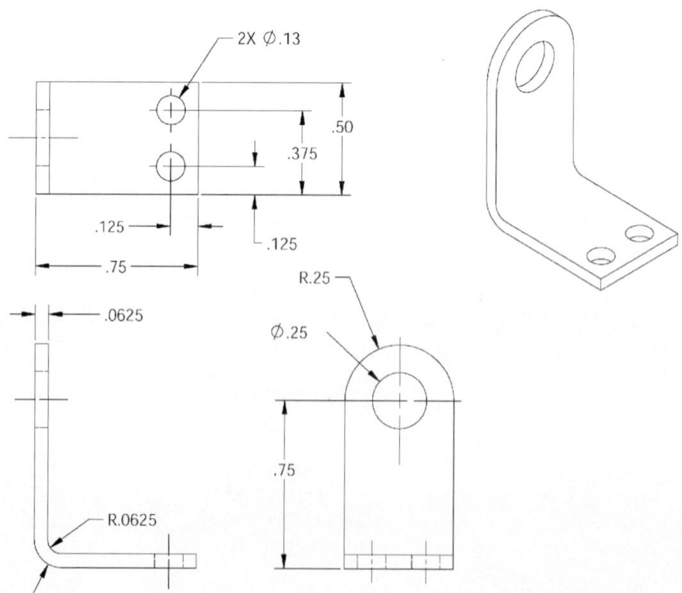

Figure 3.45

3. **Shaft Clamp** After part is modeled, add 1-mm fillets and rounds. *Note*: All dimensions are in millimeters.

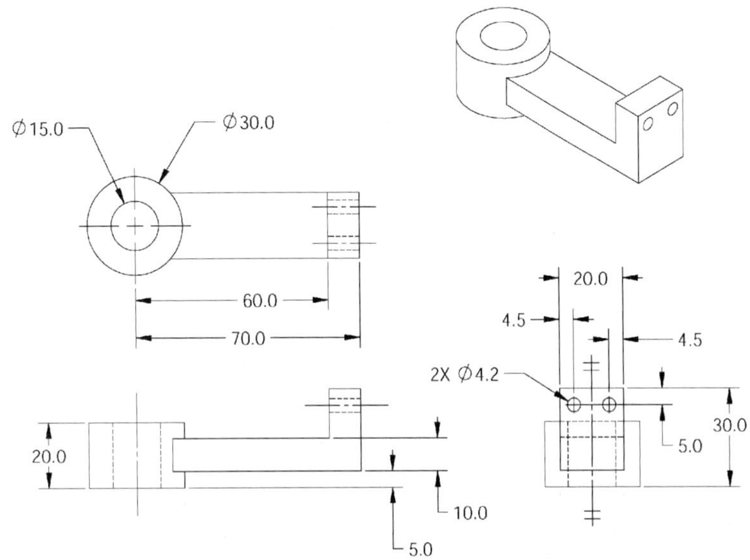

Figure 3.46

6. **U-Bracket** *Note*: All dimensions are in inches.

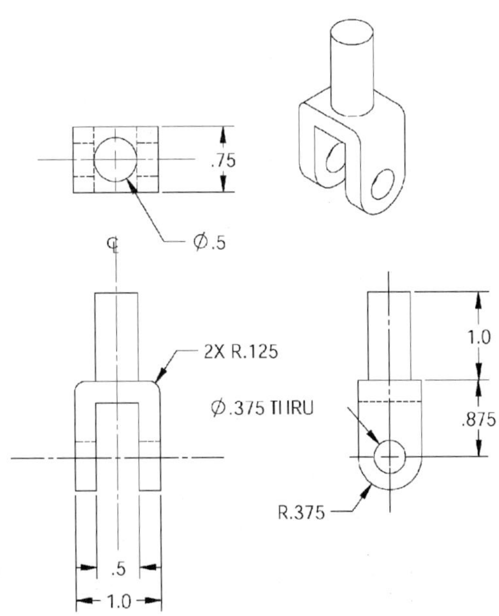

Figure 3.47

Tutorial

4 More Part Modeling

Objectives

When you have finished this tutorial, you should be able to

- Create a work plane.
- Use Grid, Snap, Object Snaps, and Ortho as sketching aids.
- Use relative coordinates to locate points.
- Use Line, Extend, Offset, and Rectangle to create sketch entities.
- Edit sketches with Undo, Trim, Mirror, Erase, and Fillet.
- Turn on and use Hot Grips and Crossing to select entities.
- Use construction lines to help constrain geometry.
- Delete and add constraints to reflect design intent.
- Extrude features using Intersect, Cut, and Join options.

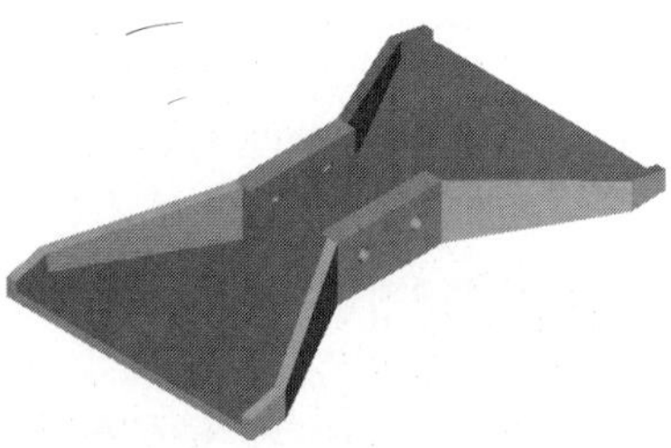

Figure 4.1

Introduction

Complicated parts are typically combinations of simpler features. You have already learned the basics of part modeling in Tutorial 2. In this tutorial you will learn more techniques for part modeling while you create the baseplate for the jack assembly. You will also use more sketching techniques to create the shapes used as profiles.

You will start modeling the baseplate part, shown in Figure 4.1, by drawing on the default coordinate system, the WCS. Then you will create a new coordinate plane that stays centered on the part and use it to sketch the second profile for the basepart. You will first sketch one-quarter of the bowtie shape of the part as seen from the top. You will mirror this sketch in two different directions to make the shape. Then you will extrude it to create the base feature.

Launch Mechanical Desktop 5

Pick: **File, New Part File**

Pick: **English (feet and inches)**

Pick: **File, Save**

Choose Save In: **MDT-Work\Jack**

Type File name: ***Baseplate.dwg***

Pick: **Save**

Load your user profile using the Options dialog box (if necessary)

Useful Commands for the Sketcher

You will try out some sketching techniques as you make this profile. Mechanical Desktop is based on AutoCAD, which provides excellent tools for 2D sketch creation. Some typical AutoCAD commands for creating and editing 2D geometry will be demonstrated here to familiarize you with different techniques. In order to focus on parametric modeling, these commands will not be covered exhaustively here. If you already know AutoCAD, these commands will be a review, but you should be patient and follow along with the tutorial. If you are new to AutoCAD, you will learn basic sketching and editing tools about which you can use Help to learn more. You will be able to apply your skills in drawing with AutoCAD further in modeling the exercises and in your engineering design projects.

Using Grid

Adding a grid to the graphics window acts as a helpful reference for your sketch. A *grid* is a background area in the graphics window covered with regularly spaced dots. You will type the Grid command to turn on the grid and set its spacing,

Command: **GRID** ↵

Specify grid spacing(X) or [ON/OFF/Snap/Aspect] <0.5000[greater]: **.25** ↵

Pick: **View, Zoom, All**

The grid fills your screen with a background grid drawn with 0.25 spacing, as shown in Figure 4.2. Note that GRID is now highlighted on the status bar to let you know that the grid is turned on.

Tip

If you choose a grid size that is too small—so that it would effectively fill your graphics window with a solid pattern of grid dots—the error message "Grid too dense to display" appears. If that happens, restart the Grid command and change to a larger value. (You can press ↵ immediately after ending a command to restart it.) If the grid values are too large, nothing appears on the screen because the grid is beyond the visual limits of the screen.

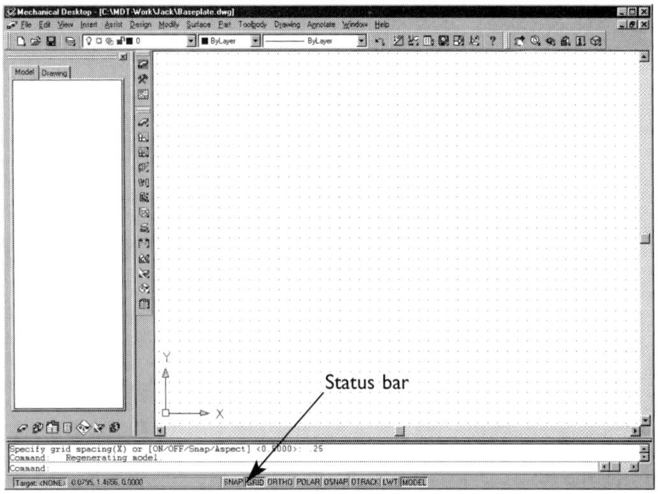

Figure 4.2

Using Snap

The Snap command restricts the movement of the cursor to make selecting locations easier. When Snap is on, your cursor jumps to the specified intervals. You will activate Snap by typing the command.

Command: **SNAP** ↵

Specify Snap spacing or ON/OFF/Aspect/Rotate/Style/Type <0.5000>: **.25** ↵

Note that SNAP is highlighted on the status bar to let you know that Snap mode is on. You should align the snap function with the grid size so that the snap spacing is the same as the grid spacing, at half the grid spacing, or some other even fraction of the grid spacing to make it useful.

Move the crosshairs around on the screen. Note that instead of the smooth movement that you saw previously, the crosshairs jump or "snap" from point to point on the snap spacing.

Tip

If you set the GRID spacing to 0.0000 and then set the SNAP, the grid spacing will automatically follow the same spacing as SNAP.

Grid and Snap Toggles

You can quickly turn the grid and snap on and off by picking their respective buttons on the status bar. If their buttons on the status bar appear highlighted (with a dark shadow around the button), they are on; buttons that are solid gray are off. Each button on the status bar acts as a ***toggle***: that is, clicking it once turns the function on; clicking it a second time turns the function off. You can toggle grid and snap on and off during the execution of other commands.

Pick: **SNAP button**

Move the crosshairs around on the screen. Note that they have been released from the snap constraint. To turn Snap on again,

Pick: **SNAP button**

The crosshairs jump from snap location to snap location again.

Pick: **GRID button**

The grid disappears from your screen, but the snap constraint remains. To show the grid again,

Tip

Pressing [CTRL]-G or [F7] also toggles the grid on and off; [CTRL]-B or [F9] toggles the snap function.

Pick: **GRID button**

Using Line

Next you will use the Line command to sketch one-quarter of the "bowtie" shape of the top view of the baseplate. Because the part is symmetrical about the horizontal and vertical centerlines, you can mirror the sketch geometry to create the remaining portions. This not only saves time (especially with more difficult shapes) but also ensures that the symmetry of the part is reflected in the sketch you make.

Often it is just as easy to sketch using fairly accurate dimensions as it is to make a rough sketch, so you will use a combination of methods to draw an accurate shape. (However, you can always change the dimension values later as needed.)

Right-click: in drawing area

Pick: **2D Sketching, Line**

Specify first point: **3,6** ↵

Tip

Do not enter spaces between coordinates as you enter them. Pressing the space bar is the same as pressing ↵ in the drawing commands.

The line should rubberband from absolute coordinate 3,6 to the current location of your crosshairs, as shown in Figure 4.3.

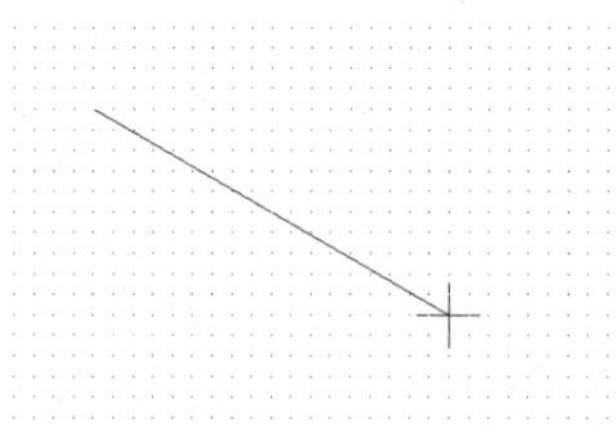

Figure 4.3

Using Relative Coordinates

Notice that the coordinates shown in the status bar currently display the location of your crosshairs. These coordinates can be set to display absolute or relative coordinates for the cursor location. You will toggle the coordinates so they display the relative length and angle of the line you are drawing, rather than its absolute coordinates.

Pick: **the coordinate display** (in the status bar)

Picking the coordinate display once turns the display off so that it appears grayed out, as shown in Figure 4.4

Figure 4.4

Pick: **the coordinate display again** (in the status bar)

Picking the coordinates again causes the display to change so that it shows the length and angle from the previous point, as shown in Figure 4.5. Use this display to help you draw the next line segments to the correct length.

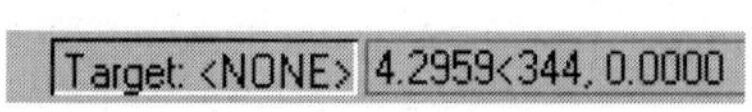

Figure 4.5

The Line command should still be active.

You should see the prompt for the next endpoint,

Tip

If the Line command is not active, press ↵ to restart it. Press ↵ again to start drawing from the last point you have selected, or type in the coordinate pair 3,6 and press ↵.

Specify next point or [Undo]: **use the Snap increment to draw segment A so that it is 3.5 inches long**

Specify next point or [Undo]: **use the Snap increment to draw segment B so that it is 1 inch long**

You will give the next length by typing in a relative distance and angle from the previous point. AutoCAD uses the format @LENGTH<ANGLE to specify relative distances.

Specify next point or [Close/Undo]: **@3<35** ↵

Reference

Engineering Design Communication

Working with Coordinates, pp. 167–171

Specify next point or [Close/Undo]: ↵

Your screen should show the line segments in Figure 4.6.

Next you will offset the side and bottom lines and then use commands to trim and extend them.

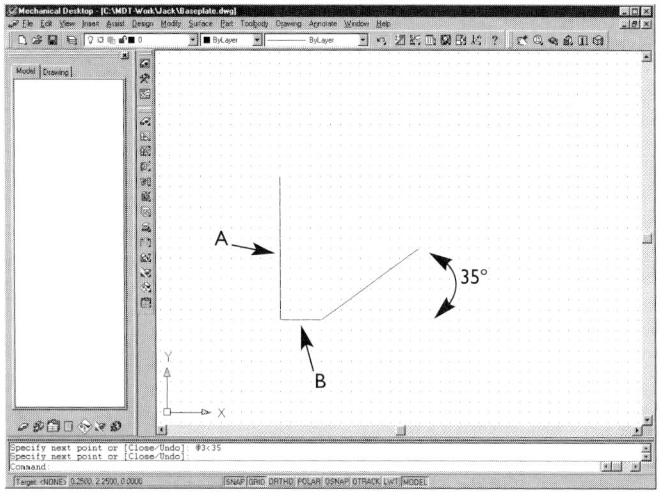

Figure 4.6

Using Offset

Reference

Mechanical Desktop On-line Help

For every command, the Help feature lists its various methods of access via menus, toolbars, and the command line. For a complete list of accelerator keys in Mechanical Desktop and AutoCAD, select Accelerator Keys in the Commands section of Mechanical Desktop Help.

The Offset command creates a new object parallel to a given object. You can select the Offset command using the Context menu. Another quick way to select the command is by typing its accelerator key (also called a command alias), which is the letter O, and pressing ↵. Some commands have different prompts depending on how they are started, so it is recommended that you start the command as instructed in the steps. You will use Offset to create parallel lines offset from some of the lines you just drew.

Right-click: in the graphics area

Pick: **2D Sketching, Offset**

To offset an object, you need to determine the *offset distance* (the distance away from the original object) or the *through point* (the point through which the offset object is to be drawn). You will specify the offset distance for both of the lines you wish to add by responding to the prompts at the command line. Refer to Figure 4.6 as you select the lines to offset.

Specify offset distance or [Through] <1.000>: **6** ↵

Select object to offset or <exit>: **select line A**

Specify point on side to offset: **pick any place in the drawing to the right of line A**

Select object to offset or <exit>: ↵

Command: ↵ **(to restart the command)**

Specify offset distance or [Through] <6.000>: **2** ↵

Select object to offset or <exit>: **select line B**

Specify point on side to offset: **pick any place in the drawing above line B**

Select object to offset or <exit>: ↵

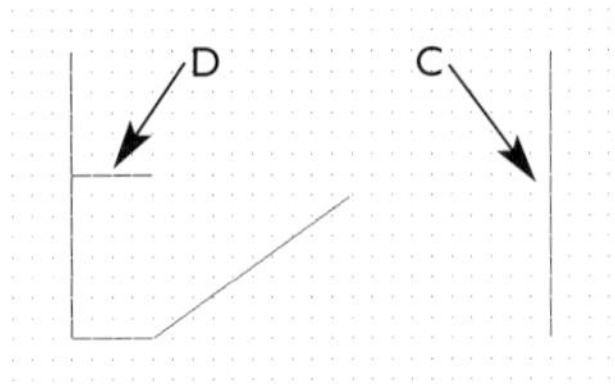

Figure 4.7

Once you have defined the offset distance, AutoCAD continues to repeat the prompt Select object to offset:, allowing you to create additional parallel lines that have the same spacing, until you press ↵ to end the command.

Now your drawing should look like Figure 4.7.

Using Undo and Redo

The Undo command allows you to undo the last operation completed. The Redo command reverses the last Undo. Undo and Redo appear as arrow icons on the Mechanical Main toolbar. You will try the icons to undo the last offset segment and then redo it.

Pick: **the Undo icon that looks like a backwards pointing arrow**

Pick: **drag down on the Undo icon to select the Redo icon on the flyout**

The offset line should have been undone and then redone.

Many commands include Undo as an option within the command, but it is also a separate command with its own options. When the full command is typed at the command line, you have options that let you undo more than one step, or mark a location where you can undo steps back to. The default option is to undo one step:

Command: **UNDO** ↵

Enter the number of operations to be undone or
[Auto/Control/Begin/End/Mark/Back]<1>: ↵

Command: **REDO** ↵

The offset line should have disappeared and reappeared again.

Next you will extend the second line you offset so that it intersects the first line offset.

Using Extend

The Extend command extends the lengths of existing entities, such as lines, polylines, and arcs, to end at a selected boundary edge. Extend has two parts: First you select the object to act as the boundary, and then you select the objects you want to extend. You can start extend from the Context menu under 2D sketching. To use its command alias, type EX ↵.

After you've selected a boundary edge, the command line options for the Extend command are Project, Edge, and Undo. You can use the Edge option to extend objects to the point at which they would meet the boundary edge if it were longer. This method is useful when the boundary edge is short and the lines to be extended would not intersect it. The Project option allows you to specify a plane of projection to be used for extending. Often 3D objects may not actually intersect the boundary edge, but appear to do so in the 2D view. This option is useful for extending objects that do not intersect the boundary in 3D space but would do so if projected onto a 2D plane. This plane can be the current viewing direction or the current User Coordinate System. You can use the Project option to extend objects as they appear in the view. The Undo option lets you undo the last object extended while remaining in the Extend command.

Refer to Figure 4.7 when making your selections. Turn Snap off if you have difficulty picking the lines with it on.

Tip

When extending a line, pick near the end of the line closest to the boundary object.

Right-click: in the graphics area

Pick: **2D Sketching, Extend**

Current settings: Projection=UCS Edge=None

Select boundary edges ...

Select objects: **pick line C** (the right most line you created using offset)

Select objects: ↵

Select object to extend or shift-select to trim [Project/Edge/Undo]: **pick line D** (the short horizontal line you offset)

Select object to extend or shift-select to trim [Project/Edge/Undo]: ↵

Command: ↵ (to restart the command)

Select boundary edges ...

Select objects: **pick the horizontal line you just created using Extend**

Select objects: ↵

Select object to extend or shift-select to trim [Project/Edge/Undo]: **pick the angled line**

Select object to extend or shift-select to trim [Project/Edge/Undo]: ↵

Now your drawing should look like Figure 4.8.

Next you will trim off the unwanted portion of the long horizontal line.

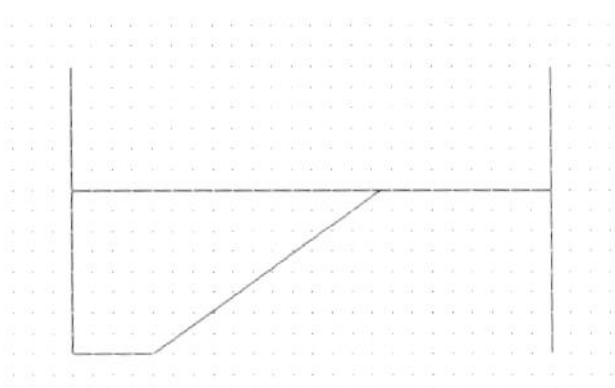

Figure 4.8

Using Trim

Trim removes part of an object in two steps. First, you select the objects that you will use as cutting edges. The *cutting edges* are drawing objects that define where another object should be cut off. Press ↵ to indicate that you have finished selecting cutting edges and want to begin the second step. After you press ↵, you select the objects to trim. When doing so, pick on the portions you want removed; the portion on the other side of the cutting edge will remain. TR is the command alias for the Trim command.

Like the Extend command, Trim has a Project option that gives you three choices for the projection method used by the command.

- View: Trims objects where they intersect as viewed from the current viewing direction
- None: Trims objects only where they intersect in 3D space
- UCS: Trims objects where they intersect in the current User Coordinate System

The Edge option lets you decide whether to trim objects only where they intersect in 3D space or where they would intersect if the cutting edge were extended. The Undo option lets you undo the last trim without exiting the command.

Refer to Figure 4.9 to understand how the cutting edge works in relation to the object to be trimmed.

When lines come together, as at a corner, you can select one or both of the lines as cutting lines, as shown in Figure 4.9, parts 1 and 3. The cursor location determines which portion of the line is removed.

Now you are ready to trim.

Right-click: in the graphics area

Pick: **2D Sketching, Trim**

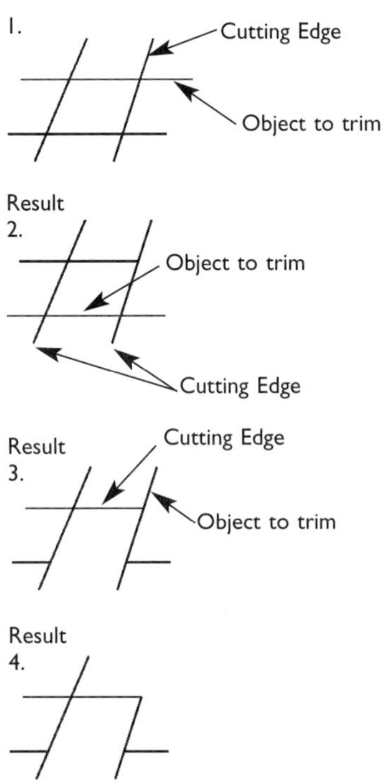

Figure 4.9

Current settings: Projection = UCS Edge = None
Select cutting edges ...
Select objects: **pick the angled line**
Select objects: ↵
Select object to trim or [Project/Edge/Undo]: **pick near the left end of the long horizontal line**
Select object to trim or [Project/Edge/Undo]: ↵

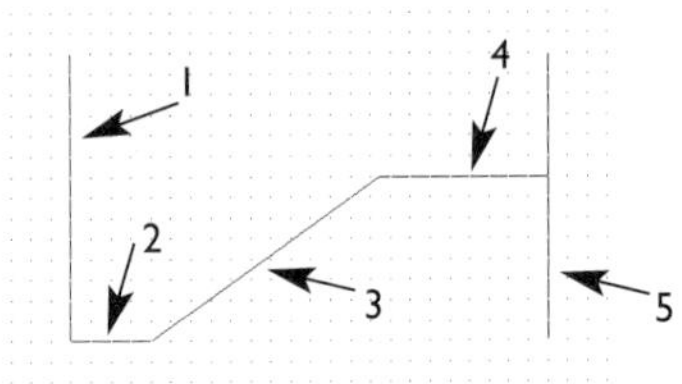

Figure 4.10

The selected end is trimmed off where it crosses the angled line, as shown in Figure 4.10.

Next you will turn on the object snap mode for Endpoint and use it to help mirror the portion you have drawn to create a finished sketch.

Using Object Snaps

You use object snaps to select locations accurately in relation to other objects in your drawing. You can use object snaps to select endpoints, midpoints, circle or arc centers, intersections, nearest objects, point entities, quadrant locations on arcs or circles, points perpendicular or tangent to objects, as well as others.

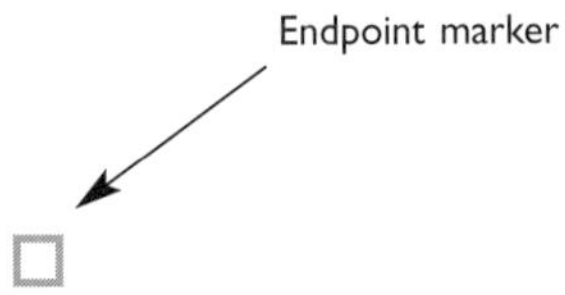

Figure 4.11

When an object snap is on, an AutoSnap marker box appears when the cursor is near a snap point (the marker for Endpoint is shown in Figure 4.11). When you pick with the marker box showing, the software will calculate that exact location to select, regardless of your actual cursor location. The Endpoint object snap, for example, locates the closest endpoint of an arc, line, or polyline vertex. You will use the Endpoint object snap as you mirror the lines you've drawn.

Right-click: **OSNAP button** (on the status bar)

Pick: **Settings** (from the context menu)

The Drafting Settings dialog box appears on the screen with the Object Snap tab uppermost. The symbols to the left of the Osnap modes indicate the shape that the cursor will change to when it is targeting each kind of Osnap. You will set the dialog box as shown in Figure 4.12.

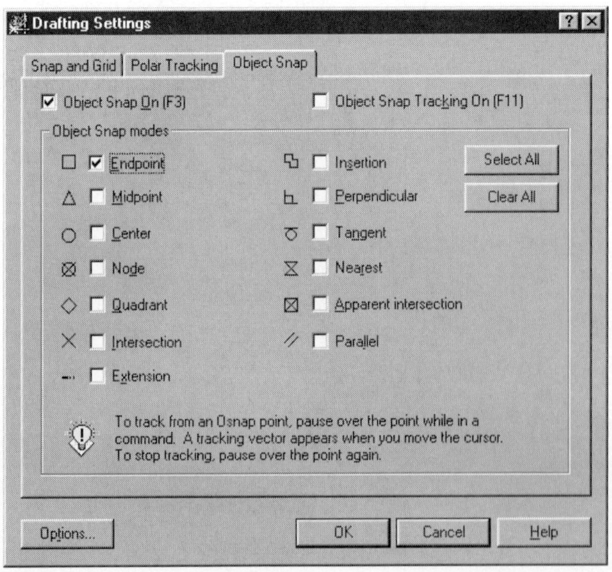

Figure 4.12

Pick: **Clear All** (to remove any current selections—in case you have different settings than the defaults)

Pick: **Endpoint** (so it is checked)

Pick: **Object Snap On** (so it is checked)

Pick: **OK** (to exit the Drafting Settings dialog box)

Now Endpoint is turned on. Anytime you are prompted to select, an AutoSnap marker box will appear when the cursor is near an endpoint. Now you are ready to use the Mirror command.

Using Mirror

The Mirror command makes a mirror-image copy of the 2D objects you select around a *mirror line* that you specify. The Mirror command allows you to delete the old objects or to keep them. The mirror line, which can be horizontal, vertical, or slanted, defines both the angle and the distance at which the mirror image is to be drawn. The mirrored image is drawn perpendicular to the mirror line. When prompted to specify the mirror line, you don't need to select an existing line; you can pick any two points from the screen to define it.

The right vertical line will serve as the first mirror line. Refer to Figure 4.10 as you pick points.

Right-click: in the graphics area

Pick: **2D Sketching, Mirror**

Select objects: **pick lines 1, 2, 3, 4**

Select objects: ↵

Specify first point of mirror line: **using the Endpoint object snap marker, pick one end of line 5**

Your cursor is now dragging a mirrored copy of the objects.

Specify second point of mirror line: **pick the other end of line 5**

Delete source objects? [Yes/No] <N>: ↵

Set Grid to off

Zoom and Pan the sketch so it is centered on the screen

Your screen should look like Figure 4.13.

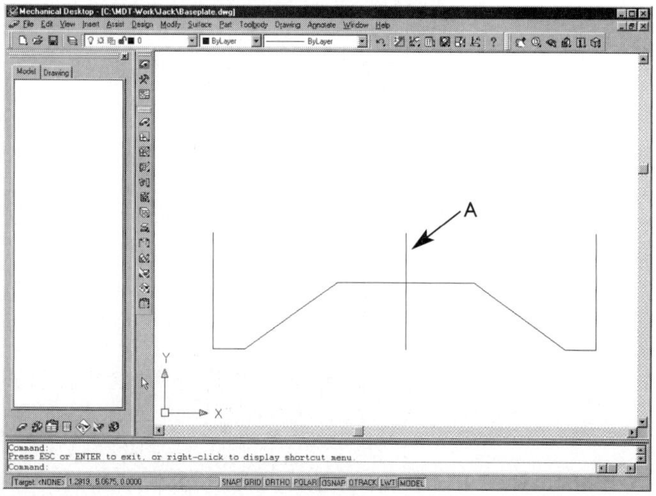

Figure 4.13

Using Erase

Next you will erase the center vertical line. It was used as a boundary for the Extend command and as the mirror line, but it is no longer needed as a part of the profile. The Erase command lets you select and erase as many objects as you wish until you exit the command.

Right-click: in the graphics area

Pick: **2D Sketching, Erase**

Select objects: **pick line A** (the center vertical line in Figure 4.13)

Select objects: ↵ (to stop selecting and apply the command to the selected objects)

Now your drawing should look like Figure 4.14.

Figure 4.14

Using Hot Grips

A quick way to make changes to your drawing is by using *hot grips*. Hot grips let you select an object (or objects) already drawn on your screen and activate certain editing commands directly from the mouse or pointing device without having to select the commands from the menu or keyboard. In most commands, you first select the command, then select the objects to be acted on. For common editing commands, such as Erase, it is convenient to be able to select the desired object and have the command take effect as soon as it is invoked.

To use the hot grips you must have a two-button mouse as your minimum hardware configuration. You will verify that hot grips are enabled using the Options dialog box. Use the menu bar to

Pick: **Assist, Options**

Pick: **Selection tab**

It becomes the uppermost tab in the dialog box, as shown in Figure 4.15. The Selection tab lets you control the tools you can use to make selections in Mechanical Desktop. You should select the box to the left of Enable Grips at this time if it is not already selected. The right side of the dialog box allows you to turn hot grips on and off and to control their color and the size of the grip box. Try using the slider control next to the image tile for the Grip size:

Pick and drag: **the slider box to the right** (to enlarge the size of the grip)

Pick and drag: **the slider box to the left** (to restore the original size of the grip)

Pick: **OK** (to exit the dialog box)

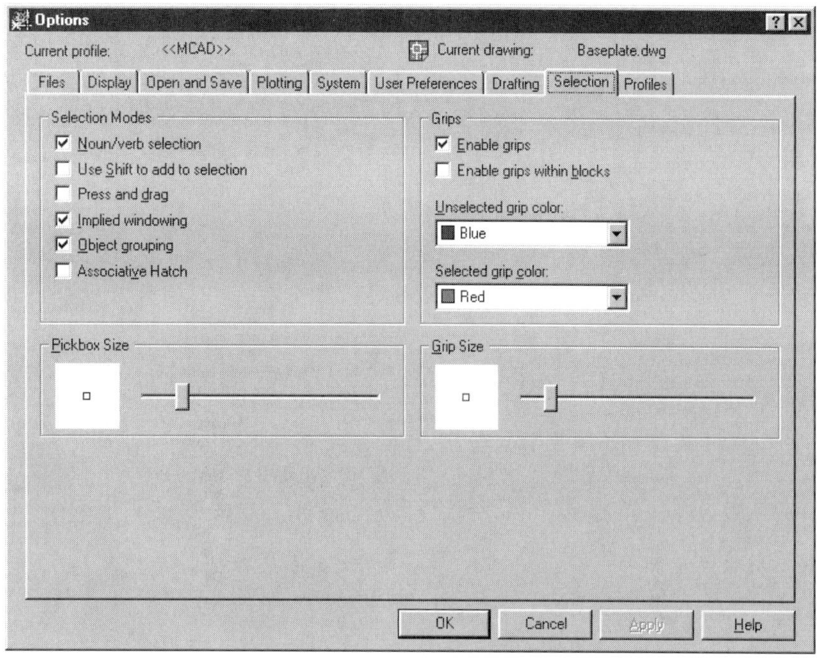

Figure 4.15

Activating an Object's Grips

You pick an object while you do not have any command active at the command prompt in order to activate its grips.

Pick the leftmost line of the sketch to activate its grips

The line becomes dashed and small boxes appear at the endpoints, as shown in Figure 4.16. These boxes are called the hot grips. You can use them to stretch, move, rotate, scale, and mirror the object.

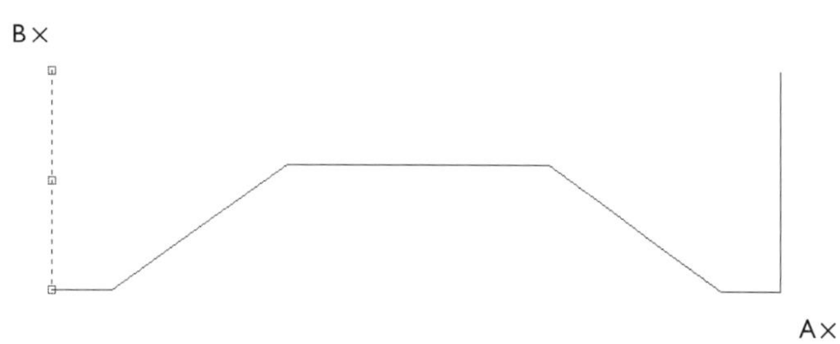

Figure 4.16

Using Crossing to Select

To select multiple objects at once, you can use implied windowing. Whenever you pick a point on the screen that is not on an object and move your pointing device, a box will start to rubberband from the point you picked. Picking a second point defines a selection rectangle, or window, that selects objects. If the box is drawn from right to left, called *implied Crossing*, the box formed selects everything that is crossed by or

Tip

You can use the Options dialog box to turn implied Windowing on and off.

contained in it. Drawing the box from left to right invokes *implied Windowing*; only objects completely enclosed by the box are selected. You can use implied Windowing and implied Crossing to select objects during any command in which you are prompted to select objects.

Select objects: **pick a point below and to the right of the sketch** (identified as A in Figure 16)

Other corner: **pick a point above and to the left of the sketch objects** (identified as B)

Note that a window box formed around the area specified by the upper left and lower right corners, and the hot grips appear for all of the entities that crossed or were included in the box. Your screen should look like Figure 4.17.

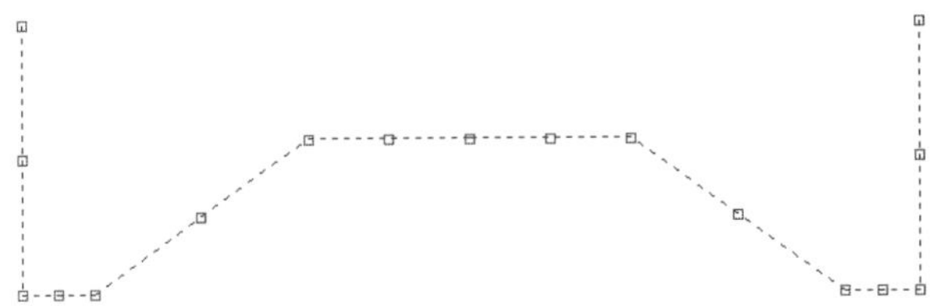

Figure 4.17

Pick: **in the grip box for the upper endpoint of the right vertical line**

The grip box turns red and filled, as shown in Figure 4.18. It will act as the *base grip*. The base grip is used as the starting point for commands invoked with hot grips active.

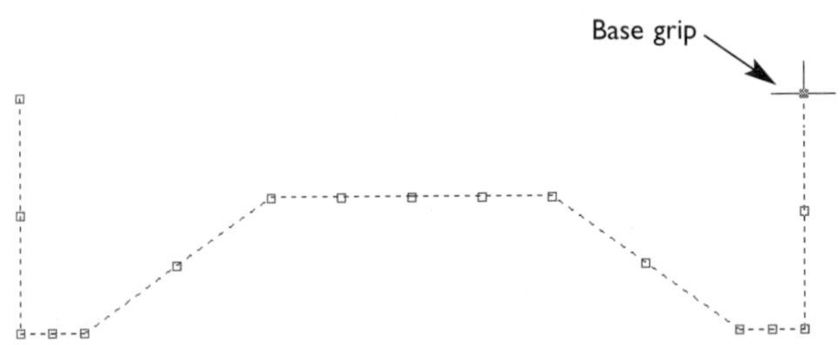

Figure 4.18

You should now see the prompt for the Stretch command in the command line area.

Command:

** STRETCH **

Specify stretchpoint or [Base point/Copy/Undo/exit]:

Right-click: **to pop-up context menu for grips**

The Hot Grips menu shown in Figure 4.19 appears. You will use the menu to select the Mirror command. The prompts will be similar to

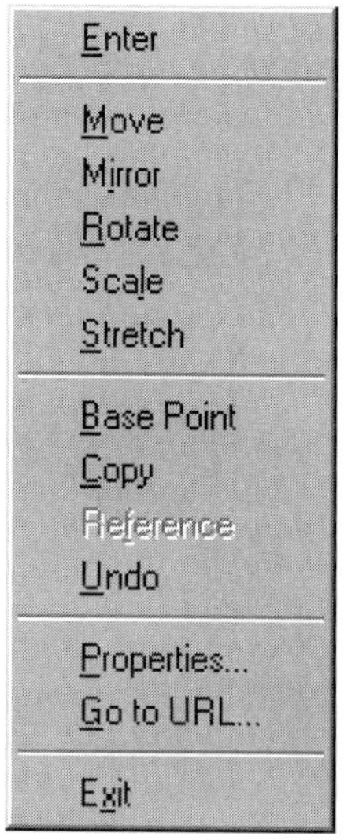

Figure 4.19

Figure 4.20

Tip

If you see a different number of required constraints, don't be concerned; you will delete the constraints in the next step and add your own.

those you used before, but you must type the command option to create a copy to mirror rather than specifying whether to delete the original objects after they have been mirrored, as you did before. Also, the base grip is assumed to be the first point of the mirror line unless you use the option Base point to specify a different first point.

Pick: **Mirror**

Specify second point or [Base point/Copy/Undo/eXit]: **C** ↵

** MIRROR (multiple) **

Specify second point or [Base point/Copy/Undo/eXit]: **press** F8 **to turn on ORTHO and pick to the left of the current location** (for the second point of the mirror line)

The mirrored copy of the objects appears. Pressing F8 toggles ORTHO on, which constrains cursor movement parallel to the grid. You can continue to select new mirror locations or, to end the command, you will press ↵.

** MIRROR (multiple) **

Specify second point or [Base point/Copy/Undo/eXit]: ↵

Press: ESC (to remove hot grips from the objects)

Your sketch is now complete. It should look like Figure 4.20.

Making the Profile

You will use your sketch as a profile, which you will extrude to create the base feature.

Right-click: in the graphics area

Pick: **Sketch Solving, Profile**

Select objects for sketch: **use implied Crossing to select all of the lines of the sketch**

Select objects for sketch: ↵

You see the message, *Solved under constrained sketch requiring 12 dimensions or constraints.*

Next you will show the constraints, then delete and add constraints as needed.

Right-click: in the graphics area

Pick: **2D Constraints, Show Constraints**

The constraint symbols appear on your screen, as shown in Figure 4.21. Note that the longer horizontal and vertical lines are actually two line segments because of the way they were made. For this profile, you will delete all of the existing constraints and add all new ones.

Enter an option [All/Select/Next/exit] <exit>: ↵

Right-click: in the graphics area

Pick: **2D Constraints, Delete Constraints**

Select or [Size/All]: **A** ↵

All of the constraints are deleted. Next you will carefully add your own constraints. You will start by setting the lower left corner of the sketch as

Tip

If your constraints are hard to read, you can make them larger by choosing Part, Part Options from the menu bar. In the dialog box that appears, choose "Constraint Size" from the Sketch Setting area and use the slider to set the size of the constraint display.

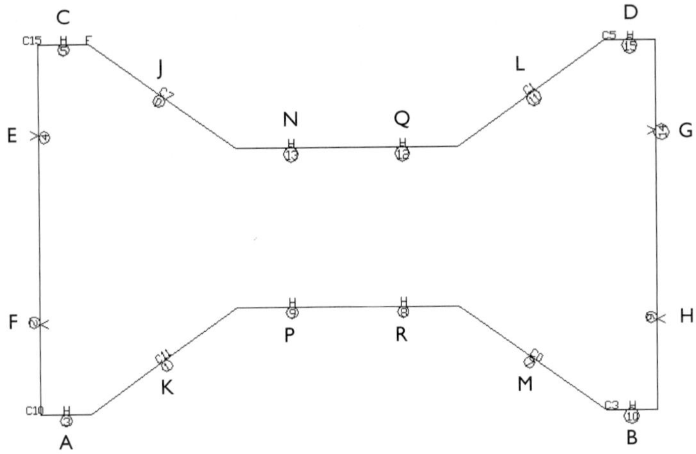

Figure 4.21

the point that will be fixed on the coordinate system. As you set this constraint, the cursor changes to a blue circle when it has targeted a point, much as the AutoSnap marker indicates the point that will be picked. When the point you want is targeted, picking will select it as the Fix point.

Right-click: in the graphics area

Pick: **2D Constraints, Fix**

Select endpoint or sketch segment to lock its position: **target the lower left corner of the sketch and pick when the cursor changes to a blue circle centered on the corner**

Select endpoint or sketch segment to lock its position: ↵

An F appears on the sketch to indicate the fixed point.

[Hor/Ver/PErp/PAr/Tan/CL/CN/PRoj/Join/XValue/YValue/Radius/Length/Mir/Fix/eXit] <exit>: ↵

Next you will add horizontal and vertical constraints to the lines, then constrain groups of lines to have equal length.

Right-click: in the graphics area

Pick: **2D Constraints, Vertical**

Select object to be reoriented: **pick all four vertical lines and then press** ↵

[Hor/Ver/PErp/PAr/Tan/CL/CN/PRoj/Join/XValue/YValue/Radius/Length/Mir/Fix/eXit] <eXit>: **H** ↵ (to add horizontal constraints)

Select object to be reoriented: **pick all 8 horizontal lines** ↵

Now you will set the short horizontal line segments to be the same size. You will set all the segments to be the same size as segment A in the lower left of the sketch. Refer to Figure 4.21 for the choices.

[Hor/Ver/PErp/PAr/Tan/CL/CN/PRoj/Join/XValue/YValue/Radius/Length/Mir/Fix/eXit] <eXit>: **L** ↵ (to add equal length constraints)

Select first object: **pick the lower right short line segment B**

Select second object: **pick the lower left short line segment A**

Select first object: **pick the upper left short line segment C**

Select second object: **pick the lower left short line segment A**

Select first object: **pick the upper right short line segment D**

Select second object: **pick the lower left short line segment A**

Now you will set the vertical segments to be the same size, then the angled lines, and finally the horizontal segments in the middle.

Select first object: **pick the upper left vertical line segment E**

Select second object: **pick the lower left vertical line segment F**

Select first object: **pick the upper right vertical line segment G**

Select second object: **pick the lower left vertical line segment F**

Select first object: **pick the lower right vertical line segment H**

Select second object: **pick the lower left vertical line segment F**

Tip

Turn Snap off if it is difficult to pick the angled lines.

Select first object: **pick the upper left angled line segment J**

Select second object: **pick the lower left angled line segment K**

Select first object: **pick the upper right angled line segment L**

Select second object: **pick the lower left angled line segment K**

Select first object: **pick the lower right angled line segment M**

Select second object: **pick the lower left angled line segment K**

Select first object: **pick the upper left horizontal center segment N**

Select second object: **pick the lower left horizontal center segment P**

Select first object: **pick the upper right horizontal center segment Q**

Select second object: **pick the lower left horizontal center segment P**

Select first object: **pick the lower right horizontal center segment R**

Select second object: **pick the lower left horizontal center segment P**

Select first object: ↵

[Hor/Ver/PErp/PAr/Tan/CL/CN/PRoj/Join/XValue/YValue/Radius/Length/Mir/Fix/eXit] <eXit>: ↵

Now you should have a good set of geometric constraints. It is time to Re-Solve the sketch, then add some dimensions.

Right-click: in the graphics area

Pick: **Sketch Solving, Re-Solve**

The sketch is resolved. You see the message, *Solved under constrained sketch requiring 6 dimensions or constraints*.

Right-click: in the graphics area

Pick: **Dimensioning, New Dimension**

Select first object: **pick on the short horizontal segment A**

Select second object or place dimension: **pick below line A to place the dimension**

Enter dimension value or [Undo/Hor/Ver/Align/Par/angle/Ord/Diameter/place]<0.9983>: **1.0** ↵

The sketch should look like Figure 4.22.

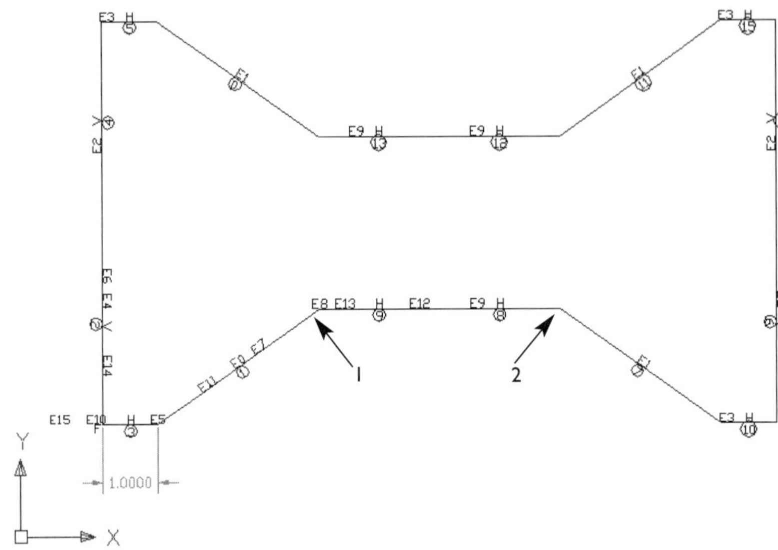

Figure 4.22

Next add the dimension for the height and width of the feature.

Select first object: **pick on the short horizontal segment A**

Select second object or place dimension: **pick on the short horizontal segment C**

Place dimension: **pick to the left of the sketch**

Enter dimension value or [Undo/Hor/Ver/Align/Par/aNgle/Ord/Diameter/pLace] <7.0000>: **7** ↵

Select first object: **pick on the left vertical line**

Select second object or place dimension: **pick on the right vertical line**

Place dimension: **pick below the sketch**

Enter dimension value or [Undo/Hor/Ver/Align/Par/aNgle/Ord/Diameter/pLace] <12.000?>: **12** ↵

Select first object: **pick endpoint 1 in Figure 4.22**

Select second object or place dimension: **pick endpoint 2**

Place dimension: **pick below the sketch**

Enter dimension value or [Undo/Hor/Ver/Align/Par/aNgle/Ord/Diameter/pLace] <3.000?>: **3** ↵

On your own, add the dimension for the 3.00 height of the center portion and the 2.00 distance that the center portion is down from the top of the part.

On your own, pan and zoom the drawing so it is centered on your screen.

When you have added the dimensions your drawing should appear similar to Figure 4.23.

Tip

You can select a dimension with hot grips, then select Move from the context menu to adjust its position.

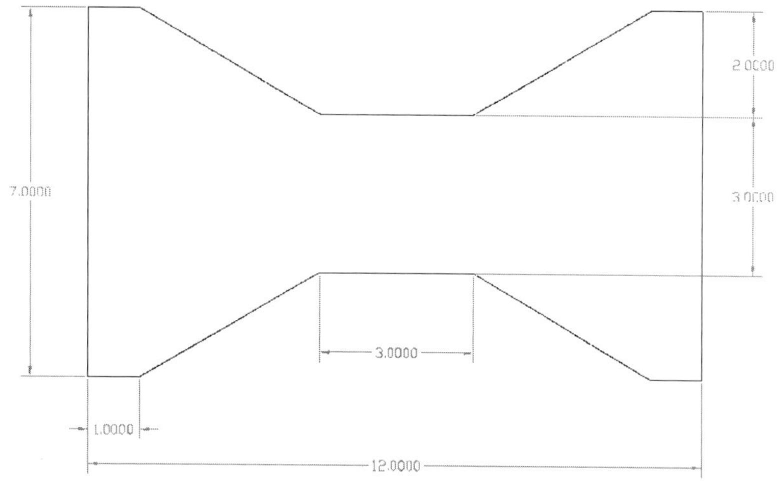

Figure 4.23

Next, you will Re-Solve the sketch again, just to make sure that you have a fully constrained sketch. You can use this command anytime you want to apply the dimensions and constraints to the profile and check to see if it is constrained properly.

Right-click: in the graphics area

Pick: **Sketch Solving, Re-Solve**

You should see the message, *Solved fully constrained sketch*. If not, review your constraints and dimensions and make changes so that your sketch resembles Figure 4.23.

Extruding the Base Feature

Now you are ready to extrude this profile as the base feature.

Right-click: in the graphics area

Pick: **Sketched & Work Features, Extrude**

The Extrusion dialog box appears. You will make the following selections in the dialog box.

Choose Operation: **Base** (this selection should be set by default and grayed out)

Set Distance: **1.5**

Set Draft Angle: **0**

Choose Termination Type: **Blind**

Pick: **OK**

Using techniques learned in previous chapters, right-click in the Browser to change the color property of the feature to red.

Pick: **View, 3D Views, Front Left Isometric**

Pick: **View, Shade, Gouraud Shaded Edges On**

Figure 4.24

The base feature you just created for the baseplate should look like Figure 4.24.

On your own, toggle the view back to wireframe.

Save your work at this time.

Next you will create a work plane as the sketch plane. You will make this a parametrically defined plane that stays aligned with the base feature you just created, even when the base feature's dimensions are changed. You will then use this work plane to sketch the shape of the front view of the baseplate. These two views will then be intersected to create the major shape of the baseplate.

Work Features

Work features, which include work planes, work points, and work axes, are special kinds of construction features that you can use to aid in placing new features on your model. *Work points* and *work axes* are always parametric features. That means they are constrained to locations relative to an existing feature through the use of geometric constraints or dimensions. *Work planes* can be either parametric or non parametric. A non-parametric work plane is one that is based on the coordinate system and stays at that location regardless of changes made to the part. Parametric work planes are dependent on the edges, faces, planes, vertices, or axes of an existing feature. Work features are useful when there is no part geometry to reference for placing or dimensioning an object. In this tutorial, you will create a parametric work plane to be used as the sketch plane. You will learn more about work points and work axes in other tutorials.

Creating a Parametric Work Plane

A work plane is an infinite plane that is related to your part. Work planes are defined by two modifiers that can be selected using the dialog box. Modifiers specify the part geometry used to orient the plane and its coordinate axes. The modifiers specify whether the plane is normal (perpendicular), parallel, or tangent, to selected edges, axes, or vertices. Nonparametric work planes can be defined by locating them on the current UCS or on any of the three planes of the WCS. At the time a work plane is created, you can also choose to have a sketch plane automatically created on it.

You will create a parametric work plane that stays centered on the existing part geometry. This is often useful when designing symmetrical parts, as features can be constrained and dimensioned from the work plane so they remain symmetrical when dimensions change.

Reference

Engineering Design Communication

Recognizing Symmetry, pp. 182–184

Right-click: in the graphics area

Pick: **Sketched & Work Features, Work Plane**

The Work Plane Feature dialog box appears on your screen. You will use it to make the settings shown in Figure 4.25. The dialog box is divided into two halves: one half for the 1st Modifier and the other for the 2nd Modifier. The lower portion of the 1st Modifier side contains the selections for creating nonparametric work planes. They are the buttons, On UCS, World XY, World YZ, and World XZ.

Select 1st modifier: **Planar Parallel** (so the button is filled)

Select 2nd Modifier: **Offset**

Set Offset: **3.5**

Select: **Create Sketch Plane** (so it is checked)

Pick: **OK**

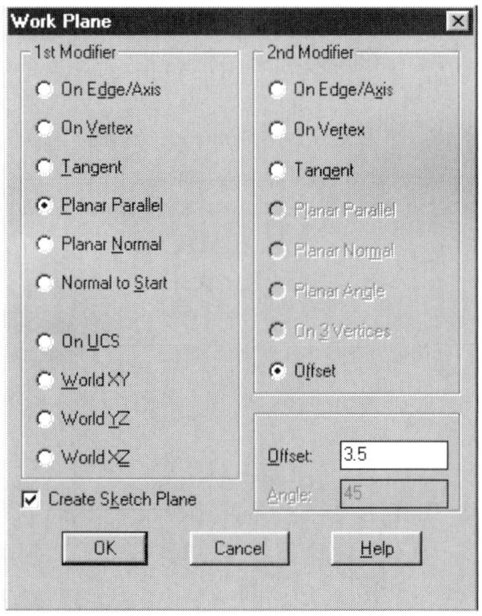

Figure 4.25

Select work plane, planar face, or [worldXy/worldYz/worldZx/Ucs]: **pick the small vertical plane on the left end of the part labeled B in** Figure 4.26.

Enter an option [Next/Accept] <Accept>: **if the correct plane is not highlighted, press left mouse to continue to the next plane, if the correct plane is highlighted press right mouse button to accept**

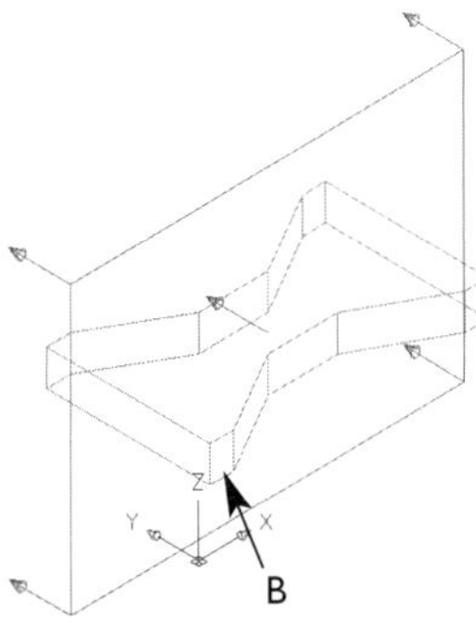

Figure 4.26

A representation of the plane appears parallel to the selected surface. Arrows point in the direction the work plane will be offset. The work plane arrows should point toward the part, as it is to be offset from surface B by one-half the width of the part to locate it in the center.

You will flip the direction for the plane as needed so that the arrows point toward the center of the part, as shown in Figure 4.26.

If the arrows do not point toward the part, use the Flip command option to toggle the offset direction.

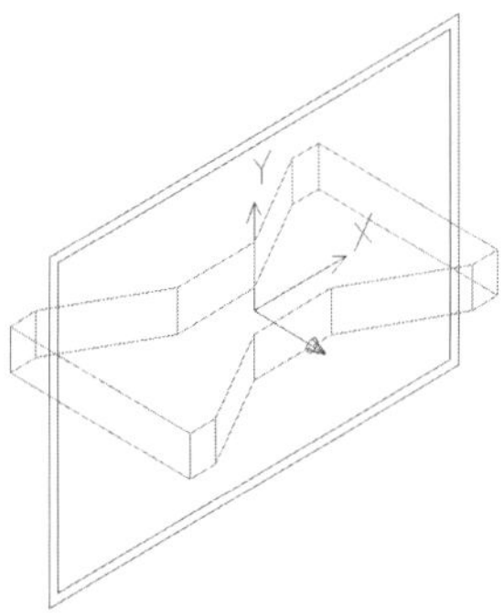

Figure 4.27

Enter an option [Flip/Accept] <Accept>: **F** ↵

Enter an option [Flip/Accept] <Accept>: ↵ **to accept when the arrows are pointed correctly toward the center of the part**

Plane=Parametric

A representation of the parametric work plane that will be created is shown on the part similar to Figure 4.27. At the prompt, you can select an edge of the part with which to align the *X*-axis, flip the direction for the positive *Z*-axis, or rotate the *XY*-coordinate plane, similar to what you did in Tutorial 2. The orientation for the *X*-direction should be as shown in Figure 4.27. If it is not, use the command options to orient it so that it appears as shown in the figure. Then

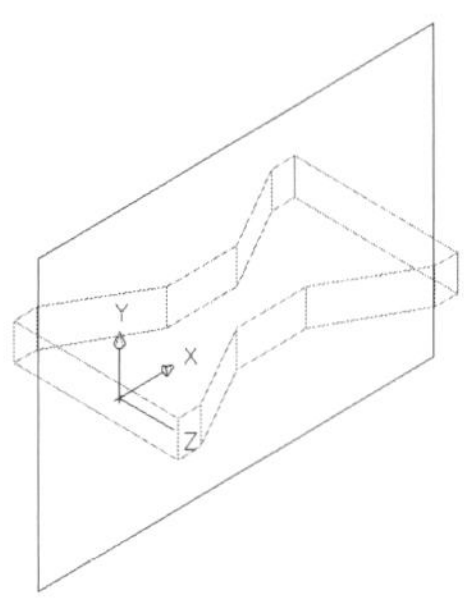

Figure 4.28

Select edge to align X-axis or [Z-flip/Rotate] <Accept>: ↵

Pick: **View, Zoom, All**

The work plane appears on your part, as shown in Figure 4.28.

Pick: **Sketch View icon**

Pick: **View, Zoom All**

Now your view is directly perpendicular to the sketch plane. The work plane fills your graphics window, as shown in Figure 4.29.

Tip

You can change your UCS icon to a wireframe appearance by typing **SHADEMODE** *and entering* **2** ⏎ *at the prompt. To return to the 3D appearance, enter 3 at the same prompt.*

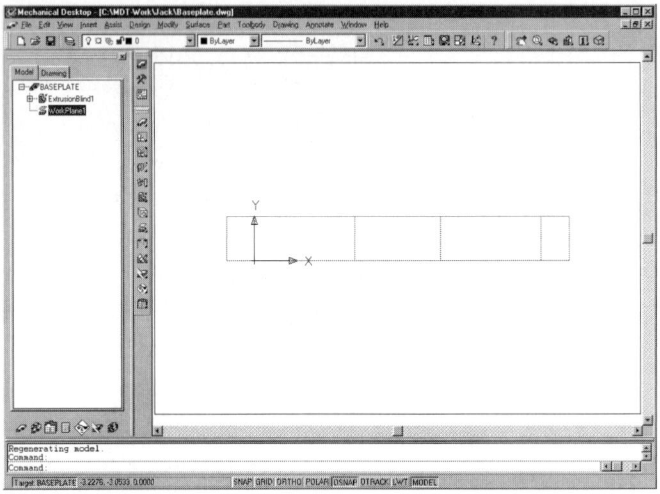

Figure 4.29

Sketching the Front View

Next you will sketch the shape you will use to generate the front view shown in Figure 4.30 on the work plane.

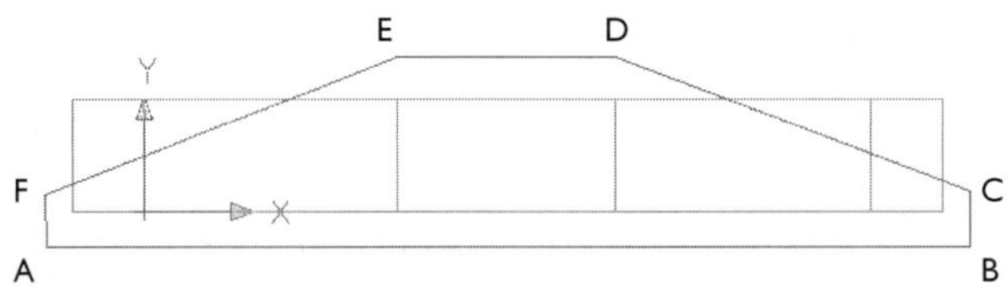

Figure 4.30

Although you could use the drawing tools you've learned to sketch the shape to the correct size, it is often easier to constrain the sketch to existing geometry when the lines are not drawn on top of each other. For this sketch, you will use Ortho to help you draw vertical and horizontal lines, but you will not use object snap to target the previous geometry exactly. Refer to Figure 4.30 while picking points.

Pick: **OSNAP on status bar to toggle it off**

Right-click: in the graphics area

Pick: **2D Sketching, Polyline**

Specify start point: **pick outside and below the lower left corner of the part (A)**

Current line width is 0.0000

Pick: **ORTHO** (on status bar to toggle it on)

Specify next point or [Arc/Close/Halfwidth/Length/Undo/Width]: **pick outside and below the lower right corner of the part (B)**

Specify next point or [Arc/Close/Halfwidth/Length/Undo/Width]: **pick a point about one-fourth of the way up the right-hand line (C)**

Set ORTHO: **OFF**

Specify next point or [Arc/Close/Halfwidth/Length/Undo/Width]: **pick just above the right middle vertex on the top edge (D)**

Set ORTHO: **ON**

Specify next point or [Arc/Close/Halfwidth/Length/Undo/Width]: **pick above the left middle vertex on the top edge (E)**

Set ORTHO: **OFF**

Specify next point or [Arc/Close/Halfwidth/Length/Undo/Width]: **pick a point outside the left vertical edge level with the short segment on the right side (F)**

Specify next point or [Arc/Close/Halfwidth/Length/Undo/Width]: **C** ↵

Your sketch should look similar to Figure 4.30.

Next you will add a construction line, as shown in Figure 4.31, that you can use to dimension the sketch from the center of the part.

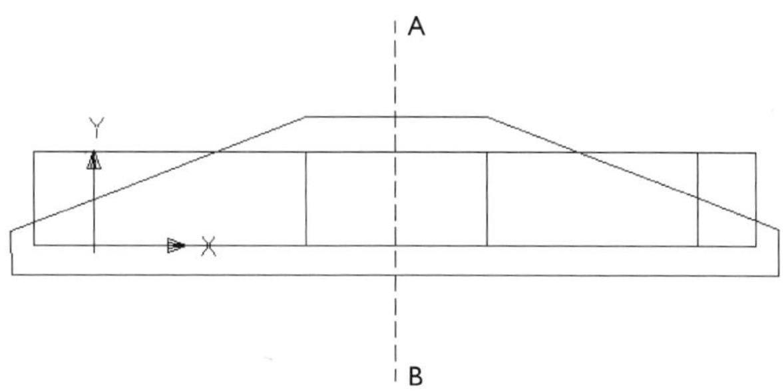

Figure 4.31

Construction Lines

Often when creating a sketch is useful to have some lines that you can use in dimensioning and creating the sketch geometry that do not become a part of the profile for extruding or otherwise creating a feature. You can use construction lines for these. (You can use construction circles in a similar way.) *Construction lines* appear in a different line pattern than the standard continuous line pattern that is used to create the profile geometry. They are included in the profile and can be edited and used for dimensioning, but they do not become part of the 3D solid. To add a construction line near the center of the sketch,

Right-click: in the graphics area

Pick: **2D Sketching, Construction Line**

Set ORTHO: **ON**

Specify next point or [Undo]: **pick point A**

Specify next point or [Undo]: **pick point B**

Specify next point or [Undo]: ↵

Next you will create a profile from the sketch.

Right-click: in the graphics area

Pick: **Sketch Solving, Profile**

Select objects for sketch: **pick the polyline**

Select objects for sketch: **pick the construction line** ↵

You see the message, *Solved under constrained sketch requiring 9 dimensions or constraints*. Now you are ready to view the constraints and add the needed constraints and dimensions.

Right-click: in the graphics area

Pick: **2D Constraints, Show Constraints**

Check out the constraints that were automatically added. You should have two horizontal and three vertical constraints only. If you have different sketch constraints, delete any other constraints; then add the following constraints on your own, so your sketch snaps to the existing geometry and looks like Figure 4.32.

Tip

You may have to use Zoom Window enlarge the lines and model edges so that you can select them.

Set Equal Length: **the 2 short vertical segments**
the two angled segments

Set Colinear: **the short segments to the model edge beside them**
the bottom segment to the model edge above it
the top segment to the model's topmost edge

Right-click: in the graphics area

Pick: **Sketch Solving, Re-Solve**

You should see the message, *Solved under constrained sketch requiring 3 dimensions or constraints*.

Using Dimension Command Options

The dimensioning command presents several options at the command line. You can use the command options when dimensioning to specify whether the dimension created will be horizontal, vertical, aligned, parallel, or angled; whether it will use ordinate dimensioning or display a diameter rather than radius; and to move a dimension's placement point. These options allow you to select entities in a sketch from which to dimension without having to select an exact location on the entity. Refer to Figure 4.32 as you are adding the dimensions.

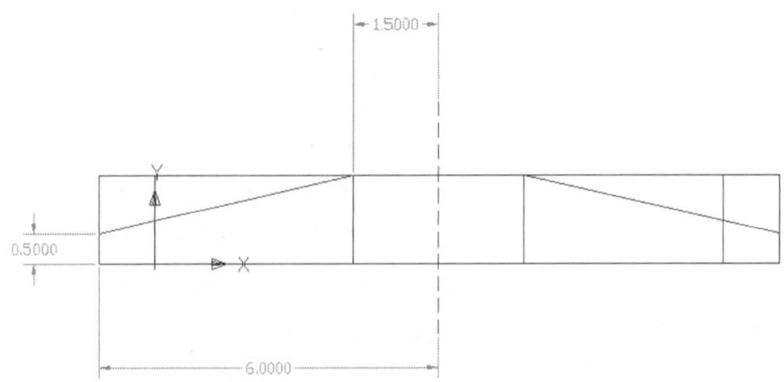

Figure 4.32

Right-click: in the graphics area

Pick: **Dimensioning, New Dimension**

Select first object: **pick the short vertical segment at the left of the sketch**

Select second object or place dimension: **pick to the left of the sketch**

Enter dimension value or [Undo/Hor/Ver/Align/Par/aNgle/Ord/
Diameter/pLace] <0.5???>:**.5** ↵

Select first object: **pick the bottom end of the vertical construction line**

Select second object or place dimension: **pick the left bottom corner point**

Specify dimension placement: **pick below the part**

The dimension may appear angled. To change it to a horizontal dimension, you will type H at the command prompt before entering the value, as follows:

Enter dimension value or [Undo/Hor/Ver/Align/Par/aNgle/Ord/Diameter/pLace] <6.3269>: **H** ↵

Enter dimension value or [Undo/Hor/Ver/Align/Par/aNgle/Ord/Diameter/pLace] <6.0814>: **6** ↵

Select first object: **pick the top end of the vertical construction line**

Select second object or place dimension: **pick the left end of the top horizontal line**

Specify dimension placement: **pick above the part**

Enter dimension value or [Undo/Hor/Ver/Align/Par/aNgle/Ord/Diameter/pLace] <6.3269>: **H** ↵

Enter dimension value or [Undo/Hor/Ver/Align/Par/aNgle/Ord/Diameter/pLace] <6.0814>: **1.5** ↵

You should see the message, *Solved fully constrained sketch*. If you do not, work on your own to add dimensions and constraints to create a working profile. Your dimensioned sketch should look similar to Figure 4.32.

Intersecting the Feature

Next you will extrude this feature and at the same time intersect it with the existing base feature. That is, the resulting part will be the intersection of the base feature and the new feature; the non-intersecting portions will be removed. This is a useful feature of Mechanical Desktop; some parametric modelers do not allow you to intersect features. You will extrude the feature using Intersect as the operation and Midplane as the termination option so that the new feature will extend in both directions from the sketch plane. The distance you will specify is the total distance in both directions for the feature.

Reference

Engineering Design Communication

Making Complex Shapes with Boolean Operations, pp. 181–182

Right-click: in the graphics area

Pick: **Sketched & Work Features, Extrude**

Choose Operation: **Intersect**

Set Distance: **7**

Set Draft Angle: **0**

Choose Termination Type: **Midplane**

Pick: **OK**

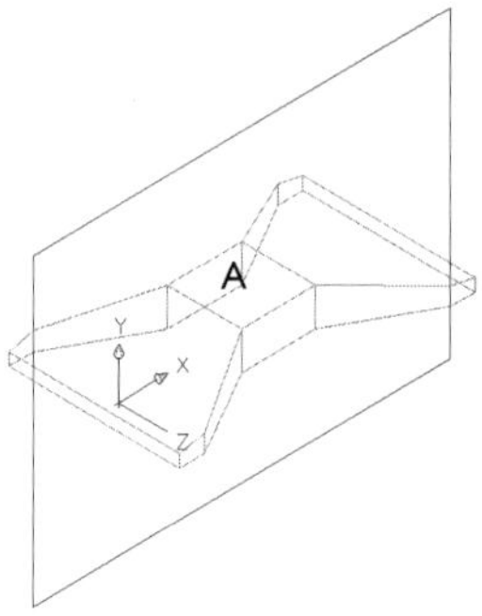

Figure 4.33

Pick: **View, 3D Views, Front Left Isometric**

The part on your screen should look like Figure 4.33. You will change the color of the new feature to match the rest of the part so it will be easier to see the new sketch lines.

Change the color of ExtrusionMidplane1 to red.

Removing Material with an Extruded Cut

Next you will make an extruded cut to remove material from the center of the part. Like intersection, the Cut operation is similar to the Boolean operation subtraction. The feature you create will be subtracted from the existing model. First, you must create a new sketch plane for the sketch geometry. This time, you will use a planar face of the model to define the sketch plane.

Right-click: in the graphics area

Pick: **New Sketch Plane**

Select work plane, planar face or [worldXy/worldYz/worldZx/Ucs]: **pick on the flat surface on the top of the part labeled A in Figure 4.33**

Enter an option [Next/Accept] <Accept>: **right-click to accept when the correct surface is highlighted**

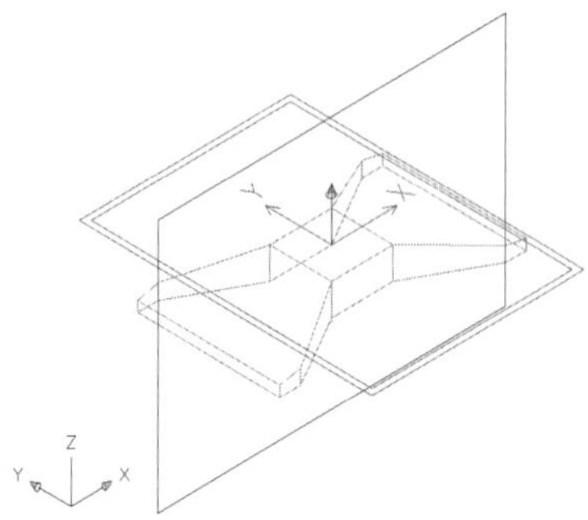

Figure 4.34

You see the message, Plane = Parametric. A representation of the plane shows in your drawing with a coordinate system showing. Use it to orient the direction for the *XY*-plane, as shown in Figure 4.34.

Select edge to align X-axis or [Z-flip/Rotate] <Accept>: **right click to accept when the coordinates are oriented like the figure**

The new sketch plane is created. Next you will change to the sketch view.

Pick: **Sketch View icon**

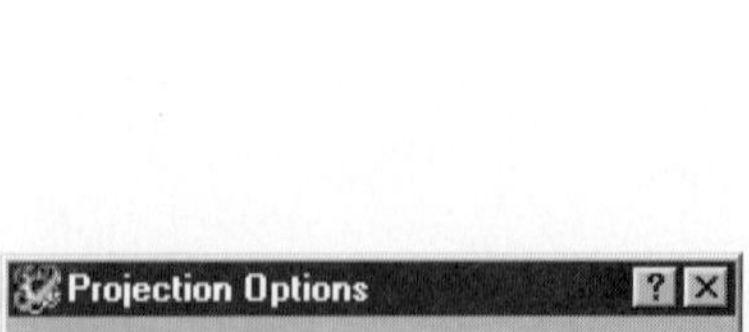

Figure 4.35

Your view is reoriented directly perpendicular to the sketch plane. The work plane in the center of the part that you used to extrude the last feature in both directions appears as a horizontal line through the center of the part, as shown in Figure 4.35. You will use feature tree in the Browser window to set the visibility of the plane to off.

Right-click: **WorkPlane1 feature** (in the feature tree)

Select: **Visible** (so it is no longer checked)

Next you will project the 3D faces of the part into the sketch plane. Projecting lines from the existing part geometry onto the sketch plane allows you to use the geometry in a new profile without re-creating it. You will use the lines you project with the Offset command to create a profile that will be used to cut away the center area.

Right-click: in the graphics area

Pick: **Sketch Solving, Project Objects to Plane**

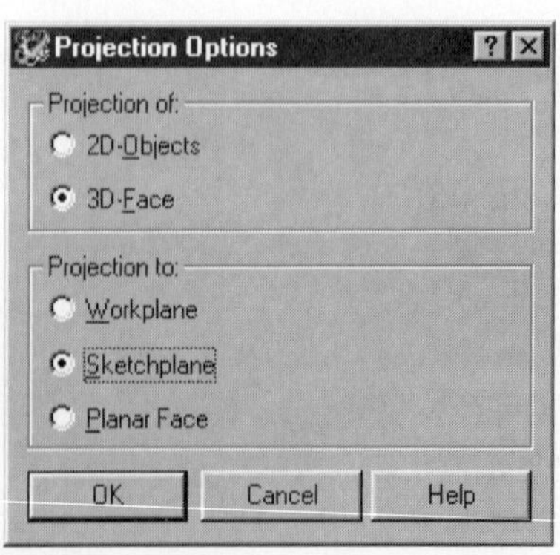

Figure 4.36

The Projection Options dialog box appears on your screen. You will use it to make the selections shown in Figure 4.36.

Set Projection of: **3D Face**

Set Projection to: **Sketch Plane**

Pick: **OK**

Select face: **pick inside the left end of the part**

Enter an option [Next/Accept] <Accept>: **press right-mouse to accept if correct surface is highlighted**

The edges of the surface are projected into the sketch plane.

On your own, restart the command and project the right surface to the sketch plane.

When you have finished,

Pick: **View, 3D Views, Front Left Isometric**

Your isometric view should show the lines projected into the sketch plane, as shown in Figure 4.37.

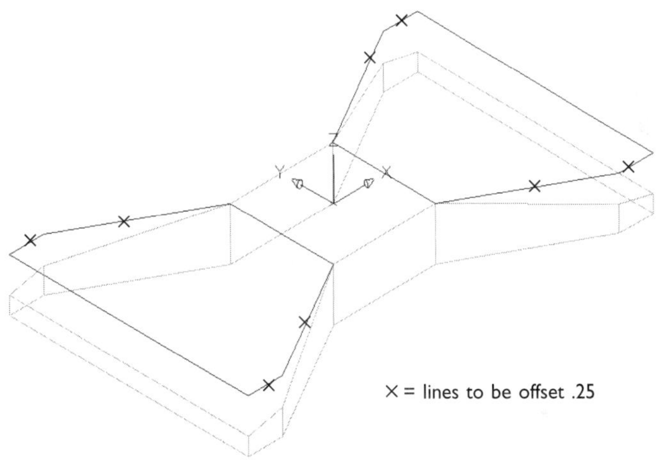

Figure 4.37

Next you will offset these lines to the inside by a distance of .25 to use as the lines of the profile. The raised edges of the baseplate should be .25 inches thick on both ends of the part, so offsetting the lines will create a sketch close to the size of the final profile. Refer to Figure 4.37 as you pick the lines to offset.

Right-click: in the graphics area

Pick: **2D Sketching, Offset**

Specify offset distance or [Through] <2.0000>:**.25** ↵

Select object to offset or <exit>: **pick on one of the angled lines** (that you created by projection)

Specify point on side to offset: **pick to the inside of the part from the original line**

Continue on your own to offset each of the lines marked in Figure 4.37 until you have sketch geometry that looks like Figure 4.38. When you have finished press ↵ to end the command.

Pick: **Sketch View icon** (so you are looking normal to the sketch plane)

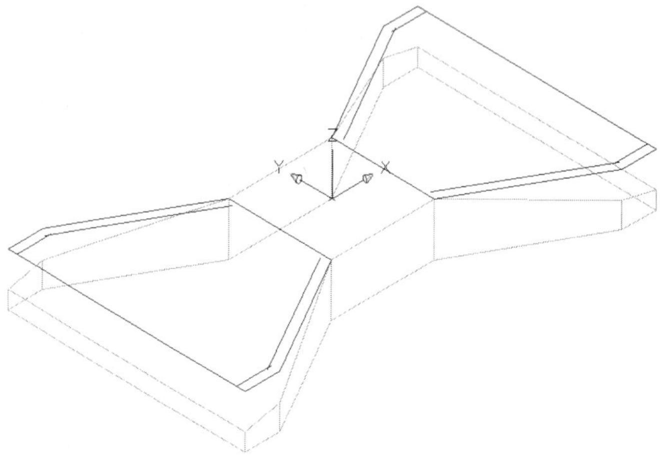

Figure 4.38

Next, you will add the lines to connect the two ends across the rectangular center section. Refer to Figure 4.39 as you pick points.

Command: **L** (the command alias for Line)

Pick: **a point slightly above the top left angled offset line**

Set ORTHO: **ON**

Specify next point or [Undo]: **about .5 inches below the point**

Specify next point or [Undo]: **pick to the right of the rectangular base and the endpoint of the right angled line**

Specify next point or [Undo]: **pick slightly above the top right angled line you offset**

Specify next point or [Undo/Close]: ↵

Repeat these steps to create the lines to connect the bottom set of angled lines.

Your sketch should look similar to Figure 4.39. It is not important if the distances between entities are not exactly the same as long as the lines intersect. You will add the relevant dimensions after you constrain the sketch.

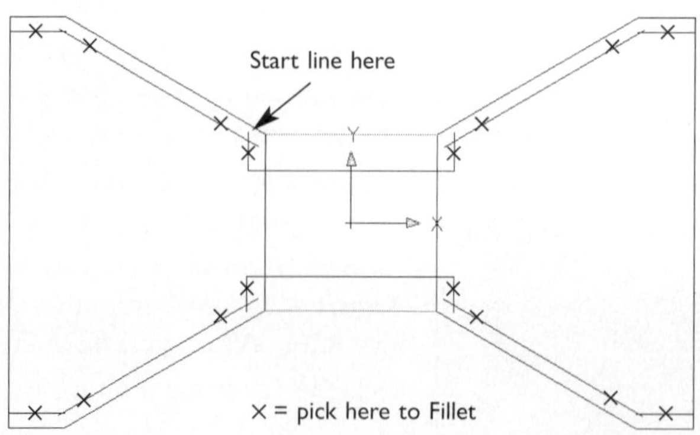

Figure 4.39

Using Fillet to Trim and Extend

An AutoCAD trick that you may find useful in creating profiles is to use a Fillet with the radius set to 0 as a way to trim and extend lines quickly. As you pick lines to fillet, be sure to pick on the portion of the line that is to remain.

Right-click: in the graphics area

Pick: **2D Sketching, Fillet**

Current settings: Mode = TRIM, Radius = 0.5000

Select first object or [Polyline/Radius/Trim]: **R** ↵ (to set the radius)

Specify fillet radius <0.5000>: **0** ↵

Select first object or [Polyline/Radius/Trim]: **pick on the top left horizontal line created by offsetting**

Select second object: **pick on the angled line that intersects it**

The lines become trimmed and form a clean corner.

Right-click: in the graphics area

Pick: **Repeat Fillet** (to restart the command)

Current settings: Mode = TRIM, Radius = 0.0000

Select first object or [Polyline/Radius/Trim]: **continue on your own to fillet each pair of lines and restart the command until all are trimmed.**

Once you have added the fillets, you can delete the lines you projected onto the sketch plane and offset from; the cut geometry is formed by the offset lines and line you added. It is easier to do this in an isometric view. Do not erase any red lines; if you cannot select a sketch line, you can delete it later. You will use the alias for the Erase Command.

Tip

One way to select lines that overlay the red model edges is to select using implied crossing and then use the Remove selection option to remove the model from the items select. To do this, at the Select objects prompt, pick to make a small crossing box on the overlapping lines, then type R ↵ and pick the red model edges to be removed. Press ↵ again to apply the erase command to the remaining items of the selection.

Pick: **Left Front Isometric view icon**

Type: **E** ↵

Select objects: **pick the lines you projected**

Select objects: ↵

When you have finished, your sketch should look similar to Figure 4.40

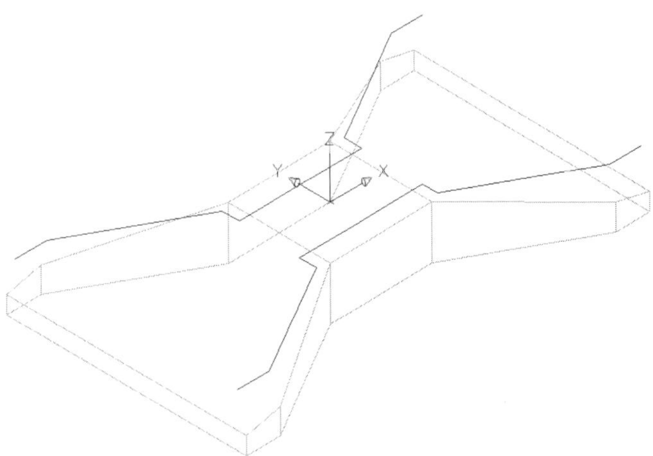

Figure 4.40

Next you will make your sketch into a profile. Because the sketch geometry is not a closed object, you will be prompted to choose an edge to close the profile. You may use existing model edges to complete a profile, which you will do in this case.

Right-click: in the graphics area

Pick: **Sketch Solving, Profile**

Select objects for sketch: **pick each of your sketch lines as shown in Figure 4.40**

Select objects for sketch: ↵

Select part edge to close the profile: **pick the line that forms the right-most edge of the part**

Enter an option [Next/Accept]<Accept>: ↵

Select part edge to close the profile: **pick the line that forms the left-most edge of the part**

Enter an option [Next/Accept]<Accept>: ↵

The sketch lines are solved as a profile. You see the message, *Solved under constrained sketch requiring 12 dimensions or constraints.*

Next you will add constraints and dimensions until the profile is fully constrained. As a rule, you should first look at the constraints that Mechanical Desktop inferred from your geometry and delete any constraints that do not reflect the design intent of the part. You may find that the lines in your profile are constrained to be parallel or colinear. If so, you should remove them.

Pick: **Sketch View icon**

On your own, delete all but the horizontal, vertical, and project (J) constraints before you proceed.

For this part, the angled lines in your sketch should be parallel to the model geometry and the area defined by the cut should run up to the rectangular base. On your own, add the following constraints:

Set Parallel: **the left two angled lines to the model edges the sketch was offset from**

Set Equal Length: **the four short horizontal segments at the ends of the part**
the four angled lines
the four short vertical segments
the two longer horizontal lines across the rectangular base

Set Colinear: **two of the short vertical segments** (one on each side) **to the edge of the rectangular base**

After setting the constraints, re-solve the sketch.

On your own, add the dimensions shown in Figure 4.41.

When you are finished, you should have constraints and dimensions similar to those in Figure 4.41. You are ready to extrude the feature.

Tip

Use the Parallel option at the command line to position the .25 dimension for the angled lines.

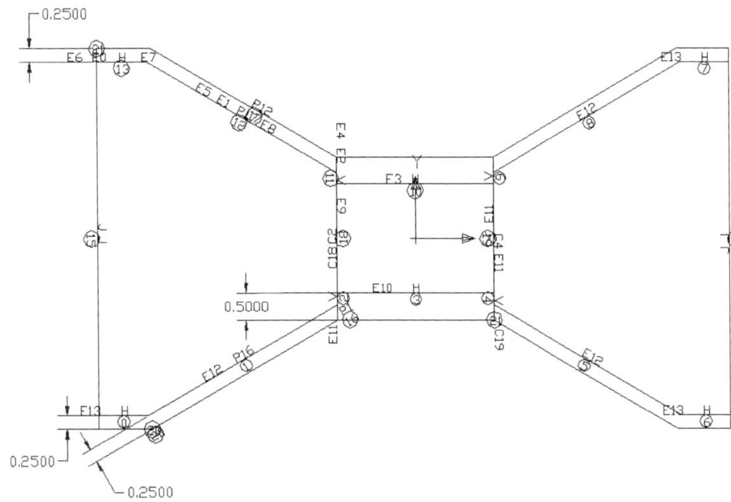

Figure 4.41

Right-click: in the graphics area

Pick: **Sketched & Work Features, Extrude**

Choose Operation: **Cut**

Set Distance: **1.25**

Set Draft Angle: **0**

Choose Termination Type: **Blind**

Pick: **OK**

Pick: **View, 3D Views, Front Left Isometric**

Tip

If you do not see the cut feature added to the part, it may be cutting in the wrong direction. To change it, right-click on ExtrusionBlind2 in the model tree. Pick Edit. You will return to the dialog box, where you can pick the Flip button and then select OK to exit. The cut should then be flipped so that it is heading into the part.

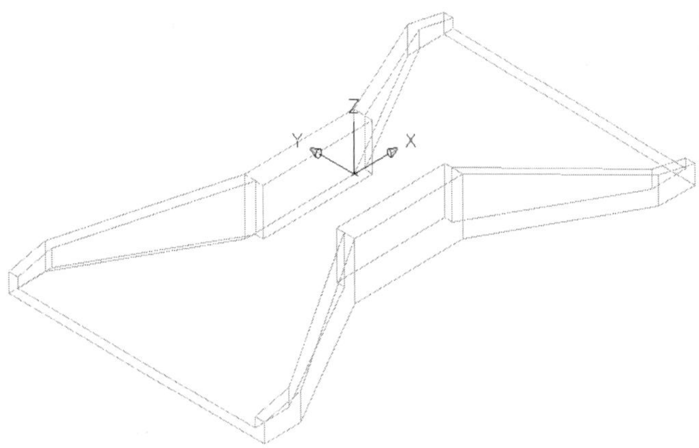

Figure 4.42

The cut should be removed from your part, as shown in Figure 4.42.

Change the color of ExtrusionBlind2 to red.

Erase any projected sketch geometry that you did not erase before (it should be black).

Sketching the Center Block with Rectangle

Next you will add the center block using the middle work plane as the new sketch plane and then extruding to both sides once again.

Right-click: **WorkPlane1** (in the feature tree)

Select: **Visible** (so it is checked)

Right-click: in the graphics area

Pick: **New Sketch Plane**

Select work plane, planar face, or [worldXy/worldYz/worldZx/Ucs]: **pick WorkPlane1**

Select edge to align X-axis or [Z-flip/Rotate] <Accept>: **right-click to accept its current setting**

The new sketch plane is created on the work plane.

Pick: **Sketch View icon**

The sketch view shows on your screen. You will use the Rectangle command to draw a box near the center of the part as shown in Figure 4.43. The sketch will be constrained to be collinear with the part geometry, but you will draw it smaller so that its edges are not over the top of the model edges, making them easier to select later.

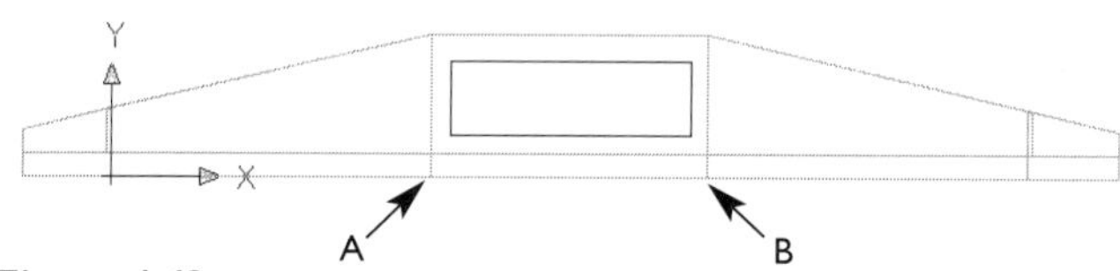

Figure 4.43

Tip

Sometimes it is easier to add colinear constraints when the sketch lines do not exactly overlap the part geometry.

Right-click: in the graphics area

Pick: **2D Sketching, Rectangle**

Pick: **to locate corners of box inside the center of the part as in Figure 4.43**

Now you are ready to solve the sketch. Because the Rectangle command forms a single closed entity (a polyline), you can use the Single Profile option. Mechanical Desktop will automatically select the polyline and solve the sketch.

Right-click: in the graphics area

Pick: **Sketch Solving, Single Profile**

Tip

When you are adding material with a Join operation, it will also work even when the sketch overlaps portions inside the part.

Tip

Sometimes it is easier to select part edges if you go to an isometric view.

On your own, add the following constraints:

Set Colinear: **the bottom line of the sketch to the bottom of the part**

the two vertical edges to the inside vertical edges labeled A and B in Figure 4.43

Now add a dimension from the top of the part to the upper horizontal line of the sketch and change its value to .35.

You should now have a fully constrained profile, as shown in Figure 4.44. You are ready to extrude the feature.

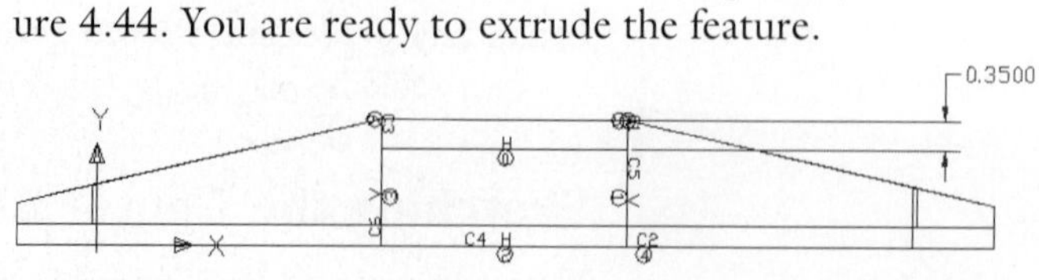

Figure 4.44

Right-click: in the graphics area
Pick: **Sketched & Work Features, Extrude**
Choose Operation: **Join**
Set Distance: **1.500**
Set Draft Angle: **0**
Choose Termination Type: **Midplane**
Pick: **OK**
On your own, turn off WorkPlane1.
Pick: **Front Left Isometric View**
Change the color of Extrusion Midplane2 to red.

The new feature should be added to your part, as shown in Figure 4.45.

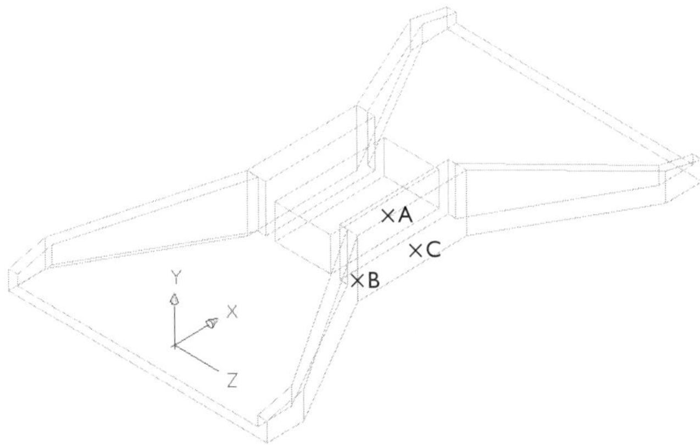

Figure 4.45

Adding the Holes

Now you are ready to add the last features to the Baseplate, the mounting holes.

Right-click: in the graphics area
Pick: **Placed Features, Hole**

The Hole dialog box appears on your screen. Make the following selections so the dialog box appears as shown in Figure 4.46.

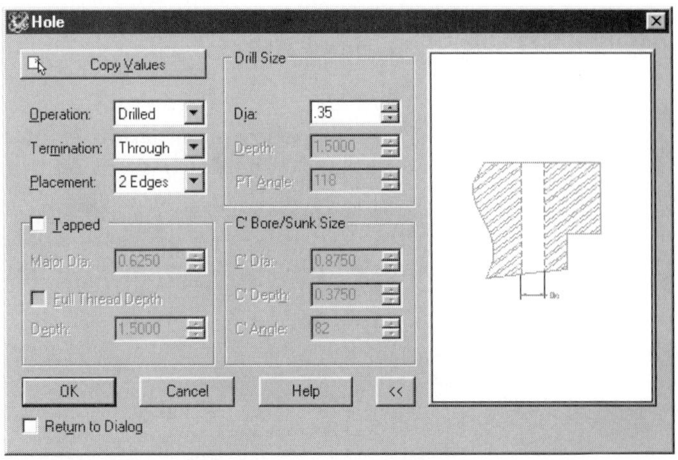

Figure 4.46

Choose Operation: **Drilled**

Choose Termination: **Through**

Choose Placement: **2 Edges**

Set Drill Size Dia: **.35**

Leave the Tapped box unchecked

Pick: **OK**

You are now ready to create through holes with a diameter of .35 inches. By selecting 2-edges as the Placement option, you will be prompted to locate the center of the hole by identifying two edges and the distance from each that the center should be. Follow the prompts and refer to Figure 4.45 to place the holes.

Select first edge: **pick the top horizontal edge of the part labeled A in Figure 4.45.**

Select second edge: **pick vertical edge labeled B.**

Computing ...

Specify hole location: **pick near the point labeled C.**

Enter distance from first edge (highlighted): **.65** ↵

Enter distance from second edge (highlighted): **.75** ↵

A hole is added to your part, as shown in Figure 4.47.

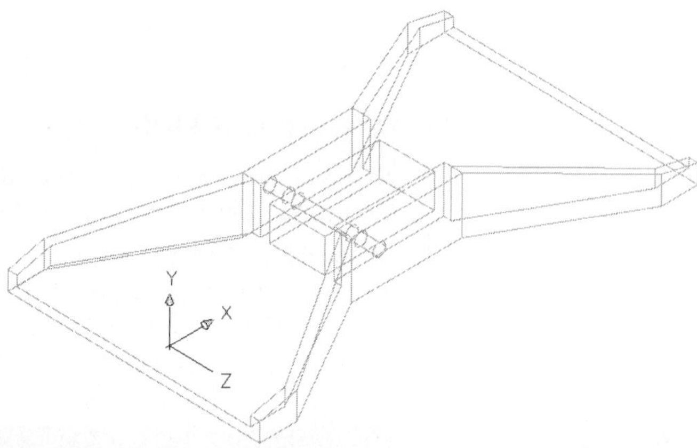

Figure 4.47

On your own, add another hole to the part on the opposite side of the rectangular base using the same settings and dimensions.

Your part should look like Figure 4.48.

On your own, change the color of the hole features to red.

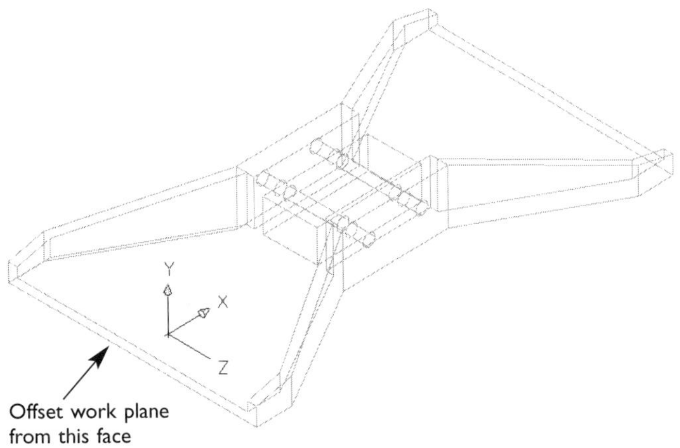

Figure 4.48

Adding a Work Plane for Assembly

When you assemble the parts for the jack, you will find it useful to have another work plane through the center of the part perpendicular to the one you created previously. As the part is 12 inches long overall, you will offset the plane 6 inches from one end.

To add this work plane,

Pick: **Sketched & Work Features, Work Plane**

Select 1st modifier: **Planar Parallel** (so the button is darkened)

Select 2nd Modifier: **Offset**

Set Offset: **6**

Select: **Create Sketch Plane** (so it is checked)

Pick: **OK**

Select work plane, planar face, or [worldXy/worldYz/worldZx/Ucs]: **pick the small vertical plane on the left end of the part**

Enter an option [Next/Accept] <Accept>: **if the correct plane is not highlighted, press left mouse button to continue to the next plane; if the correct plane is highlighted, right-click to accept**

The arrows representing the direction the work plane will be offset should point toward the part.

Enter an option [Flip/Accept] <Accept>: **F** ↵

Enter an option [Flip/Accept] <Accept>: ↵ **to accept when the arrows are pointed correctly toward the center of the part**

Plane=Parametric

Select edge to align X axis or [Z-flip/Rotate] <Accept>: ↵

Toggle to shaded mode

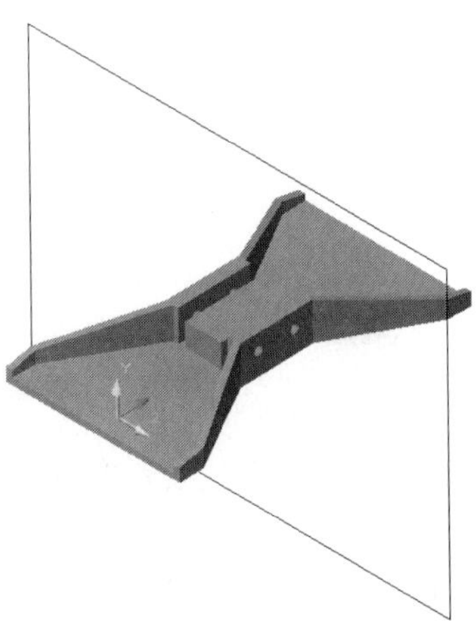

Figure 4.49

You should now see the work plane added to the part, as shown in Figure 4.49.

Use the 3-D orbiter to rotate and inspect your part.

Save your work and exit.

You have now completed the Baseplate part.

Command Summary

Action	Command Name	Description
Add Constraints	AMADDCON	Adds constraints to a 2D sketch
Adjust Constraint Size	AMOPTIONS	Manages part, assembly, surface, scene, drawing, annotation, and desktop preferences; can set size of constraint display
Construction Line	AMCONSTRLINES	Adds a construction line to the sketch that can be used for dimensioning but does not add to the 3D geometry
Create Work Plane	AMWORKPLN	Creates a work plane at a specified location
Delete Constraint	AMDELCON	Removes constraints from a 2D sketch
Erase	ERASE	Removes objects from the sketch or drawing
Extend	EXTEND	Extends the length of existing 2D lines, arcs, and polylines to a selected boundary edge
Extrude, Cut	AMEXTRUDE, C	Extrudes a feature and removes it from the model (equivalent to Boolean subtraction)
Extrude, Intersection	AMEXTRUDE, I	Extrudes a feature and retains the portions of the feature and model that intersect (equivalent to Boolean intersection)
Fillet	FILLET	Rounds or fillets the edges of 2D objects; can be used to trim lines when radius is set to 0
Grid	GRID	Controls a grid of dots on screen that can be used as a drawing aid
Hole	AMHOLE	Creates parametric hole features; opens the Hole dialog box
Hot Grips		A selection method associated with editing commands for quickly modifying existing geometry; must be enabled in the Options dialog box
Line	LINE	Draws line segments
Make a Profile	AMPROFILE	Solves 2D geometry to create a constrained sketch
Mechanical Options	AMOPTIONS	Manages settings, standards, and preferences to be used in parts, assemblies, and drawings
Mirror	MIRROR	Creates a mirror-image copy of 2D objects
New Dimension	AMPARDIM	Creates parametric dimensions to constrain sketches and define features in Model mode
Object Snap	OSNAP	Sets object snap modes
Offset	OFFSET	Creates a new 2D object (parallel line, curve, or conventric circle) at a specified distance from an existing object or through a specified point
Project Objects to Plane	AMPROJECT2PLN	Projects model geometry onto the sketch plane
Rectangle	RECTANGLE	Creates a rectangular polyline 2D object
Redo	REDO	Reverses the last Undo
Re-solve Sketch	AMRSOLVESK	Displays the number of constraints needed to fully constrain a sketch, or appends new geometry to a sketch
Set Fix Point	AMADDCON, Fix	Option of this command; sets the fixed point on a sketch

Action	Command Name	Description
Show Constraints	AMSHOWCON	Displays 2D constraint symbols on the selected sketch
Single Profile	AMPROFILE	Solves a single 2D entity to create a constrained sketch
Snap	SNAP	Controls the increment to which the cursor will jump when sketching commands are active
Trim	TRIM	Trims objects at a cutting edge defined by other objects
Undo	UNDO	Reverses the effect of the last command (or set of commands)
Visibility	AMVISIBLE	Controls the visibility of parts, features, symbols, and other entities in parts, assemblies, scenes, and drawings

Exercises

Model the parts shown in Figures 4.50 through 4.54. Additional parts for modeling may be found in the assembly exercises at the end of Tutorials 7, 9, and 10.

1. **Z-Shape** *Note*: All dimensions are in millimeters.

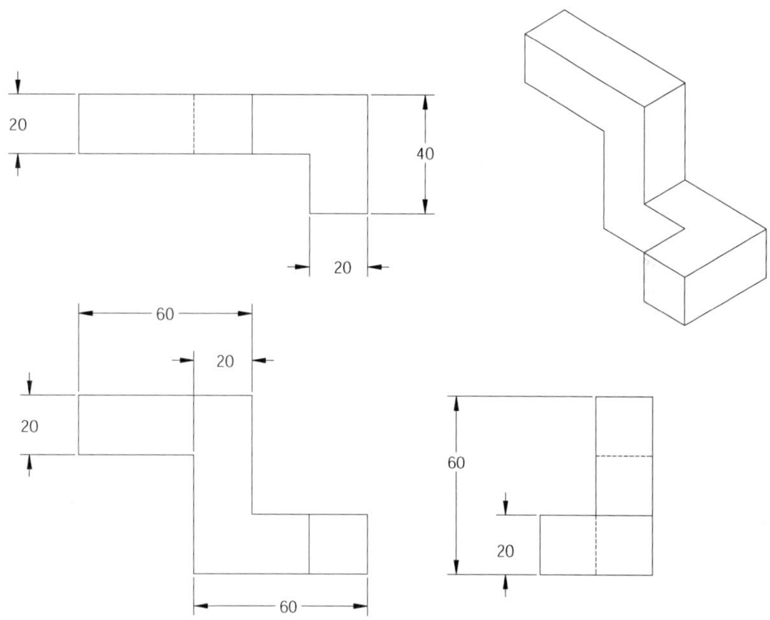

Figure 4.50

2. **Mechanical Desktop Initials** *Note*: All dimensions are in millimeters. As an additional exercise, design your own initials.

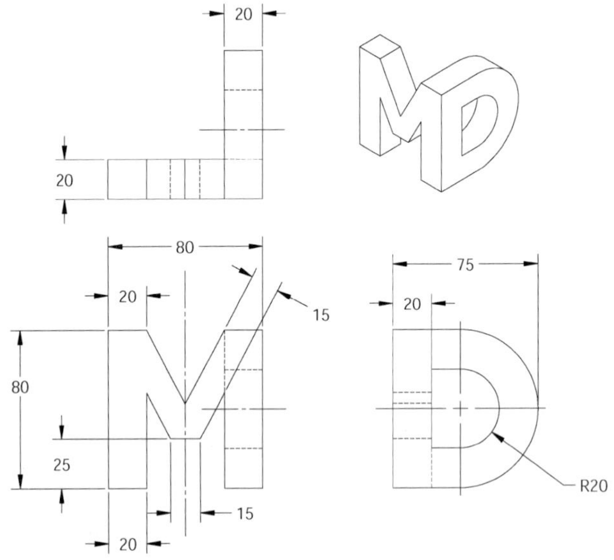

Figure 4.51

3. **Mounting Plate** *Note*: All dimensions are in inches.

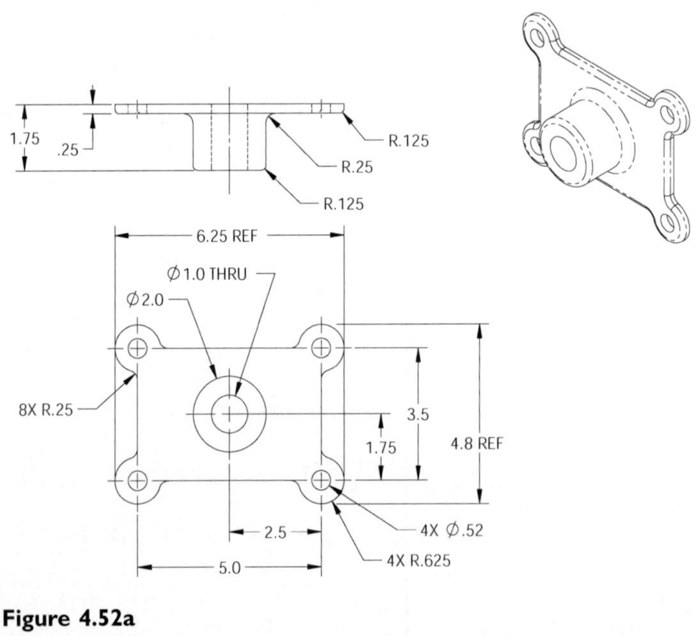

Figure 4.52a

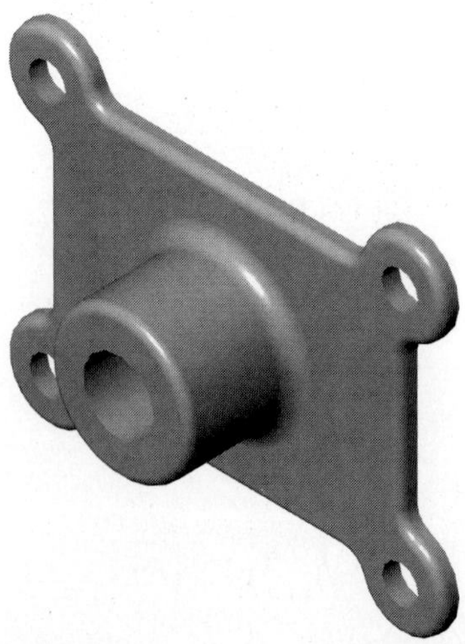

Figure 4.52b

4. **Tweezers** *Note*: All dimensions are in millimeters. After modeling the part, add 1-mm fillets and rounds to all edges.

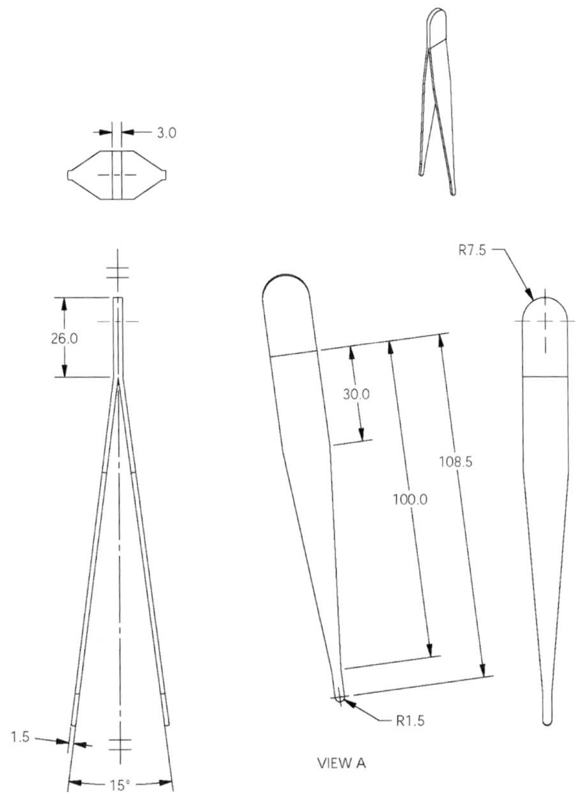

Figure 4.53

5. **Wall Bracket** *Note*: All dimensions are in inches.

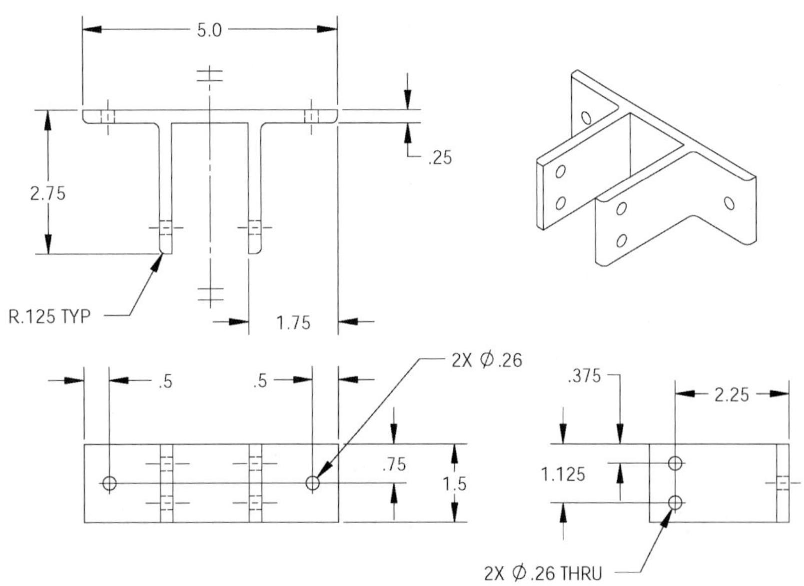

Figure 4.54

Tutorial

5 Modifying Parts

Objectives

When you have finished this tutorial, you will be able to

- Examine a model to understand parent–child relationships.
- Replay a model to see how it was made.
- Suppress and unsuppress features.
- Edit features and profiles.
- Reorder features in the model tree.
- Delete a feature.
- Create a pattern feature.

Introduction

The design of a part or an assembly is not a static process. It is an iterative one. As you gather more information and test the design ideas you are using, you will need to make changes to the models you have created. Parametric modeling provides relative ease of updating the models compared with more traditional 3D modeling processes. However, the ease with which a model can be updated for a particular purpose is often dependent on the choices that were made in the constraints, dimensions, and sketch planes, and the order in which the entities in the model were created. Often when you are making changes to a model in a work situation, it will not be a model that you created, but rather one that was created some time ago, perhaps even by an employee who is no longer at the same company. Understanding how to find information about the model and how to change it can often save you the effort of starting over from scratch. A good understanding of how models update also can help you make better choices for the dependencies among features that you create in your models.

In this tutorial you will finish creating the top part for the jack. You will start from a file that needs some additions and revisions in order to work correctly. You will investigate the part that you are provided and make changes to it. Start by opening the datafile and saving a copy to your MDT-Work\Jack directory.

Launch Mechanical Desktop 5

Pick: **File, Open**

Choose Look in: **MDT-Data directory**

Double-click: **JackTop-dta.dwg**

Pick: **File, Save As**

Choose Save In: **MDT-Work\Jack directory**

Type File name: **JackTop.dwg**

Pick: **Save**

Load your user profile using the Options dialog box (if necessary).

The top part for the jack should appear on your screen, as shown in Figure 5.1. You should see the file name *C:\MDT-Work\Jack\JackTop.dwg* at the top of your screen. Be sure you are working with a copy of the file before you proceed.

Parent–Child Relationships

Parametric models use the term *parent–child relationships* to describe how one feature relates to another. The parent feature is created first. A child feature depends on the parent feature for its location in the model. These

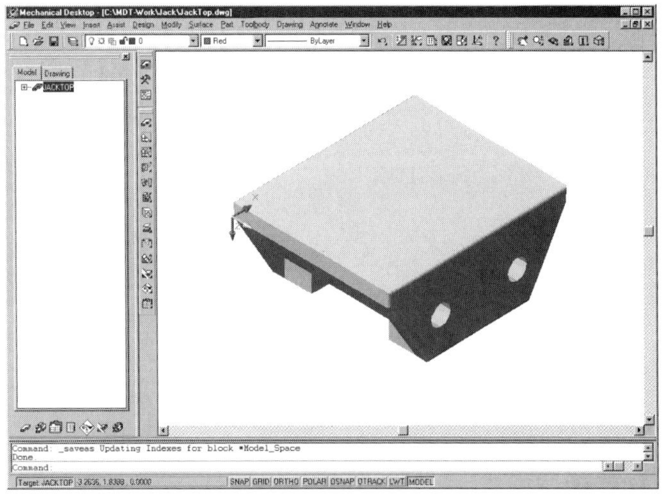

Figure 5.1

Reference

Engineering Design Communication

Parent–Child Relationships, pp. 262–263

dependencies can determine how the model updates when you make a change.

The Base Feature

The base feature is essentially the parent feature for all of the following features. Every other feature is directly or indirectly related to the base feature. As you have seen before, the model tree shows pictorially how the features are related.

Pick: **to expand the branches of the model tree**

Pick: **to expand the list below Base-block**

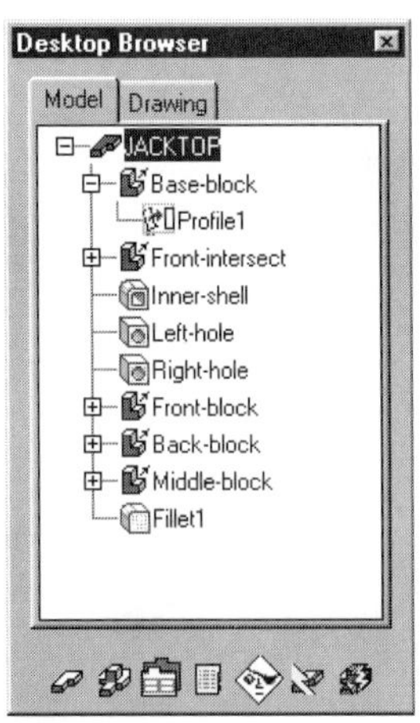

Figure 5.2

The Model Tree

Figure 5.2 shows the expanded model tree for the JackTop part. The model tree shows both dependencies between elements (in the way that the profile for a sketched feature is nested below it) and the order in which the features were created. As Mechanical Desktop regenerates the model to make changes, it follows the order shown in the model tree.

Investigating a Model

One way to investigate the model is to pick on the features in the model tree and see them highlighted on the part.

Pick: **Fillet1** (in the model tree)

It becomes highlighted on the part, as shown in Figure 5.3.

Renaming features

The features have been given names to make them easier to recognize. You should develop the habit of renaming features so that you can recognize them later. This makes it easier for someone else to work on the model. You can rename features at anytime. You will rename Fillet1 to Face-fillet.

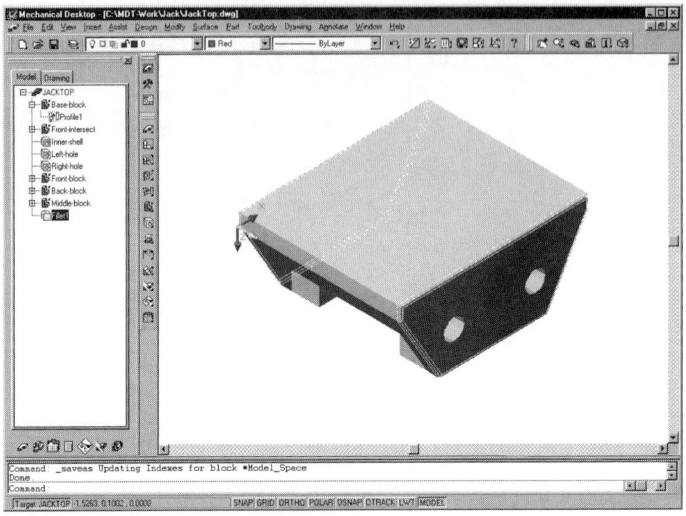

Figure 5.3

Right-click: **Fillet1** (in model tree)

Pick: **Rename** (highlight the name)

Type: **Face-fillet** (it will replace the highlighted text)

Press: ↵

The new name Face-fillet should appear in place of the previous name.

Replaying the Model

It is often helpful to get the overall picture of how the model was created by replaying the steps used to create it. Keep in mind that the model overall is the sum of the steps in its creation. The steps must always be able to be logically interpreted by the software in order to build the model shown on your screen.

Right-click: in the graphics area

Pick: **Part, Replay**

The model starts to replay from its beginning, as shown in Figure 5.4, and you see the message, Sketch plane. The selection of the sketch plane was the first choice made in this model. The prompt at this point lets you

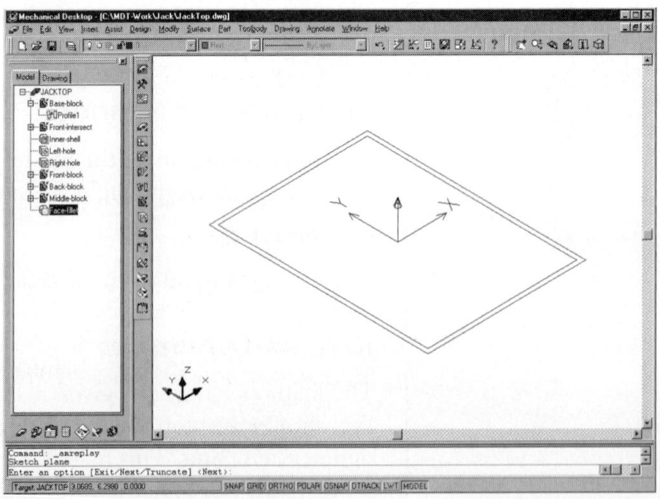

Figure 5.4

exit, go to the next step in the model, or truncate, which is to delete all the features after the current step. At each step, Mechanical Desktop reports the operation that was taken to get to that point.

Enter an option [Exit/Next/Truncate] <Next>: ↵

Closed profile

Enter an option [Display/Exit/Next/Size/Truncate] <Next>: **D** ↵

The constraints for the sketch appear on the screen.

Enter an option [Display/Exit/Next/Size/Truncate] <Next>: **S** ↵

The Constraint Display Size dialog box appears, as shown in Figure 5.5 Use the slider bar on your own to resize the constraint symbols so they are easily visible

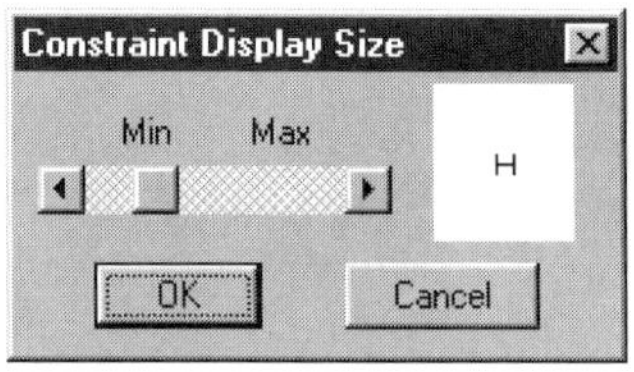

Figure 5.5

Enter an option [Display/Exit/Next/Size/Truncate] <Next>: ↵

Extrusion Blind

Enter an option [Exit/Next/Suppress/Truncate] <Next>: **S** ↵

Suppressing Features

The part should appear just as the base block that was created first. All of the other features beyond that are now suppressed. This means that, although they still exist in the database, they are not calculated for the appearance of the model at this time. You can suppress and unsuppress features as needed. When features are suppressed they appear with as dashed line in the model tree. Your model tree should look like Figure 5.6.

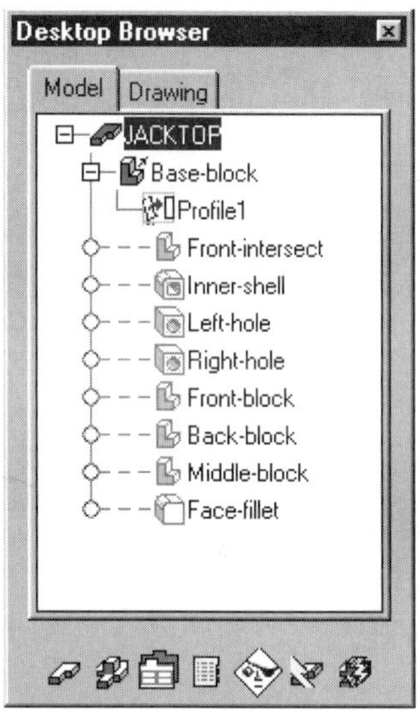

Figure 5.6

Next you will use the context menu and the model tree to unsuppress all of the features.

Right-click: **JackTop part** (in the model tree)

Pick: **Feature Suppression, Unsuppress All**

The full model is restored to your screen.

Replay the model again. This time you will continue to press ↵ to see the next step until you have completely played through the model. Notice the orientation of the sketch planes indicated by the UCS icon as you replay the model. You can use the 3D orbiter to spin the model in between steps to help you see features better. You may also want to switch to a wireframe view to see the extrusions under the part more clearly.

Right-click: in the graphics area

Pick: **Part, Replay**

Enter an option [Exit/Next/Truncate] <Next>: **press** ↵ (until model is fully replayed)

By now you should have the basic idea of how this part was created. If not, replay it again until you feel you do. Can you see some things that you would have done differently? In the next sections you will learn more about how to make changes.

Redefining the Model: Editing a Feature

As you have seen previously, you can easily select a feature to edit using the model tree. Editing a feature returns you to the dialog box that you used to create it. You can change any of the selections in the dialog box to modify the feature.

You will change the Middle-block feature so that it is extruded a distance of .9 instead of .75. First change the view so that it will show the result more clearly.

Pick: **View, 3D Views, Right**

The right-side view of the part appears on your screen, as shown in Figure 5.7.

Figure 5.7

Right-click: **Middle-block** (in the model tree)

Pick: **Edit**

The Extrusion dialog box appears on your screen with the selections that were used to create the Base-block feature. You will change the distance to .9, as shown in Figure 5.8.

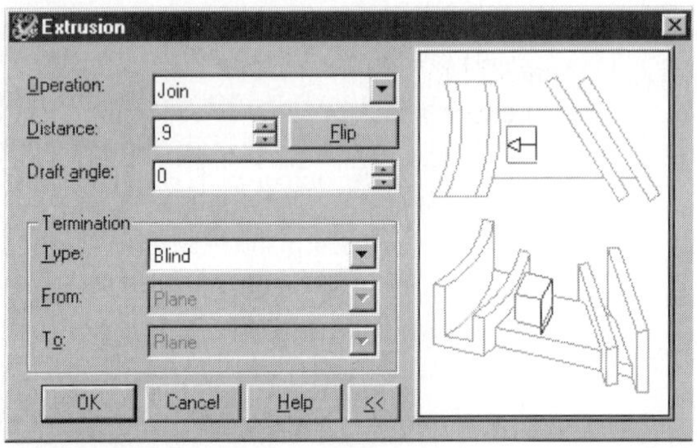

Figure 5.8

Select Distance: **.9**

Pick: **OK**

Select object: ↵

Updating a Part

As you have seen before, the part does not change immediately and the feature in the model tree is highlighted with a yellow band. Because the model is built up consecutively from the features to create the display on your screen, the change must be applied and the part must go through the steps to update the model's appearance. If you try to save the model now, you will get a message saying that you cannot until you complete the update.

Right-click: in the graphics area

Pick: **Update Part**

The model changes so that the extruded height is now .9, as shown in Figure 5.9.

Figure 5.9

Next change back to an isometric view,

Pick: **View, 3D Views, Front Left Isometric**

Redefining the Model: Editing a Profile

The profile stores the basic 2D shape that was used to create the feature. Next you will make changes to the profile for the Front-intersect feature. First you will make sure that the model tree is expanded so that the profile appears.

Pick: **the + (plus-sign) to the left of Front-intersect** (in the model tree)

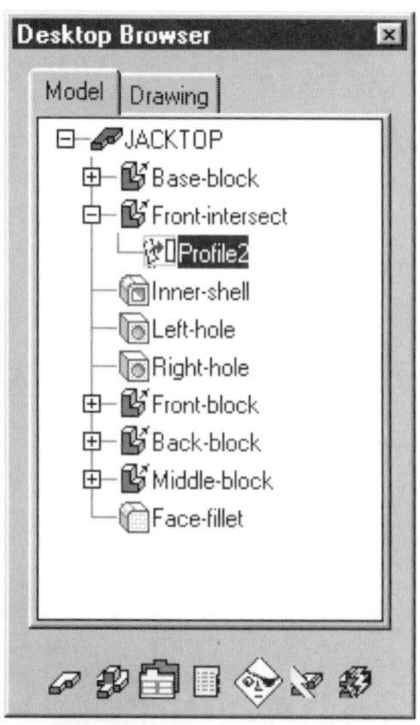

Figure 5.10

Figure 5.10 shows the model tree expanded so that you can see the profile listed.

To edit the profile,

Right-click: **Profile2** (using the model tree)

Pick: **Edit Sketch** (from the context menu)

Pick: **Sketch View icon**

Pick: **Zoom All icon**

Your view changes so that you are looking directly perpendicular to the sketch. The sketch lines and dimensions appear on your screen, as shown in Figure 5.11.

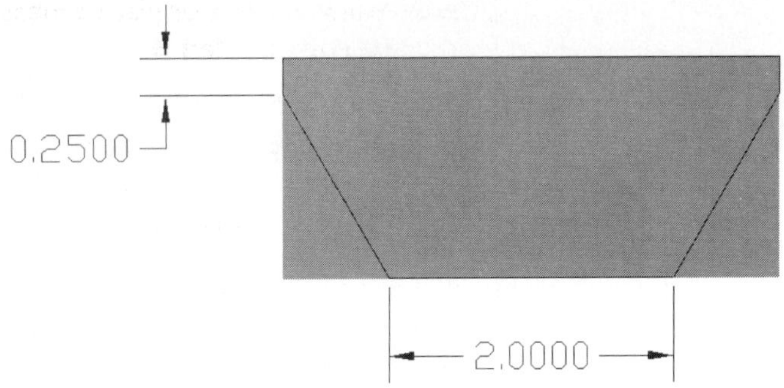

Figure 5.11

Pick: **Toggle Shading/Wireframe icon**

Next you will show the constraints that were used in the sketch.

Right-click: in the graphics area

Pick: **2D Constraints, Show Constraints**

The constraints appear on your screen, as shown in Figure 5.12.

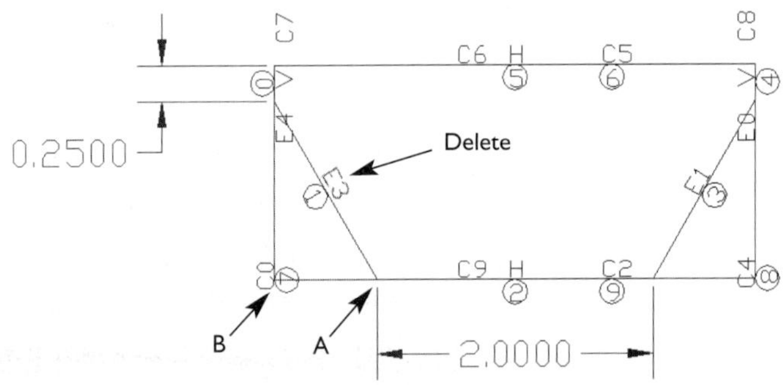

Figure 5.12

Enter an option [All/Select/Next/eXit] <eXit>: ↵

You will delete the constraint that makes angled line segment 3 equal to the other angled line, segment 1.

Right-click: in the graphics area

Pick: **2D Constraints, Delete Constraint**

Select or [Size/All]: **pick the E3 constraint shown in Figure 5.12**

Select or [Size/All]: ↵

The constraint symbol is removed.

Next you will re-solve the profile to see the effect,

Right-click: in the graphics area

Pick: **Re-Solve Sketch**

Solved under constrained sketch requiring 1 dimensions or constraints.

You will add a dimension to fully constrain the sketch.

Right-click: in the graphics area

Pick: **Dimensioning, New Dimension**

Select first object: **pick the lower left corner of the sketch labeled A in Figure 5.12**

Select second object or place dimension: **pick the lower left corner of the part labeled B**

Specify dimension placement: **pick in line with the 2.0000 dimension**

Enter dimension value or [Undo/Hor/Ver/Align/Par/aNgle/Ord/Diameter/pLace] <0.7500>: **.5** ↵

Solved fully constrained sketch

Select first object: ↵

Figure 5.13 shows how the sketch shifts to meet the new dimension requirements.

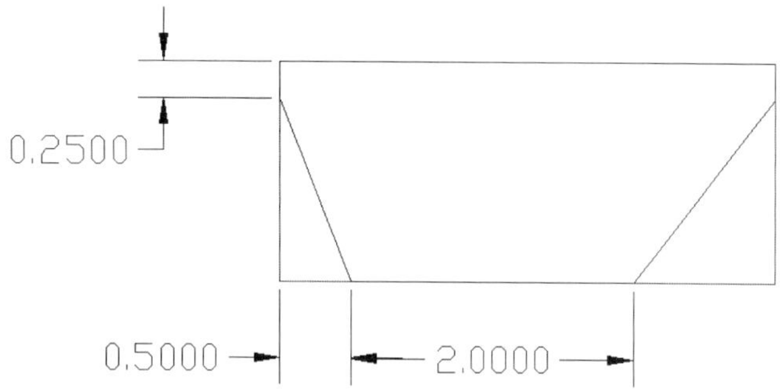

Figure 5.13

Next you will update the part to see the effect on the model.

Right-click: in the graphics area

Pick: **Update Part**

The part is updated to show the change.

On your own, switch to a left front isometric view.

Shade and spin the part.

Using Undo with Parts

You can use Undo with any command to reverse the last action taken, and you can also use it to undo changes you have made to the part. The Undo icon on the toolbar is a quick way to reverse an action. Clicking it repeatedly takes you back through the actions as executed. This is not something that you can do in all parametric modelers!

When you use Undo, Mechanical Desktop reports at the command line the command that has been undone—or reports GROUP for an action that is actually made up of several commands. Depending on the options selected, you may need to pick the icon more than once until you see the report that the command has been reversed.

Pick: **Undo icon** (until you see Group in the command line area and return to the sketch)

The changes to the view are undone; then the update to the model is undone. Finally, you return to the sketch.

Pick: **Undo icon** (until you see AMPARDIM, the Parametric Dimension command)

The dimension added to the sketch is undone.

Pick: **Undo icon** (until you see AMDELCON, the Delete Constraint command)

The deleted constraint is restored.

Now you must update the part again for the changes to take effect.

Right-click: in the graphics area

Pick: **Update Part**

The part is returned to its original appearance.

Reordering Features

Take a look at the two holes in the part. They do not go through the center block. On your own,

Spin the part to view the holes

Return to a front left isometric view

For this part to be attached to the jack, the holes must go all the way through. Let's find out how the holes were created.

Right-click: **on Left-hole** (in the model tree)

Pick: **Edit** (from the context menu)

The Hole dialog box shown in Figure 5.14 appears on the screen. This tells you that this hole was created as a placed hole feature and not as an extruded cut. You can also see that the termination for the hole is Through. Why doesn't this hole go entirely through the part?

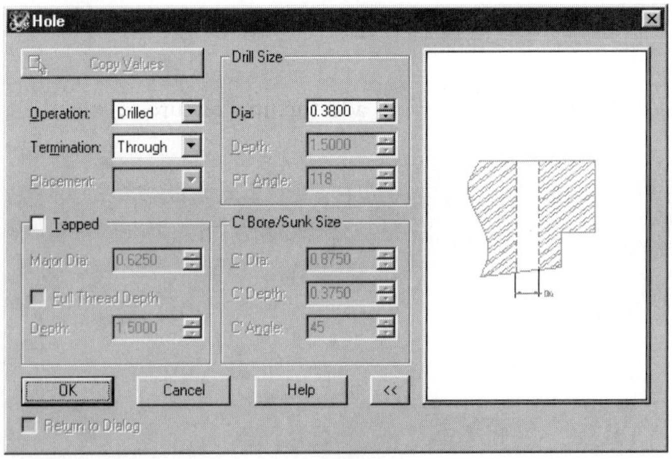

Figure 5.14

Pick: **Cancel** (to exit the dialog box without making changes)

The reason the hole doesn't go through is that the holes were added before the Middle-block and Front and Back blocks were created. To fix this, the holes need to be reordered so that they come after the Middle-block. You can reorder features by dragging them to a new location in the model tree.

You will keep your left-mouse button held down and drag the Left-hole feature so that it appears below Middle-block in the model tree. Then do the same for Right-hole.

Pick and drag: **Left-hole** (using the model tree) **to insert it after Middle-block**

Pick and drag: **Right-hole** (using the model tree) **to insert it after Left-hole**

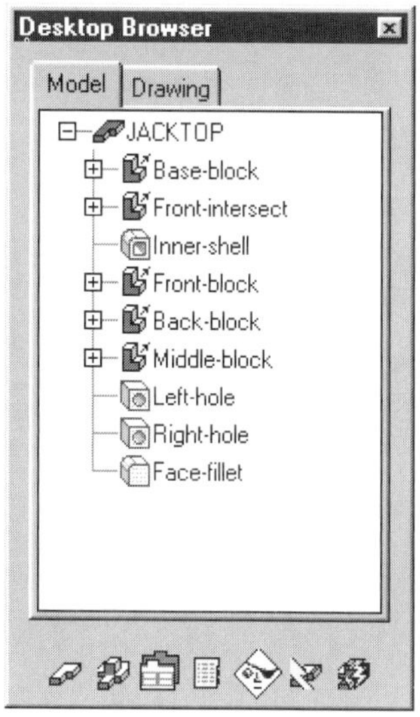

Figure 5.15

The model tree now appears as shown in Figure 5.15.

On your own,

Spin the part so that you can see the holes going through the part, as in Figure 5.16.

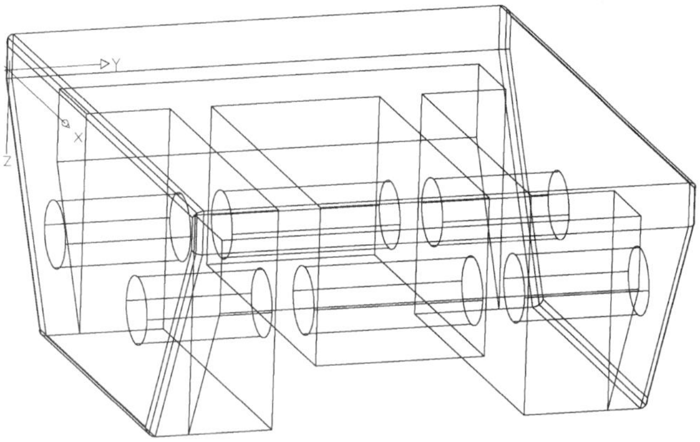

Figure 5.16

Return to a shaded front left isometric view

Adding a New Feature

Now that you have resolved the problem with the part, you are ready to add the final feature to create the pattern on the top of the part. You will follow the basic steps for feature creation.

Right-click: in the graphics area

Pick: **New Sketch Plane**

Select work plane, planar face, or [worldXy/worldYz/worldZx/Ucs]: **pick the front surface of the part labeled A in Figure 5.17.**

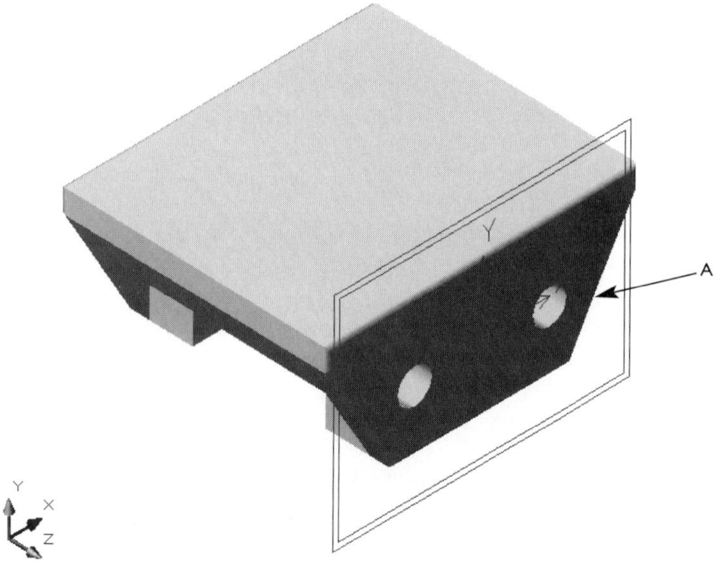

Figure 5.17

Enter an option [Next/Accept] <Accept>: **right-click if the correct face is highlighted**

Select edge to align X axis or [Z-flip/Rotate] <Accept>: **right-click to accept if the coordinates are aligned, as shown in Figure 5.17**

Now you are ready to begin sketching the new feature on this sketch plane.

Pick: **Sketch View icon**

You are now viewing the object directly perpendicular to the sketch plane.

Right-click: in the graphics area

Pick: **2D Sketching, Circle**

Specify center point for circle or [3P/2P/Ttr (tan tan radius)]: **pick just above the top line and locate the center of a circle similar to Figure 5.18.**

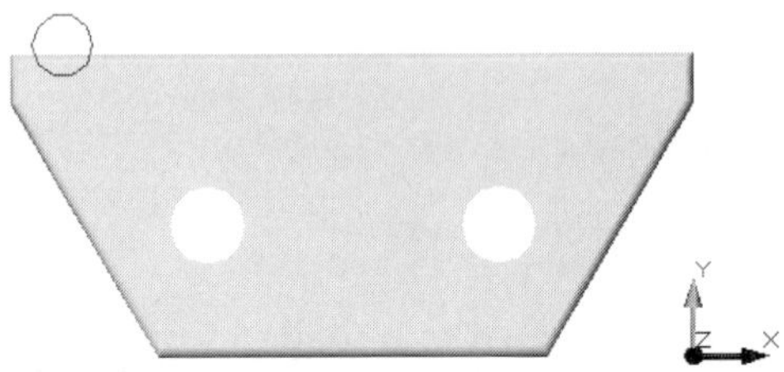

Figure 5.18

Specify radius of circle or [Diameter]: **.20** ↵

Right-click: in the graphics area

Pick: **Sketch Solving, Single Profile**

Solved under constrained sketch requiring 3 dimensions or constraints.

Right-click: in the graphics area

Pick: **Dimensioning, New Dimension**

Select first object: **pick on the circle**

Select second object or place dimension: **pick above and to the right of the circle**

Enter dimension value or [Undo/Radius/Ordinate/Placement point] <0.2000>: **R** ↵ (to enter a radial dimension)

Enter dimension value or [Undo/Diameter/Ordinate/Placement point] <0.2000>: **.20** ↵

Solved under constrained sketch requiring 2 dimensions or constraints.

Because the circle is simply a step toward an arc-shaped cut out of the top of the part, you entered its dimension as a radial dimension. Next add the dimension for the distance from the circle's center to the top of the part. (Mechanical Desktop will automatically dimension to the center of the circle when it is selected.)

Select first object: **pick on the circle**

Select second object or place dimension: **pick on the top line of the part**

Specify dimension placement: **pick to the right of the part**

Enter dimension value or [Undo/Hor/Ver/Align/Par/aNgle/Ord/Diameter/pLace] <0.08???>: **.1** ↵

Solved under constrained sketch requiring 1 dimensions or constraints.

Now add the dimension for the distance from the left side to the center of the circle.

Select first object: **pick the circle**

Select second object or place dimension: **pick on the left vertical edge**

Specify dimension placement: **pick above the part**

Enter dimension value or
[Undo/Hor/Ver/Align/Par/aNgle/Ord/Diameter/pLace] <0.0163>: **.25** ↵

Solved fully constrained sketch.

Select first object: ↵

The dimensions should look similar to those shown in Figure 5.19.

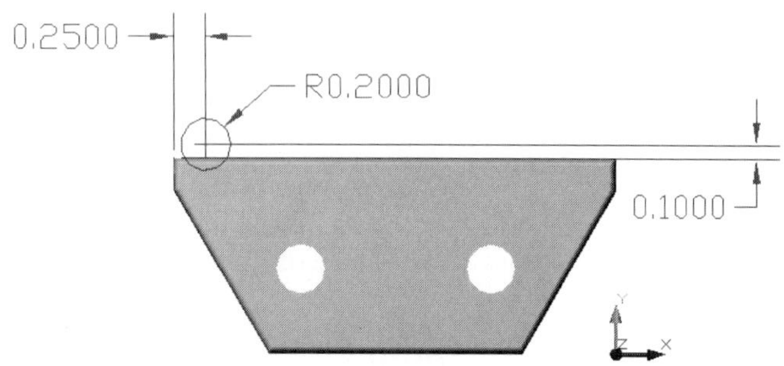

Figure 5.19

Right-click: in the graphics area

Pick: **Sketched & Work Features, Extrude**

The Extrude dialog box appears. Make the following selections.

Choose Operation: **Cut**

Choose Draft Angle: **0**

Choose Termination Type: **Through**

Pick: **OK**

The circular cut feature is completed and appears, as shown in Figure 5.20.

Figure 5.20

Creating a Pattern Feature

Next you will create a pattern of these cuts across the top of the part. One of the Placed Feature options makes a series of instances of a feature in a regular pattern, or array. Once created, the pattern feature is a single item even though it starts from an existing feature. Patterns can only be edited as a single array, not a series of individual items.

Arrays can be rectangular (arranged in a grid of columns and rows) or circular (arranged about an axis). Circular arrays can be polar (arranged in a circular pattern) or axial (in a helical pattern). As a feature is copied to make the array, the pattern can also be rotated to a particular angle. For the JackTop, you will create a rectangular patterned feature.

Reference

Mechanical Desktop On-line Help

Read more about different arrayed features by looking up Pattern Features in the Mechanical Desktop Help Index.

Switch back to the Front Left Isometric view

Right-click: in the graphics area

Pick: **Placed Features, Rectangular pattern**

Select features to pattern: **pick on the circular cut you just created**

Select features to pattern or [list/Remove]: ↵

The Pattern dialog box appears on your screen with the Pattern Control tab uppermost. Move the dialog box so you can see the dynamic preview of your selections on the part. To create the pattern, you will define the number of instances (copies) of the feature in each row or column, the distance between them, and the angle at which the pattern should be rotated. As you choose them, you will see them previewed on your part before you exit the dialog box. You will make the following selections so your dialog box looks like Figure 5.21. (The row spacing is not relevant, as you have only one row.)

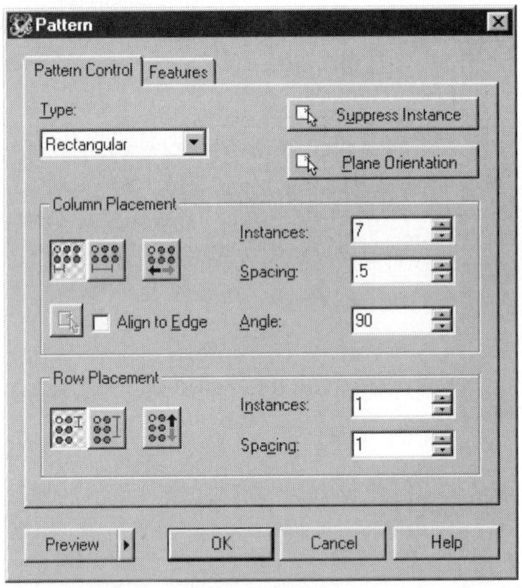

Figure 5.21

Choose Pattern Control Type: **Rectangular**

Set Column Instances: **7**

Set Column Spacing: **.5**

Set Column Angle: **90**

Set Row Instances: **1**

Set Row Spacing: **1**

Pick: **OK**

The arrayed copies of the cut feature are added to the model, as shown in Figure 5.22. Note that a single pattern feature was added to the model tree.

Figure 5.22

Save your part at this time

Using Feature Suppression

Next you will try to add a fillet to the top edges of the part.

Right-click: in the graphics area

Pick: **Placed Features, Fillet**

Use the dialog box to make the following selections.

Type: **Constant**

Radius: **0.125**

Pick: **OK**

You return to the model to select the edges to fillet. You will select just the long cut edges, not the faces of the features or the circular cuts.

Select edges or faces to fillet <continue>: **Pick on the upper sharp edges of the arrayed cut features, as shown in Figure 5.23**

Figure 5.23

Select edges of faces to fillet <continue>: ↵

Failed to compute Fillet.

Error: Blends too complex to reorder. Error: Failed to update model.

The software was unable to calculate the blend for the fillet due to the complex blends where the rounded corners meet other rounded corners. To solve this problem, you will suppress the face fillet and then add the fillet to the tops of the cuts.

Right-click: **Face-fillet** (in the model tree)

Pick: **Suppress** (from the context menu)

The Face-fillet feature should become highlighted on the part.

The highlighted features will be suppressed.

Continue? [Yes/No] <Yes>: ↵

The face fillet feature is suppressed so that it does not display, as shown in Figure 5.24.

Figure 5.24

Now you will try adding the fillets again.

Right-click: in the graphics area

Pick: **Placed Features, Fillet**

The dialog box should already contain the following selections. If it does not, make it so.

Type: **Constant**

Radius: **0.125**

Pick: **OK**

Select edges or faces to fillet <continue>: **Pick on the upper long sharp edges of the cut features as you did before**

Select edges or faces to fillet <continue>: ↵ (when all are selected).

The fillets are added to the long cut edges, as shown in Figure 5.25.

Next you will unsuppress the Face-fillet.

Right-click: **Face-fillet** (in the model tree)

Pick: **Unsuppress** (from the context menu)

The fillet is restored. But the cut edges through the filleted edge are not filleted. You will delete the Face-fillet and create a new fillet that includes the new edges.

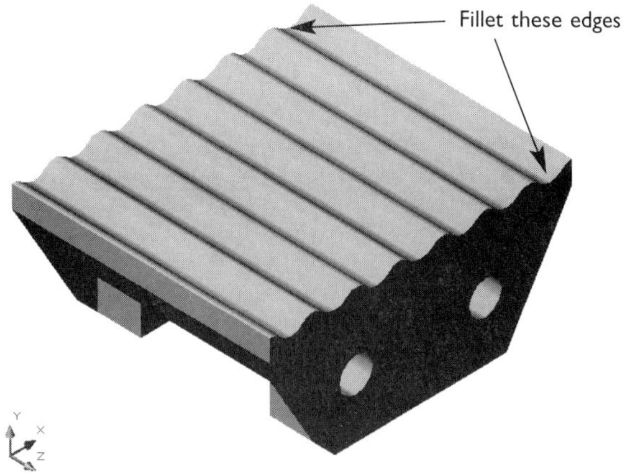

Figure 5.25

Right-click: **Face-fillet** (in the model tree)

Pick: **Delete**(from the context menu)

Highlighted features will be deleted. Continue? [Yes/No] <Yes>: [↵]

The Face-fillet is deleted from both sides of the part. Next you will create the new fillet for the same edges as were filleted before. Do not select the entire face of the part, as the hole edges should not be filleted. You may need to use the Next option to cycle through the edges to select the ones you want. You will definitely need to use the 3D Orbiter to spin the model as you select edges so you can select those on both sides of the part to be included in a single operation.

Tip

If you make a mistake while selecting edges, type U (for undo) to delete the last edge selected from the selection set. And, at any time, you can press [ESC] *to cancel the command and start again.*

Right-click: in the graphics area

Pick: **Placed Features, Fillet** (from the context menu)

Use the dialog box to make the following selections.

Type: **Constant**

Radius: **.06**

Pick: **OK**

Select edges or faces to fillet <continue>: **pick the top edges of the front and back faces shown in Figure 5.25 where the cuts meet the front surface on both sides of the part.**

Select edges or faces to fillet <continue>: [↵]

The fillet is added to the part, as shown in Figure 5.26.

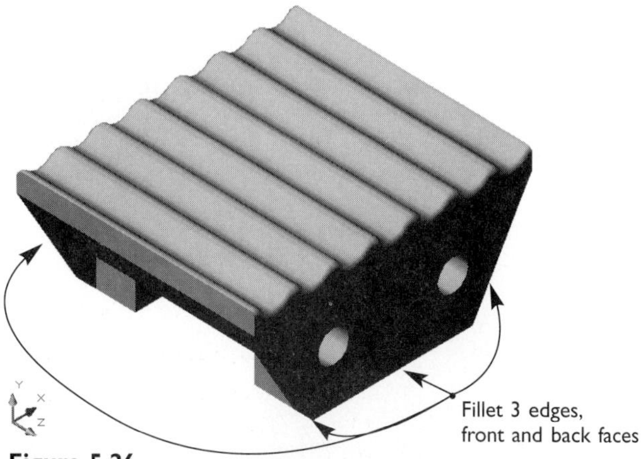

Figure 5.26

Next you will add the fillet for the edges of the front and back surfaces. The edges to select are identified in Figure 5.26.

On your own, add a .06 fillet to the 6 edges of the front and back faces.

The result is shown in Figure 5.27.

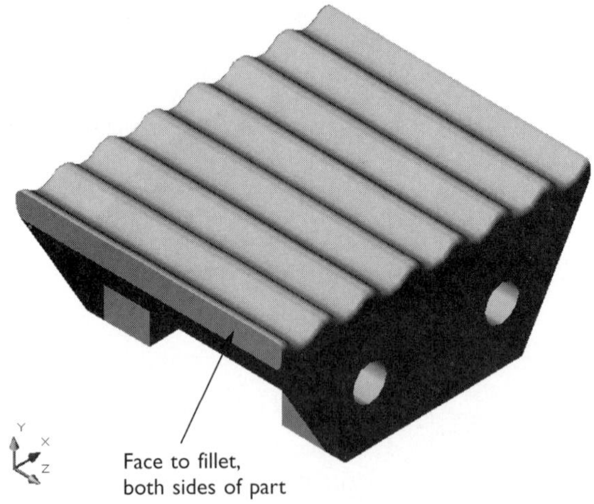

Figure 5.27

The last fillet will be added to the two side faces.

Add a .06 fillet to the face labeled in Figure 5.27 and the corresponding face on the opposite side of the part.

Your part should now look like Figure 5.28.

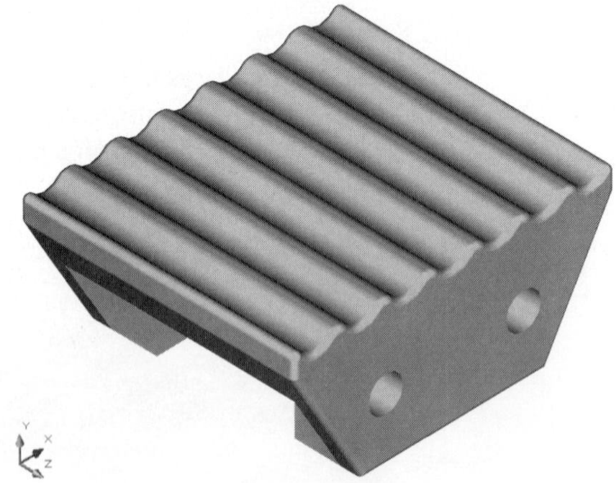

Figure 5.28

Preparing for Assembly

One final change to the JackTop part is one you will make to prepare the part to be assembled. You want the JackTop to align with the Baseplate along their centers, so you will add a work plane to use to align the part with the Baseplate's work plane.

Right-click: in the graphics area

Pick: **Sketched & Work Features, Work Plane**

The Work Plane dialog box appears.

Select: **Planar Parallel**

Select: **Offset**

Deselect: **Create sketch plane** (so it is unchecked)

Type Offset: **1.75**

Pick: **OK**

Select work plane, planar face, or [worldXy/worldYz/worldZx]: **pick the side face** (the last face you filleted)

Enter an option [Next/Accept]<Accept>: ↵

Enter an option [Flip/Accept] <Accept>: **F** ↵ (so the arrows point toward the part)

The work plane appears, as shown in Figure 5.29.

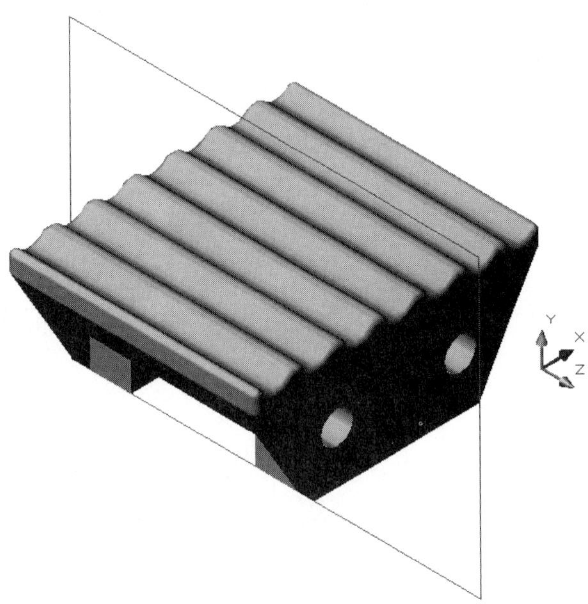

Figure 5.29

Save the part

Now you know several ways that you can use to investigate and modify a model. When you are modeling, plan ahead to create models that reflect your design intent. Think about what dimensions, features, and relationships you may later change. Try to model the part accordingly.

Command Summary

Action	Command Name	Description
3D Fillet	AMFILLET	Creates a fillet or round (placed feature) between faces of a 3D part
Draw Circle	CIRCLE	Creates 2D circle geometry
Edit Feature	AMEDITFEAT	Edits a feature by changing the operation and settings used to create it
Pattern Feature	AMPATTERN	Creates a rectangular, polar, or axial pattern of features from an existing feature
Rename (in Browser)	AMRENAME	Renames an element in the Browser window
Reorder Feature	AMREORDFEAT	Changes the order of a feature in the model tree hierarchy
Replay Part	AMREPLAY	Replays the steps used to create the features of a part
Suppress Features	AMSUPPRESSFEAT	Suppresses a feature or specifies a group of features by type to suppress
Undo	UNDO	Reverses the effect of the last command (or set of commands)
Update	AMUPDATE	Recalculates the model to show the effect of changes made to features, constraints, or dimensions on parts, assemblies, scenes, or drawing views

Exercises

1. **Reordering a Part** Open part *Reorder-1.dwg* from the MDT-Data\Exercises directory and save it to your MDT-Work directory. Make changes to the part and answer the following questions.
   (a) What happens when the Center Hole feature is changed to cut only a distance of 10?
   (b) What happens when the ends of the Base sketch are changed from a radius of 1 to 1.5?
   (c) What happens when the Join distance of the 1st Legs are changed from 10 to 20?
   (d) What happens when the 2nd Legs operation is changed from a join to a cut?
2. **Wall Bracket Problem** Open the part *WallBracket.dwg* from the MDT-Data\Exercises directory and save it to your MDT-Work directory. Using the tools you have learned in this chapter, edit the Wall Bracket part shown in Figure 5.30 so it has the following features:
   - The four mounting holes must go completely through the part.
   - The spacing of the second shaft mount must be defined so it is symmetrical about the bracket.
   - The shaft hole must go completely through both shaft mounts.
   - A .125" fillet around the outer edge of the part.

   List the steps you used to correct the part and write down the order in which the features are placed on the feature tree, or e-mail the corrected file to your instructor.

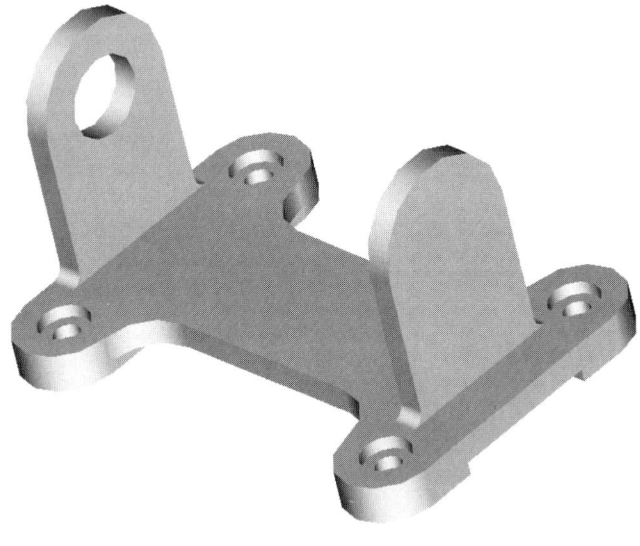

Figure 5.30

3. **Jack Modification** Open the part *JackProblem.dwg* from the MDT-Data\Exercises directory and save it to your MDT-Work directory. Modify the model (shown in Figure 5.31) as follows:

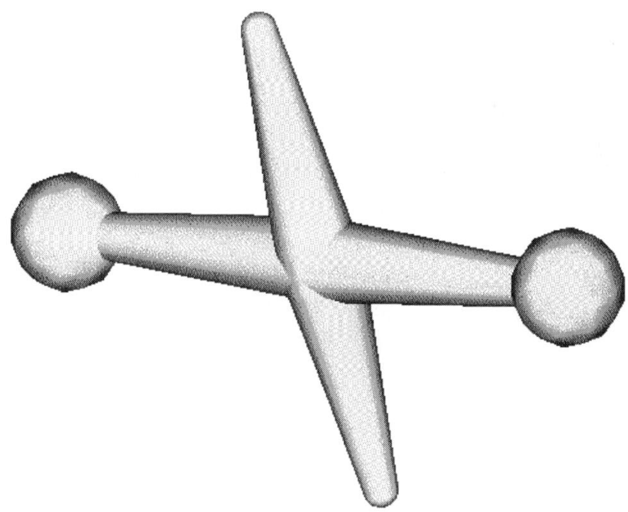

Figure 5.31

- Modify the array so that there are four instances of the leg and the sphere at the end of each leg.
- What happens if there are five instances of the leg and six of the sphere?

Tutorial

6 Creating Revolved Parts and More Part Editing

Objectives

When you have finished this tutorial, you will be able to

- Understand and control sketch assumptions.
- Create revolved features.
- Append geometry to an existing sketch.
- Edit a feature.
- Use Object Snap overrides.
- Add a work axis.
- Use the Polygon command.
- Change the default extrusion direction.
- Add a chamfer as a placed feature.

Introduction

In this tutorial, you will add to your part modeling skills by creating revolved parts, editing a sketch to modify a feature, and using an existing part to create a new one. You will also learn to control the sketching assumptions that Mechanical Desktop applies when solving a sketch. Until now, you have worked with the default settings, but it can be advantageous to customize the sketch assumptions to suit your modeling preferences. When you are finished, you will have modeled four more parts needed for the jack assembly: the stopper, pin, through bolt, and support guide parts. You will begin with the stopper part.

Launch Mechanical Desktop 5

Pick: **New Part File**

Select how to begin: **Template**

Pick: **Browse**

Select: **C:\MDT-Work** (as the directory to browse)

Pick: **MyDrawing.dwt** (If you didn't create the file in Tutorial 3, you can use the data file provided.)

Pick: **File, SaveAs**

Choose Save In: **C:\MDT-Work\Jack directory**

Type File Name: **Stopper.dwg**

Pick: **Save**

Load your user profile from the Options dialog box (if necessary)

You should be in the part modeling environment and see the file name Stopper in the title bar. Before you begin to model, you will review some of the default settings that affect the modeling process.

Setting up the Modeling Environment

Mechanical Desktop allows you to customize several aspects of your modeling environment using the Mechanical Options dialog box. To see it,

Pick: **Assist, Mechanical Options**

The dialog box appears on your screen, as shown in Figure 6.1.

The Standards tab should appear uppermost. If it is not, pick it now. You can use the settings on this tab to select the drawing standard that is applied for typical appearances in your drawing.

Pick: **the arrow for the Standard box to see the list of options**

ANSI (American National Standards Institute) should be the default standard. Notice that its icon appears as an open book. You can also

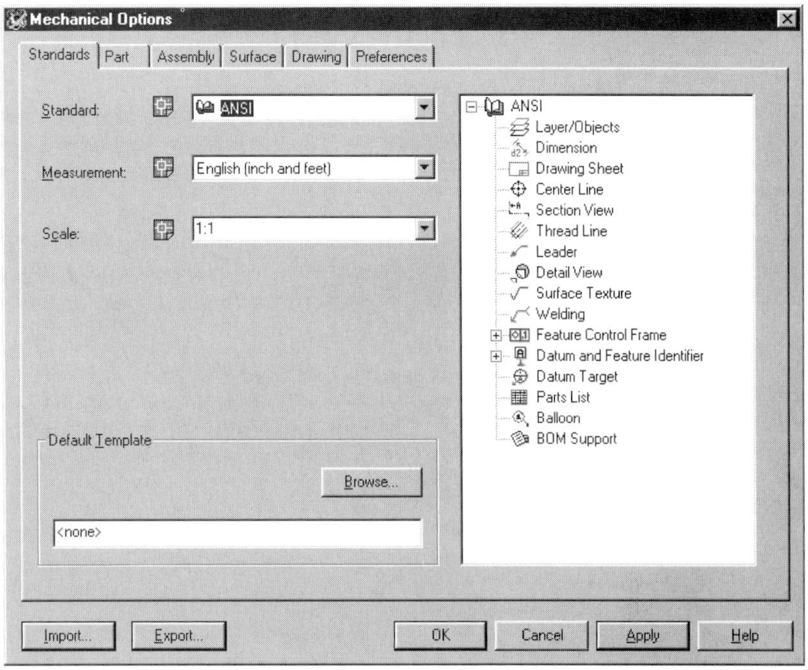

Figure 6.1

select from other international standards, such as DIN, JIS, and ISO. For now, leave ANSI selected as the standard.

The standard selected controls the default appearance for the items listed at the right of the dialog box, such as the dimension and center-line appearances.

Double-click: **Center Line**

The Center Line Properties for ANSI dialog box appears, as shown in Figure 6.2. You could use it to change the way the center lines look. For now,

Pick: **Cancel**

On the Standards tab, you can also set the default units to be English (feet and inches) or Metric, and you can preselect the scale for the model space drawing from this tab. For now, you will leave this area set to English and scale 1 = 1.

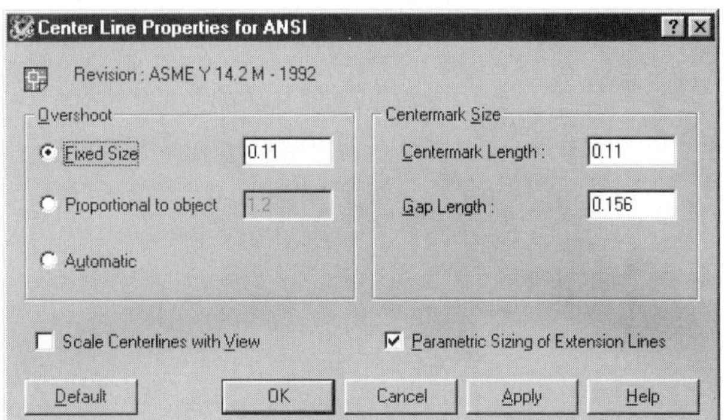

Figure 6.2

The Default template area in the lower left corner allows you to select the default template that will be used in conjunction with the present settings. Leave this area set to the current default also.

Pick: **Part tab**

The Sketch Setting area lets you set items that affect how the profile is generated from the sketch. This includes letting you choose whether to apply constraint rules. Unselecting this assumes that you do not want the software to try to generate constraints for you automatically. This is sometimes useful if you want to use your knowledge of AutoCAD to generate a very detailed and accurate sketch where the constraints might be applied incorrectly. Unselecting Assume Rough Sketch tells the software that your sketch sizes and shapes are accurate and not to make adjustments to them. You can also set the angular tolerance that is applied to the sketch as a profile is generated. The default setting of 4 degrees assumes that any line within 4 degrees of horizontal or vertical was intended to be such.

The Apply to Linetype area lets you set the profile linetype; this tells Mechanical Desktop which types of lines are intended to be incorporated in the profile that will be used to create a feature. Other linetypes, such as centerlines, are assumed to be construction geometry that will not be used to form the 3D feature.

The Constraint size lets you adjust the size at which the constraints are shown on your screen, as you saw in Tutorial 2.

Pick: **the remaining tabs, Assembly, Surface, Drawings, and Preferences**

Familiarize yourself with the controls available on each tab. You will not change any of them at this time, but you can see that this dialog box is a powerful way of controlling the appearance of your drawings and customizing Mechanical Desktop to your use.

Pick: **Cancel** (to exit the dialog box without making changes)

Sketching the Stopper Part

Now you are ready to create the Stopper shown in Figure 6.3. You will start sketching the stopper on the default coordinate system.

Figure 6.3

Pick: **Model tab** (if necessary)

Pick: **Sketch View icon**

Your view changes so that you are looking perpendicular to the *XY*-plane. The UCS icon appears in the lower left of your screen.

Right-click: in the graphics area

Pick: **2D Sketching, Rectangle**

Specify first corner point or [Chamfer/Elevation/Fillet/Thickness/Width]: **pick a point on the screen**

Specify other corner point: **pick a second point that defines the other corner of the rectangle**

You should see a rectangle on the screen. Next you will create a construction line to use for the axis of revolution for the part.

Right-click: in the graphics area

Pick: **2D Sketching, Construction Line**

Pick: **to turn Ortho on**

Specify first point: **above and to the left of the sketch**

Specify next point or [Undo]: **pick below the sketch beneath the previous point**

Specify next point or [Undo]: ↵

Your sketch should look like Figure 6.4.

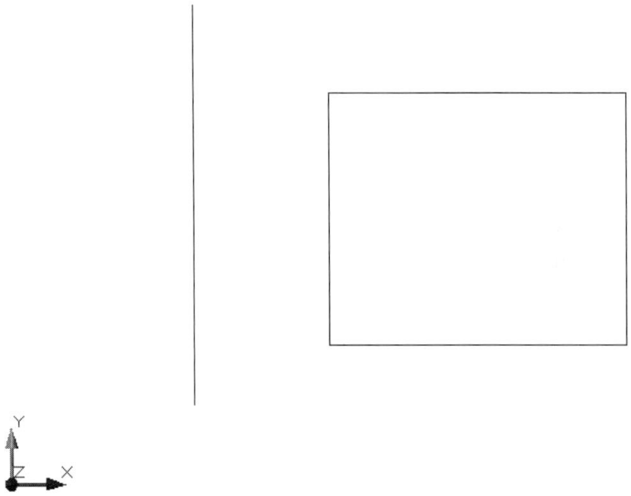

Figure 6.4

Right-click: in the graphics area

Pick: **Sketch Solving, Profile**

Select objects for sketch: **pick to the lower right of the sketched items to start a crossing box**

Specify opposite corner: **pick to included all of the sketched items including the construction line**

Select objects for sketch: ↵

Solved under constrained sketch requiring 3 dimensions or constraints.

Because you used Rectangle, your sketch should have vertical and horizontal constraints and nothing more. The construction line was created with Ortho, so it should be vertical.

On your own, review the constraints and make any changes necessary.

Now add the dimension constraints.

Right-click: in the graphics area

Pick: **Dimensioning, New Dimension**

Select first object: **pick on the right vertical line**

Select second object or place dimension: **pick to the right of the line**

Enter dimension value or [Undo/Hor/Ver/Align/Par/aNgle/Ord/Diameter/pLace] <?????>: **.25** ↵

Solved under constrained sketch requiring 2 dimensions or constraints.

Select first object: **pick on the upper horizontal line**

Select second object or place dimension: **pick above the line**

Enter dimension value or [Undo/Hor/Ver/Align/Par/aNgle/
Ord/Diameter/pLace] <?????>: **.125** ↵

Solved under constrained sketch requiring 1 dimensions or constraints.

Notice that the dimension shows rounded to .13 for the display. Internally the feature is still sized at the value you typed of .125. To check this, you can use the List command and select the line. You will see that the information displayed for the length is .125.

Select first object: **pick on the left vertical continuous line**

Select second object or place dimension: **pick the construction line**

Specify dimension placement: **pick above the sketch**

Enter dimension value or [Undo/Hor/Ver/Align/Par/aNgle/
Ord/Diameter/pLace] <?????>: **H** ↵

Enter dimension value or [Undo/Hor/Ver/Align/Par/aNgle/
Ord/Diameter/pLace] <?????>: **.25** ↵

Solved fully constrained sketch.

Select first object: ↵ (to end the command)

Tip
You may have to zoom and pan your drawing so it fits nicely on your screen.

Your dimensioned sketch should look like Figure 6.5.

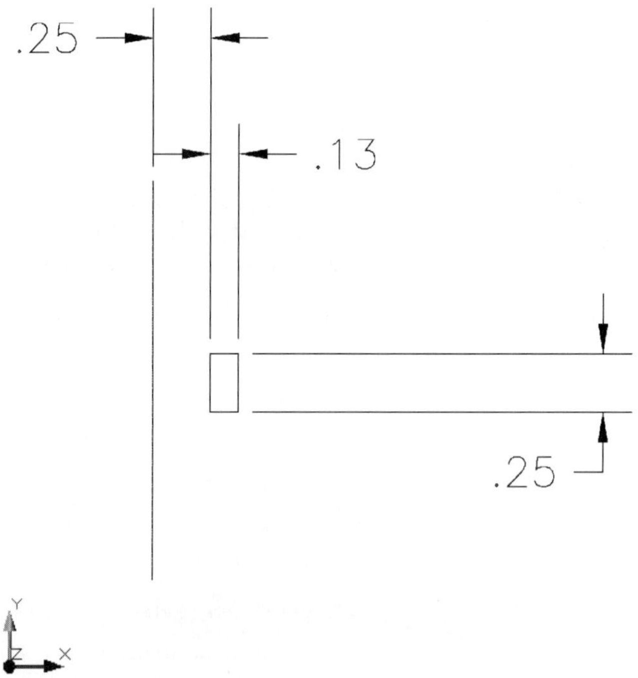

Figure 6.5

Revolving the Profile

Next you will revolve this profile to create a feature. Revolved features must have a line, part edge, axis, or construction line that can act as the axis for the revolution. You will use the construction line you drew as the revolution axis.

Reference

Engineering Design Communication

Revolved Forms, p. 185

Right-click: in the graphics area

Pick: **Sketched & Work Features, Revolve**

Select revolution axis: **pick the construction line**

The Revolution dialog box appears on your screen. Choose the different termination options to see the different ways you can specify a revolution operation. When you are done, use the dialog box to make the selections shown in Figure 6.6.

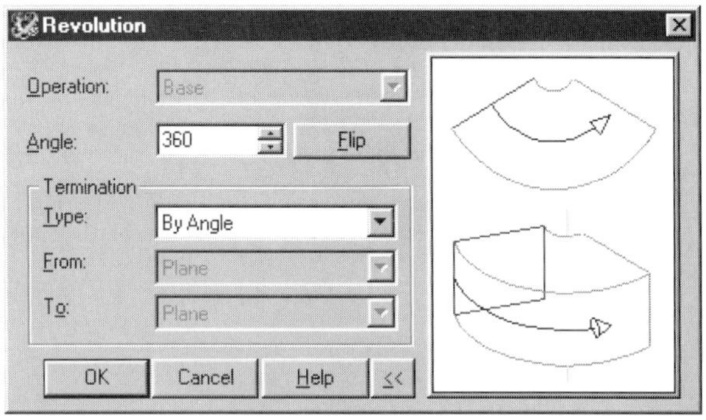

Figure 6.6

Choose Termination: **By Angle**

Set Angle: **360**

Pick: **OK**

To see the part more clearly,

Spin the model. You may have to set a new rotation center

The stopper part should look similar to Figure 6.7.

Figure 6.7

The small stopper part looks somewhat faceted when shaded. It is an accurate model, however. You can switch to a wireframe view to see that the true shape that is generated is round and not a series of flat planes.

Pick: **Toggle Shading/Wireframe icon**

Figure 6.8 shows the wireframe view of the stopper. Notice its rounded appearance.

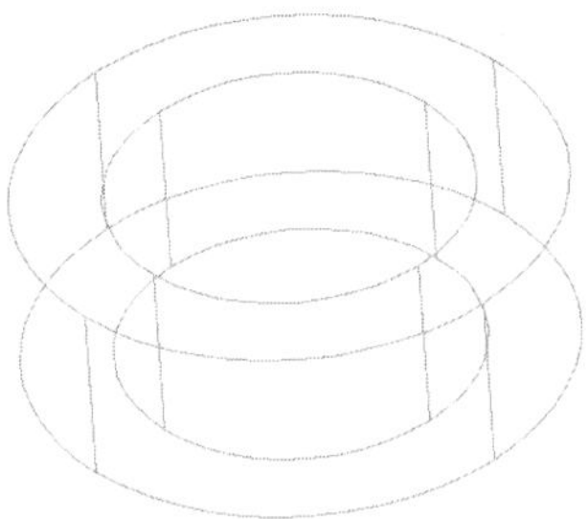

Figure 6.8

Next you will add fillets to the inside and outside edges.

Right-click: in the graphics area

Pick: **Placed Features, Fillet**

Select: **Constant**

Set Radius: **.05**

Pick: **OK**

Select edges or faces to fillet<continue>: **select the 4 circular edges** (they should appear as dashed lines as you move your cursor over the model)

Select edges or faces to fillet<continue>: ↵

The fillet is added to the four edges. Congratulations! You have already finished the stopper part.

On your own, right-click on the Stopper part in the model tree

Pick Properties and change the color for the Stopper to black

Pick: **File, Save**

You see a message stating, *"This file is being saved with drawing views that are not up to date. It is recommended that files are saved with views that accurately represent the geometry. Enter drawing mode (AMMODE). Use the AMUPDATE view option for each layout necessary and then resave the file."*

Pick: **OK**

Pick: **Drawing Tab**

Layout2 automatically updates to show four views of the part.

Double-click: **Layout1**

Layout1 automatically updates also.

Pick: **Model Tab**
Pick: **File, Save**
Pick: **File, Close**

Creating the Pin Part

Next you will create the pin shown in Figure 6.9. This time you will start from scratch instead of using your template. The pin can easily be created as a revolved feature similar to the stopper. Unlike the stopper, where the profile was revolved around an axis that was a distance away from it, creating a hole, the pin will be revolved around one of its own edges to create a solid part. You will also learn more about editing sketches by making a change to the profile and appending it to the sketch. This time you will start from scratch instead of using your template.

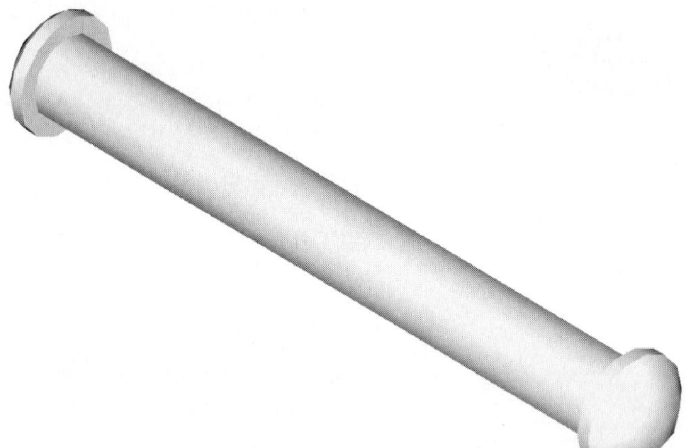

Figure 6.9

Pick: **File, New Part File**
Select How to Begin: **Start from Scratch**
Pick: **English (feet and inches)**
Pick: **File, Save**
Choose Directory: **C:\MDT-Work\Jack**
Type File Name: **Pin.dwg**
Pick: **Save**

Figure 6.10

You should now have a blank part file named *Pin.dwg* open on your screen. You will sketch the shape shown in Figure 6.10 on the default sketch plane and constrain and solve the sketch. Then you will add the rounded ends and append them to the sketch.

Turn Ortho on

Right-click: in the graphics area

Pick: **2D Sketching, Polyline**

Specify start point: **pick point A** (refer to Figure 6.10)

Current line-width is 0.0000

Specify next point or [Arc/Close/Halfwidth/Length/Undo/Width]: **pick point B**

Specify next point or [Arc/Close/Halfwidth/Length/Undo/Width]: **pick point C**

Specify next point or [Arc/Close/Halfwidth/Length/Undo/Width]: **pick point D**

Specify next point or [Arc/Close/Halfwidth/Length/Undo/Width]: **pick point E**

Specify next point or [Arc/Close/Halfwidth/Length/Undo/Width]: **pick point F**

Specify next point or [Arc/Close/Halfwidth/Length/Undo/Width]: **pick point G**

Specify next point or [Arc/Close/Halfwidth/Length/Undo/Width]: **pick point H**

Specify next point or [Arc/Close/Halfwidth/Length/Undo/Width]: **C** ↵ (to close)

You should have the half-I-beam shape for one-half of the cross section of the pin, without the rounded ends drawn. Next you will solve the sketch and use this profile to create a revolved feature. You can use Single Profile because your polyline is a single entity.

Tip

If your sketch needs more constraints, check to see that all of the lines have a horizontal or vertical constraint added to them. If the line created with the Close option has more than a 4-degree angle, it may not have been constrained to be vertical. If necessary, add these on your own.

Right-click: in the graphics area

Pick: **Sketch Solving, Single Profile**

Solved under constrained sketch requiring 5 dimensions or constraints.

You should notice a fixed point constraint for the first point selected in your sketch. That will be fine for this part. Also, you should see the short vertical segments marked as colinear. If they are not, you will have to add this constraint on your own. Next you will add constraints for the two short vertical segments to be equal length.

Right-click: in the graphics area

Pick: **2D Constraints, Equal Length**

Valid selections: line or spline segment

Select first object: **pick the upper short vertical segment**

Select second object: **pick the lower short vertical segment**

Solved under constrained sketch requiring 4 dimensions or constraints.

Select first object: ↵ (to end selecting)

Enter an option [Hor/Ver/PErp/PAr/Tan/CL/CN/PRoj/Join/XValue/YValue/Radius/Length/Mir/Fix/eXit] <eXit>: ↵ (to end command)

Next you will add four dimensions to fully define the sketch. When you are finished, your dimensioned sketch should appear similar to Figure 6.11.

Right-click: in the graphics area

Pick: **Dimensioning, New Dimension**

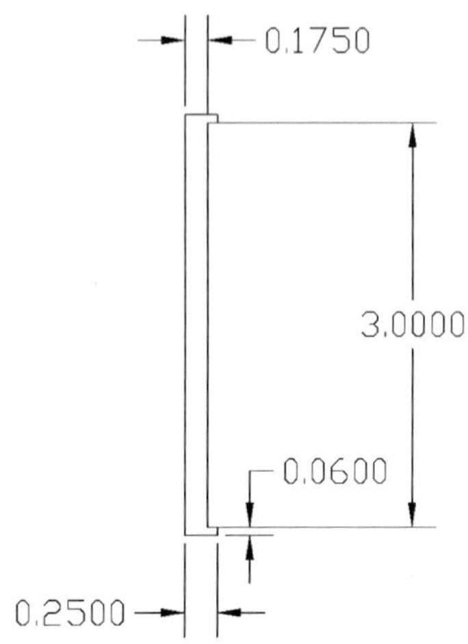

Figure 6.11

Select first object: **pick on the right long vertical line**

Select second object or place dimension: **pick to the right of the line**

Enter dimension value or [Undo/Hor/Ver/Align/Par/aNgle/Ord/Diameter/pLace] <1.5661>: **3** ↵

Solved under constrained sketch requiring 3 dimensions or constraints.

Select first object: **pick the right vertical line**

Select second object or place dimension: **pick the left vertical line**

Specify dimension placement: **pick above the sketch**

Enter dimension value or [Undo/Hor/Ver/Align/Par/aNgle/Ord/Diameter/pLace] <0.0432>:**.175** ↵

Solved under constrained sketch requiring 2 dimensions or constraints.

Select first object: **pick the bottom horizontal line**

Select second object or place dimension: **pick below the sketch**

Enter dimension value or [Undo/Hor/Ver/Align/Par/aNgle/Ord/Diameter/pLace] <0.3913>:**.25** ↵

Solved under constrained sketch requiring 1 dimensions or constraints.

Select first object: **pick the lower short vertical line segment**

Select second object or place dimension: **pick to the right of the line**

Enter dimension value or [Undo/Hor/Ver/Align/Par/aNgle/Ord/Diameter/pLace] <4.6141>:**.06** ↵

Solved fully constrained sketch.

Select first object: ↵ (to end command)

Now the sketch is fully constrained.

Next you will revolve the profile 360 degrees to create the main shape of the pin.

Right-click: in the graphics area

Pick: **Sketched & Work Features, Revolve**

Select revolution axis: **pick the long vertical left edge of the sketch**

The Revolution dialog box shows up on your screen. You will make the following selections.

Choose Termination: **By Angle**

Choose Angle: **360**

Pick: **OK**

The revolved pin feature is created. On your own,

Use the model tree to rename the feature

Type: **PinBody** (as the new name)

Change the color to magenta

Select the Front Left Isometric View

Your model should look like Figure 6.12.

Controlling the View of Curved Surfaces

Mechanical Desktop represents curved surfaces with a series of isolines, or tessellation lines. The system variable, ISOLINE, controls how many lines will be drawn to represent a surface in wireframe mode. (A different

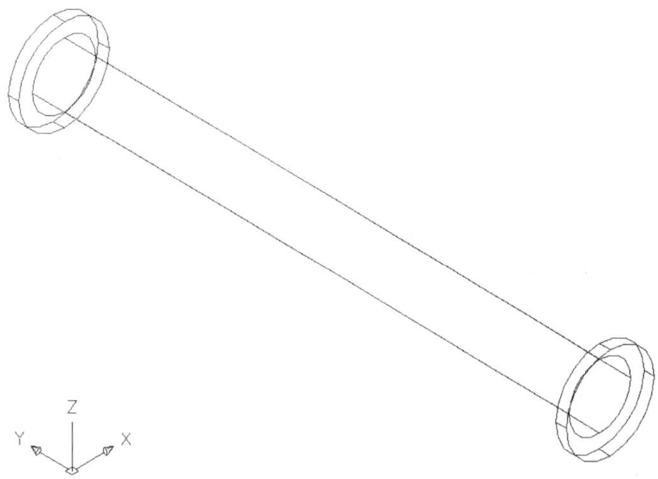

Figure 6.12

variable controls the view of shaded surfaces.) The default setting, 4, does not represent the curved surface very realistically. Setting it higher will generate a view of the feature with more tessellation lines along its curved edges. (This does not affect the model geometry, just the view of the model on your screen.)

Reference

Engineering Design Communication

Solid Model Accuracy, pp. 233-235

Command: **ISOLINES** [Enter]

Enter new value for ISOLINES <4>: **12** [Enter]

Pick: **View, Regen** (from the menu bar)

The part is regenerated with more tessellation lines, as shown in Figure 6.13.

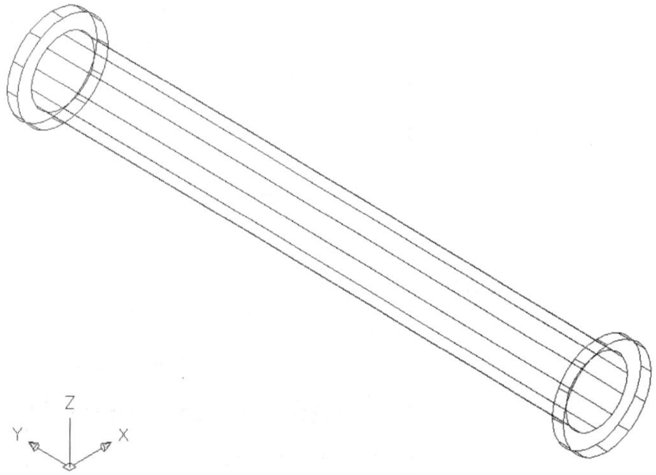

Figure 6.13

Editing a Sketch to Append New Geometry

Next you will edit the pin's sketch to add the rounded ends. Editing a sketch can sometimes be tricky, so be patient. Often it is easier to create a second feature and join it to the first one than to add something like this non-tangent rounded end, but this section will show you the basic method for adding geometry to an existing sketch. On future parts, you will then be able to choose the method that makes the most sense. In general, having more smaller features makes your model easier to update than complicated sketches that use many constraints and dimensions.

Right-click: **PinBody** (in the model tree)

Pick: **Edit Sketch** (from the context menu)

The sketch for the part reappears on your screen. You will add arcs to both ends of the sketch to create the rounded ends. Then you will extend and erase lines to form a new profile. You will then append these new sketch items to the sketch.

On your own, select the Sketch View icon and Zoom All

You will start by offsetting the top line by a distance of .13 to create an endpoint that you can use to snap to when you create the arc.

Use Zoom Window to enlarge the top portion of the sketch

Right-click: in the graphics area

Pick: **2D Sketching, Offset**

Specify offset distance or [Through] <1.0000>: **.13** ↵

Select object to offset or <exit>: **pick on the top horizontal line**

Specify point on side to offset: **pick above the top line**

Select object to offset or <exit>: ↵

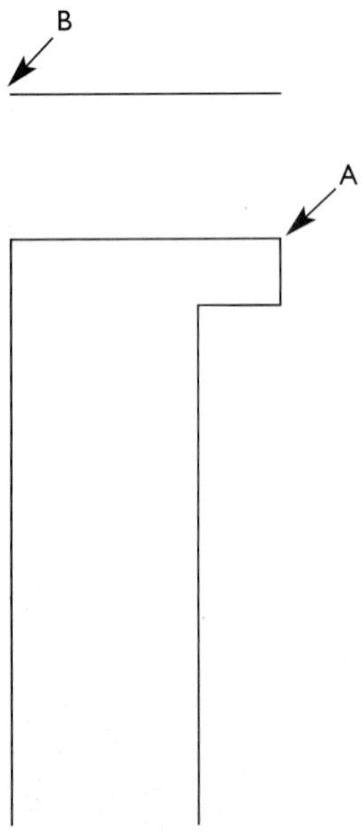

Figure 6.14

The newly offset line is created, as shown in Figure 6.14.

Using Arc

Reference

Engineering Design Communication

Arcs, p. 174

Next you will use Object Snap Endpoint as you draw an arc by selecting two endpoints from the sketch and giving the arc radius. You will choose the Arc command from the menu bar, where the submenu lists the many different methods of sketching arcs. Depending on the existing geometry and the information you have about the arc you want to draw, each of these methods is useful. Be sure to pick Start End Radius for this sketch or you won't see the following prompts.

Turn on OSNAP Endpoint

Pick: **Design, Arc, Start End Radius**

Pick: **use Osnap target to select endpoint A in Figure 6.14**

Specify second point of arc or [Center/End]: **use Osnap to target endpoint B**

Specify radius of arc: **.305** ↵

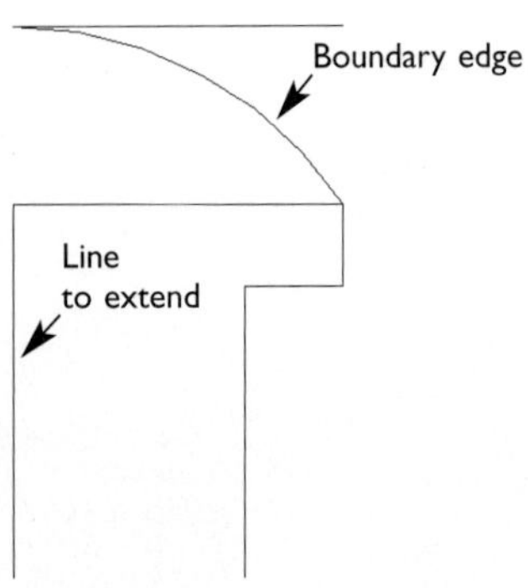

Figure 6.15

The arc is added to your sketch, as shown in Figure 6.15.

Next you will use Extend to extend the left vertical line so that it touches the new arc.

Right-click: in the graphics area

Pick: **2D Sketching, Extend**

Select boundary edges ...

Select objects: **pick the arc**

Select objects: ↵

Select object to extend or shift-select to trim or [Project/Edge/Undo]: **pick the left vertical line**

Select object to extend or shift-select to trim or [Project/Edge/Undo]: ↵ (to end the command)

The extended line in your sketch should look like Figure 6.16.

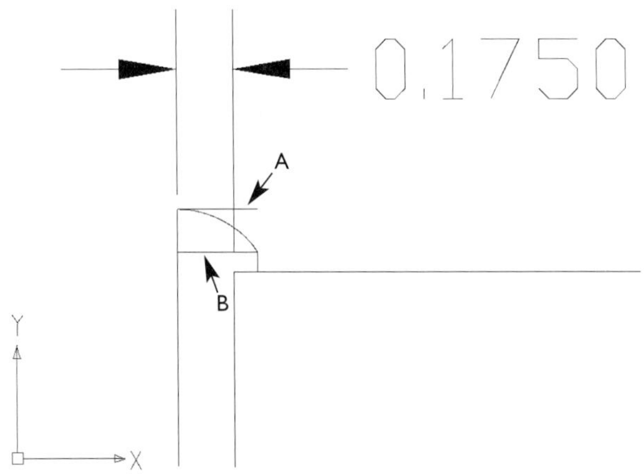

Figure 6.16

Next you will erase the line you created with offset, and the former top line of the sketch.

Right-click: in the graphics area

Pick: **2D Sketching, Erase**

Select objects: **pick line A and line B, referring to Figure 6.16**

Select objects: ↵

The two lines are removed from the screen. Your sketch should now look like Figure 6.17.

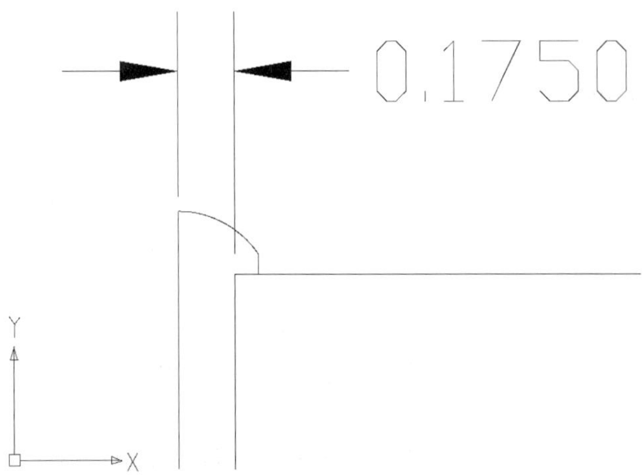

Figure 6.17

Now you will append the arc to the previous sketch to tell Mechanical Desktop to use the new shape.

Right-click: in the graphics area

Pick: **Append Sketch**

Select geometry to append to sketch: **pick the arc** ↵

Redefining existing sketch.

Solved under constrained sketch requiring 2 dimensions or constraints.

You will now need to add dimensions and constraints to define this newly added arc. First you will add a constraint that defines the center of the arc to be positioned at the same X-value as the left vertical line. This will locate its center since this is not a tangent arc. Tangent arcs that you have used in the past tutorials have their centers automatically located based on their tangency. For this arc, you must explicitly tell Mechanical Desktop where its center lies.

Right-click: in the graphics area

Pick: **2D Constraints, X Value**

Select object to be reoriented: **pick on the arc**

Valid selections: line, arc, circle or spline segment

Select object x value is based on: **pick on the left vertical line**

Solved under constrained sketch requiring 1 dimensions or constraints.

Select object to be reoriented: ↵ ↵

Next define the radius of the arc.

Right-click: in the graphics area

Pick: **Dimensioning, New Dimension**

Select first object: **pick the arc**

Select second object or place dimension: **pick above the arc**

Enter dimension value or [Undo/Diameter/Ordinate/
Placement point] <0.3050>: **.305** ↵

Solved fully constrained sketch.

Select first object: ↵ (to end command)

You have now appended the arc to the sketch. Now you are ready to update the part to see the result.

Right-click: in the graphics area

Pick: **Update Part**

Pick: **Gouraud Shaded icon**

Pick: **Left Back Isometric icon**

The rounded end is added to the top of the part, as shown in Figure 6.18.

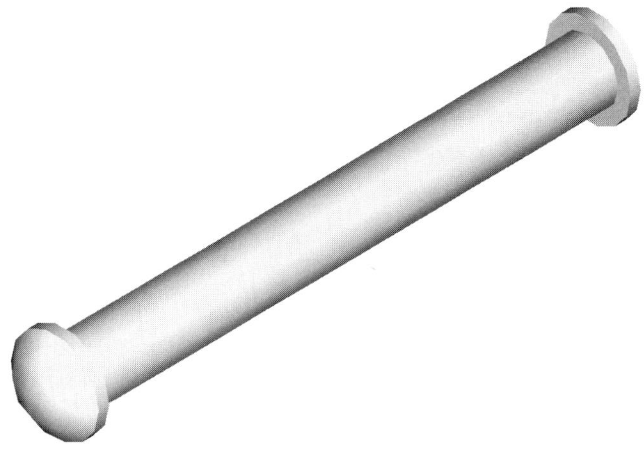

Figure 6.18

Next you will use the same method to make the rounded bottom end by editing the sketch and appending to it once more.

Pick: **Sketch View icon**

Pick: **View, Zoom, Extents** (or type **Z** ↵ **E** ↵ at the command prompt)

The view of the part fills your screen. Zoom, Extents determines the extent of objects in the drawing and magnifies the view as much as possible while still showing all objects. You will use this command again after the sketch appears by typing its command alias, Z. Then you will zoom out a bit more by specifying a scale factor at the command prompt. Scale factors less than 1 zoom out by that proportion. Scale factors greater than 1 zoom in.

Now you are ready to edit the sketch again.

Right-click: **Pinbody** (in the model tree)

Pick: **Edit Sketch**

Type: **Z** ↵ **E** ↵

Command: **Z** ↵

Specify corner of window, enter a scale factor (nX or nXP), or
[All/Center/Dynamic/Extents/Previous/Scale/Window] <real time>: **.8x**

You will mirror the arc from the top to the bottom of the part. The arc does not appear too rounded in this view, but its grips do when you select it.

Object Snap Overrides

To position the new arc, you need to select a mirror line that is exactly halfway between the two ends of the pin. To do so, you will use the Midpoint object snap to select the exact middle of the right vertical line as the first point for the mirror line. Osnap should still be on, but it is set to target endpoints. You can override this setting at any time by typing the desired object snap at the command line when a selection prompt appears. This temporary override is active for one selection only; you

Reference

Mechanical Desktop On-line Help

The command line sequences are listed in parentheses following each object snap in the AMPOWERSNAP command listing in Mechanical Desktop Help. A complete discussion of the object snap options is also included here.

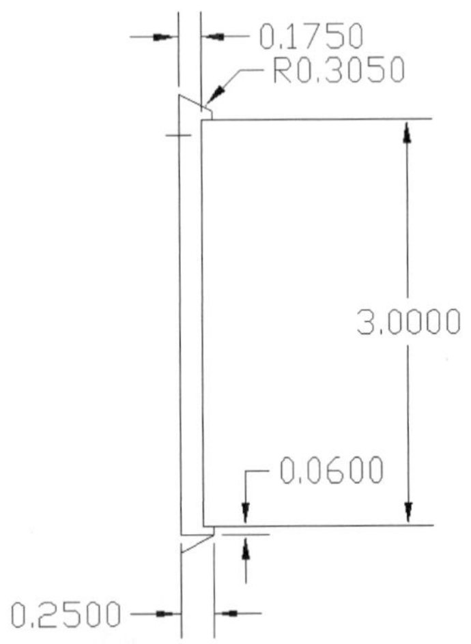

Figure 6.19

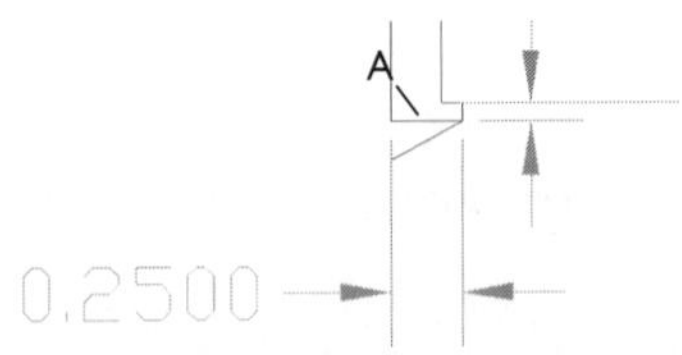

Figure 6.20

need to type it again for additional selections. When Midpoint is active, your Osnap target will look like a small triangle.

Right-click: in the graphics area

Pick: **2D Sketching, Mirror**

Select objects: **pick the top arc**

Select objects: ↵

Specify first point of mirror line: **MID** ↵

mid of: **pick on the right vertical line**

Turn on ORTHO

Specify second point of mirror line: **pick to the left of the previous point** (to define a horizontal mirror line)

Delete source objects? [Yes/No] <N>: ↵

The arc is mirrored from the top of the part to the bottom, as shown in Figure 6.19.

Pick: **View, Zoom, Window**

Enlarge the bottom section of the part so that it fills your screen, as shown in Figure 6.20.

Next you will erase the bottom horizontal line and then extend the left line so that it touches the arc. You will use command aliases to invoke the Erase and Extend commands.

Type: **E** ↵

Select objects: **pick horizontal line A, as shown in Figure 6.20.**

Select objects: ↵

The line is removed from the sketch.

Type: **EX** ↵

Current settings: Projection = UCS Edge = None

Select boundary edges...

Select objects: **pick on the arc (**you may need to use a crossing window to select it)

Select objects: ↵

Select object to extend or shift-select to trim [Project/Edge/Undo]: **pick the left vertical line**

Select object to extend or shift-select to trim [Project/Edge/Undo]: ↵ (to end command)

Now you are ready to append the sketch once again. (You could have done both ends in the same sketch, but sometimes smaller steps are easier to constrain.)

Right-click: in the graphics area

Pick: **Append Sketch**

Select geometry to append to sketch: **pick on the arc** (you may need to use a crossing window to select it)

Select geometry to append to sketch: ↵

Redefining existing sketch

Solved under constrained sketch requiring 2 dimensions or constraints.

Tip

If you have trouble constraining this sketch, review the constraints and dimensions. You may need to add the X-value constraint to the top rounded end again. You should have the constraints shown in Figure 6.21.

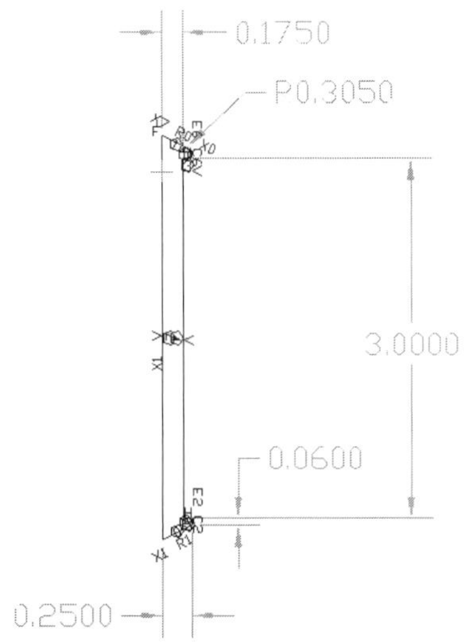

Figure 6.21

Next you will constrain and dimension this arc.

Right-click: in graphics area

Pick: **2D Constraints, X Value**

Select object to be reoriented: **pick on the arc**

Valid selections: line, arc, circle or spline segment

Select object x value is based on: **pick on the left vertical line**

Solved fully constrained sketch

Select object to be reoriented: ↵ ↵

Now you are ready to update the part.

Right-click: in the graphics area

Pick: **Update Part**

Pick: **Left Front Isometric View icon**

You should have successfully added the rounded end at the bottom of the part, as shown in Figure 6.22.

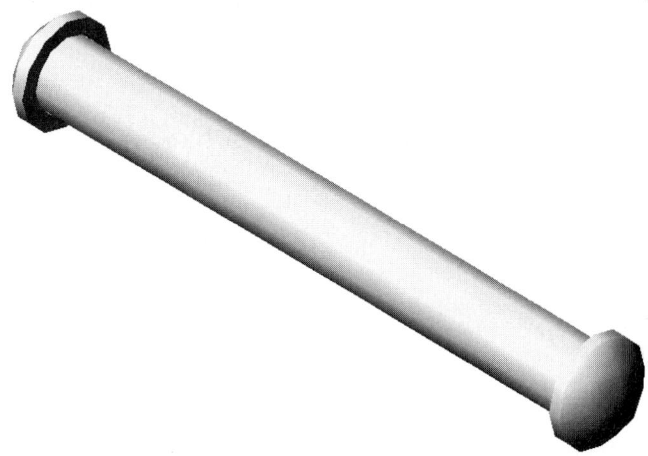

Figure 6.22

Now you have finished creating the Pin! Don't forget to save your work!

Pick: **File, Save**

Starting a New Part from an Existing One

Figure 6.23

Two more parts to go. You will be surprised how easily you can create the next one by starting out with a copy of your Pin as the basis for the SupportGuide shown in Figure 6.23.

Pick: **File, Save As**

Choose Save In: **MDT-Work\Jack directory**

Type File name: **SupportGuide**

Pick: **Save**

A copy of the Pin is open on the screen. You should see the new path and file name, *MDT-Work\Jack\SupportGuide.dwg*, in the title bar at the top of the screen. When you use Save As, it is always a good idea to check and make sure you are making the changes in the correct file.

Editing a Feature

To make the Support Guide from the Pin, consider the dimensioned view of the Support Guide shown in Figure 6.24. Remember back to creating the pin. It had a dimension of 3.00 between the ends. The Support Guide is only 2.00 long between the ends. You will edit the dimension and then add the central block and hole to complete the Support Guide part.

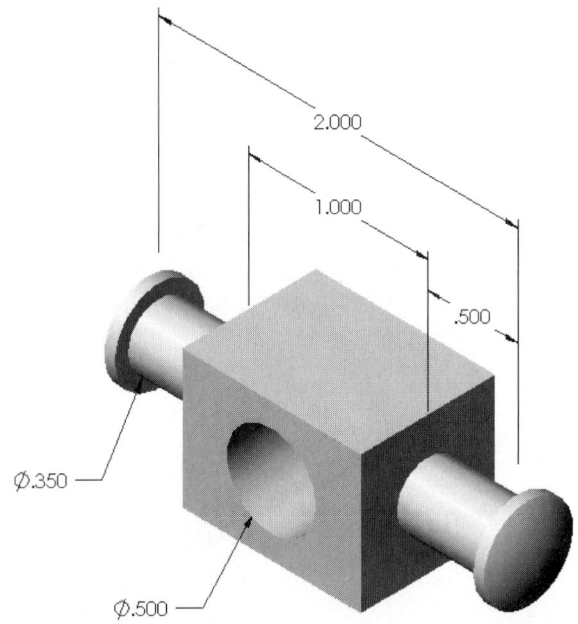

Figure 6.24

Pick: **+ sign** (to expand the model tree below PinBody)

Right-click: **Profile1**

Pick: **Edit** (from the context menu)

Pick: **Zoom All icon**

The dimensions for the feature appear on your screen, as shown in Figure 6.25, and the prompt for editing features appears.

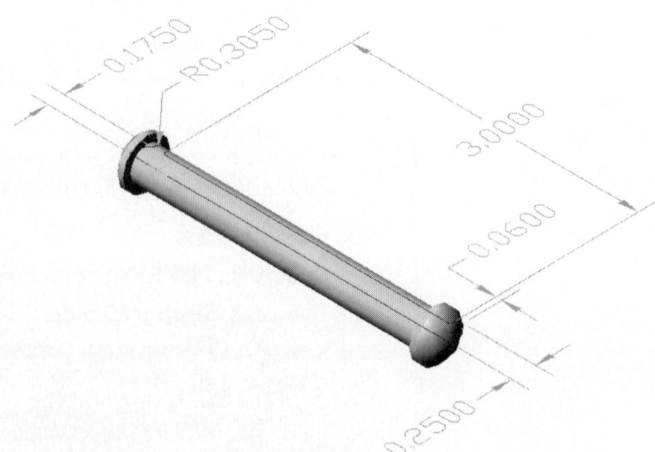

Figure 6.25

Select object: **pick on the 3.00 dimension**

Enter dimension value <3>: **2** [↵]

Select object: [↵]

The dimension value is changed, but the part still appears the same length. You must update the part for the change to take effect.

Right-click: in the graphics area

Pick: **Update Part**

The part geometry updates to the new length.

Adding a Work Axis

Next you will add the central block. The block feature should be symmetrical about the centerline of the PinBody feature, but there is no geometric entity to which you can constrain it. You will create a work axis in the center of the part so that you can use it to dimension the sketch for the block.

Right-click: in the graphics area

Pick: **Sketched & Work Features, Work Axis**

Select cylinder, cone, torus or [Sketch]: **pick on the cylindrical body of the part**

Toggle to wireframe mode (to see the axis)

The work axis is added to the part, as shown in Figure 6.26.

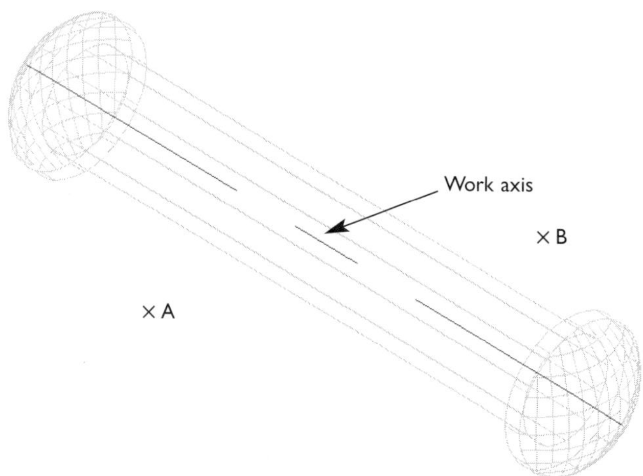

Figure 6.26

Next you will sketch a rectangle approximately in the middle of the part and extending to either side. The active sketch plane is still the WCS used to create the sketch for the PinBody. You can confirm this by selecting the Sketch View icon, but it is not necessary to be in that view. You will pick points in the isometric view and your sketch will be created on the same *X-Y* plane as the previous sketch.

Right-click: in the graphics area

Pick: **2D Sketching, Rectangle**

Specify first corner point or [Chamfer/Elevation/Fillet/Thickness/Width]:
pick near point A (as shown in Figure 6.26)

Specify other corner point: **pick near point B**

The sketched rectangle should look something like Figure 6.27.

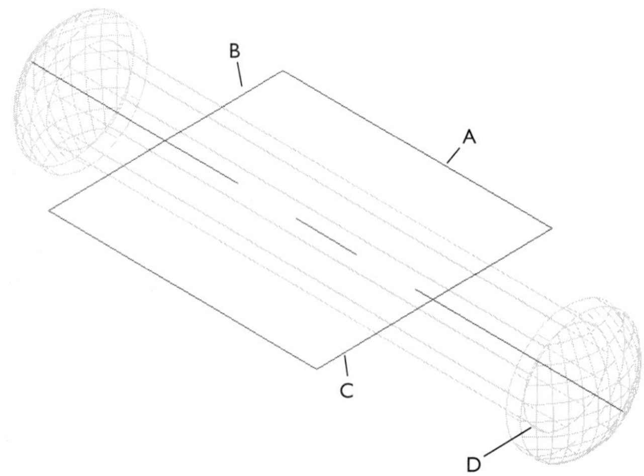

Figure 6.27

Right-click: in the graphics area

Pick: **Sketch Solving, Single Profile**

Solved under constrained sketch requiring 4 dimensions or constraints.

Select the Sketch View icon and Zoom so you can see the part well

Now you are ready to add four dimensions to size and locate this profile. As you select the Dimension command, the constraints are displayed for your review. As you used the Rectangle command, you should have nothing but Vertical and Horizontal constraints on the sketch.

Right-click: in the graphics area

Pick: **Dimensioning, New Dimension**

Select first object: **pick on the right vertical line labeled A in Figure 6.27**

Select second object or place dimension: **pick to the right of the line**

Enter dimension value or [Undo/Hor/Ver/Align/Par/aNgle/Ord/
Diameter/pLace] <????>: **1** ⏎

Solved under constrained sketch requiring 3 dimensions or constraints.

Select first object: **pick on the right vertical line A**

Select second object or place dimension: **pick the work axis**

Specify dimension placement [Undo]: **pick above the part**

Enter dimension value or [Undo/Hor/Ver/Align/Par/aNgle/Ord/Diameter/pLace] <????>: **H** (if necessary)

Enter dimension value or [Undo/Hor/Ver/Align/Par/aNgle/Ord/Diameter/pLace] <????>: **.375** ↵

Solved under constrained sketch requiring 2 dimensions or constraints.

Select first object: **pick on the top line of the rectangle labeled B**

Select second object or place dimension: **pick above the last dimension above the part**

Enter dimension value or [Undo/Hor/Ver/Align/Par/aNgle/Ord/Diameter/pLace] <????>: **.75** ↵

Solved under constrained sketch requiring 1 dimensions or constraints.

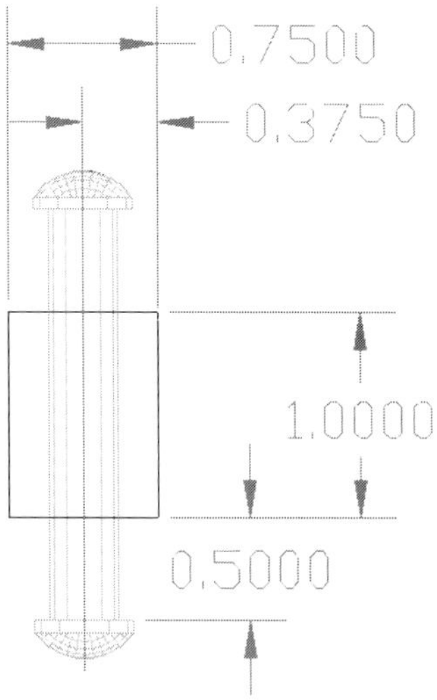

Figure 6.28

Select first object: **pick on the bottom line of the rectangle labeled C**

Select second object or place dimension: **pick on the horizontal line D representing the end of the cylinder**

Specify dimension placement: **pick to the right of the sketch**

Enter dimension value or [Undo/Hor/Ver/Align/Par/aNgle/Ord/Diameter/pLace] <????>: **.5** ↵

Solved fully constrained sketch.

Select first object: ↵

The dimensioned sketch should look similar to Figure 6.28.

Now you are ready to extrude the sketch to both sides of the sketch plane to create the center block.

Right-click: in the graphics area

Pick: **Sketched & Work Features, Extrude**

The Extude dialog box appears. You will use it to make the following selections. Remember that the extrusion distance represents the total length, even for a Midplane termination.

Choose Operation: **Join**

Choose Termination: **MidPlane**

Set Distance: **0.75**

Set Draft angle: **0**

Pick: **OK**

Figure 6.29

Pick: **Left Front Isometric icon**

Pick: **Gouraud Shaded icon**

Rename Extrusion Midplane1 feature to Block

Change the color of the SupportGuide part to magenta

The feature is created, as shown in Figure 6.29.

Next you will add the hole through the center of the block. For this feature, you will change the sketch plane to one of the outside faces of the block. (Since it is symmetric, it does not really matter which of the sides your hole goes through, but follow the steps to match the orientation in the figures.)

Pick: **Sketch View icon**

Toggle to wireframe mode

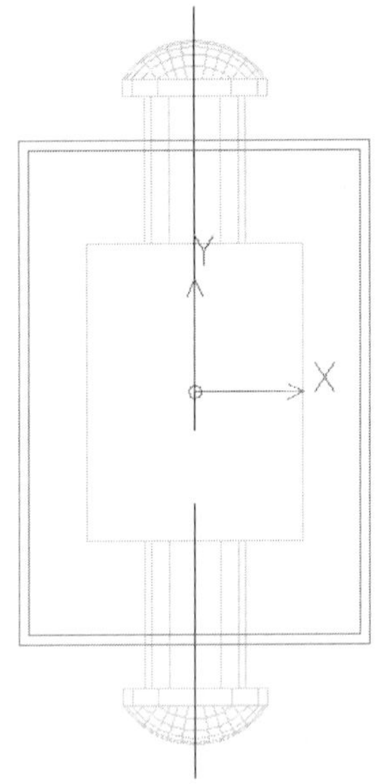

Figure 6.30

You return to the sketch view looking straight onto the block.

Right-click: in the graphics area

Pick: **New Sketch Plane**

Select work plane, planar face or [worldXy/worldYz/worldZx/Ucs]: **pick the rectangular face so it is highlighted**

Enter an option [Next/Accept] <Accept>: **right-click if the correct surface is highlighted; if not, pick to cycle to the next face**

Plane = Parametric

Select edge to align X axis or [Z-flip/Rotate] <Accept>: **press ↵ to accept** (if the coordinate system is aligned as shown in Figure 6.30.)

The new sketch plane is aligned with the rectangular surface of the block.

Right-click: in the graphics area

Pick: **Placed Features, Hole**

The hole dialog box appears on your screen. You will use it to make the following selections, as shown in Figure 6.31.

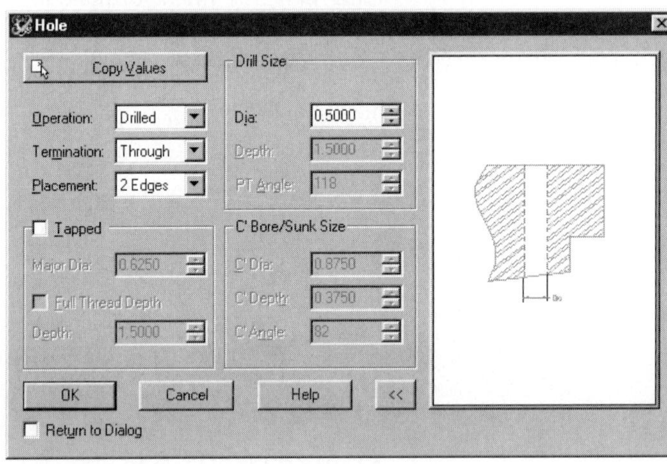

Figure 6.31

Choose Operation: **Drilled**

Choose Termination: **Through**

Choose Placement: **2 Edges**

Type Dia: **0.5000**

Deselect: **Tapped** (so it is not checked)

Pick: **OK**

You return to pick the remaining selections from the screen.

Select first edge: **pick the left edge of the rectangular block**

Enter an option [Next/Accept] <Accept>: **right-click when the correct edge is highlighted**

Select second edge: **pick the bottom edge of the block**

Specify hole location: **pick in the center of the block**

Enter distance from first edge (highlighted) <0.1332>: **.375** ↵

Enter distance from second edge (highlighted) <0.0007>: **.5** ↵

Select first edge: ↵ (to end command)

The hole should appear through the part, as shown in Figure 6.32.

Great! You have created the SupportGuide part. Shade it and spin it around to get that feeling of accomplishment. It's almost like the part is real. Now don't forget to save it.

Pick: **File, Save**

Pick: **File, Close**

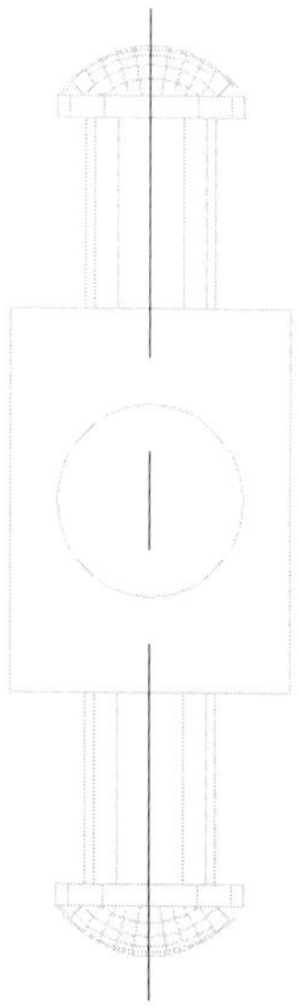

Figure 6.32

Reference

Engineering Design Communication

Parameters Control the Geometry, pp. 250-254

Creating the Bolt

Now you will need a through bolt. As you have seen, once you have made one part, it can be used for many others. The hex head bolt that you will create will be a parametric part that you can use many different times by changing its major dimensions. You will start a new part file for the Bolt.

Pick: **File, New Part File**

Pick: **English (feet and inches)**

Pick: **File, Save**

Choose Save in: **MDT-Work\Jack directory**

Type File name: **Bolt**

Pick: **Save**

A blank graphics area awaits you. The title bar displays the path and file name for the Bolt part.

You will start out by drawing the circular shape of the bolt body. This time you will not use revolution to create the cylindrical feature; rather you will extrude a circle of the desired diameter. Circular features are easy to create by revolution, but often pins, bolts, and other shapes are created by extruding circular shapes. This is because they can be easier to edit, or it can be easier to determine the sketch plane for the extruded shape than the revolved one. Also, cylindrical features are usually dimensioned by their diameter, so it can be easier to assign a diameter dimension to a sketched circle than to a revolved segment. These things all depend on the model and the design, so you must decide what method is best for you. You will use the Circle command alias as you sketch the circular shape.

Command: **C** ↵

Command: Specify center point for circle or [3P/2P/Ttr (tan tan radius)]:
pick near the center of the screen

Specify radius of circle or [Diameter]: **.25** ↵

Right-click: in the graphics area

Pick: **Sketch Solving, Single Profile**

Solved under constrained sketch requiring 1 dimensions or constraints.

Right-click: in the graphics area

Pick: **Dimensioning, New Dimension**

Select first object: **pick on the circle**

Select second object or place dimension: **pick above the circle**

Enter dimension value or [Undo/Radius/Ordinate/Placement point]
<.5000>: **.5** ↵

Solved fully constrained sketch.
Select first object: ↵

Now you are ready to extrude the sketch to create a cylindrical feature.

Right-click: in the graphics area
Pick: **Sketched & Work Features, Extrude**

Use the dialog box to make the following selections.

Choose Operation: **Base** (grayed out)
Type Distance: **17.40**
Choose Termination Type: **Blind**
Draft angle: **0**
Pick: **OK**

Pick: **Left Front Isometric icon**
Rename ExtrusionBlind1 (in the model tree) **to Shaft.**
Change the color of Bolt to blue.

Check out the part. You should have a long cylinder created. Next you will return to the sketch plane view, where you will add the sketch for the hex head.

Pick: **Sketch View icon**

Use Zoom Window to enlarge the view on your screen as needed.

Using Polygon

Reference

Engineering Design Communication

Formulas for Regular Polygons, p. 685

The Polygon command creates a regular (equilateral) polyline. There are two options for specifying the length of the edges: the Edge option, which prompts you for the endpoints of one of the edges, or the Center option, which uses a circle centered on the centerpoint you select to define the polygon. An inscribed polygon has all of its vertices on the circle (or the radial distance from the center). A circumscribed polygon has the midpoint of all edges on the circle.

You will circumscribe a hexagon about a circle to create the sketch for the head of the bolt. To do so, you will use another Osnap override in this sequence so you can select the centerpoint (CEN) of the circle.

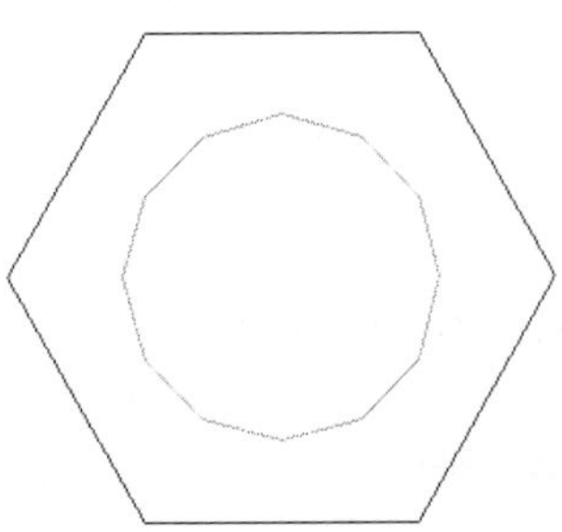

Figure 6.33

Right-click: in the graphics area

Pick: **2D Sketching, Polygon**

Enter number of sides <4>: **6** ↵

Specify center of polygon or [Edge]: **CEN** ↵

Center of: **pick on circle (you will see the Osnap circle targeting the centerpoint)**

Enter an option [Inscribed in circle/Circumscribed about circle] <I>: **C** ↵

Specify radius of circle: **.375** ↵

The sketch for the hex head is shown in Figure 6.33.

Right-click: in the graphics area

Pick: **Sketch Solving, Profile**

Select objects for sketch: **pick on the polygon** ↵

Solved under constrained sketch requiring 8 dimensions or constraints.

On your own, display the existing constraints.

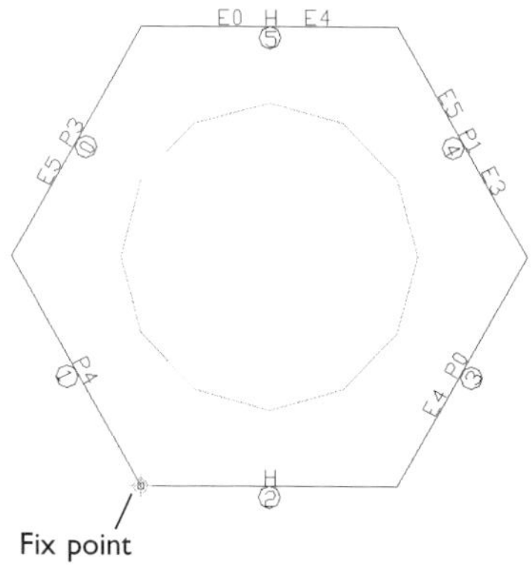

Figure 6.34

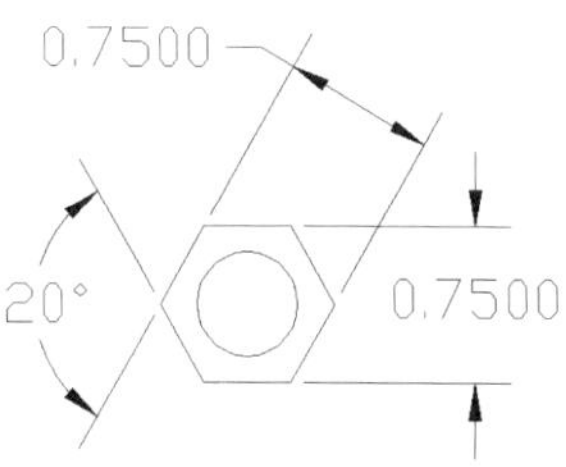

Figure 6.35

You will add constraints and dimensions until you have a fully constrained sketch. First, on you own, add a fix point constraint to anchor the sketch on the coordinate system and the equal length constraints shown in Figure 6.34. If you do not add a fix point constraint, you can dimension the location of one of the corners from the circular end view of the cylinder you have already created. Fixing the sketch on the coordinate system is a quick way of not having to locate your new sketch in relation to the previous one. It is not always the best way.

Add a fix point constraint for the lower left corner of the sketch, as shown in Figure 6.34.

Constrain four sides to be equal length, as in Figure 6.34.

Using Dimension Options: Angle, Align, Placement

Next you will add the dimensions. As you do, you will use the command line options to specify type and placement of the dimension. You will start with the angular dimension. Refer to Figure 6.35 as you follow the steps.

Right-click: in the graphics area

Pick: **Dimensioning, New Dimension**

Select first object: **pick on edge number 0**

Select second object or place dimension: **pick on edge number 1**

Enter dimension value or [Undo/Placement point] <60>: **P** ↵ (if you don't like the preview of the dimension placement)

Placement point: **pick again to place the dimension**

Placement point: ↵ (when you are satisfied with the placement)

Enter dimension value or [Undo/Placement point] <120>: **120** ↵

Select first object: **pick edge 0**

Select second object or place dimension: **pick on edge number 3**

Specify dimension placement: **pick above the part**

Enter dimension value or [Undo/Hor/Ver/Align/Par/aNgle/Ord/Diameter/pLace] <0.4086>: **A** ↵ (to align your dimension text)

Enter dimension value or [Undo/Hor/Ver/Align/Par/aNgle/Ord/Diameter/pLace] <0.7500>: **.75** ↵ (or type L to change the placement of the dimension)

Select first object: **pick edge number 2**

Select second object or place dimension: **pick on edge number 5**

Specify dimension placement: **pick right of the part**

Enter dimension value or [Undo/Hor/Ver/Align/Par/aNgle/Ord/Diameter/pLace] <0.7500>: **.75** ↵

Solved fully constrained sketch

Select first object: ↵ ↵

Your fully constrained and dimensioned sketch should be similar to Figure 6.35. Now you are ready to extrude the feature. Before you do, consider the orientation of your sketch on the coordinate system.

Switch to a left Front Isometric view

Changing the Default Extrusion Direction

Many commands for creating 3D geometry, such as Extrude, use the positive *Z*-axis as the default direction for the operation. Notice that the UCS icon, as shown in Figure 6.36, indicates that the *Z*-axis is positive in the direction of the shaft.

To extrude the hex head, you want to extrude it in the opposite, or negative, *Z*-direction. The Extrusion dialog box will not allow you to enter a negative value, but rather provides a Flip button with which you can specify a negative direction for the extrusion. When you select the Flip button, an arrow appears on screen to confirm the orientation. Move the dialog box, if necessary, so you can see the sketch as you pick Flip.

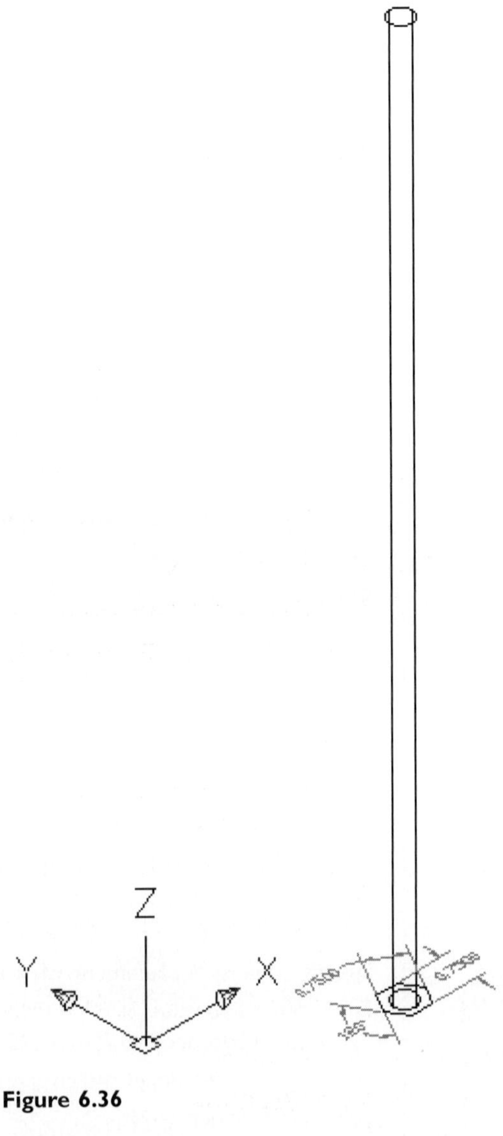

Figure 6.36

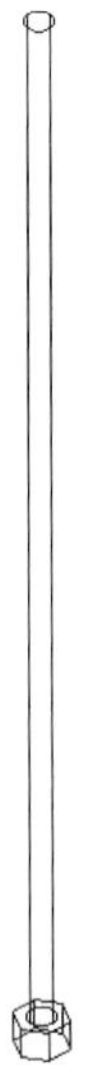

Figure 6.37

Right-click: in the graphics area

Pick: **Sketched & Work Features, Extrude**

Choose Operation: **Join**

Type Distance: **0.5000**

Choose Draft angle: **0**

Choose Termination Type: **Blind**

Pick: **Flip** (so arrow is pointing away from the shaft)

Pick: **OK**

The hex head is extruded, as shown in Figure 6.37.

Rename ExtrusionBlind1 to HexHead

Shade and spin the part

Save the part so far

Adding the Chamfer

To complete the bolt, you will add another kind of placed feature, a chamfer. Chamfers are commonly added to the ends of bolts to make them easier to insert into a hole. As with other dialog boxes, you can use the image tile in the Chamfer box to see the three options available for specifying a chamfer: equal distance, two distances, or distance plus an angle. When you are done, you will use the Equal distance.

Pick: **Zoom, Window** (to enlarge your view of the end of the shaft)

Right-click: in the graphics area

Pick: **Placed Features, Chamfer**

Choose Operation: **Equal distance**

Type Distance1: **.05**

Pick: **OK**

Select edges or faces to chamfer: **pick the circular edge at the end of the shaft**

Select edges or faces to chamfer: ↵

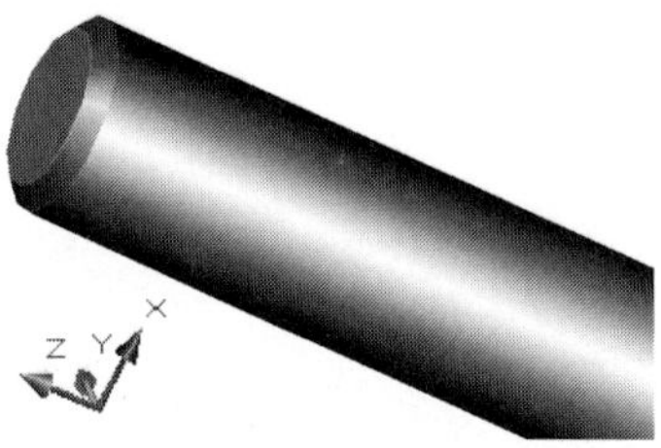

Figure 6.38

The chamfer appears, as shown in Figure 6.38.

Congratulations! You have finished this tutorial and completed four more parts for the Jack assembly. Save your work and close this part file.

Pick: **File, Save**

Pick: **File, Close**

Command Summary

Action	Command Name	Description
3D Chamfer	AMCHAMFER	Creates a beveled edge (placed feature) on 3D objects
Append Sketch	—	Adds 2D geometry to an existing profile
Create Polygon	POLYGON	Creates a regular (equilateral) polyline object; object can be inscribed or circumscribed about the circle defined by the radius
Draw an Arc	ARC	Creates 2D arc geometry
Extend	EXTEND	Extends existing 2D geometry to a specified boundary edge
Mechanical Options	AMOPTIONS	Manages settings, standards, and preferences to be used in parts, assemblies, and drawings
Regenerate View	REGEN	Regenerates the display of the model to reflect current settings and the model data
Revolve	AMREVOLVE	Creates a solid feature by revolving a profile about an axis
Set a Zoom Factor	ZOOM, <value>X	Zooms the view to a specified number of magnifications
Set View of Curved Surfaces	ISOLINES	System variable that controls the number of tesselation lines used to represent a curved surface in wireframe display
Work Axis	AMWORKAXIS	Creates a work axis from a sketched location or from an existing circular edge
Zoom, Extents	ZOOM, E	Zooms to the maximum magnification possible so that the extents of the drawing (visible or not) fit on screen

Exercises

Model the parts shown in Figures 6.39 through 6.43.

1. Candle Holder *Note:* All dimensions are in inches.

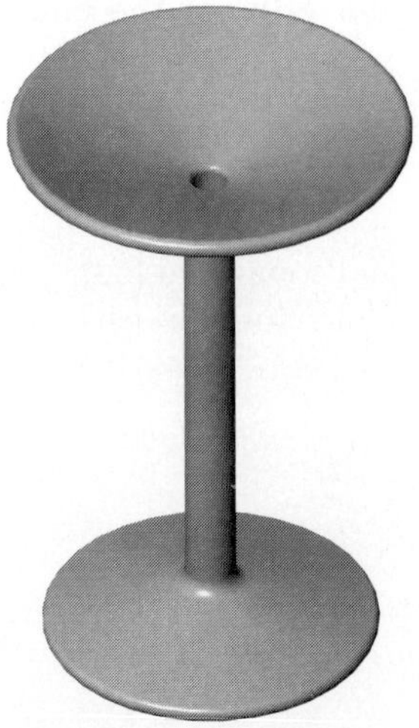

Figure 6.39a

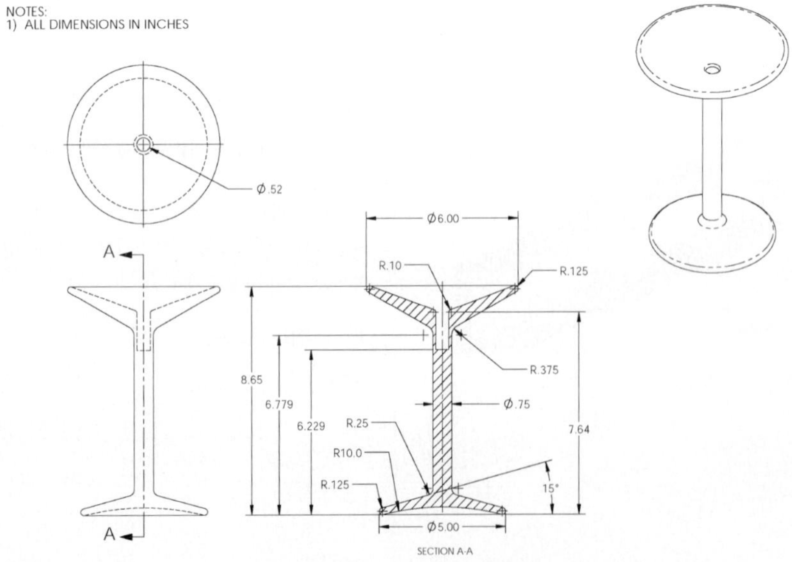

Figure 6.39

Figure 6.40a

2. **Door Stopper Body** *Note:* All dimensions are in inches.

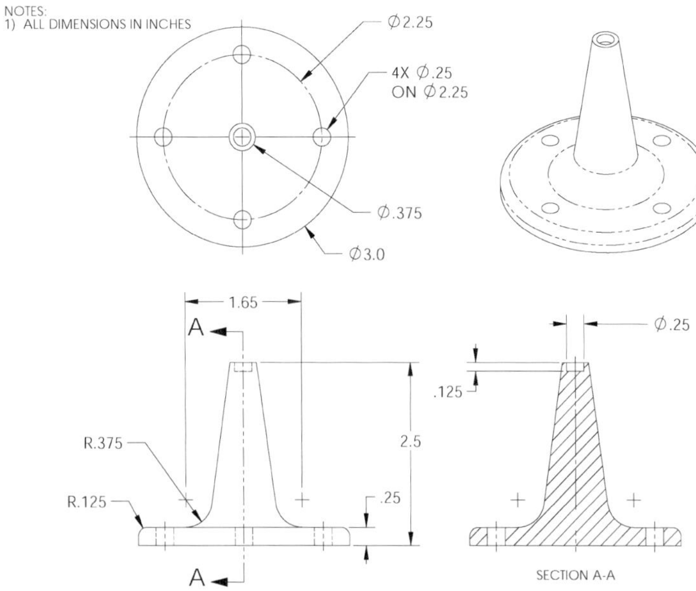

Figure 6.40

3. **Nozzle** *Note:* All dimensions are in millimeters. Add .25-mm fillets to outer edges.

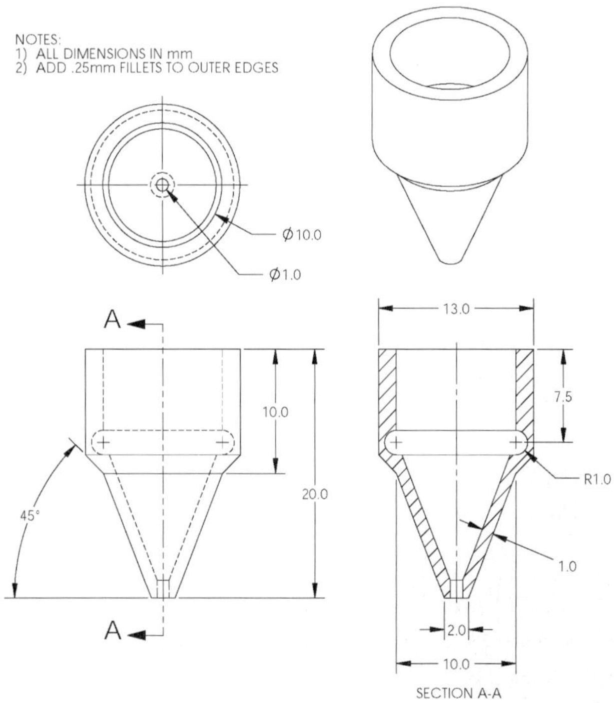

Figure 6.41

4. Shaft Guide *Note:* All dimensions are in inches.

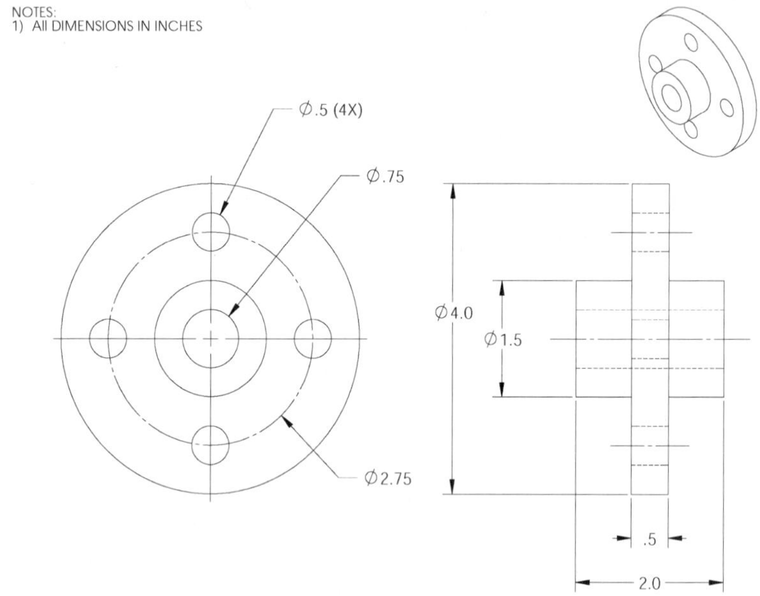

Figure 6.42

5. Top *Note:* All dimensions are in millimeters. All radii 1.0 unless otherwise specified. As an additional exercise, design your own top.

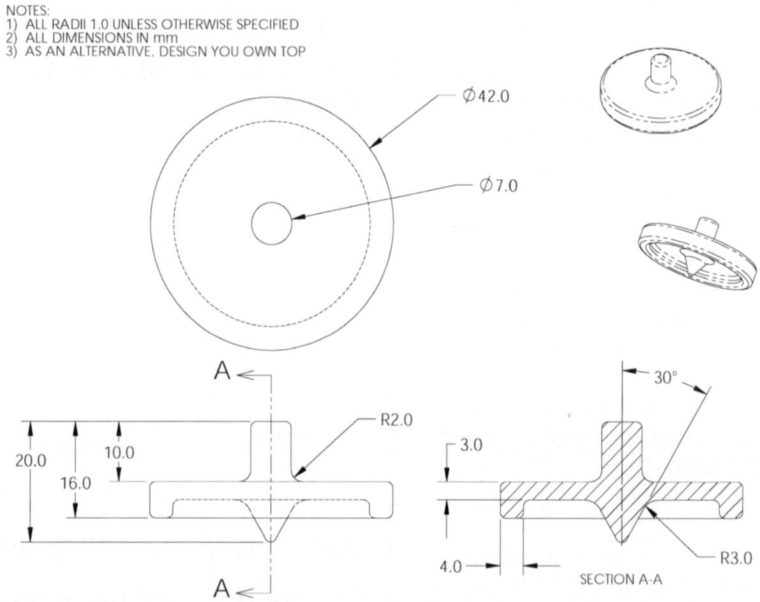

Figure 6.43

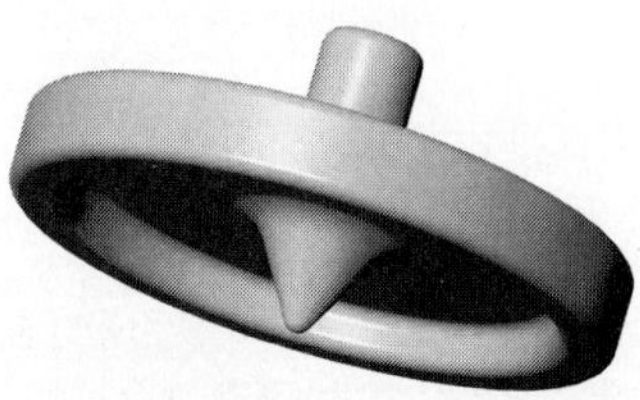

Figure 6.43a

Tutorial 7

Creating an Assembly

Objectives

When you have finished this tutorial, you will be able to

- Add external parts to an assembly.
- Add and edit assembly constraints.
- Use the 3D Manipulator.
- Add standard parts.
- Create an exploded assembly scene.
- Analyze a model for interference.
- Determine mass properties of a part.
- Localize external parts in an assembly.

Introduction

In this tutorial, you will begin working in Mechanical Desktop's assembly environment. Since Tutorial 2, you have been working in the Part environment, where you can create individual parts and layouts for part drawings. In the assembly environment, as you saw in Tutorial 1, you can combine part models into assembly models to analyze and refine how parts fit together, and to create drawings and scenes that document and illustrate the assembly.

Assembly models can include parts that have already been created, but you may also make parts, called local parts, while you are in the assembly environment. In addition, you can add standard parts from a variety of sources to your assembly model. You will create an assembly of the jack you have been working on that contains all three of these types of parts.

Checking for Data Files

Before you begin, check to see that you have copied the folder of completed parts (Jackparts) provided with the data files into your working directory. Although you could use your parts for the assembly, the datafiles will ensure that you achieve the results shown in the figures.

Tip

Right-click and drag the files to the new location in Windows Explorer, then pick "Copy here" from the context menu to copy the files.

Copy (don't move) the Jackparts directory from MDT-Data to MDT-Work

You should have the following parts available in *C:\MDT-Work\Jackparts:*

Tip

If you wish to use the parts you created, make a copy of them (don't just move them) in a folder called Jackparts in your MDT-Work directory. If you encounter difficulty, you can always start again using the data files provided.

- Baseplate
- SupportGuide
- JackTop
- Pin
- Stopper
- ThickLeg
- ThinLeg
- Bolt

Launching Mechanical Desktop 5 Power Pack

For this tutorial, you will use launch Mechanical Desktop so that it loads the Power Pack. You should have a Mechanical Desktop 5 Power Pack icon on your desktop (or on your Start menu). You will use this icon to launch Mechanical Desktop Power Pack. If you do not have Power Pack available, you can complete most of the tutorial using the standard Mechanical Desktop.

Double-click: **Mechanical Desktop 5 Power Pack icon**

Pick: **to minimize Mechanical Desktop Power Pack Today**

Load your user profile (if necessary)

You should see Mechanical Desktop Power Pack at the top of your screen, as shown in Figure 7.1. (If not, exit and launch the software again.) The Power Pack adds easy access to over 500,000 standard parts that you can add to your models.

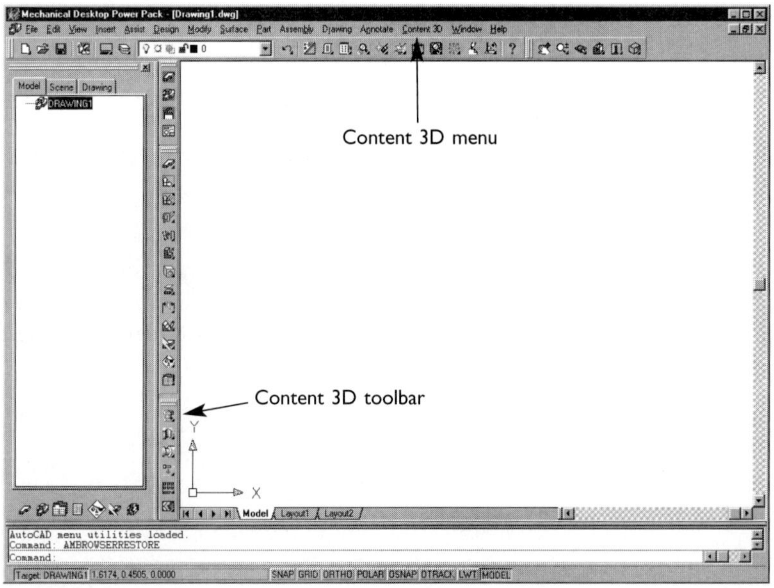

Figure 7.1

On your screen, you should now see an additional menu and toolbar, Content 3D, which is useful for adding standard parts and performing engineering analyses, as you will see in Tutorial 10. Additional items also appear on the Mechanical Main toolbar.

Starting a New Assembly

You will open a new file for the jack assembly and save it to your working directory.

Pick: **File, New**

Select how to begin: **Start From Scratch**

Pick: **English (feet and inches)**

Pick: **File, Save**

Choose Save In: **C:\MDT-Work**

Type File name: **JackAsmb** (as the file name)

Pick: **Save**

The file name and path, *C:\MDT-Work\JackAsmb.dwg*, should appear in the title bar at the top of the Mechanical Desktop window.

In many parametric modelers, assemblies and parts are stored in two different file types. In Mechanical Desktop, both kinds of models plus drawing information are stored in the *.dwg* format. Unless you specify a part file, the new file opened at startup will open an assembly environment. Notice that the Desktop Browser now has the assembly environment tabs: Model, Scene, and Drawing.

Switch to a front left isometric view of the default coordinates.

Pick: **Front Left Isometric View icon**

You should see the UCS icon displayed so that you see *X*-, *Y*-, and *Z*-axes.

Assembling the Jack Parts

To begin the jack assembly, you will add parts by attaching them as externally referenced parts. Later, you will create a new local part within the assembly and export it so it may be used in other designs if needed.

External and Local Parts

"External" and "local" refer to where the data for a part (or subassembly) is stored. *External* parts are stored in files separate from the assembly file. These are the individual part files you created so far. External parts may be used in more than one assembly. Typically, you will store each part in a separate drawing file and then attach these external parts to form an assembly.

If you create a part in an assembly, that is a *local* part. It is stored in the assembly file and cannot be used in a different assembly unless you export it to a separate, external part file. You may find it convenient at times to create local parts within an assembly, or to localize an external part so it is stored with the assembly. Localizing an external part breaks the link to the original part file. That way if you change the part in the subassembly, that part will not change externally and in all the other places it has been used.

Because parts and assemblies can have the same bidirectional associativity as the model and drawing, it is important to organize your part files so it is clear which assemblies they belong to.

Reference

Engineering Design Communication

Parametric Assemblies, pp. 292–293

Organizing Your Parts

When you attach an external part to an assembly, Mechanical Desktop keeps track of that part's location on your hard drive. That is, the part itself is not stored with the assembly, but the location of the part file in your computer system is. When a part is updated from an assembly, the software must be able to locate the part on the hard drive to make the changes to the part model. Conversely, when you open an assembly file, Mechanical Desktop goes to the external part files for data needed to build the assembly model. It is important for you to keep the files that you use in parametric modeling organized. Mechanical Desktop helps you do this using the metaphor of an assembly catalog.

The Assembly Catalog

The assembly catalog is how you tell Mechanical Desktop where the parts that you will use in the assembly are stored. It can also act as a reference for you to see where the parts came from as they were added to the assembly. If you move a part to a different location on your drive using the operating system (for example, by using Windows Explorer), you will have to tell Mechanical Desktop where to find that external part file. It cannot keep track for you. However, you can use the assembly catalog to help you find and reload parts.

To open the assembly catalog and begin adding parts to the assembly,

Right-click: in the graphics area

Pick: **Assembly Menu**

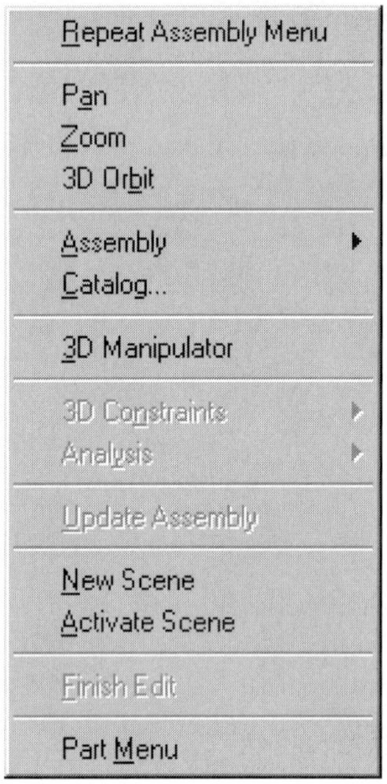

Figure 7.2

The context menu changes to show the assembly menu choices, as seen in Figure 7.2. Notice the selection, Part Menu, at the bottom of the context menu. You can toggle between the part and assembly context menus by picking on these selections.

Pick: **Catalog** (from the context menu)

The Assembly Catalog appears on your screen, as shown in Figure 7.3. Make sure that the External tab is uppermost.

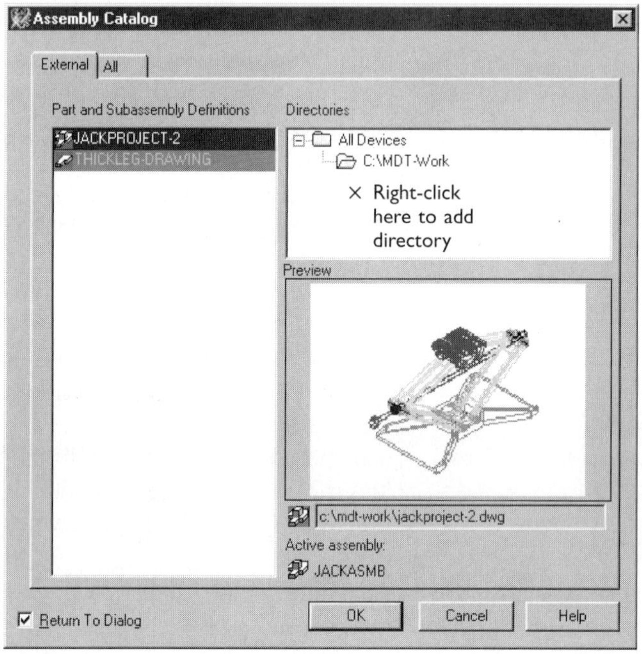

Figure 7.3

You can access additional context menus while you are in the Assembly Catalog dialog box by right-clicking in either the Directories area or the Part and Subassemblies Definitions area. You will right-click in the Directories area to tell Mechanical Desktop where the parts that you want to assemble are located.

Right-click: **in the Directories area of the Assembly Catalog**

Pick: **Add Directory**

The Browse for Folder dialog box appears. Use it to select the directory where your parts for the jack are stored. This should be *C:\MDT-Work\Jackparts*. (If you do not have this directory, stop and create it as instructed at the beginning of this tutorial.) When you have the correct directory selected,

Pick: **OK**

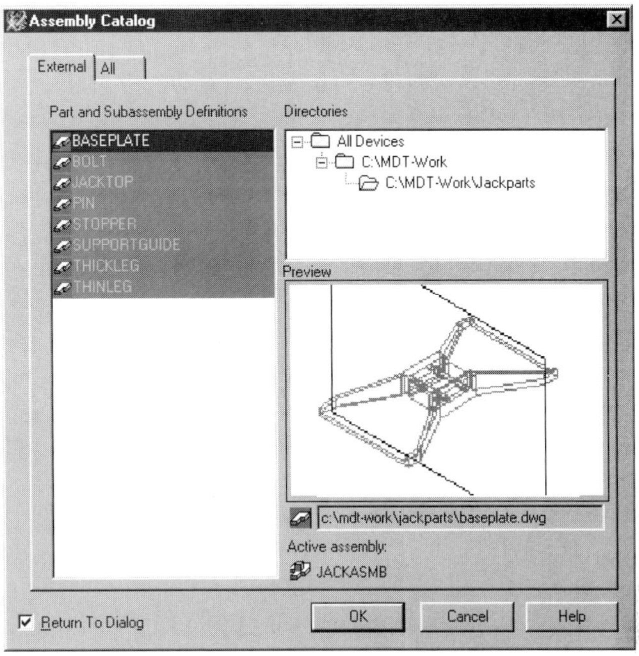

Figure 7.4

The directory is added to the list at the right of the Assembly Catalog. All of the parts stored in the selected directory are listed at the left, as shown in Figure 7.4.

Next you will attach parts to the assembly.

Adding an External Part

To add an external part to the assembly,

Right-click: **Baseplate** (in the list at the left of the Assembly Catalog)

Pick*:* **Attach** (from the context menu)

You return to your drawing for the following selections.

Specify new insertion point: **0,0,0** ↵

Specify insertion point for another instance or <continue>: ↵

The insertion point specifies where the part instance will be located on the coordinate system in the assembly. An instance is like a copy; you can attach several instances of the same external part file. This instance of the Baseplate part will be located at the origin of the new coordinate system.

The dialog box returns to your screen. You will exit the dialog box and look at the results.

Pick: **OK**

The Baseplate appears on your screen, as shown in Figure 7.5.

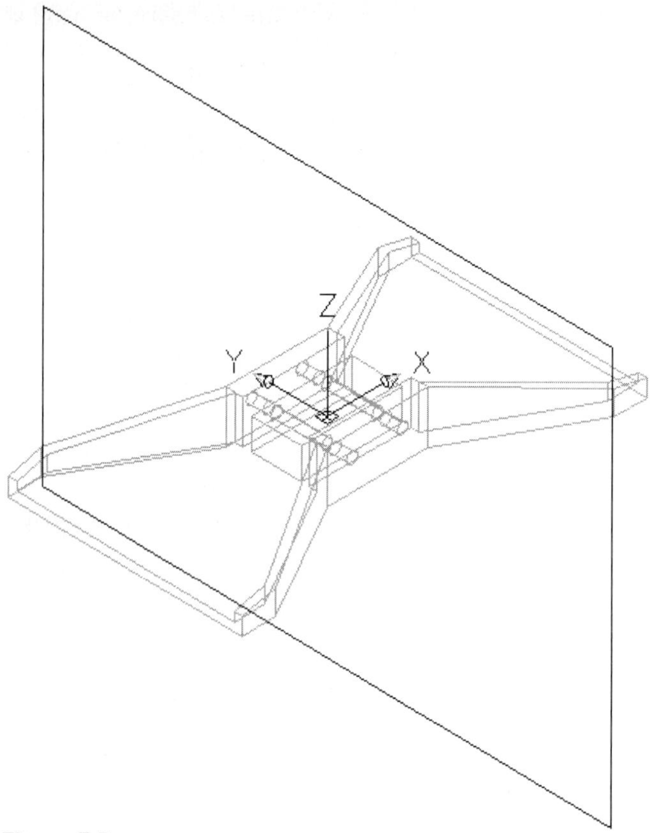

Figure 7.5

Next you will bring in the remaining parts.

Right-click: in the graphics area

Pick: **Catalog**

The assembly catalog returns to your screen. This time you will double-click on the part name to attach it to the assembly.

Double-click: **JackTop**

Specify new insertion point: **pick above the Baseplate**

Specify insertion point for another instance or <continue>: ↵

You return to the dialog box,

Double-click: **ThickLeg**

Specify new insertion point: **pick to the right of the Baseplate**

Specify insertion point for another instance or <continue>: **pick to the left of the Baseplate**

Specify insertion point for another instance or <continue>: ↵

Double-click: **Pin**

Specify new insertion point: **pick above and to the left of the Baseplate**

Specify insertion point for another instance or <continue>: **pick below the other pin, continue until you have four total pins**

Specify insertion point for another instance or <continue>: ↵

Double-click: **Stopper**

Specify new insertion point: **pick below the Baseplate**

Specify insertion point for another instance or <continue>: ↵

Double-click: **ThinLeg**

Specify new insertion point: **pick to the left of the parts**

Specify insertion point for another instance or <continue>: **pick to the right of the parts**

Specify insertion point for another instance or <continue>: ↵

Double-click: **SupportGuide**

Specify new insertion point: **pick below the ThinLeg**

Specify insertion point for another instance or <continue>: **pick below the other ThinLeg part**

Specify insertion point for another instance or <continue>: ↵

Double-click: **Bolt**

Specify new insertion point: **pick to the left of the parts**

Specify insertion point for another instance or <continue>: ↵

Pick: **OK** (to exit the Assembly Catalog)

Save your work

Now you should have the Baseplate, JackTop, two ThickLegs, two ThinLegs, two SupportGuides, four pins, the bolt, and the stopper on your screen, similar to Figure 7.6.

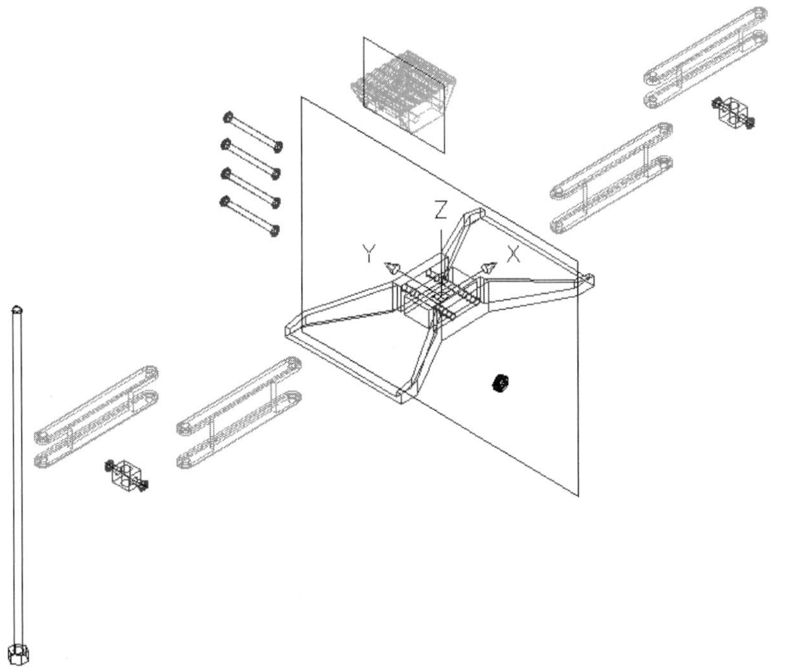

Figure 7.6

Zoom your drawing as needed so that the parts show on the screen clearly.

The parts are placed in the assembly, but they are not fit together. Assembly constraints are used to align parts with one another. First you will use assembly constraints to position the JackTop so that it always stays centered above the Baseplate.

Assembly Constraints

There are four ways that you can constrain parts to fit with one another: mate, flush, angle, and insert (Figures 7.7 through 7.10).

Mate

The Mate constraint allows you to select points, axes, planes, or nonplanar faces on different parts and mate or match them together so that the selected items are parallel and either touch facing each other or are offset by a distance.

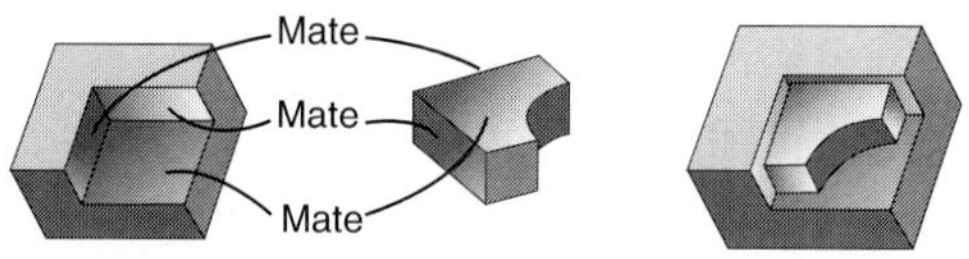

Figure 7.7

Flush

The Flush constraint allows you to select two items and constrain them so that they are parallel and face the same direction. They can be offset from one another by a distance that you specify.

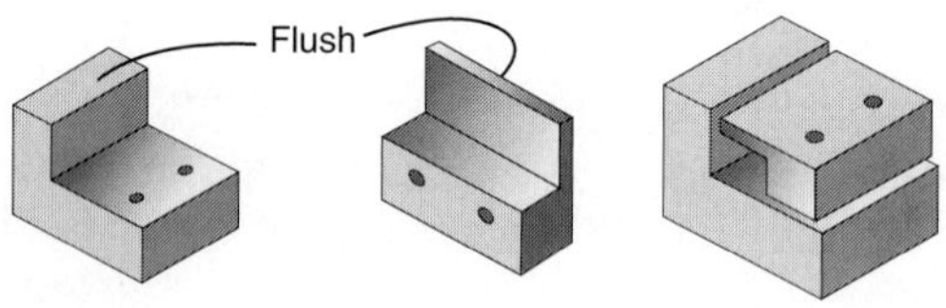

Figure 7.8

Insert

The Insert constraint makes the selected circular edges or faces align on the same axis. It also forces their faces to lie in the same plane or be separated by an offset distance that you specify.

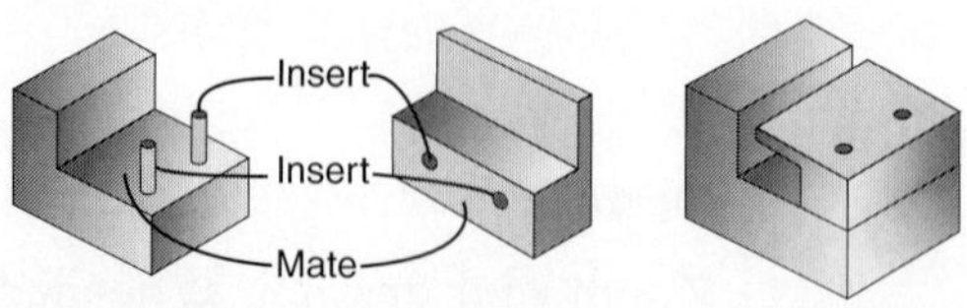

Figure 7.9

Angle

The Angle constraint allows you to control the angle between two planes, two vectors, or a combination of a plane and a vector.

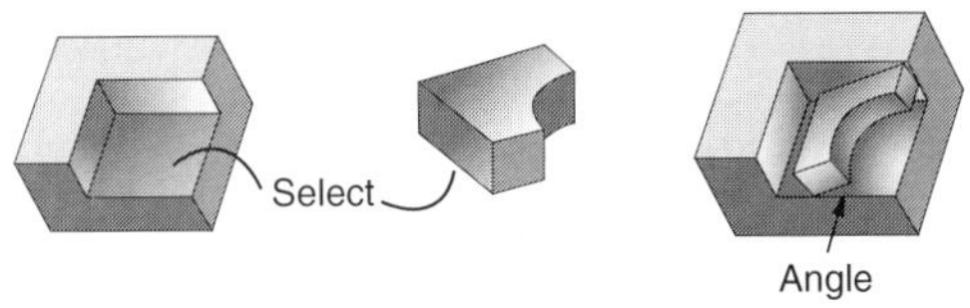

Figure 7.10

You will use the Mate constraint first to mate the center work planes that have been added to the Baseplate and JackTop parts. You will see how useful work planes are in assembling parts.

Figure 7.11 shows the Baseplate and JackTop parts arranged so that you can see the middle work planes clearly and the locations on the object you will select in the next steps.

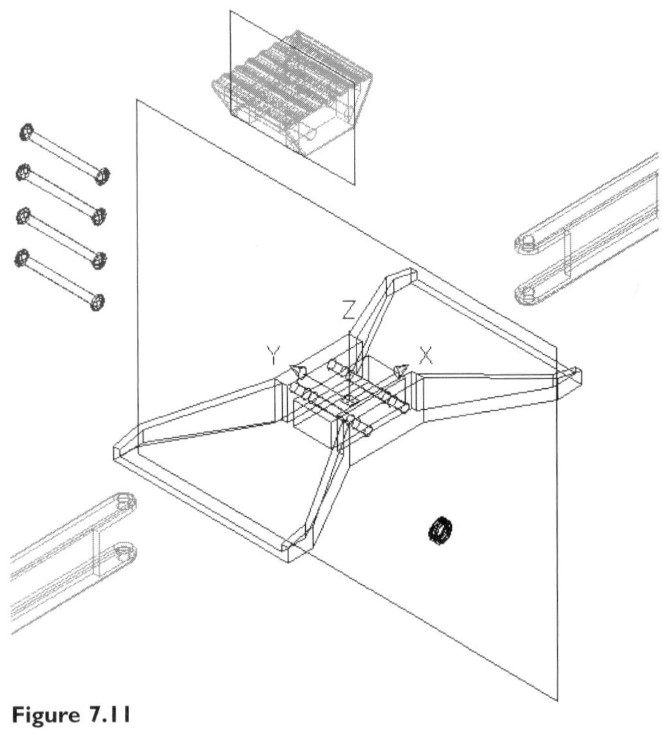

Figure 7.11

Right-click: in the graphics area

Pick: **3D Constraints, Mate** (from the context menu)

Select first set of geometry: **pick on the work plane through the middle of the Baseplate** (as shown in Figure 7.11)

First set = Plane

The middle work plane should be selected. An arrow appears as shown in Figure 7.12, indicating the direction the surface will be mated.

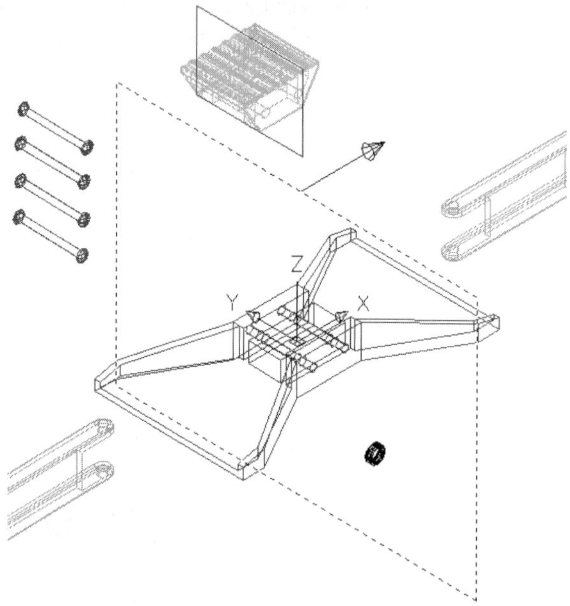

Figure 7.12

Enter an option [Clear/Flip] <accEpt>: ↵ (if the arrow is oriented correctly, if not type F ↵ to flip the direction and then press ↵)

Select second set of geometry: **pick the middle work plane on the JackTop part**

Second set = Plane

Enter an option [Clear/Flip] <accEpt>: **F** ↵ (if necessary to flip the arrow so that it points toward the arrow on the other part, as shown in Figure 7.13.)

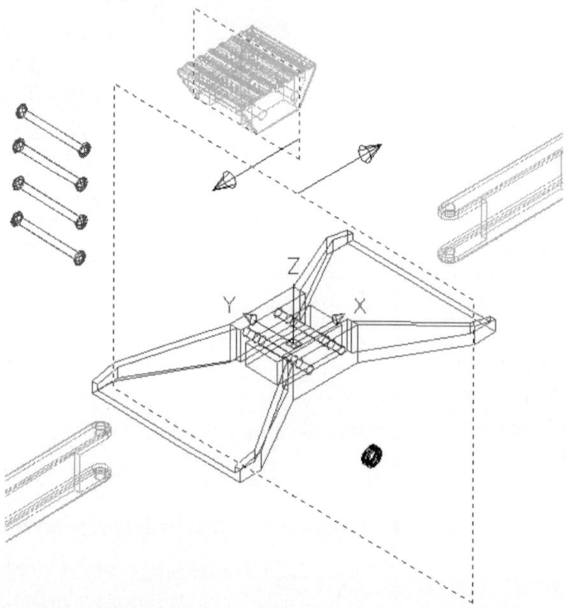

Figure 7.13

Enter an option [Clear/Flip] <accEpt>: ↵ (to accept when the arrow is facing correctly)

Enter offset <0.0000>: ↵

The JackTop part moves so that the work planes touch and they are aligned along their centers (or along the Y-axis). Next you will add another mate constraint. This time you will specify to mate two faces—one on the JackTop and the other on the Baseplate so that these surfaces are parallel and face each other. You will give an offset distance between them that you can use to control the height of the top part above the Baseplate. The first mate constraint specified that the planes are located parallel to and facing each other, but not that they aligned on all sides. Adding this new constraint will limit the JackTop so its bottom surface will remain parallel to the top of the Baseplate. Later you will constrain the ThinLeg to the JackTop to position it in the other direction.

When your mouse is over the object before you pick the surface to constrain, the entire object will become highlighted. After you pick, you will see the line or face you picked highlighted and you can use the mouse buttons and command options to select the right surface.

Pick: **Zoom Window to enlarge the view of the Baseplate and JackTop**

Right-click: in the graphics area

Pick: **3D Constraints, Mate**

Select first set of geometry: **pick on the top center surface of the Baseplate, as shown in Figure 7.14**

First set = Plane

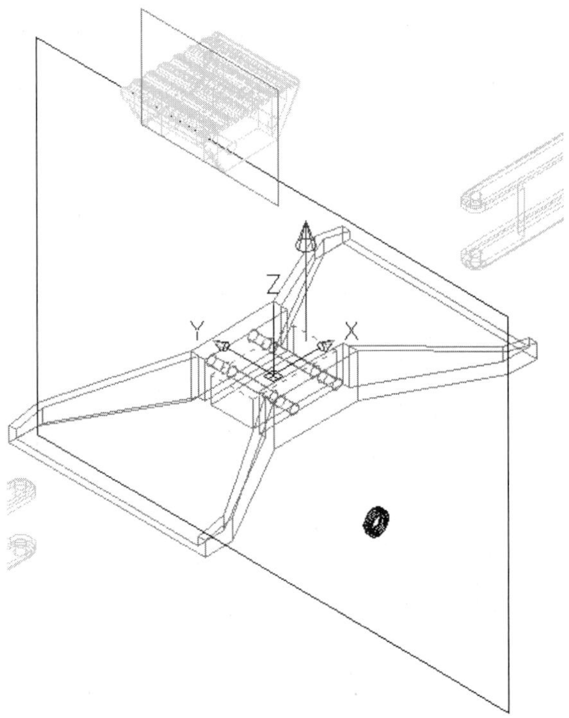

Figure 7.14

Tip

You can type C ↵ at the command prompt to select again if a line or surface on the wrong feature is highlighted. If you do not see the First set = Plane at the prompt, an edge or axis may have been selected instead of the surface. If so, clear the prompt and try again or type an option letter at the prompt, Clear/fAce/Point/cYcle. Typing A ↵ will switch to selecting a face rather than an edge. You can use the other options to select a point, or to cycle through the selections using the pick button on your mouse.

The top surface should be selected. An upward arrow appears as shown in Figure 7.14, indicating the direction the surface will be mated. If a line is highlighted, type A to highlight a face. Use the left mouse button, or type Y, to cycle through the faces.

Tip

Use the 3D Orbiter to orient the Jack-top so you can pick on the lower surface. You may need to launch the 3D Orbiter once before you can use it while a command is active. You can also Zoom in on the part so you can better pick the wire that defines the surface.

Enter an option [Clear/Next/Flip/cYcle] <accEpt>: ↵ (if the arrow is oriented correctly, if not type F ↵ to flip the direction and then press ↵)

Select second set of geometry: **pick the center face of the JackTop shown in Figure 7.15**

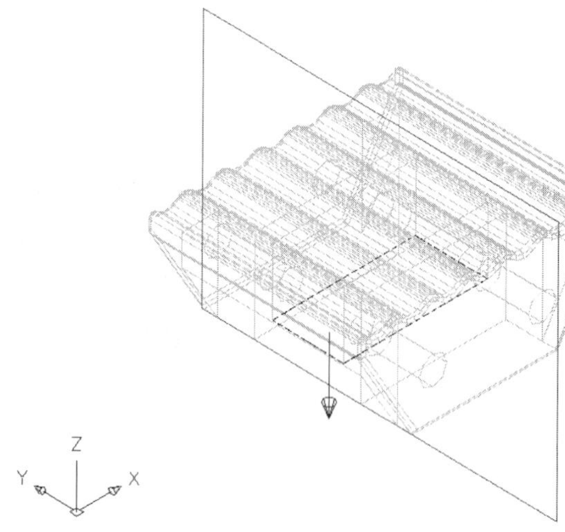

Figure 7.15

Second set = Plane

Enter an option [Clear/Next/Flip/cYcle] <accEpt>: ↵ (if the correct surface is highlighted if not type N ↵ to select the Next surface, continue until the correct surface is selected)

Enter an option [Clear/Flip] <accEpt>: ↵ (to accept when the arrow is facing correctly)

Enter offset <0.0000>: **7** ↵

The JackTop part moves so that it is above the middle plane of the Baseplate and is 7 units above it, as shown in Figure 7.16.

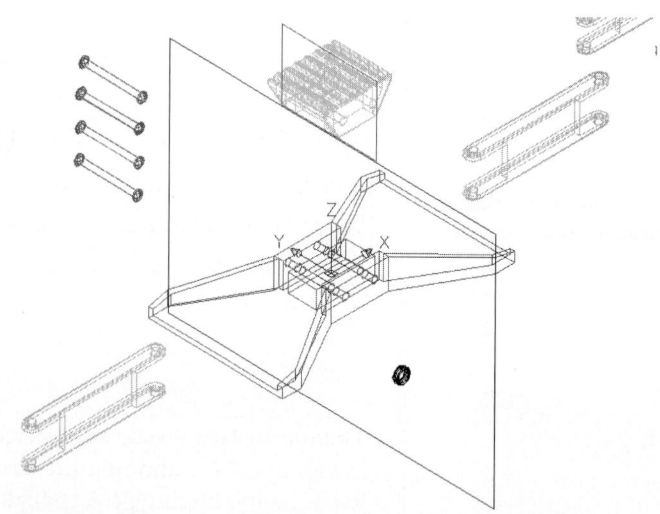

Figure 7.16

Pick: **Zoom Realtime and enlarge the left-hand ThickLeg part** (as shown in Figure 7.17)

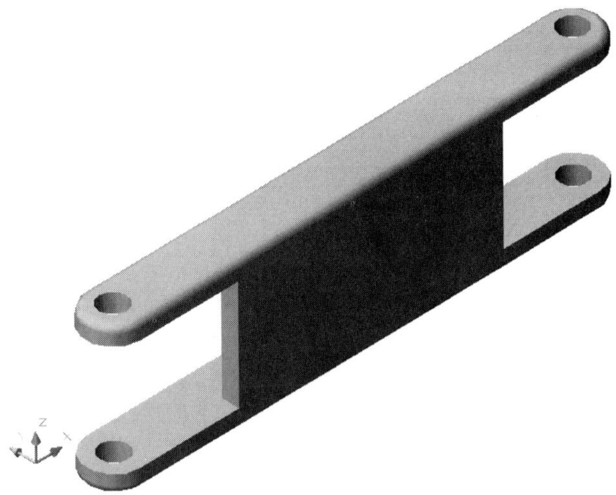

Figure 7.17

You will use the 3D Manipulator to move some of the parts so they are closer to their final orientations.

Using the 3D Manipulator

You can use the 3D Manipulator to position parts in a way that makes them easier to select in the assembly. The 3D Manipulator displays handles that you can grab to rotate or move the part. The central handle is used for free move. It allows you to grab the part and relocate it by picking with the mouse. You can also move or rotate a part by clicking and dragging on the handles on the axes. If you click and drag along the selected axis, movement will be constrained along that axis. If you click and drag in the direction of a rotation, the part will rotate about the perpendicular axis. You can rotate freely by dragging, or enter an exact angle to rotate to.

You will use the dialog box to rotate the legs of the jack so the midplates of the ThickLegs are on the bottom and the midplates of the ThinLegs are on top, as they will be when you assemble them. You will rotate all of them about the X-axis.

Toggle shading on so you can see the plates more clearly as you rotate

Right-click: in the graphics area

Pick: **3D Manipulator**

Select objects: **pick on the ThickLeg part**

Select objects: ↵

Middle handle is FREE MOVE, other handles are ROTATE and MOVE

Select handle to manipulate (Selecting no handle shows option dialog) or

[Undo/Mode (follow)] <Exit>: **pick and drag to the right on the handle at the positive end of the Z-axis**

ROTATE mode, rotating around X-axis [RETURN for options]: **if you see this message, press ↵; if not, press ESC to try again**

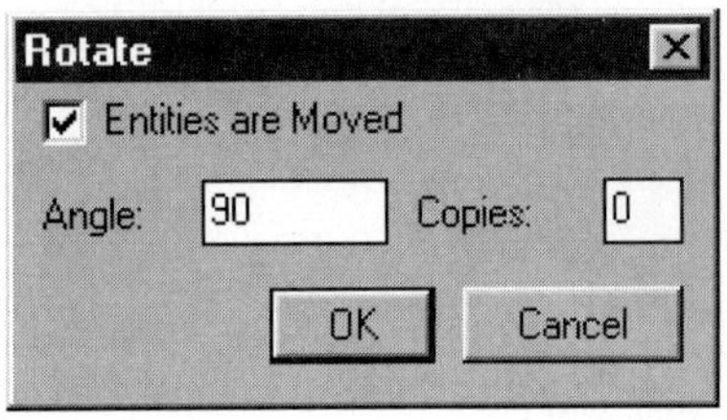

Figure 7.18

The Rotate dialog box appears, as shown in Figure 7.18. You will enter 90 to rotate the leg 90 degrees about the X-axis so the midplate is on the bottom.

Type Angle: **90**

Pick: **OK**

The part moves and you are able to select again, undo, or exit.

Select handle to manipulate (Selecting no handle shows option dialog) or

[Undo/Mode (follow)] <Exit>: ↵

Repeat these steps on your own to rotate all four legs so your screen looks like Figure 7.19. The midplates of both ThinLegs should be on top (a rotation of −90, or 270, degrees). Zoom your model as need to see the parts easily.

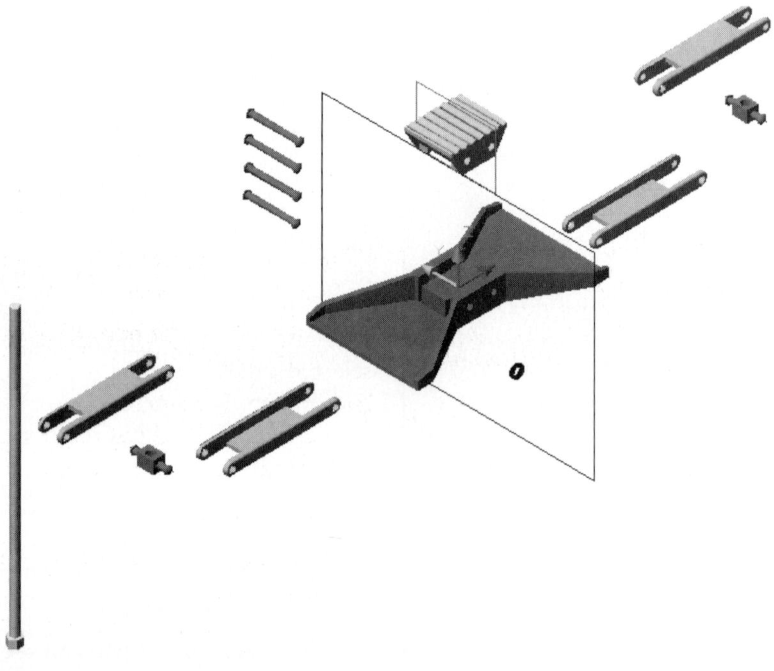

Figure 7.19

Save your work at this time

Insert Constraint: Insert and Mate

This time you will use the Insert constraint to align the parts and mate a surface at the same time.

You will use the Insert constraint first to align the axis of the cylinder of the SupportGuide pin and the holes through the ThinLeg. During the same command, you will align the square face so that it touches the inside flat surface of the ThinLeg part. The circular edges you select will define both the axis and the face to be aligned.

Zoom up the left SupportGuide and Thinleg

Right-click: in the graphics area

Pick: **3D Constraints, Insert** (from the context menu)

Select first circular edge: **pick on the circle on the square face** (where the pin cylinder meets it, as shown in Figure 7.20)

First set = Plane/Axis

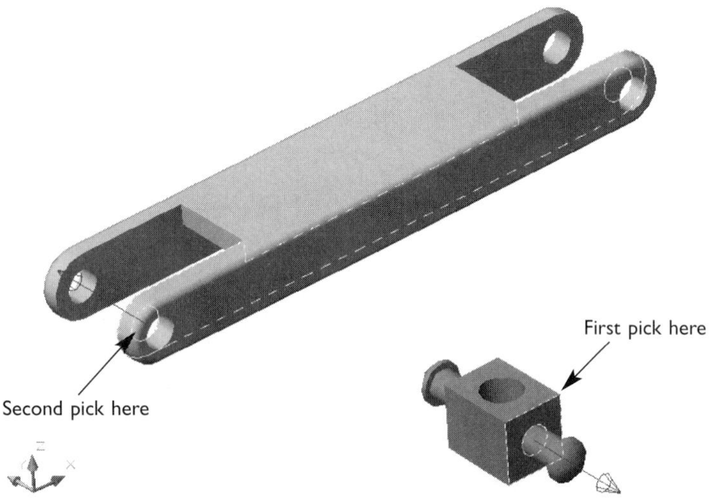

Figure 7.20

A line appears, establishing the axis direction. If it appears as shown in Figure 7.20, you will accept it. If not, you will type C at the command options to Clear the selection and start again.

Enter an option [Clear/Flip] <accEpt>: [Enter] (if you have the correct surface and direction selected)

Select second circular edge: **pick the inside right inner hole, as shown in Figure 7.20**

Second set = Plane/Axis

Enter an option [Clear/Flip] <accEpt>: [Enter] (if you see the arrow pointing inward, if not use F for Flip)

Enter offset <0.0000>: [Enter]

The Insert part moves so that its faces and holes are aligned with the ThinLeg, as shown in Figure 7.21.

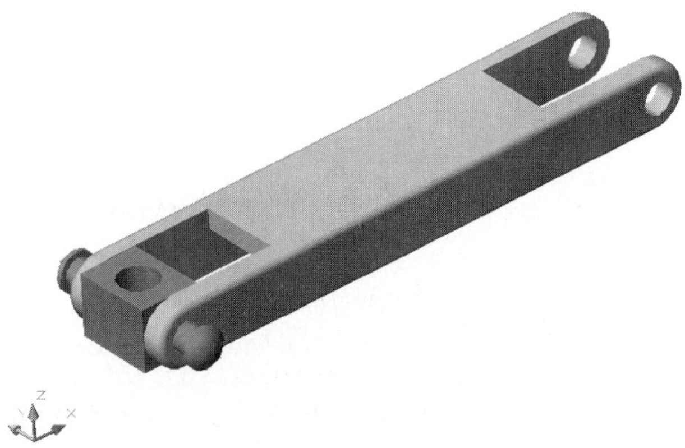

Figure 7.21

Degrees of Freedom

When a part is constrained along one or two directions, it can still move in other ways. The number of axes along which a part can move is termed its *degrees of freedom* (DOF). Typically in a parametric assembly, the more constraints you add between two parts, the fewer the degrees of freedom that part has. To view the degrees of freedom for a part graphically,

Right-click: the **SupportGuide you just mated** (from the model tree)
Pick: **DOF Symbol**

The directions along which the insert part can move are shown as arrows on the screen, as in Figure 7.22. As you can see, you can still rotate the SupportGuide part along the insert axis.

Tip
If it is hard to see the symbol, zoom in or out so the rotation arrow moves off the part.

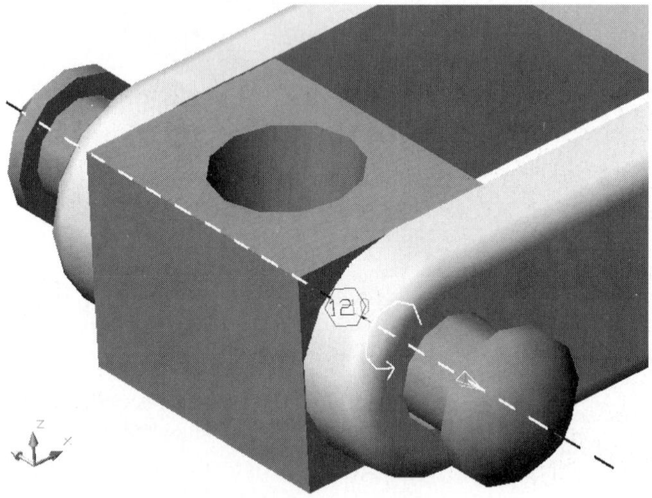

Figure 7.22

Checking the DOF symbol can help you determine whether a part is constrained to move in only the allowed directions. If you need to fix the position further, you can add more constraints. In this case you will add the bolt through the SupportGuide later and that will constrain it further. At this point the SupportGuide should still be able to rotate so that the bolt can align through the holes, so you will not constrain it further.

On your own, insert the Support Guide to the ThinLeg part on the right side of the assembly and save your work.

Tip
Picking the right geometry the first time generally depends on how unambiguous the first pick is. Zoom or use the 3D Orbiter so you can target the circle or line where it is not on top of other lines.

Mating the ThickLeg

Next you will use the Mate constraint to align the ThickLeg with the holes in the Baseplate. Refer to Figure 7.23 for the locations to select.

Zoom your view so you can select the parts easily.

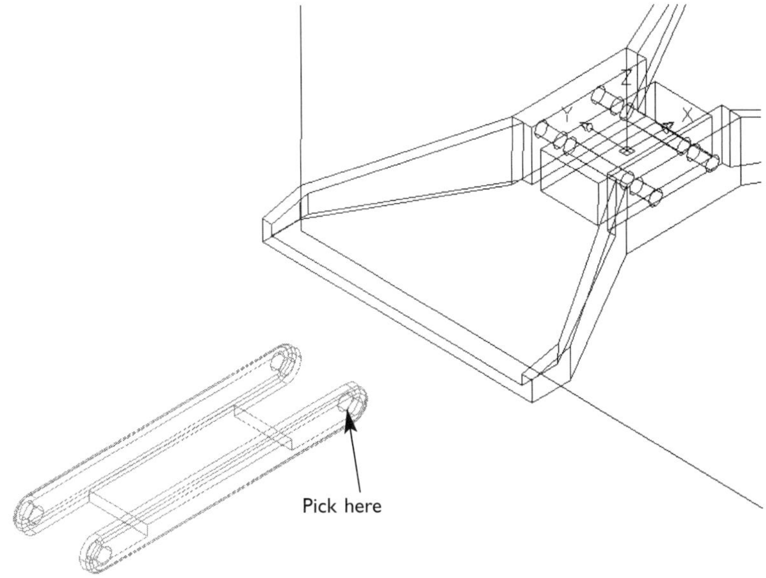

Figure 7.23

Switch to a wireframe view.

Right-click: in the graphics area

Pick: **3D Constraints, Mate**

Select first set of geometry: **pick on the inner right-hand hole through the outside of the ThickLeg**

First set = Axis, (arc)

The axis should appear aligned through the ThickLeg holes, as shown in Figure 7.24.

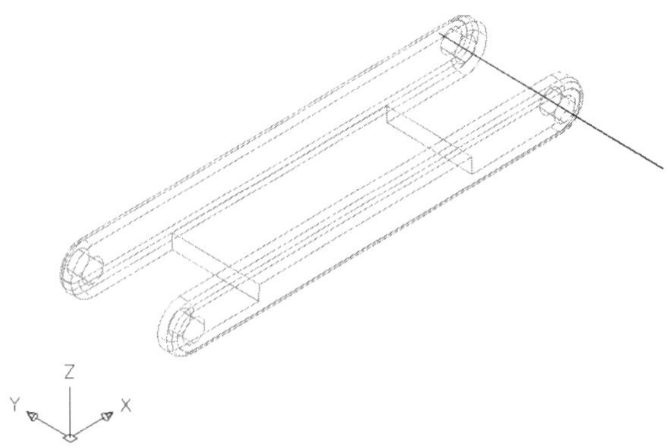

Figure 7.24

Select first set or [Clear/fAce/Point/cYcle] <accEpt>: **right-click** (if the axis is shown correctly; if not type C [↵] to clear and select again)

Select second set of geometry: **pick on the rounded holes through the near side at the left end of the middle block of the Baseplate, as shown in Figure 7.25**

Second set = Axis, (arc)

The second axis should be shown through the holes in the Baseplate, as shown in Figure 7.25.

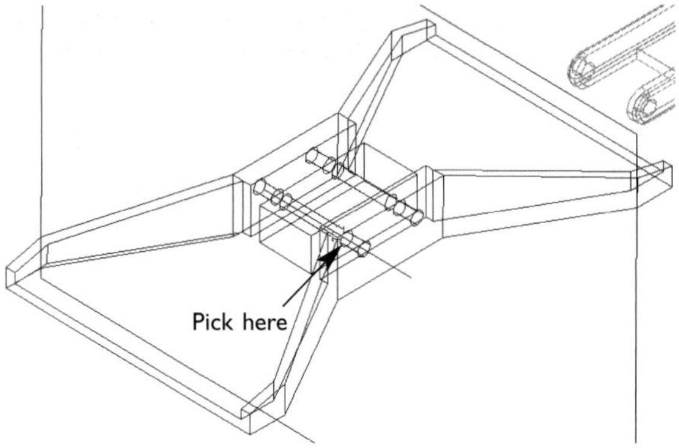

Figure 7.25

Select second set or [Clear/fAce/Point/cYcle] <accEpt>: **right-click** (if the axis is shown correctly; if not type C ↵ to Clear and select again)

Enter offset <0.0000>: ↵

The axis of the holes through the ThickLeg aligns with the holes in the Baseplate and the surfaces are mated, as shown in Figure 7.26

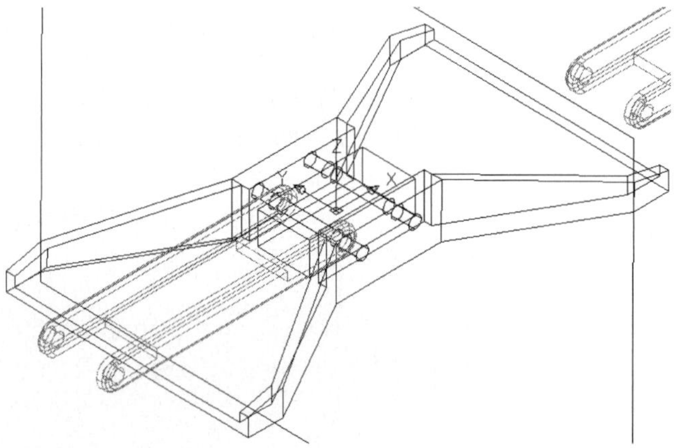

Figure 7.26

Pick: **Top view icon**

Your view switches to show the top view of the parts, as shown in Figure 7.27. The parts are aligned on the axis, but the surfaces of the ThickLeg and Baseplate parts do not necessarily match up. The Mate constraint aligns faces or axes so they touch; it does not constrain the faces fully. You will add an additional mate constraint to match the surface of one part to the surface of the other.

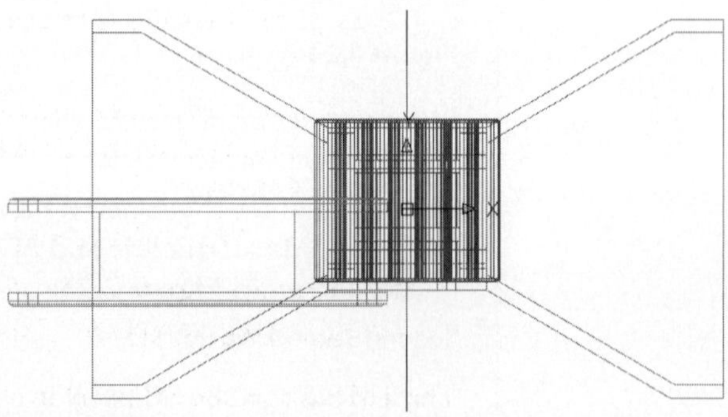

Figure 7.27

Before you do, you will fit the entire assembly to your screen. Then you will use the 3D Manipulator to move the ThickLeg part away from the sub-assembly to make it easier for you to select its features. You will change the location for the 3D Manipulator so that it aligns with the ThickLeg entity.

Pick: **Front Left Isometric view icon**

Pick: **Zoom Window icon and enlarge the assembly so that it fills your screen**

Right-click: in the graphics area

Pick: **3D Manipulator**

Select objects: **pick on the ThickLeg part**

Select objects: ↵

Middle handle is FREE MOVE, other handles are ROTATE and MOVE

Select handle to manipulate (Selecting no handle shows option dialog) or

[Undo/Mode (follow)] <Exit>: **pick near the center of the ThickLeg**

The Common Options dialog box appears on your screen, as shown in Figure 7.28.

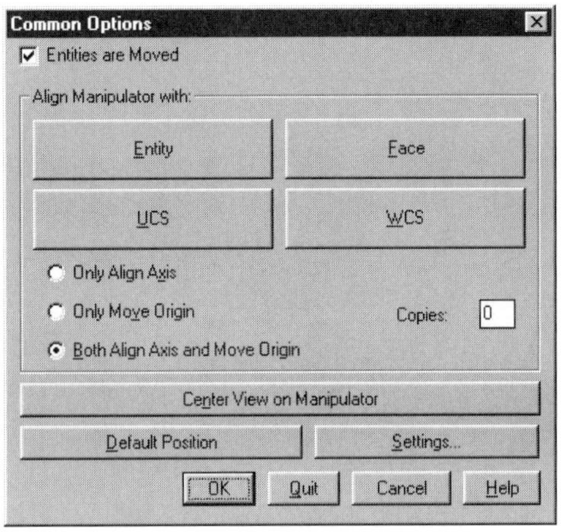

Figure 7.28

Choose Align Manipulator With: **Entity**

Select object: **pick on the ThickLeg**

Middle handle is FREE MOVE, other handles are ROTATE and MOVE

Select handle to manipulate (Selecting no handle shows option dialog) or

[Undo/Mode (follow)] <Exit>: **pick on the handle at the center of the axes** (It can be difficult to see this handle because it is in the center of the part.)

FREE MOVE handle selected; move cursor in arbitrary direction: **move the cursor to position the ThickLeg left of the Baseplate and press the pick button**

Middle handle is FREE MOVE, other handles are ROTATE and MOVE

Select handle to manipulate (Selecting no handle shows option dialog) or

[Undo/Mode (follow)] <Exit>: ↵

The ThickLeg is moved away from the other parts, as shown in Figure 7.29.

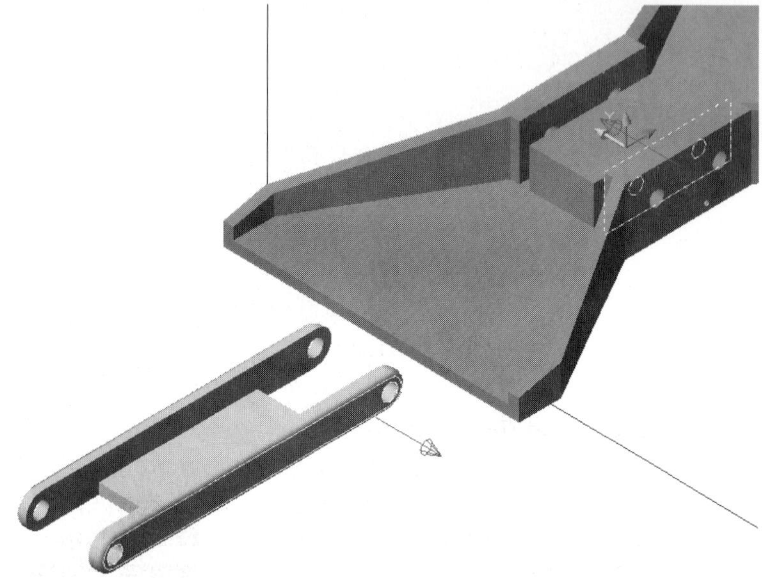

Figure 7.29

Now you are ready to mate the surfaces of the ThickLeg and the Baseplate. Refer to Figure 7.29 for the next selections. You will type A ↵ at the command options to select to use faces as the type of 3D entity to mate.

Use Zoom Window to enlarge the ThickLeg and left end of the Baseplate

Right-click: in the graphics area

Pick: **3D Constraints, Mate**

Select first set of geometry: **pick the right outside rounded side of the ThickLeg, as shown in Figure 7.29**

First set = Axis, (line)

Select first set or [Clear/fAce/Point/Next/cYcle] <accEpt>: **A** ↵

First set = Plane

Enter an option [Clear/aXis/Point/Next/Flip/cYcle] <accEpt>: ↵ (if the correct surface is highlighted and the arrow points outward from the surface, if not, type C ↵ and select again)

Select second set of geometry: **pick on the inner flat face of the Baseplate, as shown in Figure 7.29**

Second set = Point, (line)

Select second set or [Clear/aXis/fAce/Next/cYcle] <accEpt>: **A** ↵

Second set = Plane

Enter an option [Clear/aXis/Point/Next/Flip/cYcle] <accEpt>: ↵ (if the correct surface is highlighted; if not use the command options, N for Next to go on to the next surface, or C to clear and select again)

Enter offset <0.0000>: ↵

Pick: **Top View icon**

Pick: **Gouraud shaded icon**

Pick: **Left Front Isometric icon**

The ThickLeg pops back into place in the subassembly and now the surfaces are correctly mated, as shown in Figures 7.30a and 7.30b.

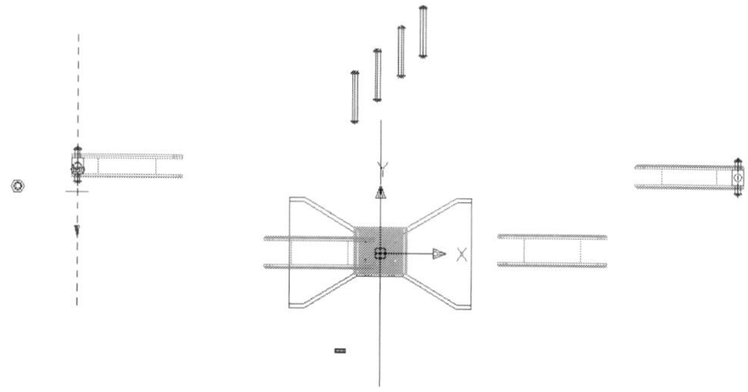

Figure 7.30a

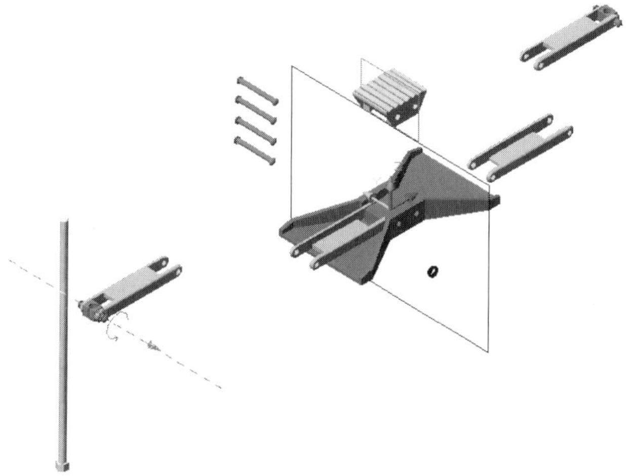

Figure 7.30b

Constraining the ThinLeg and SupportGuide

Next you will mate the axis of the holes through the ThinLeg with the holes through the ThickLeg. Because the ThinLeg and SupportGuide are constrained at this axis, you will see an additional prompt that confirms first the part you wish to constrain before you select the axis.

Use Zoom Window so you can see the holes in both parts, as shown in Figure 7.31

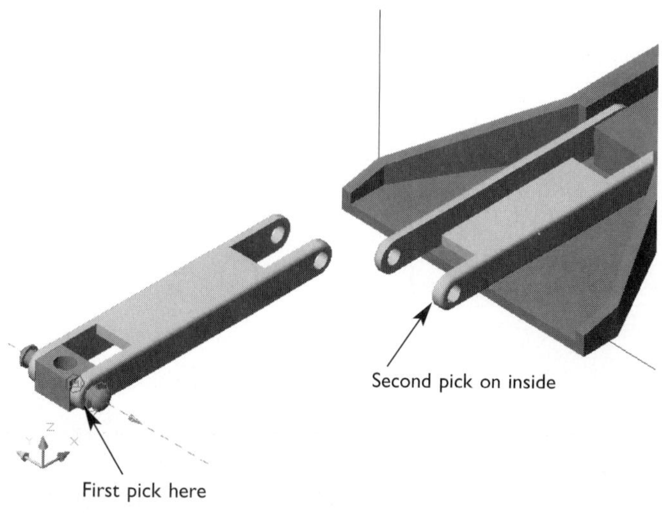

Figure 7.31

Right-click: in the graphics area

Pick: **3D Constraints, Mate**

Select first set of geometry: **pick on the holes through the ThinLeg where the ends of the SupportGuide extend through, as shown in Figure 7.31**

Enter an option [Next/Accept]<Accept>: ↵ **(if the ThinLeg part is highlighted)**

First set = Axis, (arc)

Select first set or [Clear/fAce/Point/cYcle] <accEpt>: ↵ (if the axis of the holes is selected)

Select second set of geometry: **pick the holes through the inside free-end of the left ThickLeg as shown in Figure 7.31**

Second set = Axis, (arc)

Select second set or [Clear/fAce/Point/cYcle] <accEpt>: ↵ (if the axis of the holes is selected)

Enter offset <0.0000>: ↵

The holes align, but the surfaces of the two parts do not necessarily match. Next you will mate the surfaces of the ThickLeg and Thinleg so that they line up correctly. First move the ThinLeg to the side so you can select the surface easily. (You do not need to move the SupportGuide as it should not be in the way.)

Right-click: in the graphics area

Pick: **3D Manipulator**

Pick: **the ThinLeg**

Pick: **the center handle**

Pick: **away from the Baseplate**

Tip

If you don't see the 3D Manipulator handles when a part is zoomed up, pick the part again and use the dialog box to align the handles with the entity as you did before.

[Undo/Mode/Exit]<Exit>: ↵

Zoom your view so that you can see the parts clearly

Right-click: in the graphics area

Pick: **3D Constraints, Mate**

Select first set of geometry: **pick the outside flat surface of the ThinLeg**

First set = Point, (line)

Select first set or [Clear/aXis/fAce/Next/cYcle] <accEpt>: **A** ↵ (if necessary)

First set = Plane

Enter an option [Clear/aXis/Point/Next/Flip/cYcle] <accEpt>: ↵ (if the correct surface is selected. If not use the prompts until you have the correct surface selected and then press ↵)

Select second set of geometry: **pick the inside flat surface of the ThickLeg**

Second set = Point, (line)

Select second set or [Clear/aXis/fAce/Next/cYcle] <accEpt>: **A** ↵

Second set = Plane

Enter an option [Clear/aXis/Point/Next/Flip/cYcle] <accEpt>: ↵ (when the correct surface is selected)

Enter offset <.0000>: ↵

The ThinLeg moves back into place and mates with the ThickLeg part, as shown in Figure 7.32. Fit the view on your screen as necessary.

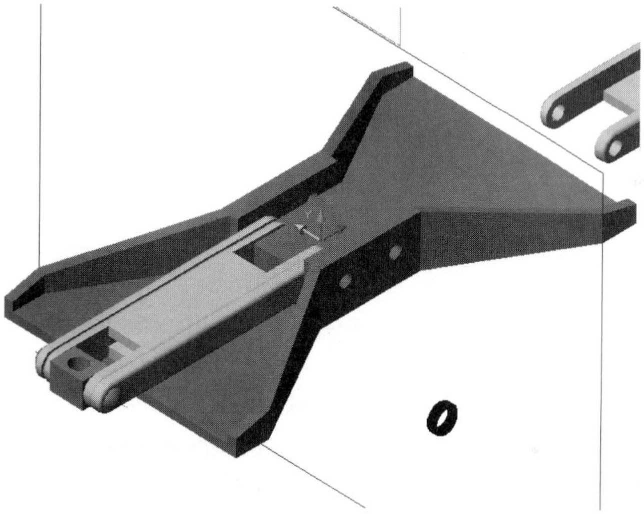

Figure 7.32

Using Move to Reposition a Part

In addition to the 3D Manipulator, you can use the Move or Rotate commands to reposition parts for easier selection while adding constraints. After you have added the new constraint, the parts will "pop" back into position.

Command: **M** ↵

Select objects: **pick on the ThinLeg part**

Select objects: ↵

Specify base point or displacement: **pick anywhere on the ThinLeg**

Specify second point of displacement or <use first point as displacement>: **pick slightly below and to the left of the JackTop part**

The ThinLeg moves to that location. On your own, use Zoom Window to enlarge the view, as shown in Figure 7.33

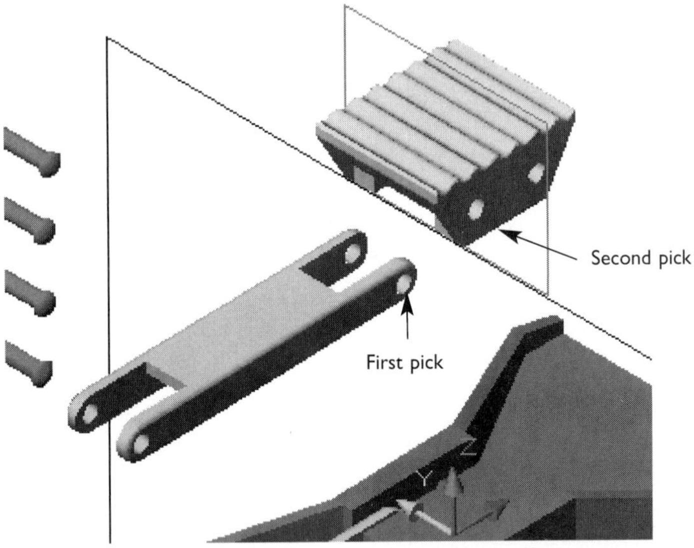

Figure 7.33

Constraining the ThinLeg to the JackTop

Next you will mate the holes through the ThinLeg with the holes through the JackTop.

Toggle to wireframe mode

Right-click: in the graphics area

Pick: **3D Constraints, Mate**

Select first set of geometry: **pick on the hole through the free end of the ThinLeg**

First set = Axis, (arc)

Select first set or [Clear/fAce/Point/cYcle] <accEpt>: ↵ (when you have the axis through the holes selected)

Select second set of geometry: **pick on the left hole through the JackTop**

Second set = Axis, (arc)

Select second set or [Clear/fAce/Point/cYcle] <accEpt>: ↵ (when you have the axis of the holes selected)

Enter offset <0.0000>: ↵

The top end of the ThinLeg rotates so that its holes are aligned with those on the JackTop. But the correct surfaces do not necessarily line up. Figure 7.34 shows the appearance of the assembly thus far.

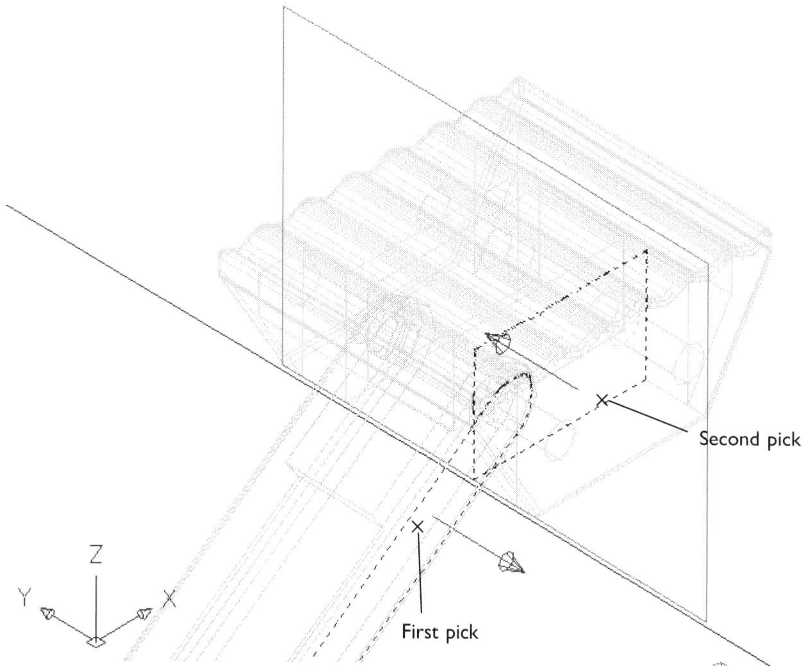

Figure 7.34

You will add another mate constraint to match the surfaces.

Right-click: in the graphics area

Pick: **3D Constraints, Mate**

Select first set of geometry: **pick on the flat surface of the ThinLeg indicated in Figure 7.34**

First set = Plane

Enter an option [Clear/Next/Flip/cYcle] <accEpt>: ⏎ (if the correct surface is selected)

Select second set of geometry: **pick on the flat surface inside the JackTop indicated in Figure 7.34**

Second set = Plane

Enter an option [Clear/Next/Flip/cYcle] <accEpt>: ⏎ (if the correct surface is selected)

Enter offset <0.0000>: ⏎

Use Zoom so you can see the whole assembly

The left side of the jack is assembled except that the hole through the center of the SupportGuide part does not align horizontal so that the threaded rod can fit through it. In the actual jack, the SupportGuide part on the other side combined with the equal length sides will keep the hole oriented horizontally. You will add an Angle constraint to tell Mechanical Desktop to keep the axis of the hole aligned with the bottom of the Baseplate.

Tip

If you have trouble selecting the part, cancel the command and try picking it first so that its grips appear. Then right-click and pick the Move command from the context menu.

Tip

If you use Crossing to select multiple objects, or if you select an unwanted object by mistake, you can type R ⏎. The Remove objects: prompt will appear and you can then pick the objects you want to remove from the selection set. To continue adding objects to the selection set, type A ⏎.

Command: **M** ⏎

On your own, follow the prompts and move the SupportGuide part on the left side of the assembly slightly to the left of the other parts so that you can select it easily.

Your model should look similar to Figure 7.35.

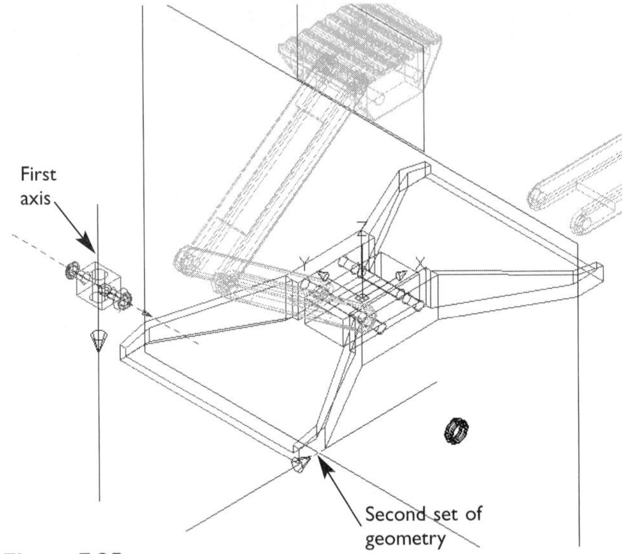

Figure 7.35

Right-click: in the graphics area

Pick: **3D Constraints, Angle**

Select first set of geometry: **pick the hole through the SupportGuide** (to select its axis)

First set = Axis/Vector, (arc)

Select first set or [Clear/fAce/Point/Flip/cYcle] <accEpt>: ↵ (when the correct axis shown in Figure 7.35 is selected)

Select second set of geometry: **pick the short horizontal edge of the Baseplate** (shown in Figure 7.35)

Second set = Axis/Vector, (line)

Enter an option [Clear/Next/Flip/cYcle] <accEpt>: ↵ (when the correct edge is selected)

Enter angle <0>: ↵

On your own, use Zoom Window to enlarge your view and then shade the part

The hole should be oriented so that it is horizontal, as shown in Figure 7.36. Notice how the DOF symbol changes.

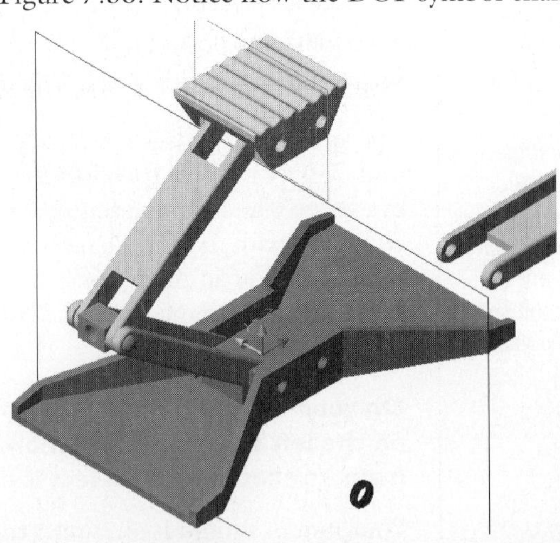

Figure 7.36

Right-click: **SupportGuide** (in model tree)
Pick: **DOF Symbol** (to shut it off)

Editing Constraints

Any of the constraints that you add are parametric. If you make mistakes, or the design changes, you can always delete or edit the constraints that you have chosen. Next you will change the distance that the JackTop is offset from the Baseplate to see the effect of editing the constraints. You will select the constraint to edit by using the model tree.

Figure 7.37 shows the JackTop part in the model tree with the constraints that are applied to it listed below it. You will right-click on the constraint mating a plane to a plane. As you pick on the constraints shown in the model tree, arrows appear in the graphics window, pictorially identifying how the constraint is applied.

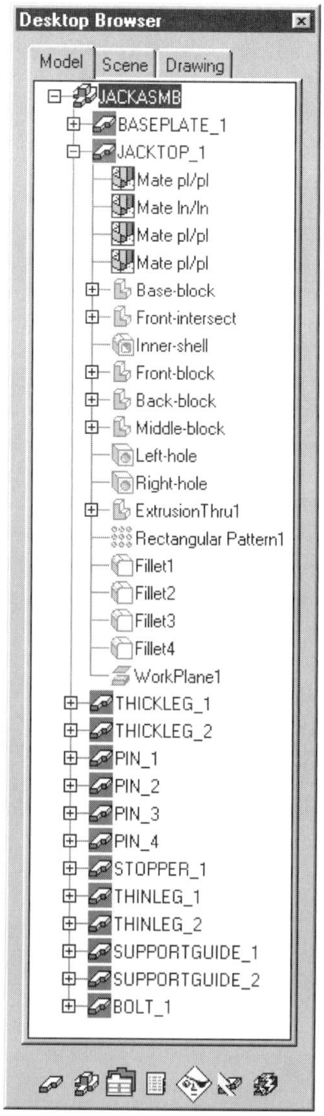

Figure 7.37

Pick: **to expand the model tree for the JackTop** (if needed)
Pick: **each of the different mates in turn** (notice how they become highlighted in the assembly)
Right-click: **on the Mate/pl/pl that defines the distance between the JackTop and the Baseplate**

The surfaces of the part and arrows appear, identifying the constraint, as shown in Figure 7.38.

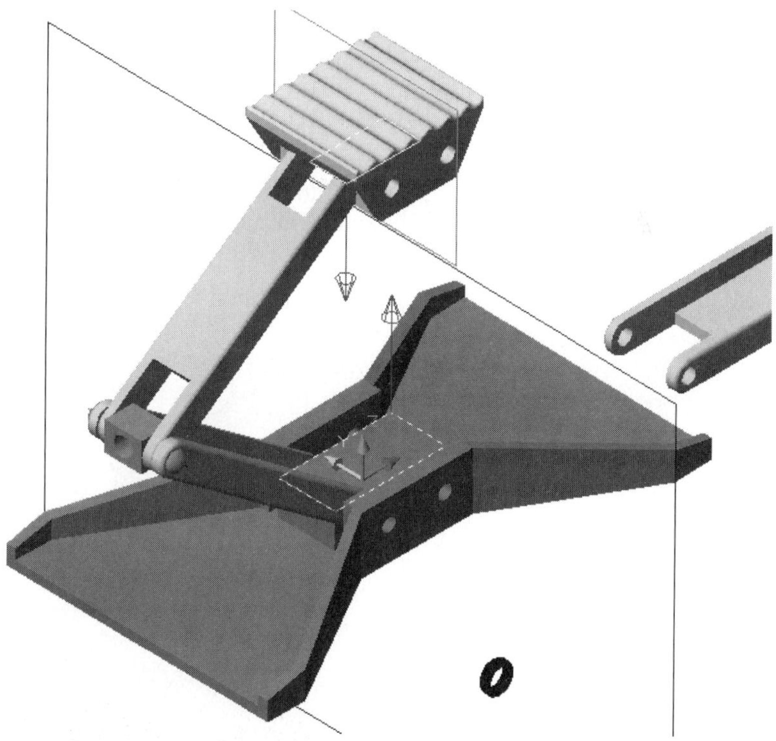

Figure 7.38

Pick: **Edit** (from the context menu)

The Edit 3D Constraint dialog box appears. The offset value is currently set to 7.00. You will change the value by clicking on the upward pointing scroll arrow, as shown in Figure 7.39.

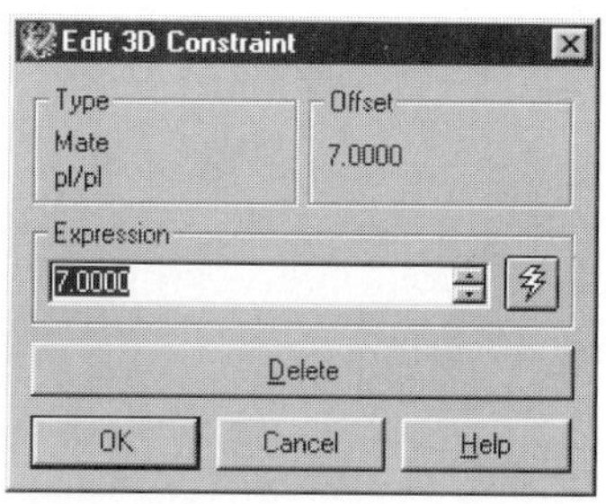

Figure 7.39

Pick: **the upward arrow to the right of the Expression box**

Continue picking on the value until you cannot change it any longer

As you pick to increase the offset, the assembly is modified so that the JackTop moves upward.The constraints between the other parts causes them to move also. The upper whole number value that you can move this part to should be 12.00. Why is it unable to move further? You probably realize that the length of the ThickLeg and ThinLeg will determine how far up it will move.

On your own, pick on the downward pointing arrow to the right of the box until you have set it to −5

What happened? This time the JackTop part went right through the Baseplate and the arms continued to move downward. Mechanical Desktop does not detect when two solids collide; it only applies the constraints that you have defined.

On your own, set the offset distance to 5.75

Pick: **OK** (to exit the dialog box)

Constraining the Four Pins

Next you will insert two pins through the holes in the JackTop, and then two more pins through the two sets of holes in the Baseplate. As you have seen, the software does not detect that one solid passes through another. You will add the pins to the right side of the assembly even though you do not have the legs assembled on that side. You will complete the steps mostly on your own.

Right-click: in the graphics area

Pick: **3D Constraints, Insert**

Using what you have learned, select on the circular face *inside* the end of the pin. (Remember, Insert matches the axis and the face so you need to select the face that will be flush to the outside of the JackTop.) Then select to mate it to the hole exiting the JackTop. Use a similar procedure for the pins through the Baseplate. When you have completed these steps, your assembly should look like Figure 7.40.

Tip

It may be easier to select the circular edges of the parts if you switch to wireframe mode.

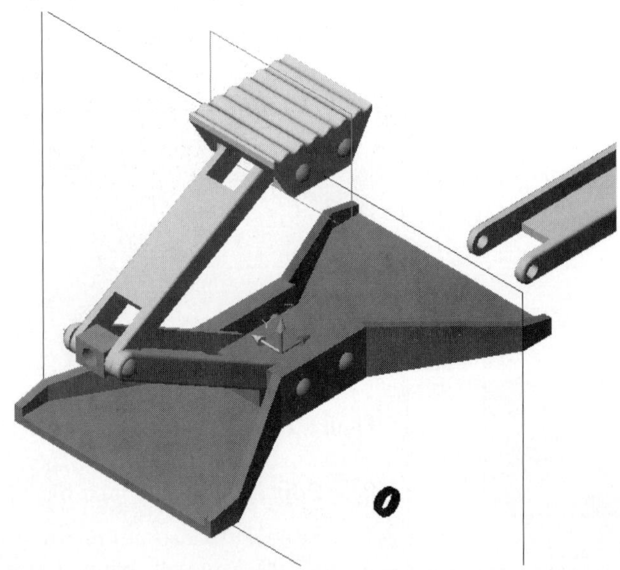

Figure 7.40

Tip

If you have trouble adding a constraint, review the list of constraints in the Browser window to be sure you have not added any that you do not need. An existing constraint may be preventing you from adding a constraint.

Next you will add the parts to the right side of the assembly on your own. Use the methods you have learned to constrain the parts to finished the jack assembly. (Refer to the list of constraints in the Browser window for help with choosing constraints.)

Constrain the ThickLeg2 part to the Baseplate

Constrain ThinLeg2 to ThickLeg2

Constrain ThinLeg2 to the JackTop

Constrain the axis of the hole through the SupportGuide to remain parallel to the Baseplate

Finally, you use the Mate constraint to align the axis of the stopper to the hole through the SupportGuide part at the left side of the assembly. Then add another Mate constraint, mating the circular face of the stopper to the square surface at the end of the SupportGuide part.

Mate the Stopper axis to the axis of the hole through the SupportGuide

Mate the Stopper's circular face to the square face of the SupportGuide

Your assembly should now look like Figure 7.41.

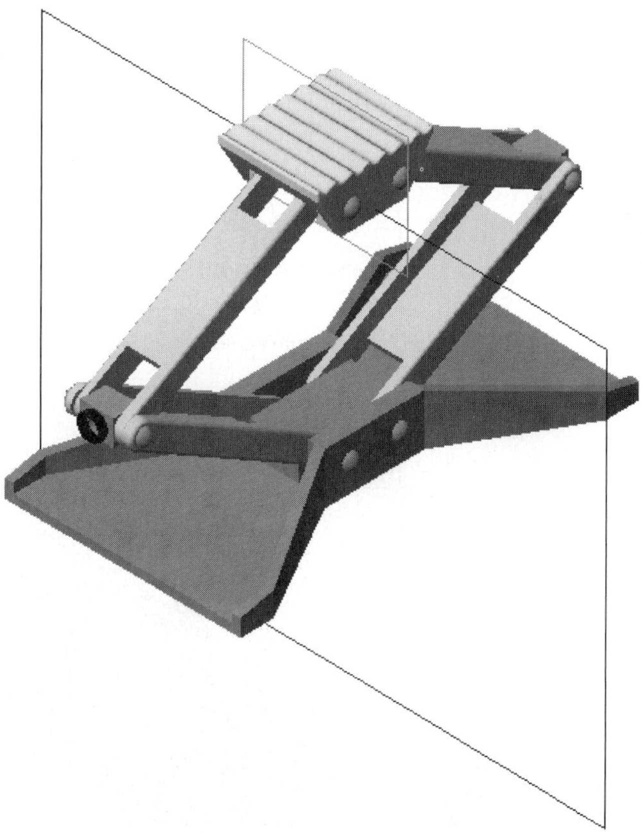

Figure 7.41

Inserting the Bolt

Now you are ready to add the bolt through the center of the holes in the SupportGuides. You will use the Insert constraint to align the bolt and match its face to the surface of the Stopper, as shown in Figure 7.42.

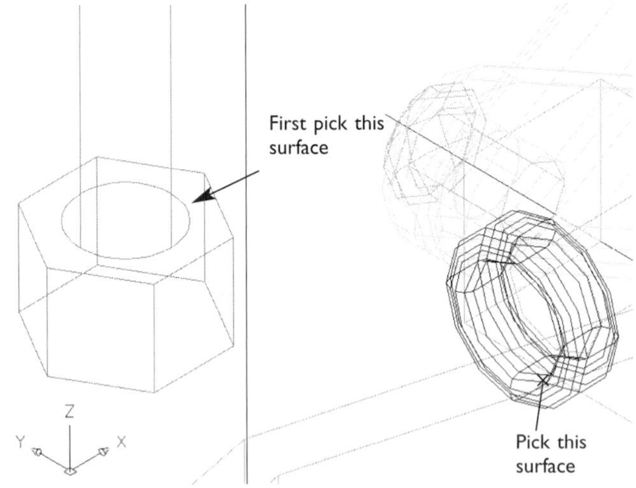

Figure 7.42

Right-click: in the graphics area

Pick: **3D Constraints, Insert**

Select first circular edge: **pick on the rounded edge where the bolt shaft meets the bolt head**

First set = Plane/Axis

Enter an option [Clear/Flip] <accEpt>: ↵ (if the correct surface is selected)

Select second circular edge: **pick on the round surface on the flat side of the Stopper**

Second set = Plane/Axis

Enter an option [Clear/Flip] <accEpt>: ↵ (if the correct surface is selected)

Enter offset <.0000>: ↵

The bolt should be inserted in your assembly, as you see in Figure 7.43.

Save your work

Figure 7.43

Adding Stock Parts (Power Pack)

Next you will add a standard part using the Mechanical Power Pack features. If you do not have Power Pack, skip this section and continue with the next section, "Viewing the Assembly."

Reference

Engineering Design Communication

Standard Parts, pp. 303–304

Mechanical Desktop with Power Pack provides a variety of stock parts that are typically ordered from outside companies to complete your designs. These stock parts include fasteners, specially formed holes, shafts, bearings, standard steel shapes, and others.

You use the Content 3D menu to select the stock parts. You will replace the pins that you added in the assembly with a generated shaft part. First, save your file to a new name so your assembly with the pins remains unchanged when you delete the pins.

Use SaveAs to save your assembly to MDT-Work and name it *JackAsmb-Local.dwg*

On your own, delete the 4 pins from the assembly

Switch to wireframe view and Zoom up on the center of the Baseplate

Pick: **Content 3D, Shaft Generator**

Specify start point or [Existing shaft]: **CEN** ↵

Center of: **pick the rounded edge of the left hole in the Baseplate**

Specify centerline endpoint: **CEN** ↵

Center of: **pick the same rounded edge on the back of the Baseplate**

Specify point for new plane <parallel to UCS>: ↵

New part created.

Reference

Mechanical Desktop On-line Help

See the AMSHAFT3D command in Mechanical Desktop Help for more information about generating shaft features.

The 3D Shaft Generator dialog box pops up, as shown in Figure 7.44. You can use it to select many different features that you can use in generating shaft models. The Outer Contour tab is selected by default.

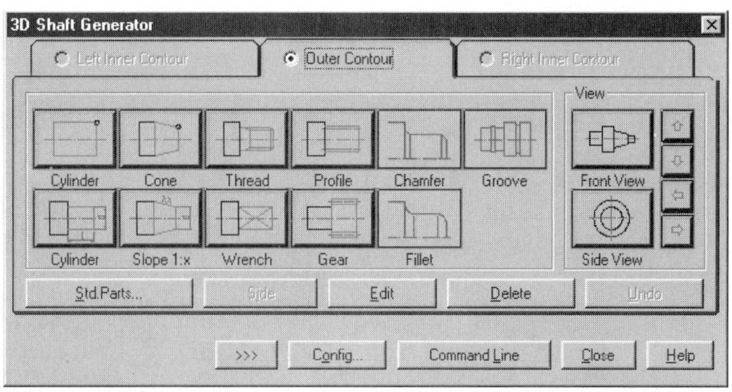

Figure 7.44

Pick: **the lower left Cylinder button** (as the type of shaft model to generate)

Specify length or [Associate to/Equation assistant] <2>: **3** ↵

Specify diameter or [Associate to/Equation assistant] <1.5>: **.35** ↵

You return to the Shaft Generator dialog box. Now you can select internal and snap groove type features for the ends.

Pick: **Left Inner Contour tab** (to see the choices available)

Pick: **Right Inner Contour tab** (to see the choices)

Pick: **Close**

The cylindrical shaft is added to the model.

On your own, use the shaft generator to add pins to the remaining holes

Pick: **Front Left Isometric icon**

Tip

If you do not have Power Pack available, continue on from here.

On your own, use Zoom Window to enlarge the assembly in your view

Pick: **Gouraud Shaded icon**

Viewing the Assembly

The Scene tab in the Desktop Browser lets you make views of your assembly that are quick and easy to use. Saving some scenes that you can move to easily helps you visualize the parts and to show them to others easily.

Pick: **Scene tab** (so that it appears uppermost)

Making a New Scene

To create a new scene, you will right-click and use the context menu.

Right-click: **in the blank area of the Scene tab in the Desktop Browser**

Pick: **New Scene**

Enter new scene name of the active assembly (JACKASMB) <SCENE1>: **ASSEMBLED** ↵

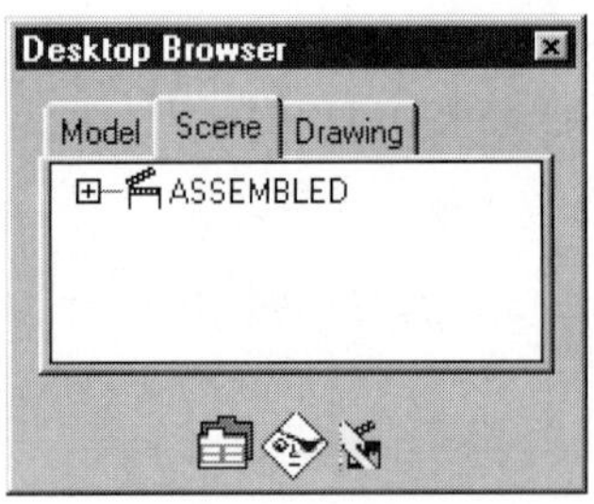

Figure 7.45

The new scene name Assembled appears in the model tree on the scene tab, as shown in Figure 7.45.

You will not make any changes to this scene. It will show the fully assembled parts the way you currently have them.

Now you will create another scene that you will change to an exploded view.

Right-click: **in the blank area of the scene tab**

Pick: **New Scene**

Enter new scene name of the active assembly (JACKASMB) <SCENE1>: **EXPLODED** ↵

The scene name Exploded appears below the name Assembled in the tree for the Scene tab. It is now the current scene; Assembled is grayed out, indicating that it is not active.

Explode Factor

Next you will change the explode factor for this scene to create a view that shows the parts as though they haven't been assembled together yet or are exploded away from one another.

Right-click: **Exploded** (in the tree on the Scene tab)

Pick: **Explode Factor**

Use the explode factor dialog box that appears to set the explosion distance for the scene to 10, as shown in Figure 7.46.

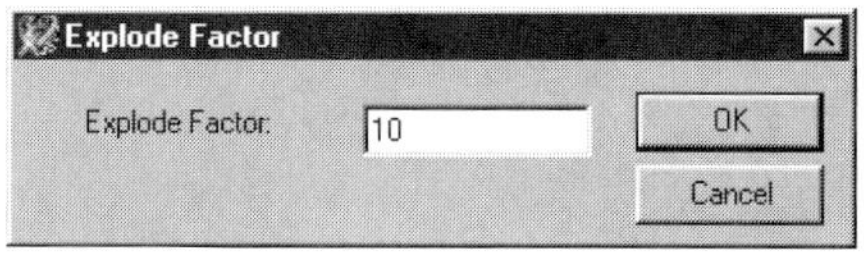

Figure 7.46

Type Explode Factor: **10**

Pick: **OK**

Pick: **Zoom All**

Mechanical Desktop automatically explodes the assembly by 10, but as you can see in Figure 7.47, the way Mechanical Desktop explodes the assembly may not be desirable.

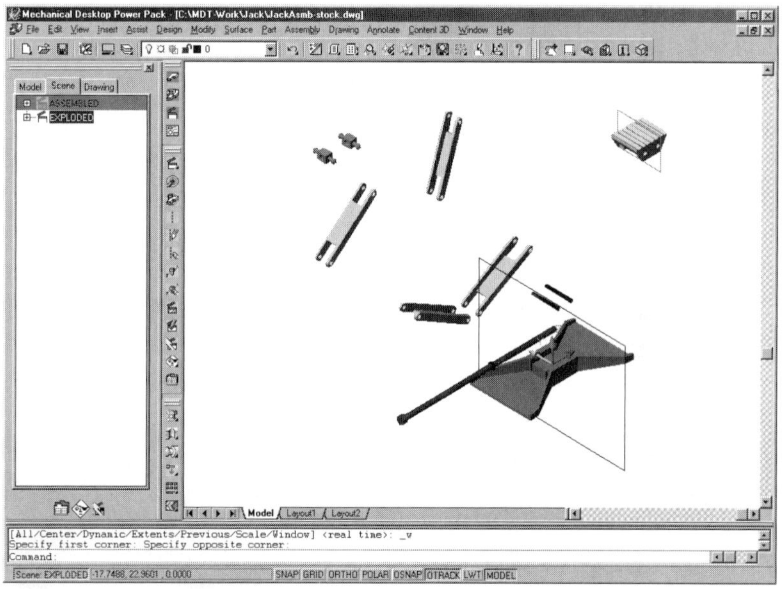

Figure 7.47

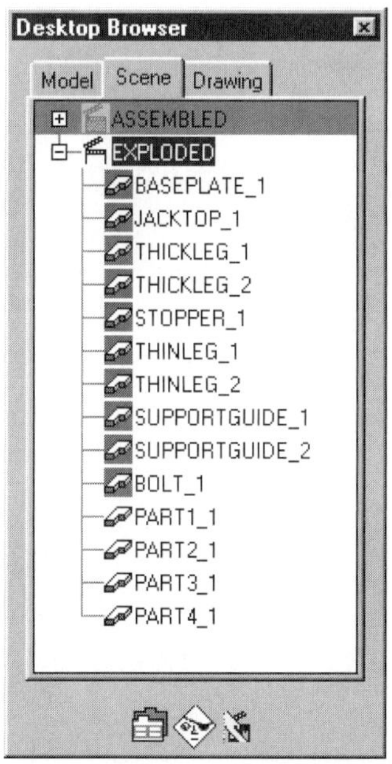

Figure 7.48

Using Tweak

The Tweak command is another way to create an exploded view so that all of the individual parts are positioned the way you would like them.

Pick: **Undo icon to undo the explode factor** (or use the dialog box to reset it to 0)

Expand the model tree under Exploded in the Browser window

You will see all of the parts in the assembly under the exploded scene, as shown in Figure 7.48.

You will change the view to prepare to tweak the first part.

Pick: **Front View icon**

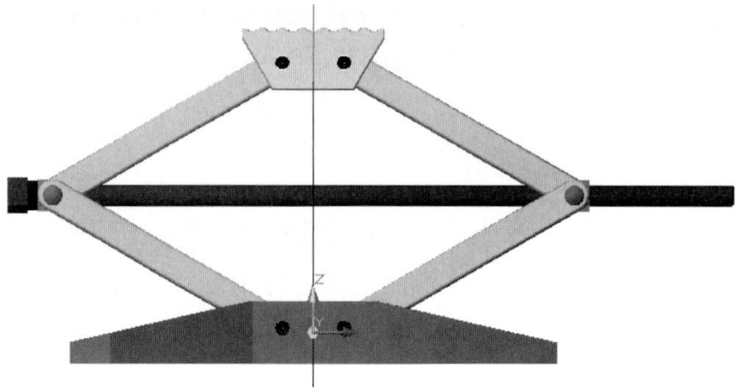

Figure 7.49

Your screen should look like Figure 7.49.

Right-click: **JackTop in the Browser window under Exploded**

Pick: **New Tweak**

The Tweak Part/Subassembly box appears on your screen, as shown in Figure 7.50. With tweak you are able to move, rotate, or transform a part or subassembly to position it for an exploded view. For this assembly you will only move the parts. In general, you want to move the parts far enough away from each other that they do not overlap in the isometric view.

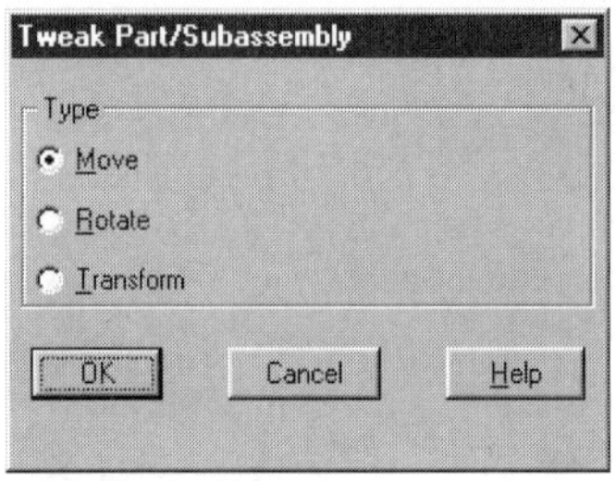

Figure 7.50

Reference

Engineering Design Communication

Assembly Drawings, pp. 473–474; Sketching Assemblies, pp. 102–103

Select: **Move**

Pick: **OK**

Other parts disappear except for the base plate and the top part. Mechanical Desktop prompts you for reference geometry. This means that you must select an edge to indicate the direction you wish to move the JackTop. In general, you want to keep parts lined up with the other parts where they assemble. If possible, move any single part along only one axis direction.

Select reference geometry: **pick a vertical line on the base plate**

An arrow should appear, as shown in Figure 7.51.

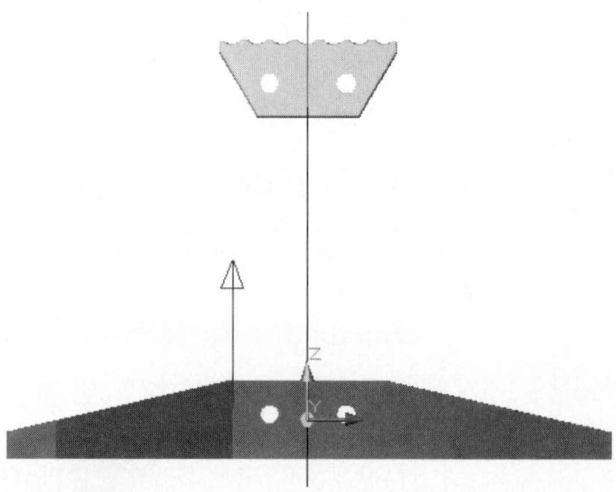

Figure 7.51

Now you will move the JackTop part a distance of 6 away from its mated position.

Enter distance <.0000>: **6** ↵

Pick: **Zoom All**

The assembly should now appear on your screen, as shown in Figure 7.52.

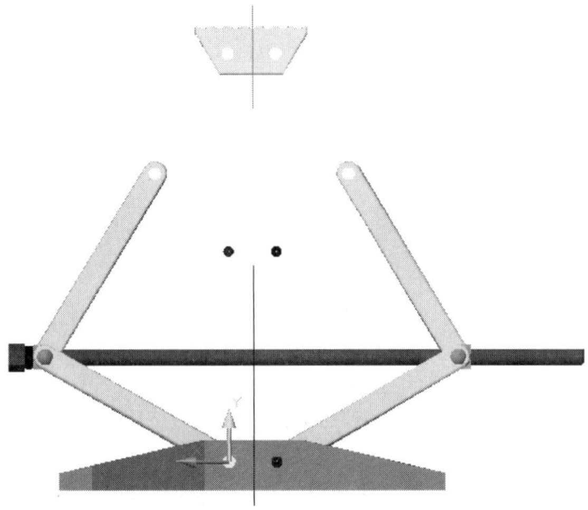

Figure 7.52

The JackTop part is moved a distance of 6 above its original position. (To move in the opposite direction, you would enter a negative number.) You can also access the tweak command by right-clicking in the graphics area.

Right-click: in the graphics area

Pick: **New Tweak**

Select part or subassembly to tweak: **pick the bolt**

Pick: **Move** (in the Tweak Part dialog box)

Pick: **OK**

Select reference geometry: **pick a horizontal line on any of the parts**

Enter distance <.0000>: **depending on which way the arrow is pointing, enter 20 or −20 to position the bolt to the left of the assembly**

Tip

Switch to the isometric view after moving a part to see if it overlaps another part and needs to be moved farther away.

You will tweak the rest of the parts on your own so that your assembly looks like Figure 7.53. For the pins in the Baseplate, switch to a right side view to explode them along the Z-axis.

Tweak the rest of the parts on your own so the exploded view is similar to Figure 7.53

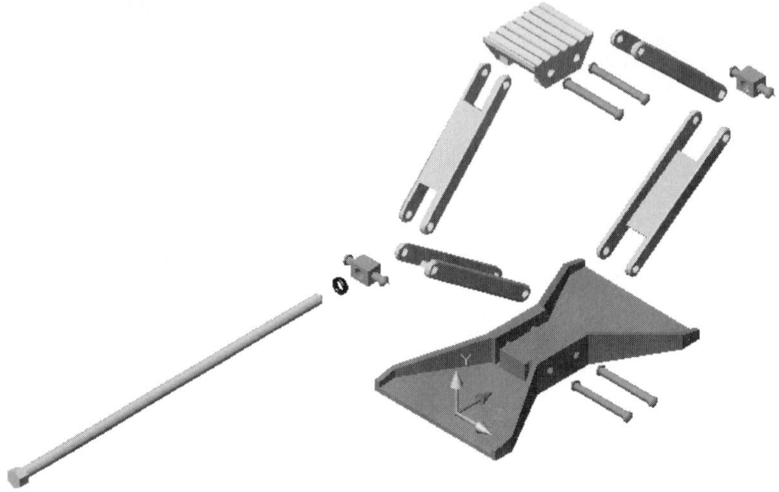

Figure 7.53

Activating a Scene

To switch between scenes, you use the model tree on the Scene tab.

Right-click: **Assembled** (from the tree)

Pick: **Activate Scene**

The original assembled view is restored.

Next you will return to the Model Tab in the Desktop Browser to see how MDT can be used to analyze fit and mass properties of your model.

Pick: **Model tab** (so that it appears uppermost in the Browser)

Analyzing the Model

Creating a solid model can provide many benefits for analyzing the design. Using MDT, you can analyze mass properties, check for interferences, and determine the minimum distance in 3D between different locations on parts.

Checking for Interference

Though the software does not automatically detect when two solids collide, you can use the analysis functions built into MDT to determine whether parts interfere (or overlap one another).

You will switch to wireframe so it is easier to view the interference.

Toggle to wireframe mode

Right-click: in the graphics area

Pick: **Analysis, Check Interference**

Nested part or subassembly selection? [Yes/No] <No>: ↵

Select first set of parts or subassemblies: **pick on the right cylinder through the JackTop part**

Select second set of parts or subassemblies: ↵

Select second set of parts or subassemblies: **pick on the JackTop part**

Parts/subassemblies do not interfere.

Mechanical Desktop reports that the parts do not interfere, as they shouldn't. To see how Mechanical Desktop reports interference, you will modify the JackTop part so the holes are smaller.

Activating and Editing a Part

Before you can make changes to parts in the assembly environment, you need to activate the part. Part commands then become available on the context menus for that part, and any part commands will only act on the active part. Because your part files are external parts, any changes you make to the assembly can also be saved with the part file. You will change the JackTop part and save your changes to see how this bidirectional associativity works.

Right-click: **JackTop in the Desktop Browser**

Pick: **Activate Part**

The rest of the parts in the Desktop Browser are dimmed out. You will edit the hole features in the JackTop to change their diameters to .30 instead of .38.

Expand the JackTop listing in the Desktop Browser.

Right-click: **Left-hole**

Pick: **Edit**

Set Dia: **.30**

Pick: **OK**

Select object: ↵

Right-click: **Right-hole**

Pick: **Edit**

Set Dia: **.30**

Pick: **OK**

Select object: ↵

Right-click: in the graphics window

Pick: **Update Part**

Now you will activate the assembly again and save your changes.

Right-click: in the graphics window

Pick: **Assembly, Activate Assembly**

Enter assembly name to activate or [?] <JackAsmb>: ↵

Pick: **Save**

The External file Save dialog box appears and Mechanical Desktop reports that it will save changes to the external file before saving the assembly file, as shown in Figure 7.54.

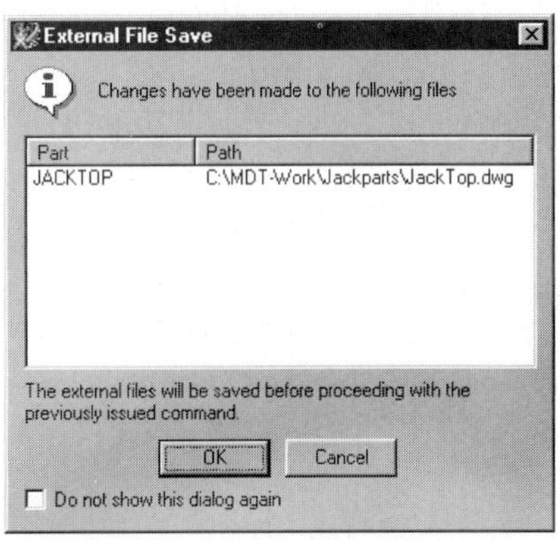

Figure 7.54

Pick: **OK**

Checking for Interference with the New JackTop

You are now ready to check for interference again.

Right-click: in the graphics area

Pick: **Analysis, Check Interference**

Nested part or subassembly selection? [Yes/No] <No>: ↵

Select first set of parts or subassemblies: **pick on the left pin through the JackTop part**

Select first set of parts or subassemblies: ↵

Select second set of parts or subassemblies: **pick on the JackTop part**

Comparing 1 parts/subassemblies against 1 parts/subassemblies.

Interference 1:

PIN_4

PIN

JACKTOP_1

JACKTOP

Create interference solids? [Yes/No] <No>: **Y** ↵

Highlight pairs of interfering parts/subassemblies? [Yes/No] <No>: ↵

A new solid is created in a contrasting color (red) to show you where the parts overlap. If you had selected several parts at once, you could use the option to highlight pairs and Mechanical Desktop would highlight each pair of interfering parts.

You will move the solid away from the model so you can see it more clearly.

Command: **M**

Select objects: **LAST** ↵

Select objects: ↵

Specify basepoint or displacement: **pick on the solid**

Specify second point of displacement: **pick above the JackTop**

The interference solid is shown in Figure 7.55.

You can list the properties of this solid to learn more about how the parts interfere. Or use the Distance or dimensioning commands to determine how much you must change the parts to keep them from interfering. If you are modeling a force fit, you can verify that the amount of overlap is correct.

Remove the interference solid before you continue.

Pick: **the interference solid** (so the grips appear)

Type: **E** ↵ (for Erase)

You will now return the JackTop holes to their intended size and save again so your assembly does not have parts that interfere. On your own,

Activate the JackTop part

Edit the Left-hole diameter to be .38

Edit the Right-hole diameter to be .38

Turn WorkPlane1 visibility to off (if you have not already done so)

Update the part

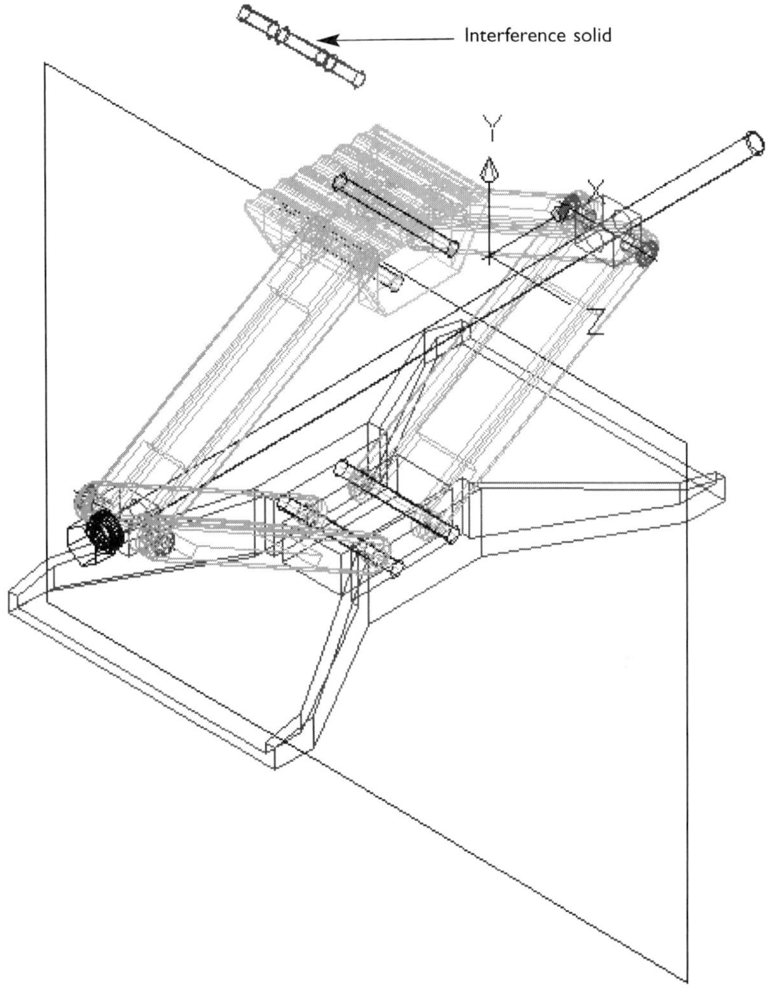

Figure 7.55

Activate the Baseplate and turn WorkPlane2 off (if you have not already done so)

Activate the assembly

Save the assembly and the changes to the Jacktop external part

Mass Properties

Reference

Engineering Design Communication

Determining Mass Properties, pp. 376–383

Next you will analyze the mass properties for the model.

Right-click: in the graphics area

Pick: **Analysis, Mass Properties**

You return to your drawing to select the items for analysis.

Enter an option (parts/subassemblies) [Name/Select] <Select>: ↵

Select part and subassembly instances: **pick the Baseplate part**

Select part and subassembly instances: ↵

The Mass Properties dialog box appears.

Pick: **Material button** (near the top right of the dialog box)

The Select Material dialog box appears on your screen. You will assign the material MILD_STEEL to the Baseplate.

Pick: **MILD_STEEL** (from the drop-down menu of materials)
Pick: **Assign button**
Pick: **OK**

Now use the mass properties dialog box that reappears to select the coordinate system that will be used for the analysis, the assembly units, and the mass units. You will select the options shown in Figure 7.56.

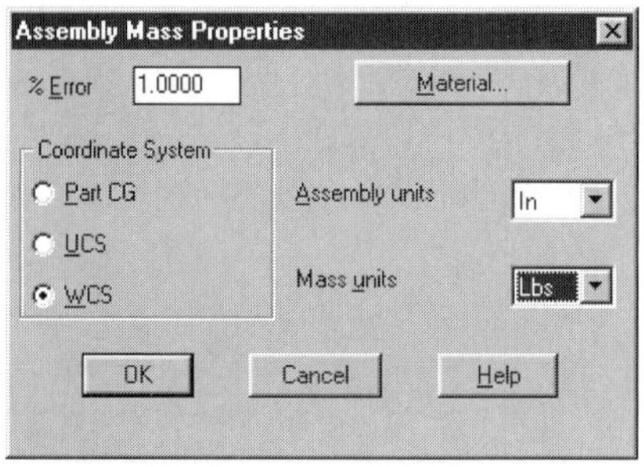

Figure 7.56

Select: **WCS**
Choose Assembly units: **Inches, Pounds**
Pick: **OK**

The mass properties for the Baseplate are shown, as you see in Figure 7.57.

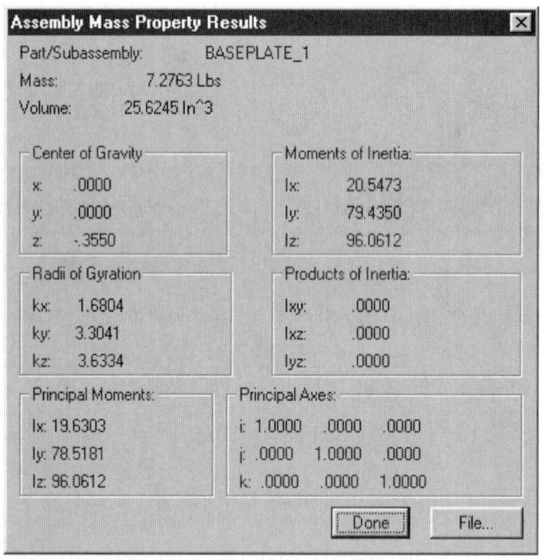

Figure 7.57

Pick: **Done**

Localizing the Assembly's Parts

To simplify file management when sending an assembly to a manufacturer or others, you can convert all of the external parts to local parts so they will be stored in a single file. You will do this now with your jack assembly.

Right-click: in the graphics window
Pick: **Catalog**
Pick: **All tab**
Pick: **Baseplate** (in the list of external files in the left)

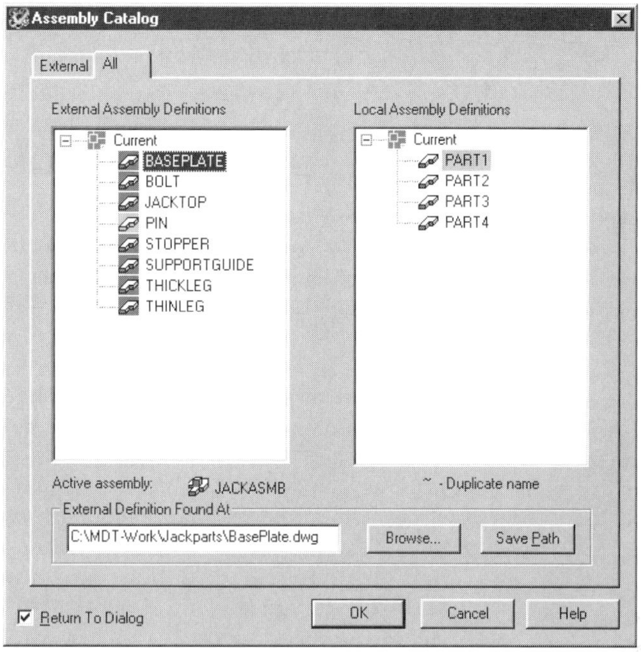

Figure 7.58

When you select a file, the location of that file on your hard drive appears in the area at the bottom of the dialog box, as shown in Figure 7.58.

Right-click: **Baseplate**
Pick: **Localize**

The part shifts to the right side of the dialog box under Local Assembly Definitions, as shown in Figure 7.59. The part as currently defined is now stored in the assembly file. Any change made to this part in the assembly will no longer automatically update the external part file.

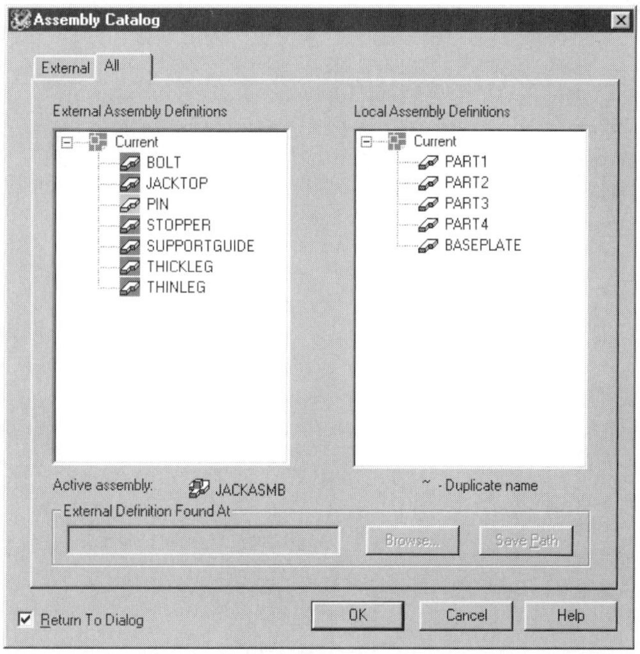

Figure 7.59

Repeat the steps to localize all of the external parts on your own

Pick: **OK** (when you are finished)

Save your work

Congratulations! You have completed your assembly model of the jack. In the next tutorial, you will use it to make an assembly drawing.

Command Summary

Action	Command Name	Description
Activate Part	AMACTIVATE	Activates a local or external part in an assembly so it may be edited
Angle Constraint	AMANGLE	Creates an angular assembly constraint between parts, subassemblies, or assemblies
Assembly Catalog	AMCATALOG	Attaches and removes, converts, and organizes local and external parts in an assembly
Delete Part	AMDELETE	Deletes a part, part instance, or scene
Delete Tweaks	AMDELTWEAKS	Removes tweaks in a scene so part moves back to original position
Edit Assembly Constraints	AMEDITCONST	Changes offsets in or deletes assembly constraints
Insert Constraint	AMINSERT	Constrains two circular edges or faces in an assembly so that they share the same axis and the faces are coplanar
Interference	AMINTERFERE	Analyzes selected parts and reports any interference between them
3D Manipulator	AMMANIPULATOR	Dynamically rotates, copies, or moves parts freely or along 3D axes
Mate Constraint	AMMATE	Creates a mate constraint between points, planes, axes, or nonplanar faces
Mass Properties of a Part	AMPARTPROP	Lists the mass properties of the currently active part
Generate Shaft	AMSHAFT3D	Creates rotationally symmetrical shaft parts and inner and outer shaft contours
Tweak	AMTWEAK	Positions a part or subassembly in an exploded scene
Set Explode Factor	AMXFACTOR	Controls the explosion factor for a scene
Move	MOVE	Moves selected geometry to a new location on the coordinate system

Exercises

1. **Cabinet Assembly** Assemble the cabinet parts provided in *MDT-Data\Exercises\CabinetParts* to create the assembly shown in Figure 7.60.

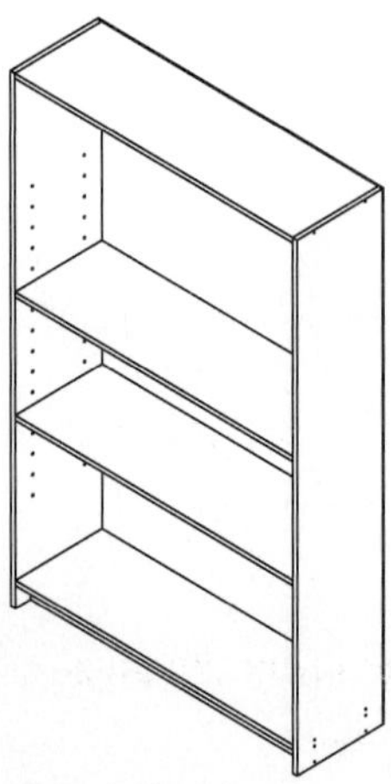

Figure 7.60

In Exercises 2 and 3, create the assemblies shown in Figures 7.61 and 7.62 using the part dimensions given.

2. M6 Hardware Assembly

ITEM NO.	QTY.	PART NO.	DESCRIPTION	MATERIAL
1	1	81006018	M6 X 18 SHCS	18-8 STAINLESS
2	1	81006100	M6 LOCK WASHER	18-8 STAINLESS
3	1	81006000	M6 FLAT WASHER	18-8 STAINLESS

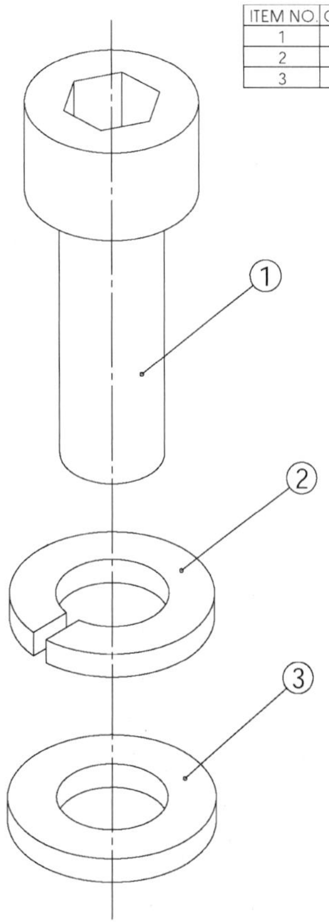

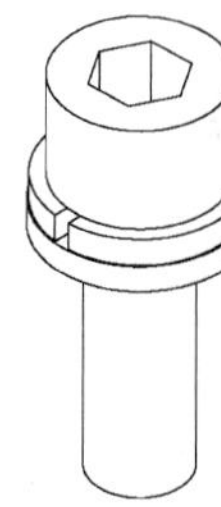

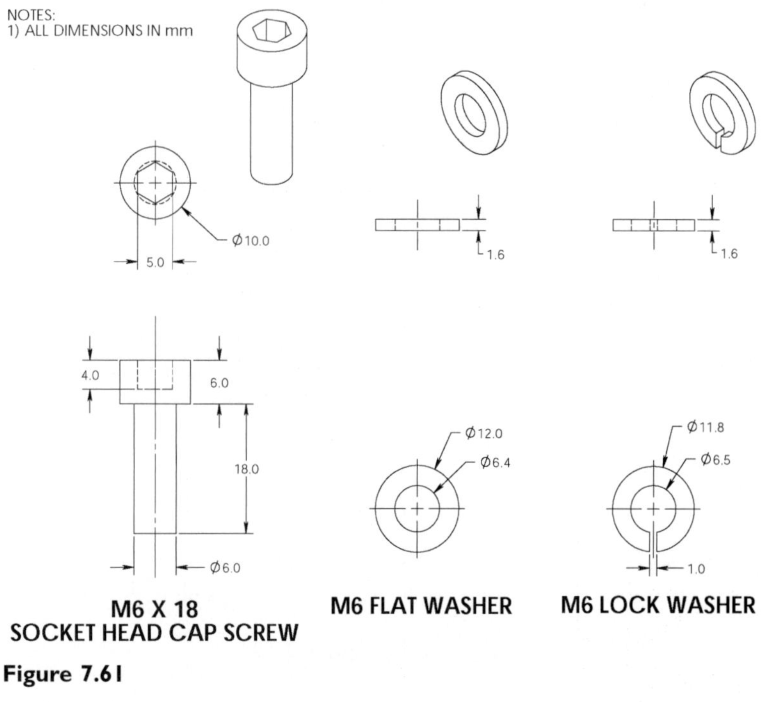

Figure 7.61

3. Adjustment Latch Assembly

ITEM NO.	QTY.	PART NO.	DESCRIPTION	MATERIAL
1	1	60001025	BASE	6061 T-6 AL
2	1	60001023	PIN	18-8 STAINLESS
3	1	60001024	ADJUSTMENT LATCH	6061 T-6 AL
4	4	81006018	M6 X 18mm SHCS	18-8 STAINLESS
5	4	81006100	M6 LOCK WASHER	18-8 STAINLESS
6	4	81006000	M6 WASHER	18-8 STAINLESS

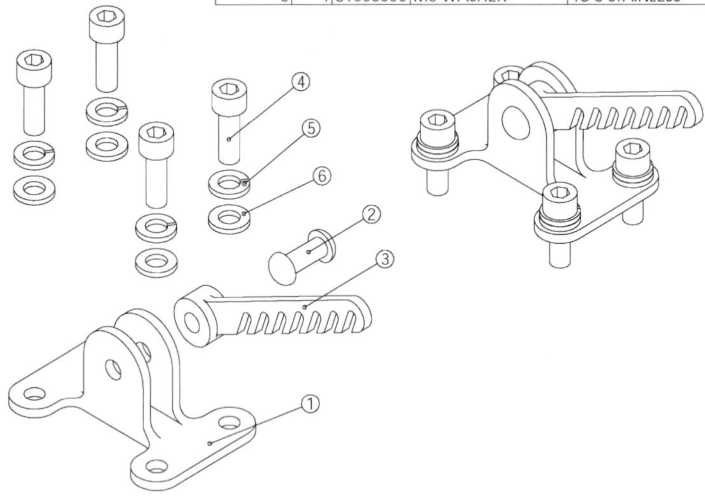

NOTES:
1) ALL DIMENSIONS IN mm

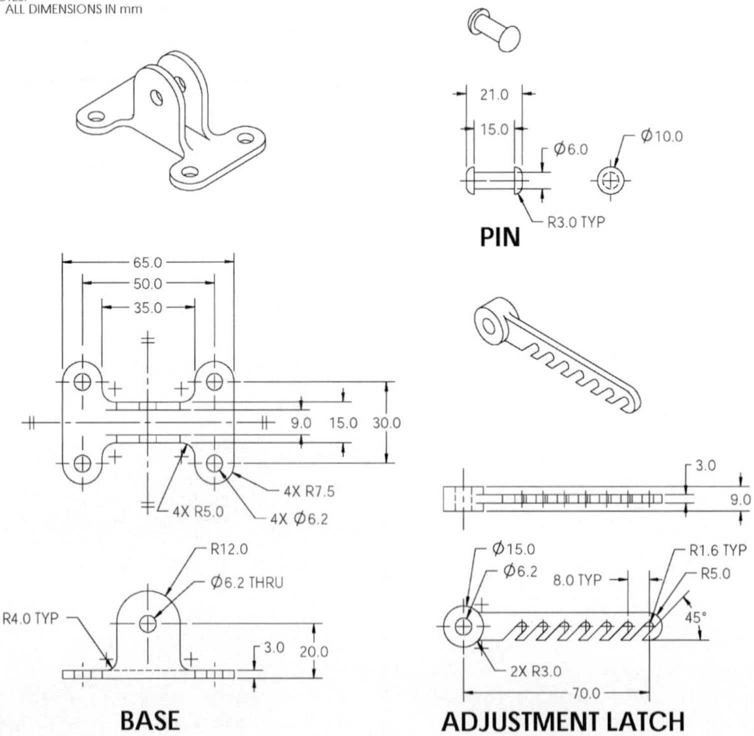

Figure 7.62

Tutorial

8 Working Drawings from Models & Assemblies

Objectives

When you have finished this tutorial, you will be able to

- Use standard drawing formats.
- Select scenes to add to your drawing.
- Add ball tags and a parts list.
- Add drawing notes.
- Make a dimensioned and toleranced part drawing.
- Add section views to a drawing.

Introduction

In this tutorial, you will learn to create part and assembly drawings of the jack assembly. Mechanical Desktop provides many standard drawing sheet and border arrangements that you will use to set up your layouts. You will first create a large-format assembly drawing of the jack assembly that includes ball tags and a parts list. You will also make a dimensioned and toleranced part drawing from one of the jack parts using a standard drawing sheet.

Starting

You will start by opening the data file *JackAsmb-dta.dwg* in the MDT-Data directory and saving it to a new file name. (If you wish, you may begin from the *JackAsmb-Local* that you created in Tutorial 7, but your screens and prompts may not match the steps exactly. If you encounter difficulties in the steps, you can begin the tutorial again starting with the data file.)

Start by launching Mechanical Desktop from the icon that includes the Power Pack. If you do not have Mechanical Desktop Power Pack available, you can complete the tutorial using Mechanical Desktop commands.

Launch Mechanical Desktop 5 Power Pack

Pick: **File, Open**

Navigate to the MDT-Data directory

Pick: **JackAsmb-dta.dwg**

Pick: **OK**

Pick: **File, SaveAs**

Choose Save in: **MDT-Work**

Type File name: **JackAsmb-Drawing.dwg**

Pick: **Save**

Load your user profile (if necessary)

The jack assembly appears on your screen, as shown in Figure 8.1. The new file name *JackAsmb-Drawing.dwg* appears in the Title bar. The Model tab appears uppermost in the Desktop Browser. You should also see the Scene and Drawing tabs. The parts in this assembly are local parts.

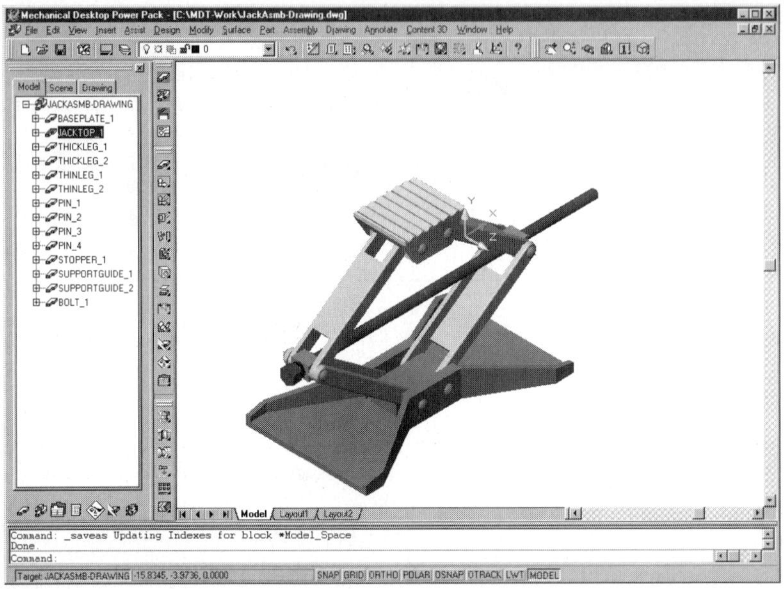

Figure 8.1

Creating a Layout for the Assembly Drawing

Tip

It is easier to start your new drawing using an ANSI-C template than to change the sheet size later, but there may be times when you are not aware of what sheet size to use for the drawings when you start the model. Many companies standardize on only one sheet size, such as C or D, in order to save on setup effort and cost and to make a set of drawings fit together nicely.

You will modify one of the default layouts so that it represents an ANSI standard 22 × 17-inch drawing sheet, and add a border and title block before you add the views of your assembly. You have created a similar layout in Tutorial 3, but this time you will use the Mechanical Desktop Power Pack features. If you do not have Power Pack, refer to Tutorial 3 as necessary throughout this tutorial.

Pick: **Drawing tab** (in the Desktop Browser)

Your screen switches to the Drawing tab, where you should see Layout1 shown in Figure 8.2.

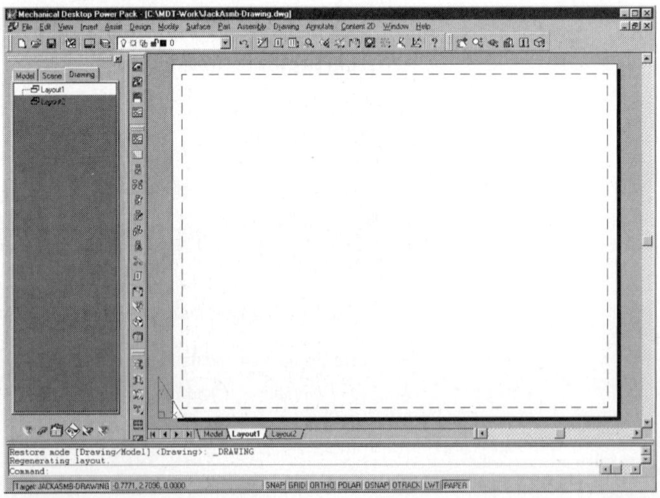

Figure 8.2

Setting Up the Sheet Size and Adding a Title Block

Most drawings are printed on standard sheet sizes with a border and title block. The AMTITLE command lets you both select the sheet size and

Reference

Mechanical Desktop On-line Help

You can view the help on Customizing Borders and Title Blocks to see how to replace the standard borders and title blocks with a customized one of your own design.

Reference

Engineering Design Communication

Standard Sheet Sizes and Title Blocks, pp. 618–620

add a standard border and title block to the drawing. This saves time and gives the drawing a finished look. Although most companies have their own approved title blocks that you will be required to use, Mechanical Desktop includes a number of standard title blocks that use features of the software to make it easy to change and add to the title block. You will select one of the standard title blocks from the Annotate menu.

Pick: **Annotate, Drawing/Title Revisions, Drawing Title/Borders**

The Drawing Borders with title Block dialog box appears, where you will make the selections shown in Figure 8.3.

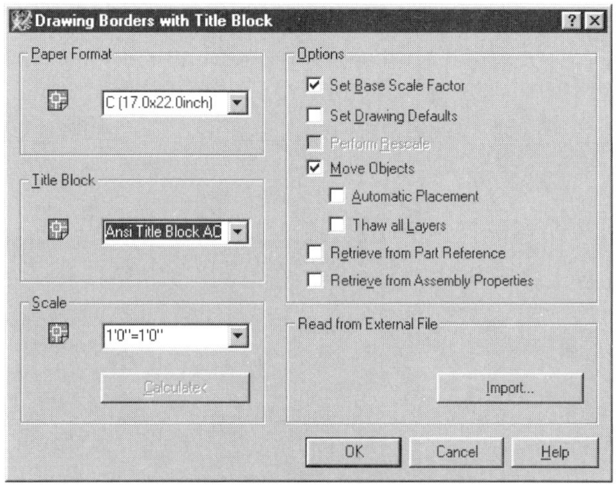

Figure 8.3

Choose Paper Format: **C** (17.0 × 22.0 inch)
Choose Title Block: **ANSI Title Block AC**
Pick: **OK**

The Page Set Up dialog box shown appears on your screen next. You will make selections as shown in Figure 8.4.

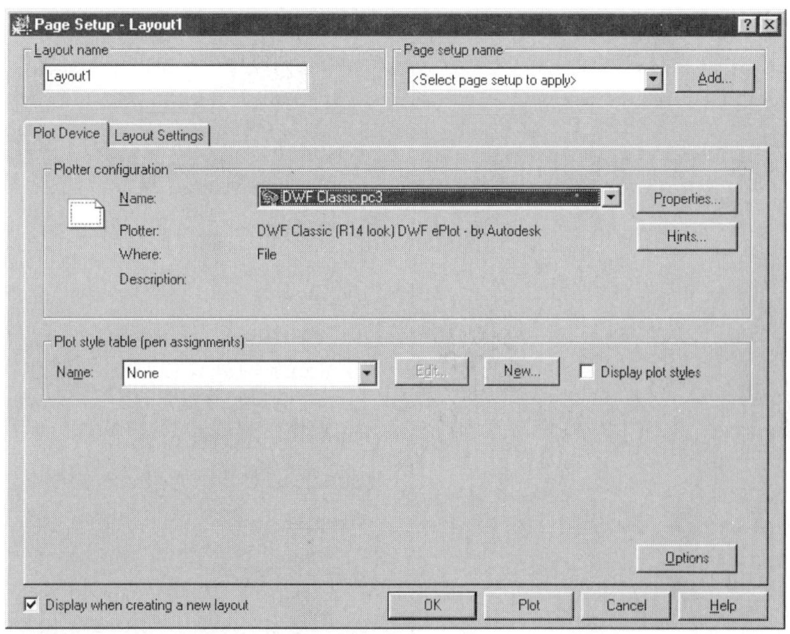

Figure 8.4

Pick: **Plot Device tab** (so it is uppermost)

Pick: **DWF Classic.pc3** (by expanding the list for Plotter Configuration Name)

The *DWF Classic.pc3* plotter configuration file was designed to create Drawing Web Format (DWF) files that resemble AutoCAD Release 14 drawings. The selection of this plotter configuration over any other plotter configurations that you may have available is not significant, except that it already contains different standard sheet sizes. If you select a plotter configuration file that does not allow larger sheets (such as that for a laser printer), you will not be able to select larger sheet sizes. You can change which plotter configuration is selected later if you want.

Pick: **Properties button**

The Plotter Configuration Editor–*DWF Classic.pc3* dialog box appears. You will use it to select ANSI-C as the paper size, as you see in Figure 8.5.

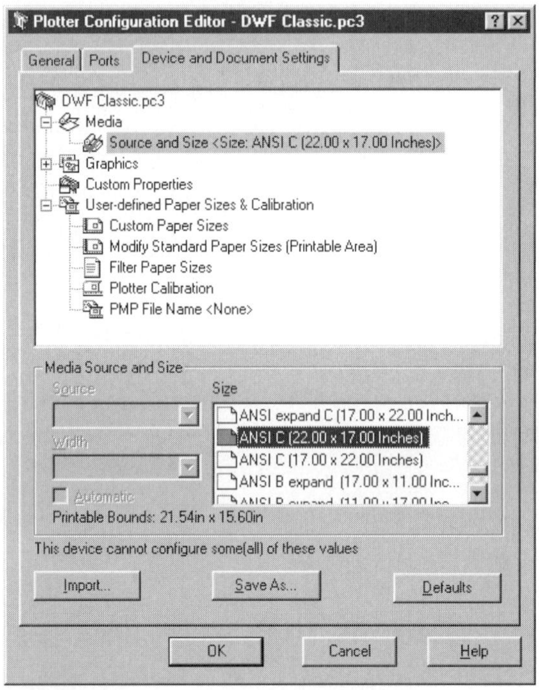

Figure 8.5

Pick: **Device and Document Settings tab** (so that it appears uppermost)

Pick: **Source and Size** (under Media) (if necessary)

Pick Size: **ANSI C** (22.00 × 17.00 Inches) (by scrolling through the list at the lower right)

Pick: **OK**

The Changes to a Printer Configuration File dialog box may appear on your screen. You will use it to save the changes to the printer configuration file. (If you cannot save changes to the files on your network, you can substitute a different directory where you have permission to save the file.)

Pick: **OK** (to save the changes to the configuration file, if necessary)

Pick: **OK** (to exit the Page Setup Dialog box)

The title block appears and you are prompted for an insertion point. The default is the 0,0,0 point on the sheet, or the bottom left corner of the printable area of the sheet.

Specify insertion point: ↵

The Change Title Block Entry dialog box appears. This dialog box allows you to enter text in the various title block areas to indicate the revision number of the drawing, its drawing number, who created it, and so on. You will accept the default file drawing name and enter your own name, as in Figure 8.6.

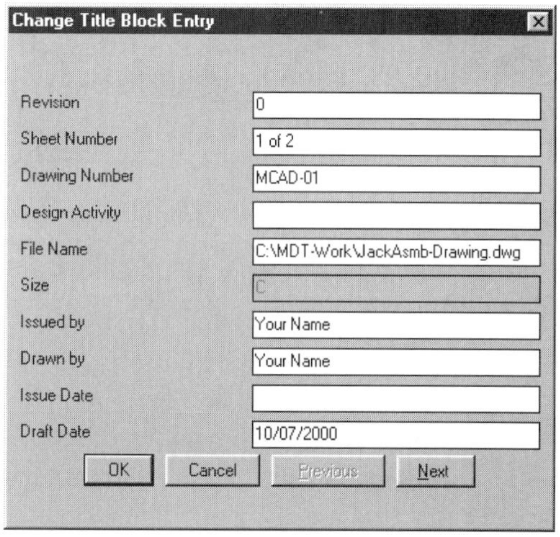

Figure 8.6

Pick: **Next** (to see the next page of information)

Pick: **OK**

Select objects for centering

Select objects: ↵

The title block appears in the drawing, as shown in Figure 8.7.

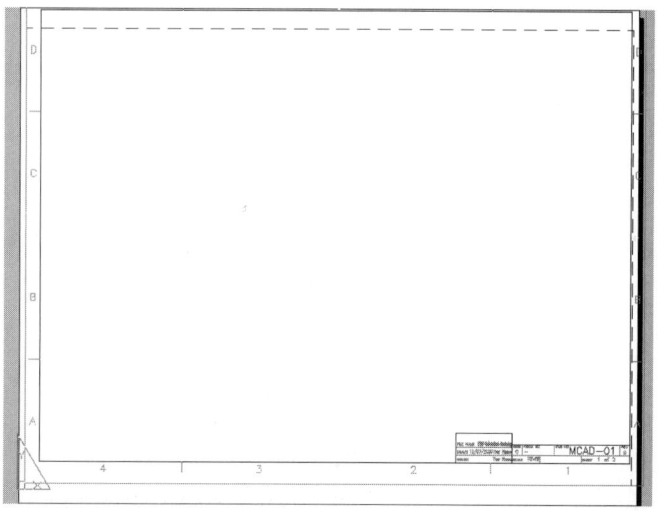

Figure 8.7

Tip

The border and title block are not designed perfectly for every printer. You may have to scale them down and move them into position to make them fit perfectly on your sheet. You will eventually want to design your own custom border and title block to replace the standard ones.

The layout returns to your screen. Move your crosshairs around the layout and look at the coordinate display to verify that the new size will allow plotting on a 22 × 17-inch sheet.

Command: **SCALE** ↵
Select objects: **pick anywhere on the lines or text of the title block**
Select objects: **pick anywhere on the border**
Select objects: ↵
Specify base point: **END** ↵
endpoint of: **pick on the lower left end of the bottom line of the border**
Specify scale factor or [Reference]: **.915** ↵

Command: **M** ↵
Select objects: **pick the title block**
Select objects: **pick the border**
Select objects: ↵
Specify base point or displacement: **END** ↵
endpoint of: **pick the lower left end of the border**
Specify second point of displacement or <use first point as displacement>: **.75,0** ↵

The drawing border should now appear centered in the available drawing area, as shown in Figure 8.8.

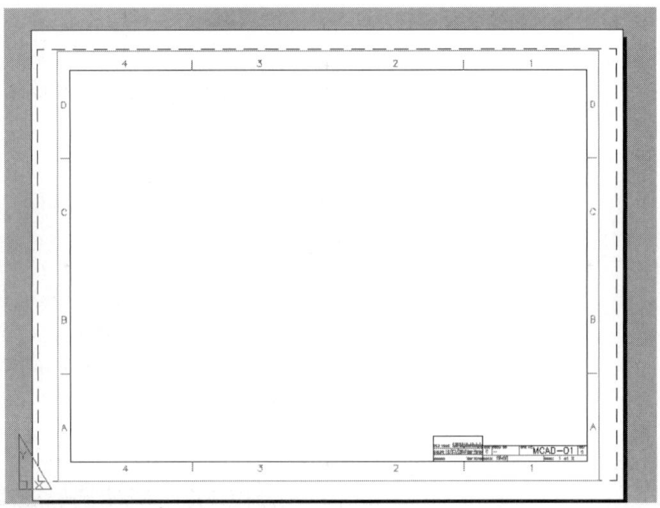

Figure 8.8

Next you will see how you can edit the information once it has been entered into the title block.

Pick: **Zoom Window icon**
Enlarge the title block so that it fills the screen, as shown in Figure 8.9.

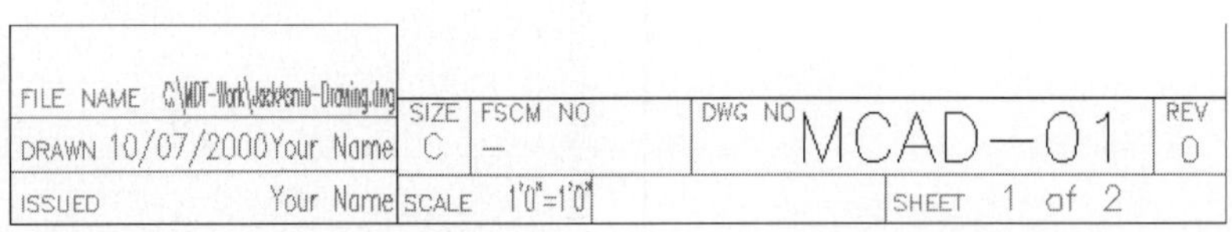

Figure 8.9

The ATTEDIT command allows you to edit block attributes, which is how this information is added to the title block. You will type it at the command line.

Command: **ATTEDIT** ↵

Select block reference: **pick on the lines or text of the title block** (not the border)

The Edit Attributes dialog box appears on your screen, showing the information displayed in the title block. Use it to make any necessary corrections to the attributes you have typed.

Pick: **OK**

Pick: **Zoom Previous icon**

Tip

You could also select Drawing/Title Revisions, Drawing Title/Borders from the Annotate menu as you did before, then select Change Title Block in the Edit Drawing Border/Title Block dialog box to open a dialog box to edit the title block entries.

Now you are ready to add views to create an assembly drawing.

Exploded Assembly Drawing

As you saw in Tutorial 7, you can create scenes to store useful views of your model. When you are creating drawings for printing, you can select a scene to use in quickly creating printable views that show the parts clearly. Exploded isometric views are often printed and used as assembly drawings, which are given to manufacturing personnel as a guide for making the device.

Using Scenes in Drawing Views

You will add a drawing view that uses the Exploded scene created in Tutorial 7 to position the parts in the drawing.

Pick: **Zoom All icon**

Right-click: in the graphics area

Pick: **New View**

The Create Drawing View dialog box appears on your screen. Before you can create an isometric view, you must first establish a base view in the drawing that you can use to develop the isometric from. You can later delete this view if you do not want to use it. You will make the following selections, as shown in Figure 8.10.

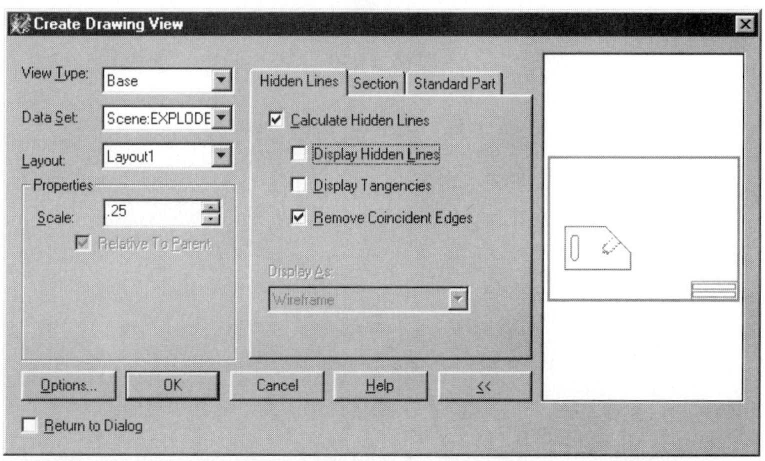

Figure 8.10

Choose View Type: **Base**

Choose Data Set: **Scene: EXPLODED**

Choose Layout: **Layout1**

Type Scale: **.25**

Select: **Calculate Hidden Lines**

Deselect: **Display Hidden Lines** (so it is unchecked)

Deselect: **Display Tangencies** (so it is unchecked)

Select: **Remove Coincident Edges** (so it is checked)

Pick: **OK**

Your model appears so you can specify the orientation of the views. You are first prompted for a plane that will define the front view of the model. You can specify this plane by picking a face on the part, a work plane, or by using the current UCS, a View, or orientations of the World Coordinate Sytem (WCS). You will use the WCS to define the orientation.

Select planar face, work plane or [Ucs/View/worldXy/worldYz/worldZx]: **Z** ↵

Define X axis direction:

Select work axis, straight edge or [worldX/worldY/worldZ]: **X** ↵

Adjust orientation [Z-flip/Rotate] <Accept>: ↵

Specify location of base view: **pick in the lower left quadrant of the layout** (select again if necessary to fit the view on the sheet)

Specify location of base view: ↵

An orthographic base view of the exploded jack assembly appears on the screen, similar to Figure 8.11. The hidden lines are not displayed. Exploded orthographic views such as this one are often plotted as assembly drawings. In this case, you will add an exploded isometric view and later delete this view.

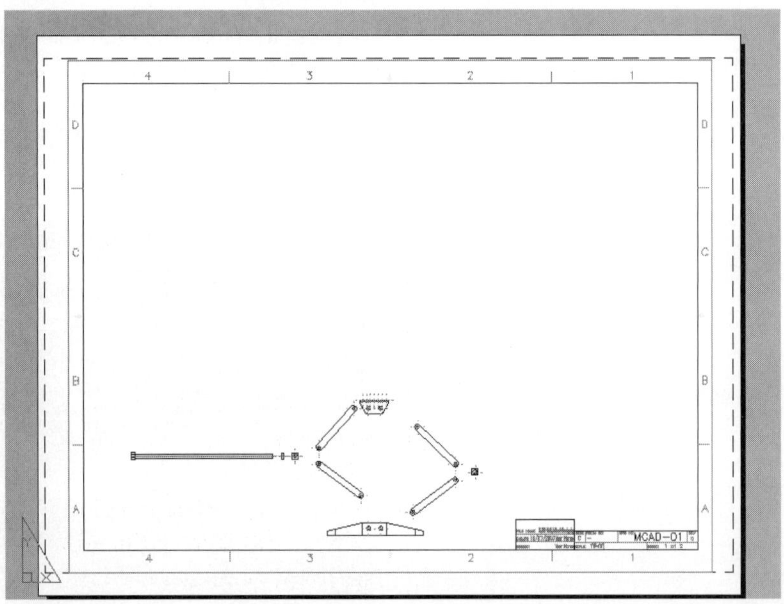

Figure 8.11

To create the isometric view,

Right-click: in the graphics area

Pick: **New View**

The Create Drawing View dialog box appears on your screen. Now that you have created a base view, isometric view is one of the options under View Type. The Data Set and Layout choices are grayed out.

Select View Type: **Iso** (metric)

Select Scale: **1.5**

Select: **Relative to Parent** (so it is checked)

Select: **Calculate Hidden Lines**

Deselect: **Display Hidden Lines** (so it is unchecked)

Deselect: **Display Tangencies** (so it is unchecked)

Select: **Remove Coincident Edges** (so it is checked)

Pick: **OK**

Select parent view: **pick the base view**

Specify location for isometric view: **pick above and to the left of the ortho view to get a left front iso view** (even it if does not fit inside border)

Specify location for isometric view: ↵

Move the view on your own to reposition it inside the border

The isometric view is added, as shown in Figure 8.12.

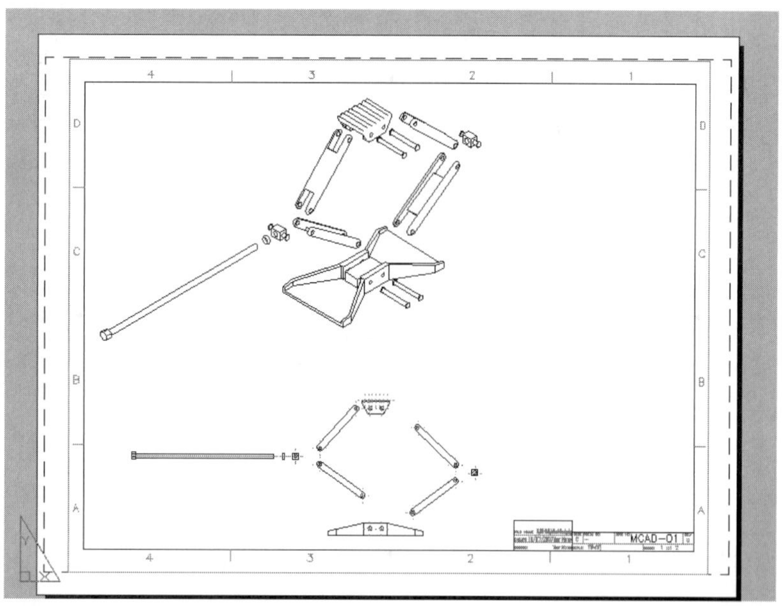

Figure 8.12

Next you will delete the base view of the exploded assembly and add an orthographic view of the fully assembled jack.

Right-click: in the graphics area

Pick: **Delete View**

Select view to delete: **pick the orthographic base view**

Enter an option [Next/Accept] <Accept>: ↵

The Delete Dependent Views message box appears. You will select no so that the isometric view remains.

Pick: **No**

The orthographic view is deleted from the drawing layout.

Next you will add a fully assembled view of the jack. This will be a new base view that uses the Assembled Scene for the position of the parts.

Right-click: in the graphics area
Pick: **New View**
Select View Type: **Base**
Select Data Set: **Scene:ASSEMBLED**
Select Layout: **Layout1**
Select Scale: **.25**
Select: **Calculate Hidden Lines**
Deselect: **Display Hidden Lines** (so it is not checked)
Deselect: **Display Tangencies** (so it is not checked)
Select: **Remove Coincident Edges**
Pick: **OK**

Select planar face, work plane or [Ucs/View/worldXy/worldYz/worldZx]: **Z** ↵
Define X axis direction:
Select work axis, straight edge or [worldX/worldY/worldZ]: **X** ↵
Adjust orientation [Z-flip/Rotate] <Accept>: ↵
Specify location of base view: **pick in the lower left quadrant of the layout** (select again if necessary to fit the view on the sheet)
Specify location of base view: ↵

The assembled view of the jack is added to the drawing sheet, as shown in Figure 8.13.

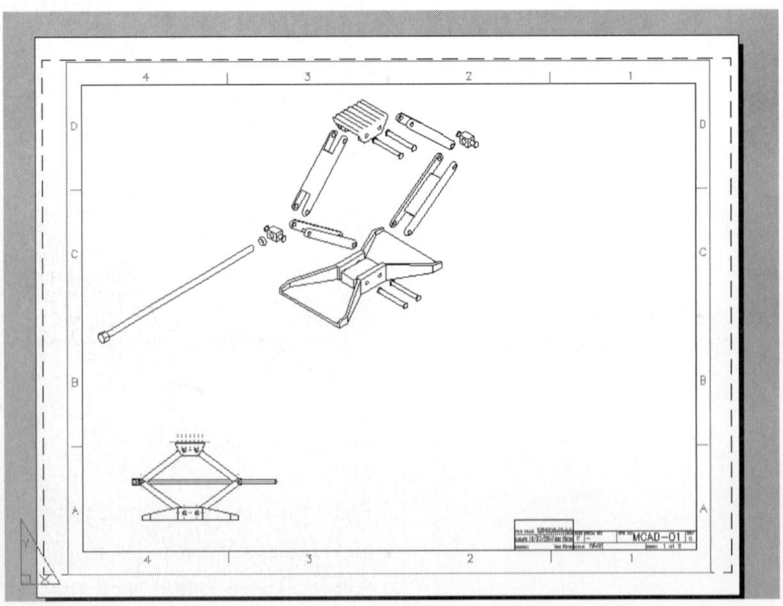

Figure 8.13

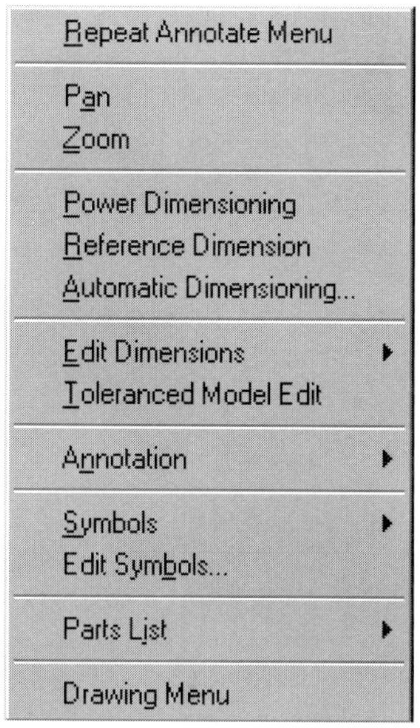

Figure 8.14

Reference

Mechanical Desktop On-line Help

To learn more about the BOM database, look up the AMBOM command in Mechanical Desktop Help.

Adding a Parts List

Next you will add a parts list, which is also called a bill of materials (BOM). A parts list is a table of information about the parts in the drawing, which allows you to include information about the material the part is to be made from or about standard parts that will be purchased for the assembly. Parts are identified by number in the parts list, and balloon numbers on the parts identify them in the drawing.

Right-click: in the graphics area

Pick: **Annotate Menu**

The context menu switches to show the annotation menu selections shown in Figure 8.14. You will use it to select the command to create a bill of material.

Pick: **Parts List, BOM Database**

Creating the BOM Database

The BOM dialog box appears on your screen. It is similar to a spreadsheet where the information about your parts is stored. The headings Item, Qty, Name, Material, Note, and Vendor appear across the input boxes. The icons at the top of the dialog box start commands that you use to work with the BOM, such as print, add parts, delete and add columns and rows, insert and delete items, change assembly properties such as the listed material, expand or collapse subassemblies in the list, sort list items, add balloons to the drawing, and export and import BOM tables. Some of these icons are labeled in Figure 8.15.

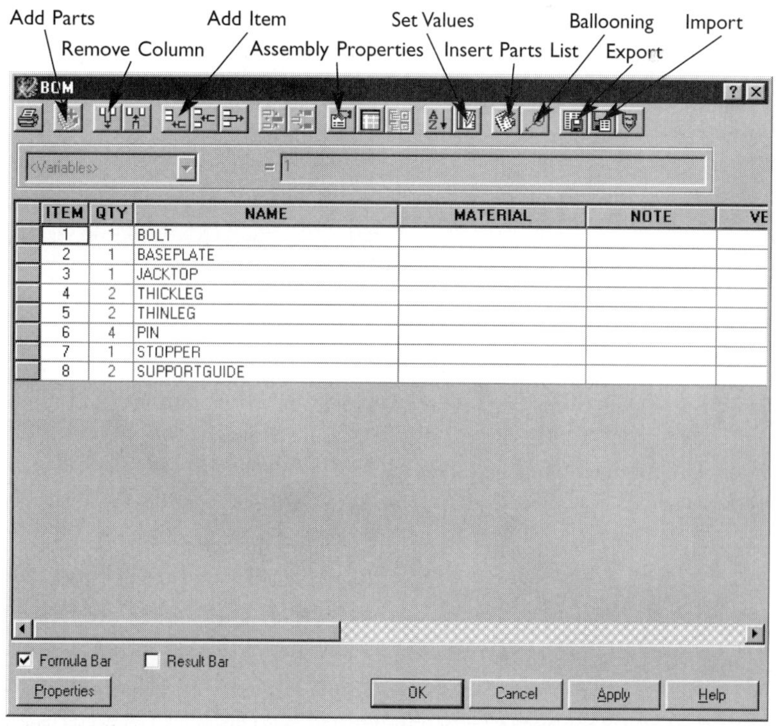

Figure 8.15

Figure 8.16

Tip

You can double-click in the empty box and type in the value.

You can customize the BOM to include the columns you need to describe your assembly. For this drawing, you need only the part number, quantity of each, the part name, and the material the part will be made from.

Next you will add materials information to the BOM for all parts. The pin and bolt will not be manufactured but will be purchased or taken from inventory, so you will enter STOCK as the material of the pin and bolt.

Right-click: **in the empty box across from Pin**

Pick: **Set Values**

The Set Value dialog box shows up on the screen. Make the following selections, as shown in Figure 8.16.

Choose Column: **Material**

Type Value: **STOCK**

Pick: **OK**

The text STOCK appears as the material for the pin. On your own,

Set the value for Material for all of the parts, as shown in Figure 8.17

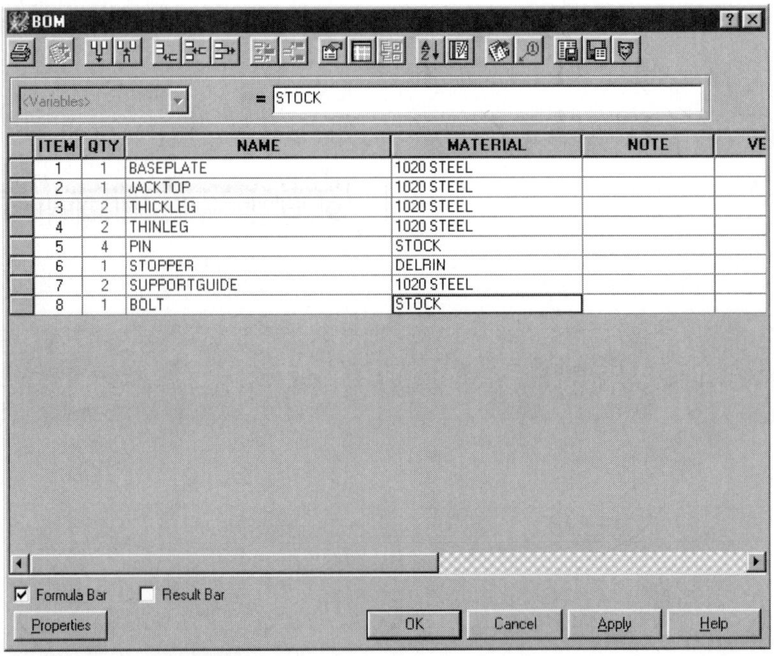

Figure 8.17

Pick: **OK** (when finished)

Creating the Parts List

Now that you have created the BOM database, you are ready to add a parts list to your drawing.

Right-click: in the graphics area

Pick: **Parts List, Parts List**

The Parts List dialog box appears, which presents the information in the BOM database and offers options for formatting the parts list.

First, you will delete the unused columns Note and Vendor. You will delete the two columns as a group.

Click and drag on the Note and Vendor headings so they are both highlighted.

Pick: **Remove Column icon**

Do you really want to remove these columns?

Pick: **Yes**

The two empty columns are deleted. Next you will resize the Name and Material headings.

Position your cursor near the division between the column headings until it appears as a double-headed arrow, then press the pick button and, holding the button down, drag to resize the column smaller.

Now you will look at the properties for the BOM and set the color for the text used in it.

Pick: **Properties** (at bottom of dialog box)

Set Color Swatch adjacent to Heading Text to blue (default is yellow).

Set Color Swatch adjacent to Data Text to blue.

Pick: **OK**

Finally, you will change the attachment point to Top right, as shown in Figure 8.18, to make it easier to position the parts list.

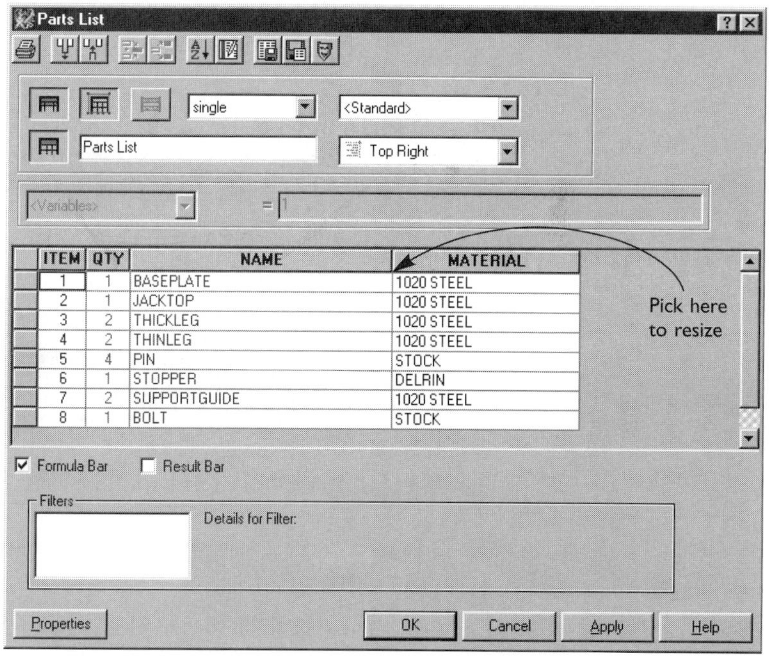

Figure 8.18

Choose: **Top Right** (from the drop-down menu)

Pick: **OK**

BOM table name is JackAsmb-Drawing

Specify location: **END** ↵

Pick: **the top right inside corner of the border**

Parts list was attached to the geometry.

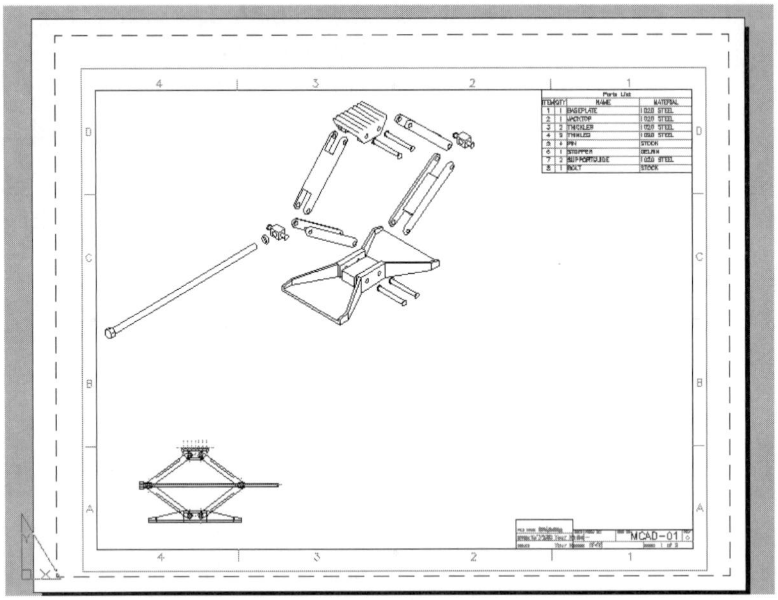

Figure 8.19

The parts list is added to the drawing, as shown in Figure 8.19.

Balloon Numbers

Tip
You can rename Parts List in your model tree to rename the heading for the parts list added to the drawing.

Tip
Move your isometric view left if it is too close to the parts list. (Return to the Drawing context menu for Move View.)

A balloon number, also called a ball tag, is a circled number that identifies the item on the drawing as it is named in the parts list. Mechanical Desktop generates these balloon numbers automatically or semiautomatically for you. First you will try the automatic selection. Then you will use hot grips to edit their placement.

Right-click: in the graphics area

Pick: **Parts List, Balloons**

Select part/assembly or [Auto/Collect/Manual/One/Renumber]: **A** ↵

Select pick object: **pick a crossing box around the isometric view of the parts starting in the lower right**

Select pick object: ↵

The balloon numbers appear in a horizontal arrangement that is attached to the part at the arrow end and to the cursor location for the circled numbers. You will use the command prompt to specify a vertical arrangement.

Align Standalone/Horizontal/<Vertical>: **V** ↵

Align Standalone/Vertical/<Horizontal>: **pick to the left of the isometric view**

The balloon numbers appear in the drawing, as shown in Figure 8.20, but they are too large for the available space. Next you will edit the properties of the balloon numbers to a better size. You will use the Mechanical Options dialog box, which has settings for many of the Mechanical Desktop functions.

Pick: **Annotate menu, Annotation Options**

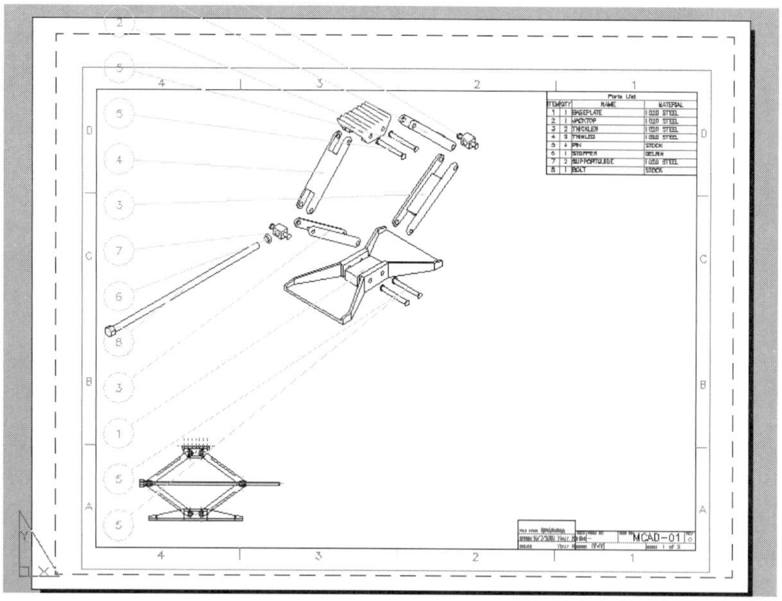

Figure 8.20

Pick: **Standards tab** (in the dialog box that appears)

The Standards tab appears, as shown in Figure 8.21.

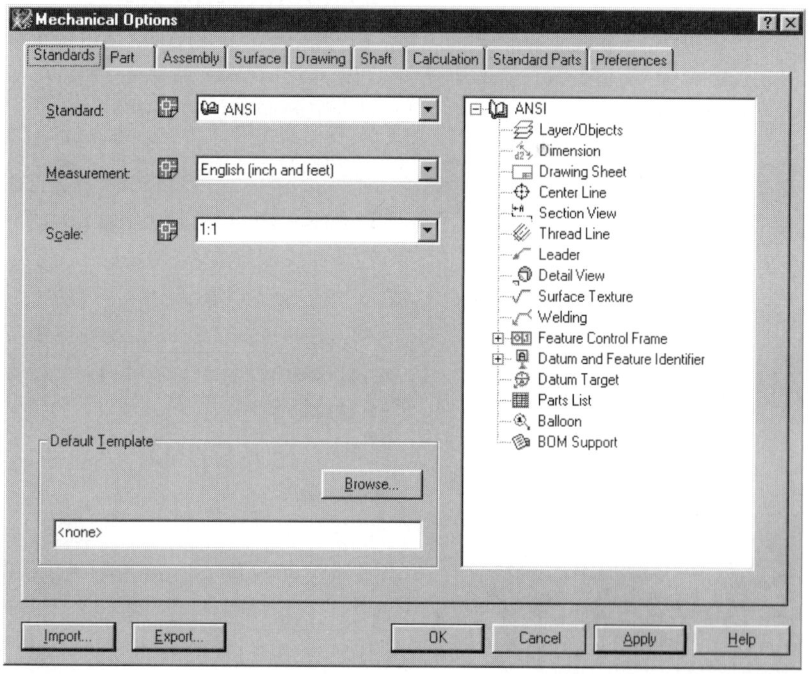

Figure 8.21

The window on the right contains a tree menu showing the drawing entities controlled by ANSI standards. (You could view a similar list for other standards listed in the menu at left.) You will look at the setting for balloons.

Right-click: **Balloon** (in the tree menu at the right)

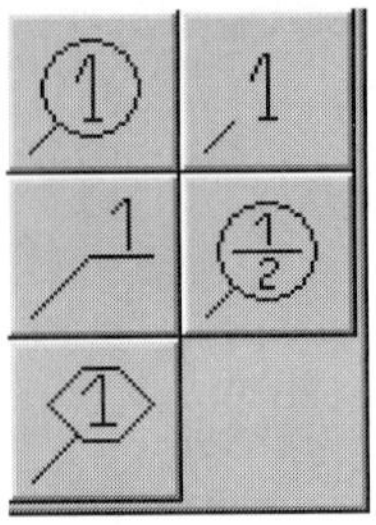

Figure 8.22

Pick: **Properties**

The Balloon Properties for ANSI dialog box shows up on your screen. It indicates that the settings used here are based on the ASME Y14.34M standard revised in 1996. In the dialog box you can select a custom symbol or different balloon type by picking on the image tile, as shown in Figure 8.22. You can set the text height (which also sizes the circle drawn around it) and text color, set the arrow style, and edit the columns to use as the selection for the number. You will change the text height so the dialog box is set as in Figure 8.23.

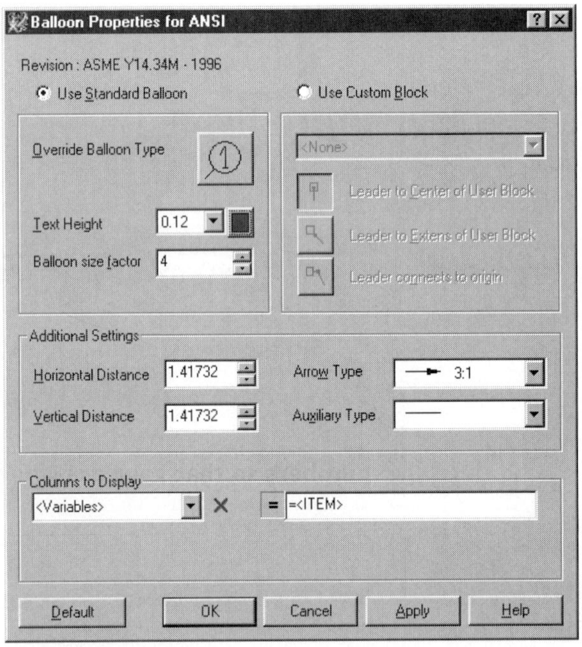

Figure 8.23

Choose Text Height: **.12**

Pick: **color box next to text height**

Pick: **blue**

Pick: **Apply**

Pick: **OK**

Pick: **OK**

The balloon numbers are resized. They still do not fit in the drawing neatly because of the default spacing that was used to make them.

Next you will use their hot grips to rearrange their placement and erase duplicate numbers.

While the command prompt is blank,

Pick: **on the balloon number at the top of the screen** (you may have to scroll as it is off the top of the drawing sheet)

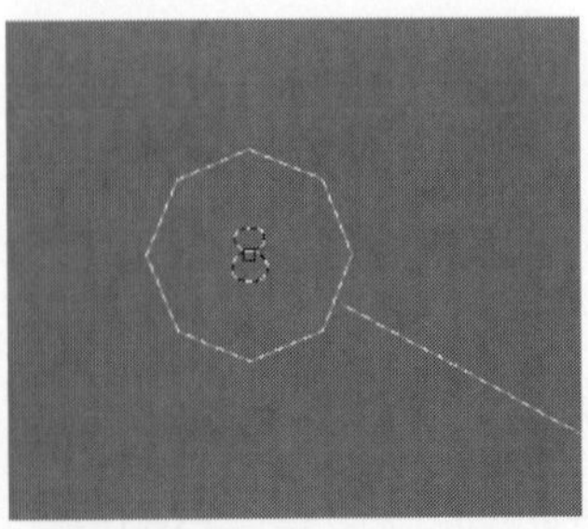

Figure 8.24

Figure 8.24 shows one of the ball tags selected so that you see its hot grips.

Pick: **in the small grip box inside the circled number**

The grip becomes active and changes color. The circle now appears attached to the cursor. On your own,

Position the balloon number so that it is located nearer the part it identifies

The preference is for the angled leader lines from balloon numbers to all be at essentially the same angle, but not horizontal or vertical. As you are placing the balloon numbers, try to keep this in mind. At times there is no good arrangement that shows the lines all at the same angle. The balloon numbers do not need to line up in a row, although sometimes this looks nice. Nor do all of the balloon numbers need to be on the same side of the drawing. Try to avoid placing them where the leaders cross other parts. In general, try to make it easy for readers to find the number of the item they are interested in. Also, it is not considered necessary to identify more than one part where parts appear similar and cannot be mistaken for other parts. You will use the Erase command to delete the duplicate balloon numbers.

Command: **E** [Enter]

Select objects: **pick the duplicated balloon numbers that are in the less desirable locations**

Select objects: [Enter]

The duplicated numbers are erased.

Continue on your own to reposition any of the other balloon numbers so that your drawing looks like Figure 8.25

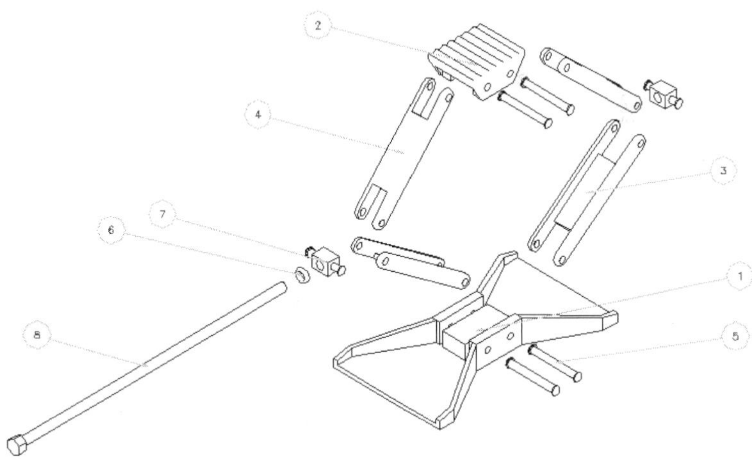

Figure 8.25

Congratulations! You have created an excellent assembly drawing that would have taken many hours to hand draw. It would also have taken many hours to draw as a 2D CAD drawing. All of this was done from your assembly model.

Adding a Shaded Image

To make your drawing even one bit nicer, you can insert a shaded image of the part. This shaded image must have been previously created using the commands you learned in Tutorial 1 to create a shaded view for use in other drawings or printing in other applications. You can use the file you created in Tutorial 1, or use the *jack.tif* file in the MDT-Data directory.

Pick: **Insert, Raster Image** (from the menu bar)

The Select Image File dialog box appears on your screen.

Select File: **jack.tif** (from the data files)

The Image dialog box appears. You will accept the default selections, as shown in Figure 8.26.

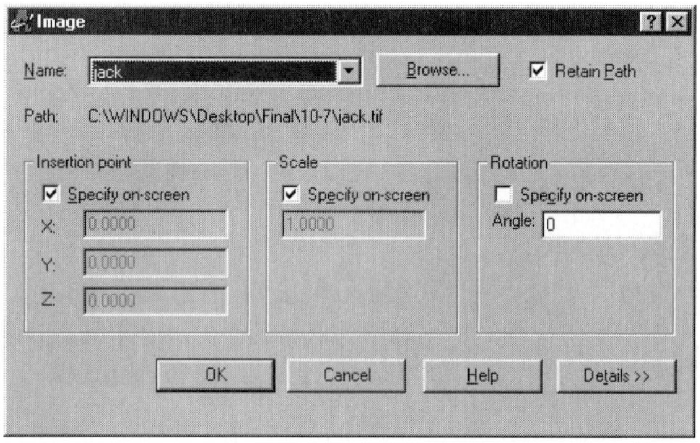

Figure 8.26

Pick: **OK**

You return to the drawing to specify the insertion point and the scale factor for the drawing.

Specify insertion point <0,0>: **pick in the upper left of the drawing area**

Base image size: Width: 1.000000, Height: 0.750000, Inches

Specify scale factor or [Unit] <1>: **move your cursor to size the drawing so that it fits in the available space**

The shaded image is added to the drawing. You can use hot grips to select the image, then move or scale if further if you want to adjust it more. Your completed drawing should look like Figure 8.27.

Save your finished assembly drawing

Printing the Assembly

You can print your assembly drawing even if you do not have a large-format printer. In the plot dialog box, choose your plot device (printer). If it supports 22 × 17 paper, your layout is ready to print. If not, choose your printer, choose to print Layout, and set the scale to 1 = 2 for an 8.5 × 11 sheet of paper. Make sure to UNselect the box at the top of the dialog box for "Save changes to layout." You do not want the changes in the plot dialog box to change your sheet size or other settings for the layout when you return to the drawing tab.

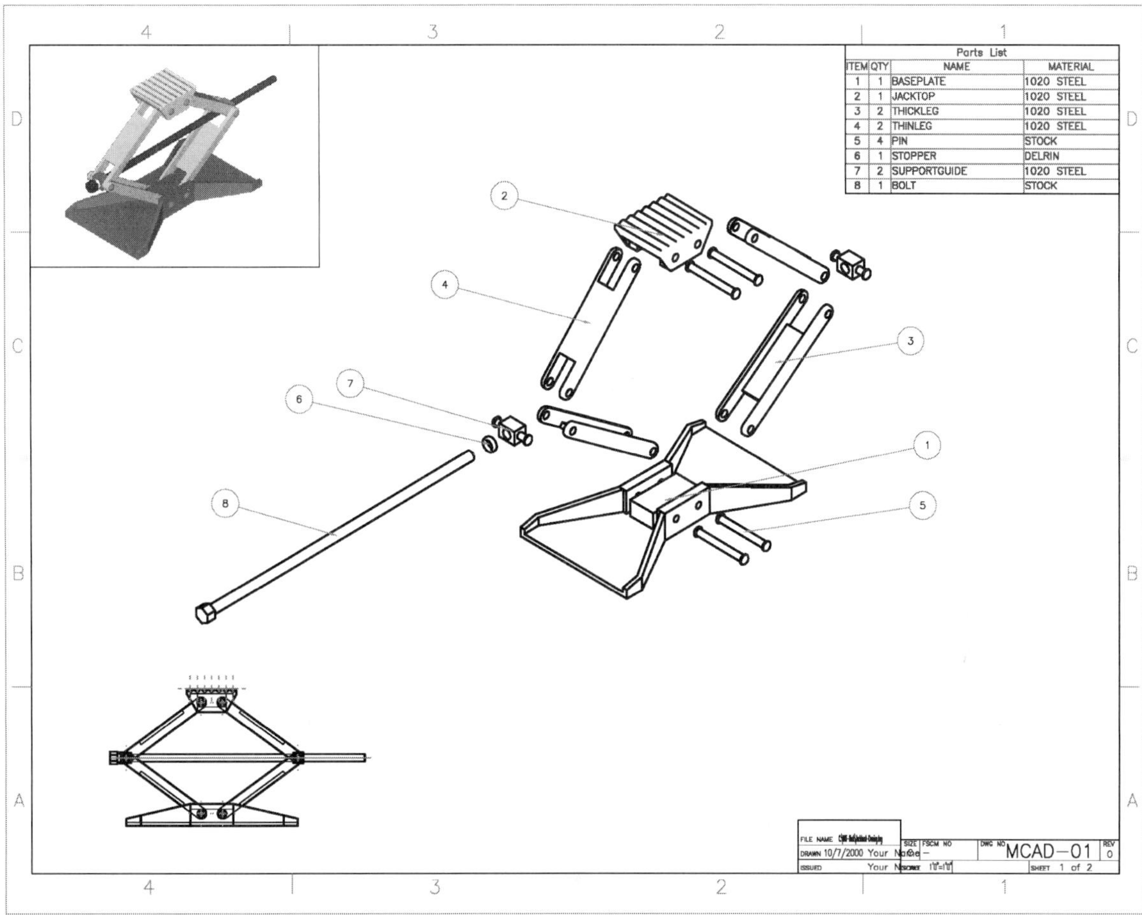

Figure 8.27

Reference

Engineering Design Communication

Part Drawings, pp. 476–477

Dimensioned Part Drawing

A full set of working drawings includes part drawings in addition to the assembly drawing you just created. You can create any number of dimensioned part drawings in the same drawing file. To do so, you will set up a new layout and make it active. Then you will add orthographic drawing views of the Baseplate part to this sheet.

Right-click: **Layout2** (in the Drawing tree)

Pick: **Rename**

Type: **Base-ortho** [Enter]

Layout 2 is renamed to Base-ortho. Next, you will briefly switch to the model tab to activate the Baseplate part. Activating the part allows you to specify the active part as the source of the drawing views you will add in the next steps. You will also change the UCS here so that you can see the World UCS as you specify the front view for the part.

Pick: **Model tab**

Right-click: **Baseplate**

Pick: **Activate Part**

Pick: **Assist, New UCS, World**

Toggle to wireframe view

The World UCS is now active.

Pick: **Drawing tab**

Double-click: **Base-ortho** (to activate the layout)

You switch to the blank drawing layout where you will layout views of the baseplate. Notice that because this layout is one of the two created by default, its size is 8.5 × 11 with no border or title block. You will set up the layout as you did for Layout1. (Refer back to that section if you need help.)

On your own,

Use Page Setup with the DWF Classic.pc3 and use Properties to set the size to ANSI-C

Add a standard ANSI-C title block and border

Add the name Baseplate, Sheet 2 of 2, and the same information as in Layout1 to the title block when prompted

Resize the border and title block to fit on the sheet

Different drawing views

Next you will add a set of orthographic views of the Baseplate part to the drawing.

Right-click: in the graphics area

Pick: **Drawing menu**

Pick: **New View**

Use the Create Drawing View dialog box to make the following selections.

Choose View Type: **Multiple**

Choose Data Set: **Active Part**

Choose Layout: **Base-ortho**

Choose Scale: **.5**

Select: **Calculate Hidden Lines**

Select: **Display Hidden Lines**

Deselect: **Display Tangencies** (so it is not checked)

Select: **Remove Coincident Edges**

Pick: **OK**

Your model appears so you can specify the orientation of the views. You are first prompted for a plane that defines the front view of the model. You could pick a plane on the Baseplate to define this view, but the Baseplate is aligned with the WCS, making it easy to use that.

Select planar face, work plane or [Ucs/View/worldXy/worldYz/worldZx]: **Z** ↵

Define X axis direction:

Select work axis, straight edge or [worldX/worldY/worldZ]: **X** ↵

Adjust orientation [Z-flip/Rotate] <Accept>: ↵

Regenerating layout.

Specify location of base view: **pick in the lower left quadrant of the sheet** (repick if needed)

Specify location of base view or [Done] <next view>: ↵

Specify location of projected view or [New parent view]: **pick above the previous view** (repick if needed)

Specify location of projected view or [Done] <next view>: ↵

Specify location of projected view or [New parent view]: **pick to the right of the base view** (repick if needed)

Specify location of projected view or [Done] <next view>: ↵

Specify location of projected view or [New parent view]: ↵

Three orthographic views of the base are added to the drawing, as shown in Figure 8.28.

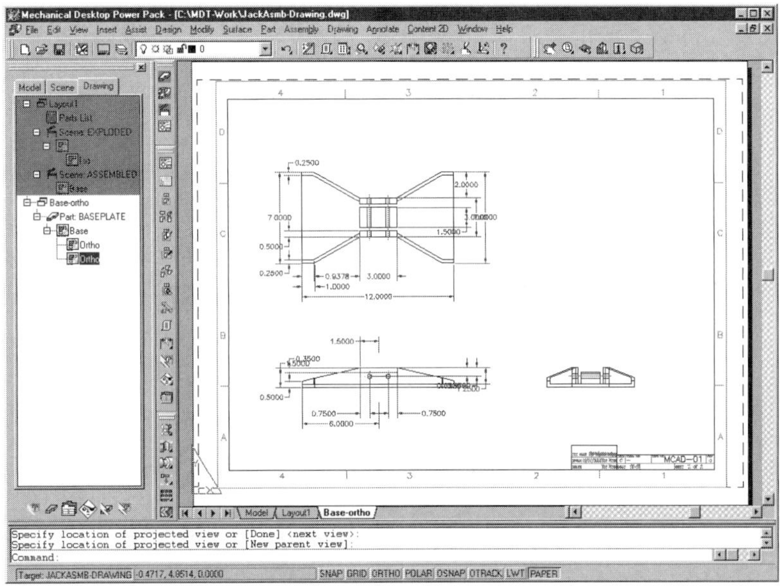

Figure 8.28

The hidden lines are generated automatically, as are the centerlines. The views should be located relatively far apart. Drawings are more easily read when there is plenty of white space and room for dimensions and notes.

The parametric dimensions—those you used to model the part—appear in the views automatically. Good placement practices and techniques cannot be provided entirely by the rules of the software. You must clean up the dimensions so that they are easy to read.

Good Dimensioning Technique

When creating a dimensioned drawing, keep in mind its purpose, which is usually to communicate to someone the specifications for an acceptably manufactured item. If your drawing can be interpreted unambiguously, it serves the purpose. Organizations like ANSI also publish standards for the sizes of dimension features, like arrowheads and text, that guide you in creating a drawing that is legible and prints clearly and describe how dimensions and tolerances are to be applied and interpreted. In addition, there are many rules of thumb to help guide you in placing the dimensions where they can be read easily. Some of these rules are as follows:

- Place dimensions where the feature shows its shape clearly and true size.
- Do not cross a dimension line with another dimension line or extension line.
- It is acceptable for extension lines to cross one another.
- Place dimensions off drawing views.

Reference

Engineering Design Communication

Developing Drawing Views, pp. 485–489

Tip

If you wish to use the style you created in Tutorial 3, you can substitute your ThickLeg-Drawing.dwg file in MDT-Work for the file to import from.

- Do not duplicate dimensions.
- Group dimensions together to make them easy to read.
- Use a meaningful number of decimal places in the dimension value (often untoleranced inch values are shown to two decimal places and metric values are shown to whole millimeters or one decimal place).

As you look at the drawing in Figure 8.28, you can see that many of these guidelines are violated. You will start cleaning up the dimensions by modifying the dimension style. Then you will rearrange the dimensions for better placement.

The dimensions currently use the default style STANDARD. You will use the style you created in Tutorial 3, MECHANICAL, to change the dimension characteristics. This file is stored in the datafile *ThickLeg-Style-dta.dwg*. Mechanical Desktop allows you to import dimension styles from other drawings so you do not have to re-create them.

Pick: **Assist, Format, Style Importer**

Navigate to MDT-Data

Double-click: **ThickLeg-Style-dta.dwg**

The Dim Style Importer dialog box appears, as shown in Figure 8.29, where you will choose the Mechanical style from the list at left.

Pick: **Mechanical**

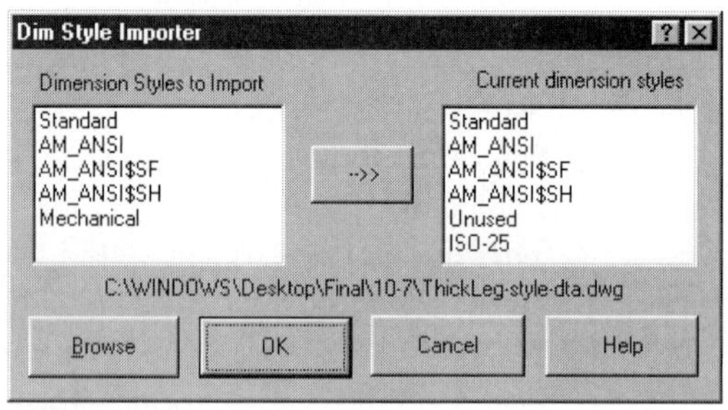

Figure 8.29

Pick: **the arrow button**

Pick: **OK** (after Mechanical appears in the list on the right)

Next you will make MECHANICAL the current dimension style. (Refer to Tutorial 3 if necessary to review working with dimension styles.)

Pick: **Annotate, Edit Dimensions, Dimension Style**

The Dimension Style Manager pops up on your screen, as shown in Figure 8.30. The available dimension styles are listed in the left side of the dialog box.

Select: **MECHANICAL**

Pick: **Set Current**

Pick: **Close**

The dimensions that already exist do not change. Changing the style will only change the dimensions added after it is changed. Next you will update all of the existing dimensions so that they use the current style.

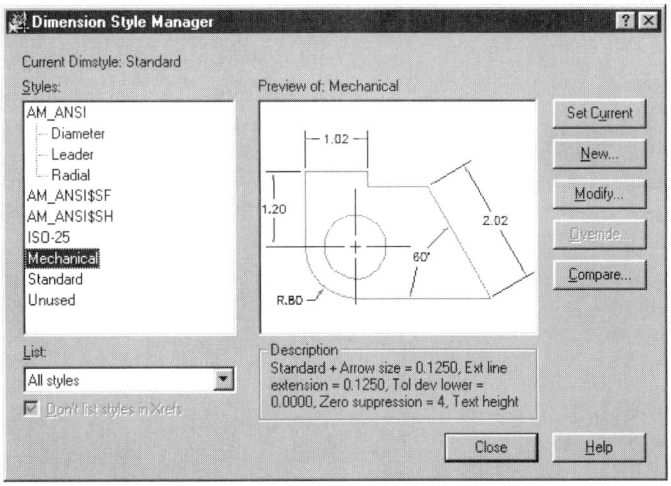

Figure 8.30

Command: **DIM** [↵]

Dim: **UPDATE** [↵]

Select objects: **ALL** [↵]

Select objects: [↵]

Dim: [ESC]

The dimensions update to show only two decimal places and smaller text, arrowheads, and other features. Figure 8.31 shows the drawing as it appears at this time.

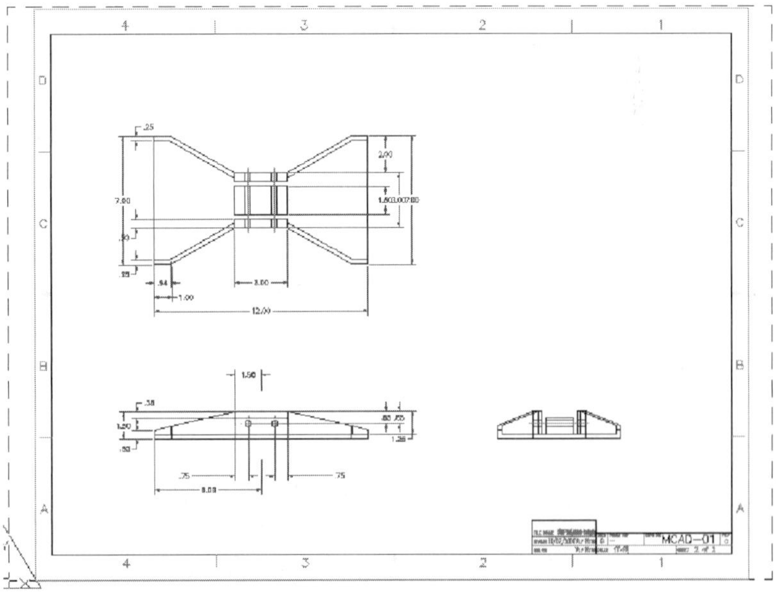

Figure 8.31

The smaller dimension features will make it easier to place the dimensions on the sheet so that they don't overlap and show clearly. Don't forget that you are working with a 22 × 17-inch sheet. Although the values look small on your screen, they will be perfectly legible when printed.

Use Zoom Window to enlarge the top and front views

Start by using the dimensions' hot grips and dragging the longest dimensions further away from the object outline. This will help you read them better as well as give better placement. You will hide some of the extraneous dimensions later.

Turn ORTHO on (to make it easier to move the dimensions along a single axis).

Pick: **the 7.00 vertical dimension at the right of the top view so that its grips appear**

Pick: **the middle grip near the dimension value 7.00 so that it becomes hot**

Move your cursor to drag the dimension line and value to the right

Pick: **to place the dimension line as shown in Figure 8.32**

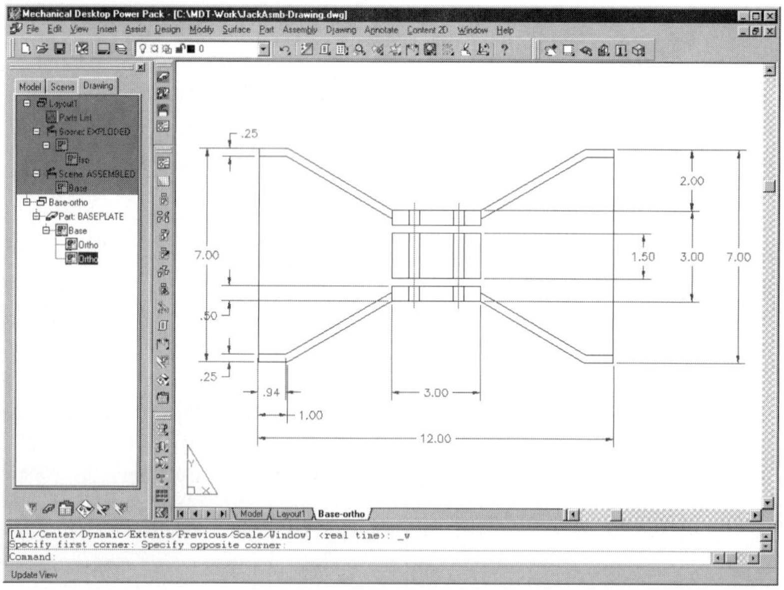

Figure 8.32

Press: ESC (to deactivate the grips)

Next you will use grips to rearrange the 2.00 and 3.00 dimensions so that they are in a rank the next in from the 7.00 dimension.

Pick: **the 2.00 vertical dimension near the right of the top view so that its grips appear**

Pick: **the middle grip near the dimension value 2.00 so that it becomes hot**

Move your cursor to drag the dimension line and value to the right

Pick: **to place the dimension line as shown in Figure 8.32**

Press: ESC (to deactivate the grips)

On your own, use the grips to locate the 3.00 dimension as shown in Figure 8.32

On your own,

Use the grips to reposition the remaining dimensions for the front and top views so that your drawing looks like Figure 8.33

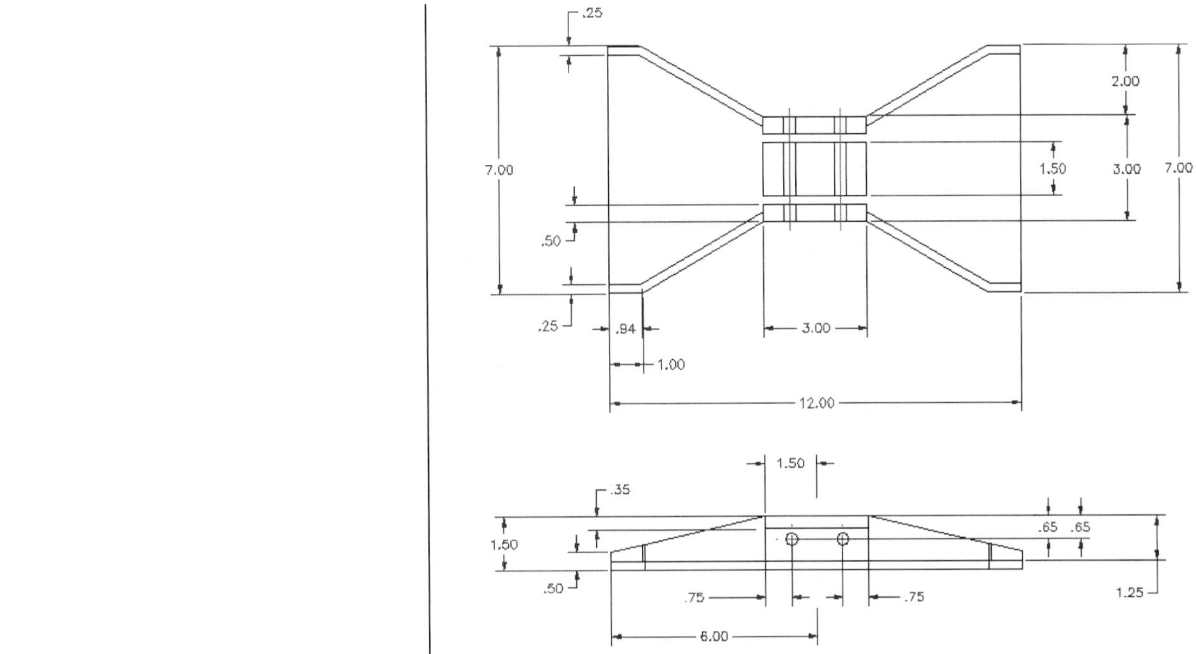

Figure 8.33

Notice the 7.00 dimension attached to the upper left corner of the top view in Figure 8.33. Its extension lines overlap onto the drawing lines because it is attached to the far corner instead of the close corner of the feature. There is nothing wrong with extension lines that cross part edges, but they should show a gap of about 1/16 between the feature and the extension line. Extension lines over the top of object lines are not okay. You will use hot grips to move the dimension attachment points.

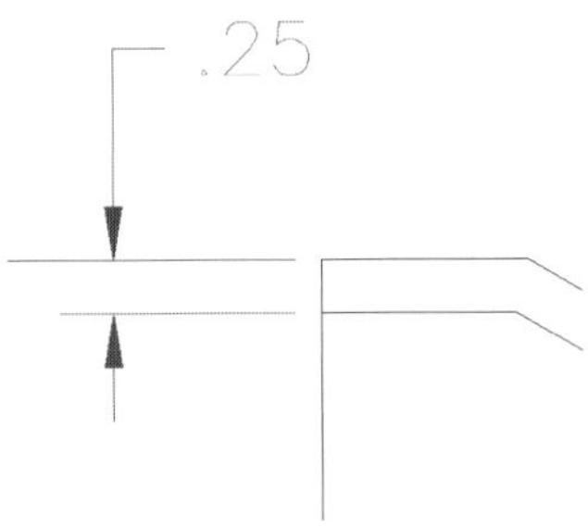

Figure 8.34

Turn on OSNAP Endpoint.

Pick: **the 7.00 vertical dimension near the left of the top view so that its grips appear**

Pick: **the right grip on the upper extension line so it becomes hot**

Specify stretch point or [Base point/Copy/Undo/eXit]: **pick on the top corner of the object to relocate the extension line attachment**

Press: ESC ESC (to deactivate the grips)

The extension line should appear, as shown in Figure 8.34.

Repeat these steps for other extension lines, as necessary.

Turn OSNAP off

Next you will hide the extraneous dimensions.

Pick: **Drawing, Drawing Visibility**

Use the Desktop Visibility dialog box that appears for the following selections. Make sure that none of the boxes appear selected, as shown in Figure 8.35.

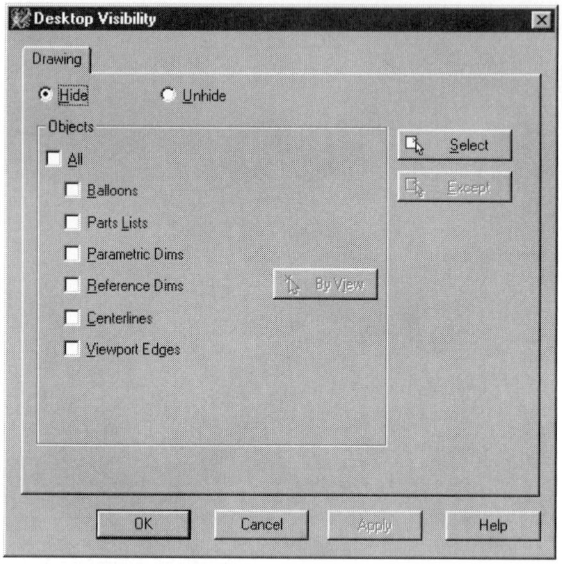

Figure 8.35

Pick: **Select**

Select objects to hide: **in the top view, pick the .94 dimension, the lower .25 dimension, and the left 7.00 dimension**

Select objects to hide: **in the front view, pick the horizontal 1.50 dimension, and the rightmost .65 dimension**

Select drawing objects to hide: ↵

Pick: **OK**

The selected dimensions are removed from the screen.

Next you will move the 6.00 dimension from the front view to the top view. Notice that you can also use this command to flip or reattach dimensions. Read about these options using Help.

Right-click: in the graphics area

Pick: **Annotate Menu**

Pick: **Edit Dimensions, Move Dimension**

Enter an option [Flip/Move/movemUltiple/Reattach] <Move>: ↵

Select dimension: **pick the 6.00 dimension in the front view**

Select destination view: **pick on the top view**

Select location: **pick above the 12.00 dimension**

Select location: ↵

Select dimension: ↵

Move the 1.25 and 1.50 dimensions from the front view to the side view on your own

Move the 3.00 dimension from the front view to the top view on your own

Rearrange the dimensions further using hot grips, if necessary. You can use the grips to move the dimension value alone by selecting its box as the base grip. Your drawing should look like Figure 8.36.

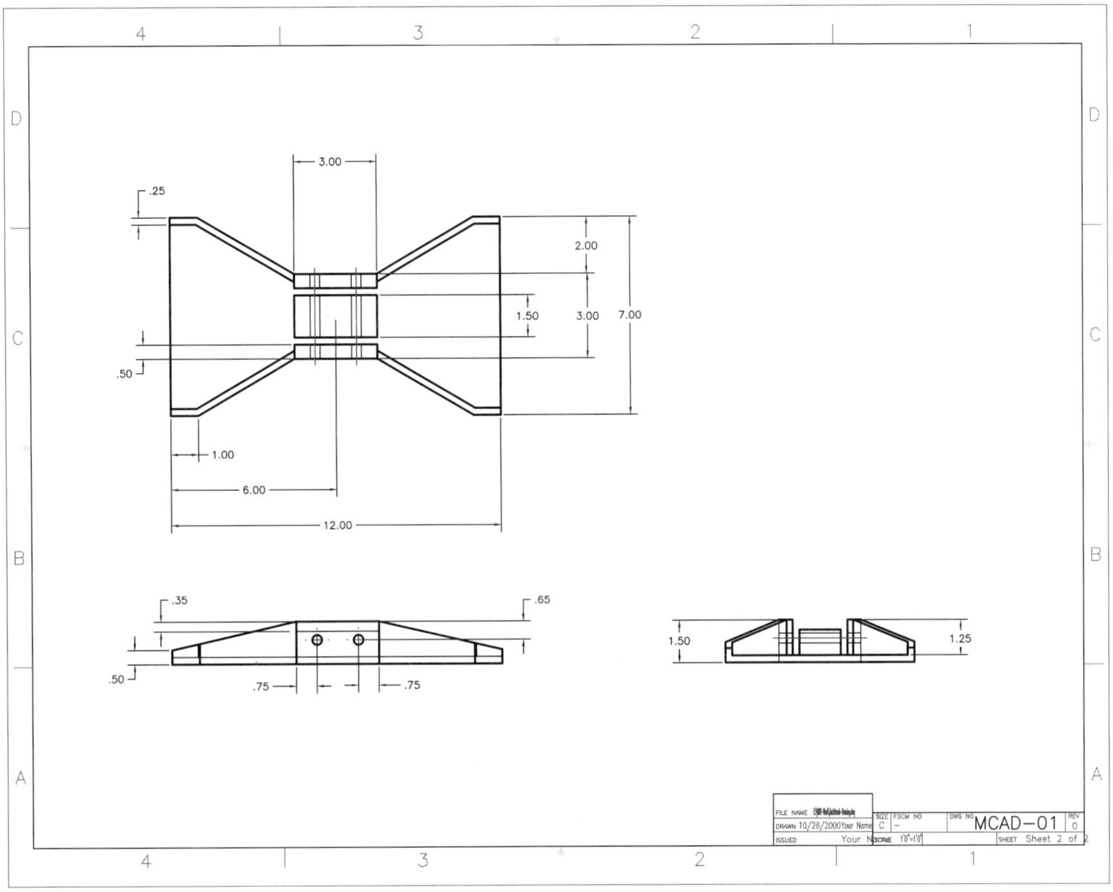

Figure 8.36

Reference

Engineering Design Communication

Nonparametric Dimensions, p. 274

Adding a Reference Dimension

The dimensions that automatically appeared in your drawing are called *parametric dimensions*. This indicates that they drive the sizes of the features they are associated with. Changing the value of a parametric dimension will change the size of the feature that it relates to. Sometimes in the process of creating a model, the dimensions that you want to show on the drawing are not available as parametric dimensions—although the better you plan your model, the less often you will need to add nonparametric dimensions. A *reference dimension* in Mechanical Desktop is one that gets its size value from the features in the drawing views and can update if the model changes but cannot be used to change the sizes of features in the model (it is associative, but not bidirectional). You will add a reference dimension for the size of the slot in the top view. Refer to Figure 8.37 as you pick.

Right-click: in the graphics area

Pick: **Reference Dimension**

Select first object: **pick the upper right corner of the slot labeled A in Figure 8.37**

Select second object or place dimension: **pick the lower right corner of the slot labeled B**

Specify dimension placement: **pick to the right of the part**

Specify placement point or [Undo/Hor/Ver/Align/Par/aNgle/Ord/reF/Basic]: ↵

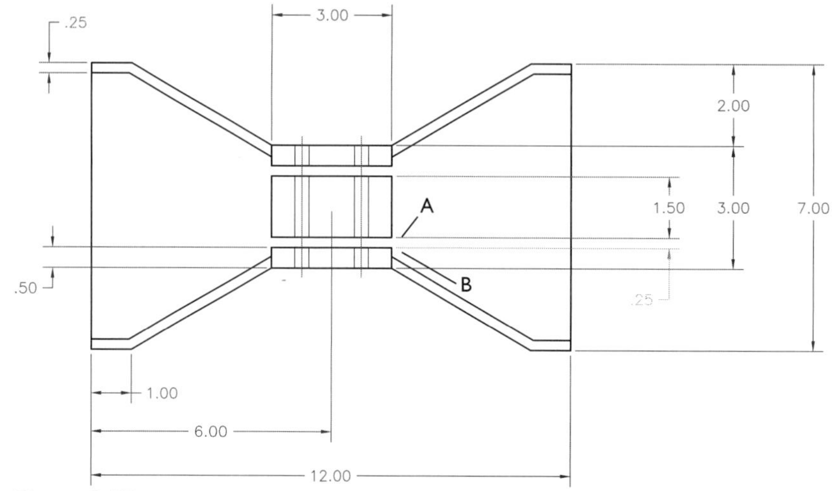

Figure 8.37

Select first object: ↵

The reference dimension appears, as shown in Figure 8.37. Notice that it is a different color than the parametric dimensions. This is to remind you that it does not drive the part geometry; rather it gets its size information from the part.

Use hot grips to move the .25 reference dimension value down so it does not overlap another extension line

Next you will add a vertical centerline through the top view. You will later add text as a note specifying that the part is symmetric about that centerline.

Right-click: in the graphics area

Pick: **Annotation, Centerline**

Select edge: **pick the left vertical edge of the top view**

Select mirrored edge or <ENTER>: **pick the right vertical edge of the top view**

Specify first trim point: **pick above the view** (as shown in Figure 8.38)

Specify second trim point: **pick below the view**

The centerline is added to the drawing. You can use the LTSCALE command to adjust the lengths of the dashes in line patterns. You will type it at the command prompt.

Command: **LTSCALE**

Enter new linetype scale factor <1.0000>: **.03** ↵

Notes

You can add notes to your drawing using standard AutoCAD commands for adding text. Notes can be free-standing notes on the drawing or associated with specific drawing geometry. You will create a new layer on which to create the notes and set it to the current layer.

Pick: **Layer icon**

Pick: **New**

Type: **TEXT** (as the new layer name)

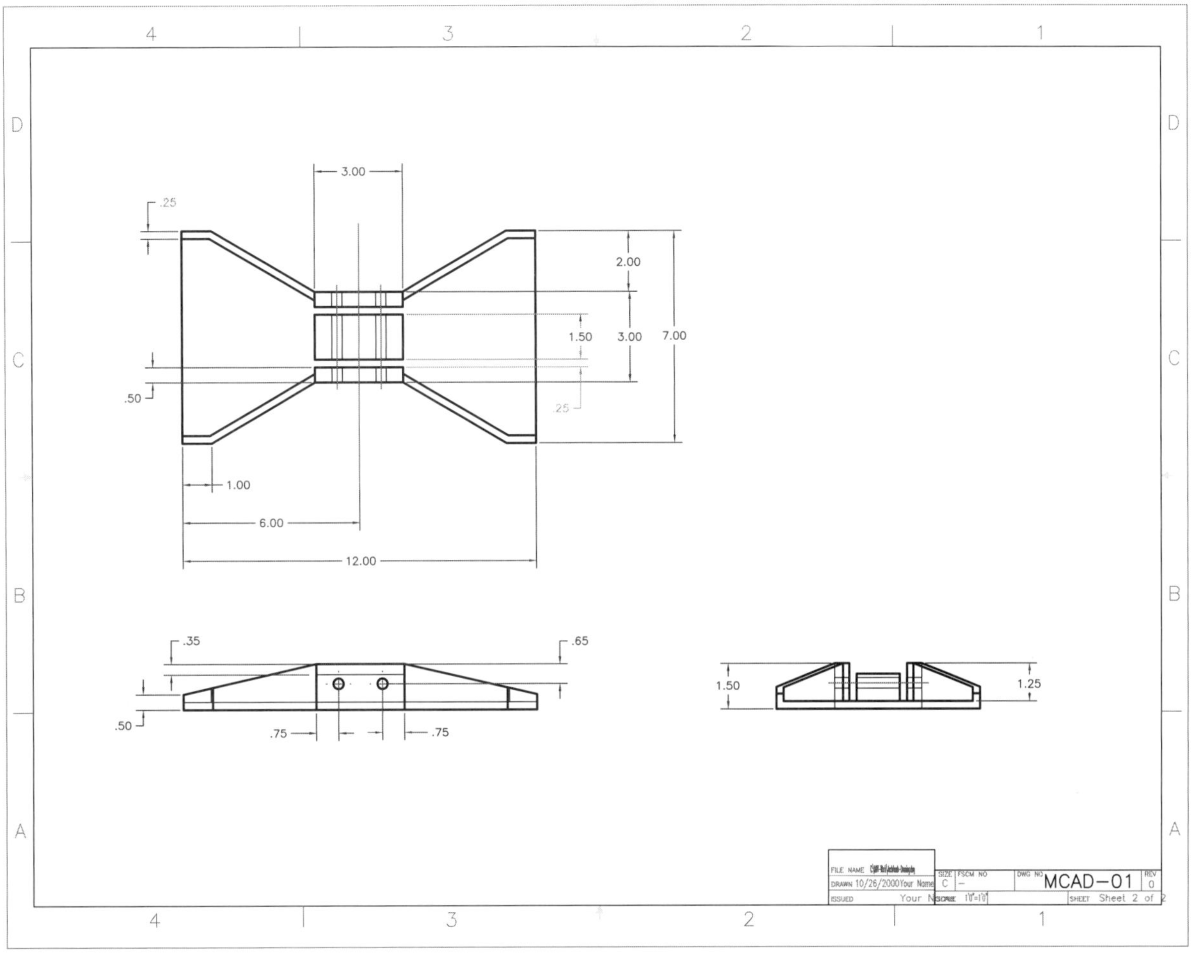

Figure 8.38

Pick: **on the color tile across from layer text**

Pick: **RED** (as the color for the layer)

Pick: **OK**

Pick: **Current**

Pick: **OK**

Layer Text should now be the current layer.

Annotations with Leaders

Use annotation when you need to make specific notes that apply to certain features or items. For these notes, it is important to identify where on the part the note applies so they include a leader from the text to the part. Annotation text and geometry (the leader line and arrow) can then be associated with entities in the drawing views so that they will move with the view.

Next you will add text and a leader specifying that the part is symmetrical about the centerline.

Right-click: in the graphics area

Pick: **Annotation, Leader**

Specify first leader point, or [Settings]<Settings>: ↵

The Leader Settings dialog box appears on the screen with the Annotation tab uppermost, as shown in Figure 8.39.

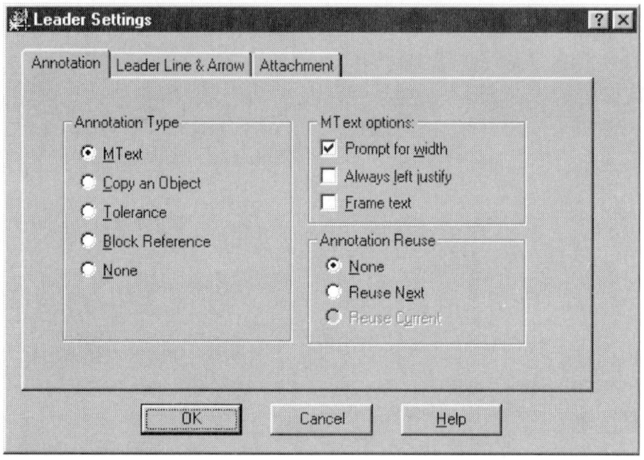

Figure 8.39

The Annotation tab allows you to select to add MText, Copies of an Object, Tolerances, Blocks, or nothing to the end of the leader. MText should be selected, which indicates that you may enter an annotation with multiples lines of text. Do not make any changes to this tab.

Pick: **Leader Line and Arrow tab**

The settings on this tab allow you to control the appearance of the line and arrow style use for the leader. Do not make any changes to this tab at this time, as shown in Figure 8.40.

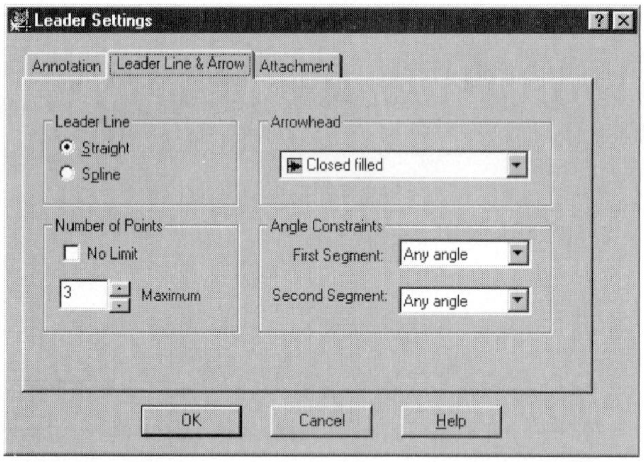

Figure 8.40

Pick: **Attachment tab**

You use the settings on the attachment tab to determine how text, blocks, or other annotation attaches to the leader. (This is in the general sense of annotation, not the MDT sense of attaching entities to the parametric drawing views.) Do not make any changes at this time. The dialog box should look like Figure 8.41.

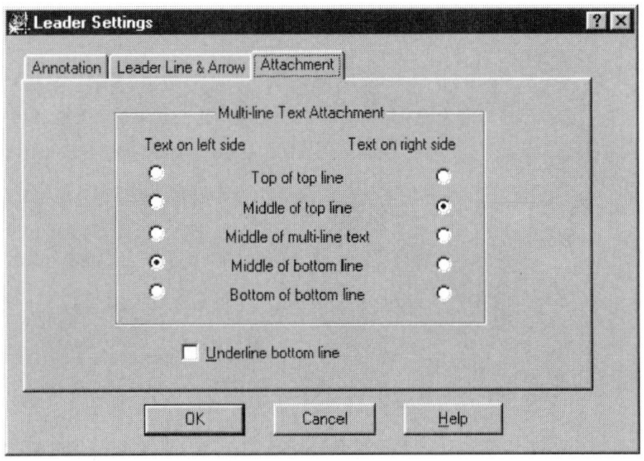

Figure 8.41

Pick: **OK**

Specify first leader point or [Settings]<Settings>: **pick on the upper end of centerline in the top view**

Turn Ortho off

Specify next point: **pick to draw an angled leader line, as shown in Figure 8.42**

Turn Ortho on

Specify next point: **pick to draw a horizontal tail for the leader**

Specify text width <.0000>: ↵

Enter first line of annotation text <Mtext>: **PART IS SYMMETRICAL ABOUT CENTERLINE** ↵

Enter next line of annotation text: ↵

The leader and text are added to the drawing, as shown in Figure 8.42.

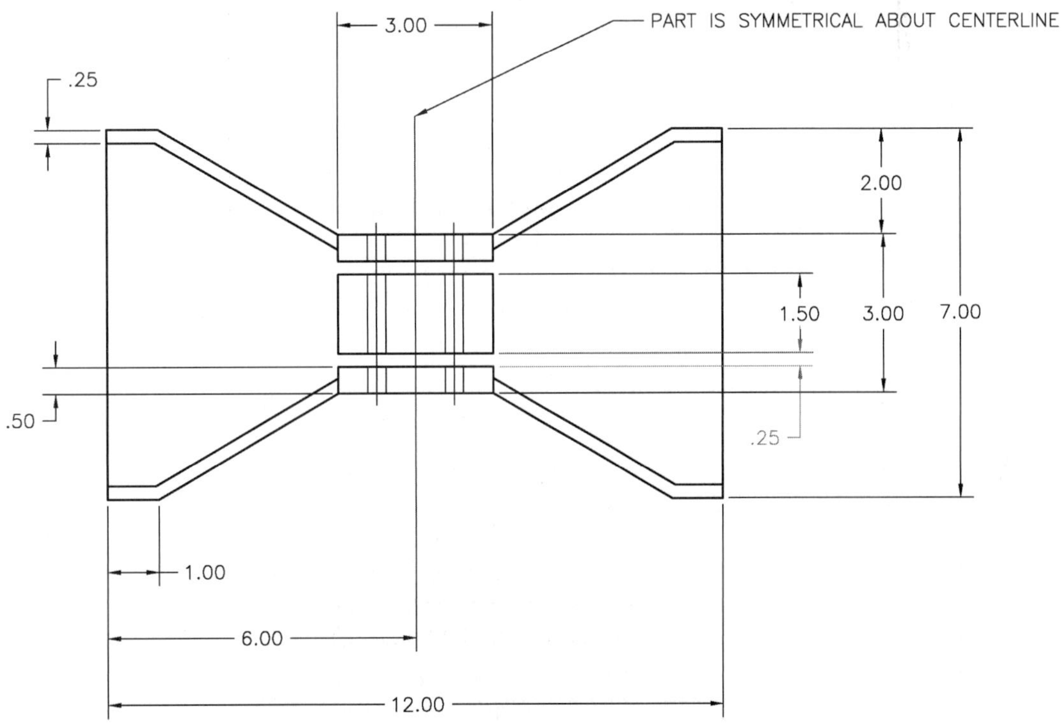

Figure 8.42

Associating an Annotation

The text and leader that you just created are a free-standing note on the sheet. If you were to move the view or make a change to the part that results in the centerline being shifted over, the leader and note would no longer align with the part. Specific notes like this one should be associated with the drawing view so that if the view changes the note moves with it. You must select on a drawing line to attach the annotation.

Right-click: in the graphics area

Pick: **Annotation, Create Annotation**

Select objects to associate to view.

Select objects: **pick the leader line and the text**

Select objects: ↵

Select point in view to attach annotation: **pick on the center of the top middle horizontal line in the top view**

Annotation containing 2 objects attached to view

Moving a Drawing View

Next you will move the top view down slightly so that you can see that annotation is attached to the view.

Right-click: in the graphics area

Pick: **Drawing Menu**

Pick: **Move View**

Select view to move: **pick near the center of the top view**

Specify new view location: **pick slightly below the previous location**

Specify new view location: ↵

Hooray! The drawing view moves and the leader and text move with it. You have created a specific note that is associated with the drawing view. Now even if the part geometry changes the note will remain in a reasonable location in the drawing.

On your own,

Use Zoom All to show the entire drawing sheet

Your drawing should look like Figure 8.43.

General Notes

General notes are notes that apply to the entire drawing rather than to specific features. They are typically located in the lower right of the drawing near the title block, or sometimes in the title block area, although there is no specific standard. You should group general notes together so that they are easily read rather than placing them here and there on the sheet where space permits.

Move the drawing views to make some space for the general notes (if necessary)

Right-click: in the graphics area

Pick: **Annotate Menu**

Pick: **Annotation, Paragraph Text**

Specify first corner: **pick near point A, as shown in Figure 8.43** (in line

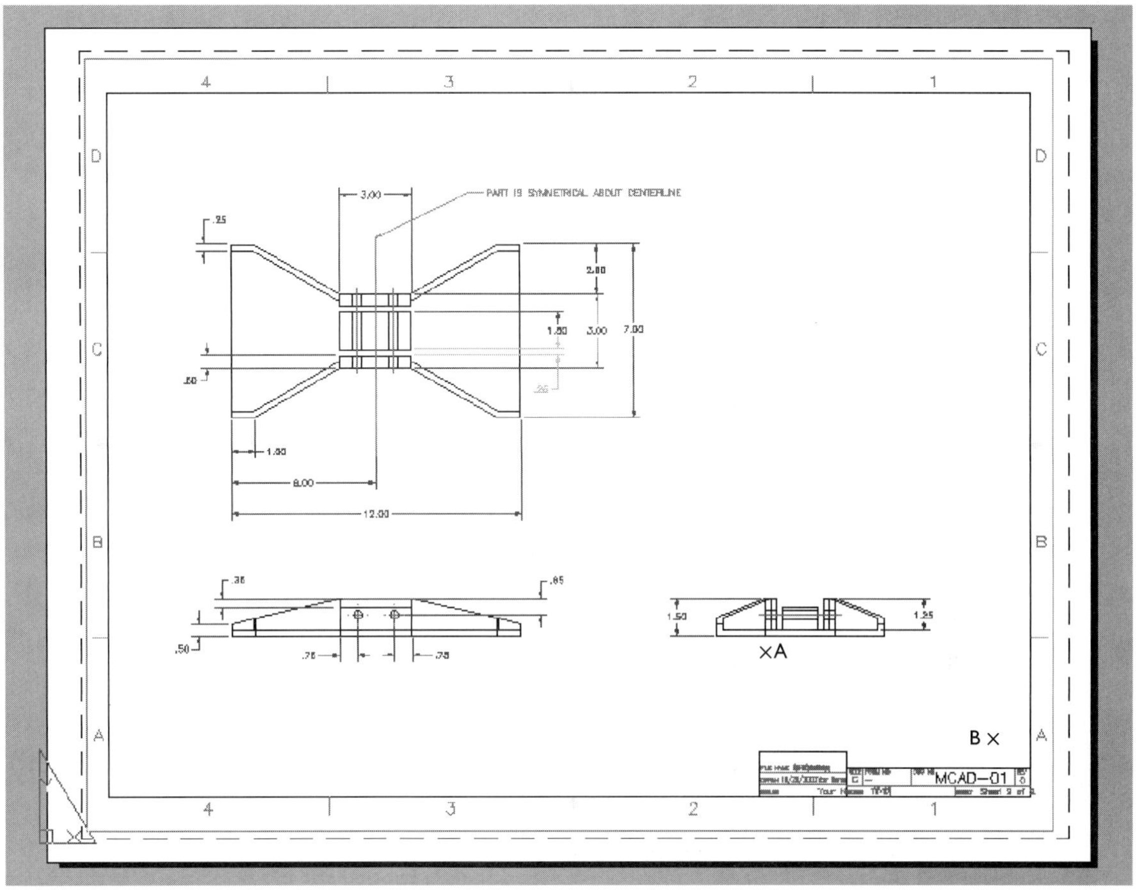

Figure 8.43

with the left corner of the title block and still below the right-side view)

Specify opposite corner or [Height/Justify/Line spacing/Rotation/Style/Width]: **pick near point B above the right line of the title block**

The Multiline Text Editor dialog box appears.

Pick: **Properties Tab**

The Properties tab appears uppermost. You will use it to pick the text style that you created previously, named MyText, as shown in Figure 8.44.

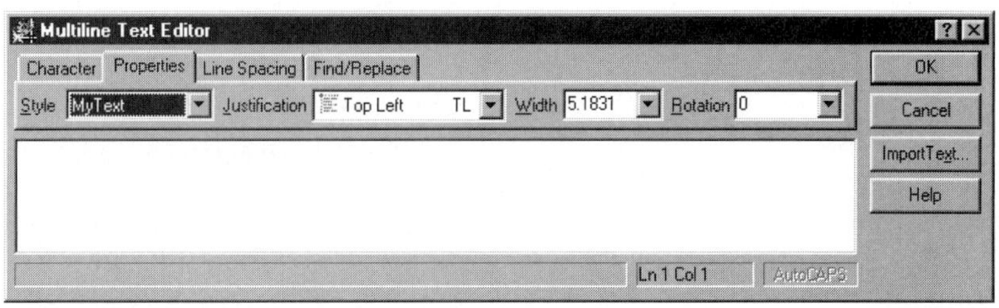

Figure 8.44

Choose Style: **MyText**

Pick: **Character Tab**

The Character tab returns to the top of the stack. You will enter the following note in the input box; use the Symbol menu to select and enter the plus/minus symbol.

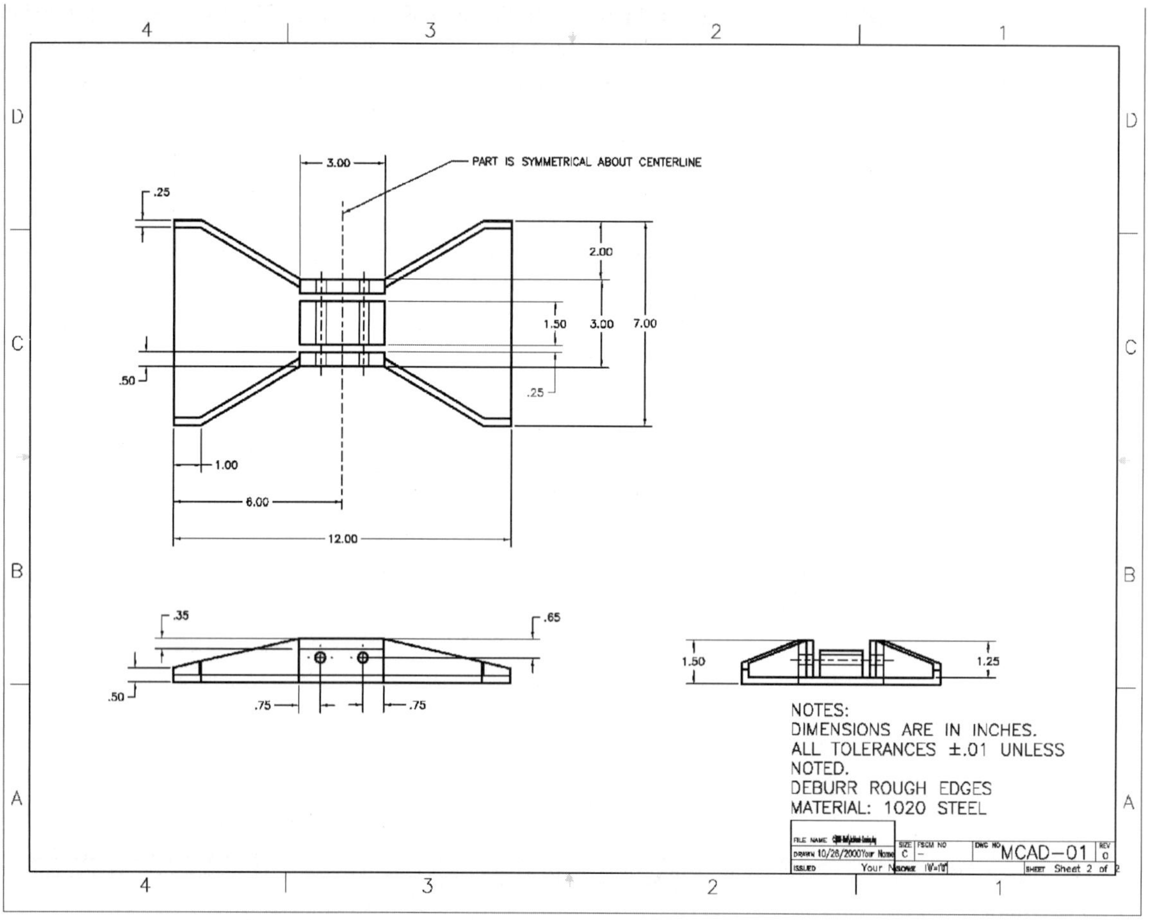

Figure 8.45

Type:

NOTES:

DIMENSIONS ARE IN INCHES

ALL TOLERANCES ±.01 UNLESS NOTED

DEBURR ROUGH EDGES

MATERIAL: 1020 STEEL

Pick: **OK**

The general notes are placed in your drawing, as shown in Figure 8.45.

Specifying Tolerance

A general tolerance note will not always ensure that parts will fit together accurately. For example, the ThickLeg parts must be able to slide in the two slots in the Baseplate part. In order for them always to slide, the Baseplate slot must always be larger than the ThickLeg part that fits it. However, all parts are manufactured with some amount of variation. Stating a tolerance allows you to specify an allowable variation that will still allow the parts to fit.

The width of the slot is not currently dimensioned in the model, but its nominal size is .25 inches. The ThickLeg feature that fits it was also modeled at a size of .25 inches. For an RC6 sliding fit, the tolerance for

Reference

Engineering Design Communication

Tolerancing, pp. 514–524

the hole would be from .250 inches to .2514 inches. The ThickLeg part would then be allowed to vary between .249 inches and .2481 inches. This way the ThickLeg tab will slide in the slot. Even though the slot and tab will vary in size, the smallest hole size is always larger than the biggest tab that will fit into it.

You will create a new dimension style to use for the toleranced dimensions so that you can see the precision required for the tolerances.

Pick: **Annotate, Edit Dimensions, Dimension Style**

The Dimension Style Manager appears. You will use it to create a new style from a copy of the MECHANICAL style, but this style will use four decimal places.

Pick: **New**

In the Create New Dimension Style dialog box that appears,

Type New Style Name: **MECH-TOL**
Start With: **MECHANICAL**
Use for: **All dimensions**
Pick: **Continue**
Pick: **Primary Units tab** (so that it appears uppermost)
Choose Precision: **0.0000**
Pick: **OK**
Pick: **Set Current**
Pick: **Close**

Mech-Tol is now the current dimension style.

Next you will edit the reference dimension for the width of the slot to include a tolerance.

Right-click: in graphics area
Pick: **Edit Dimensions, Power Edit**
Select objects: **pick the .25 reference dimension**
Select objects: ↵

The Power Dimensioning dialog box appears on your screen. There are two buttons in the top right that you can pick to select a type of fit and to select the tolerance display. You will pick both buttons in turn so you can see this great feature. You will first pick the button labeled lc3.

Pick: **Add Fit button** (from the upper right of the dialog box)

Lc3 shows as the type of fit selected. This is a locational clearance fit, which is generally used for stationary parts that are assembled You will switch to a running and sliding clearance type of fit.

Pick: **... across from the lc3 value, as shown in Figure 8.46**

The Fits dialog box appears on the screen. It has a great representation of the fit concept in the image tile at the right. We will be basing the tolerance on the hole size.

Pick: **Hole tab** (if necessary)
Pick: **RC (from the list at left)**
Pick: **6 (from the numerical list)**

Notice that the nominal size for the hole is still shown as .2500, but the RC6 fit states an upper deviation of +.0014 and lower deviation of 0, as shown in Figure 8.47. The limits of size for the hole feature are shown

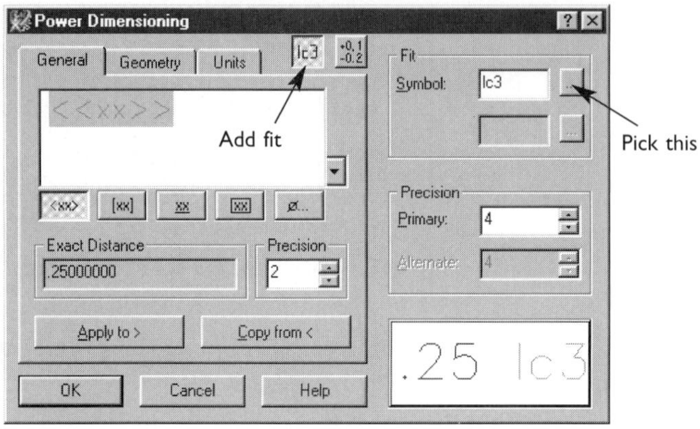

Figure 8.46

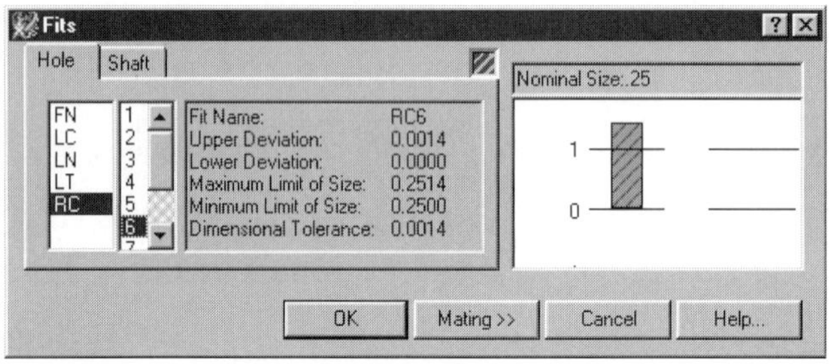

Figure 8.47

and the tolerance of .0014. Pretty cool! Now you don't have to look up the fit in the back of a textbook!

Pick: **OK**
Pick: **Add Tolerance button**

The Tolerance selections appear in the right side of the dialog box. You will type in the values shown in Figure 8.48 for the upper and lower deviation.

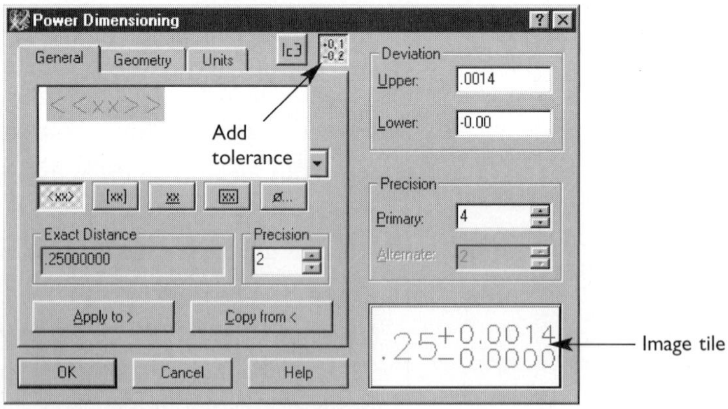

Figure 8.48

Type Upper Deviation: **.0014**
Lower Deviation: – **0.00**
Pick: **Image tile** (colorful picture below the deviations, as shown in Figure 8.48)

The Select Tolerance Type dialog appears, as shown in Figure 8.49. To select the method for displaying the tolerance, pick the corresponding image tile, then set the precision.

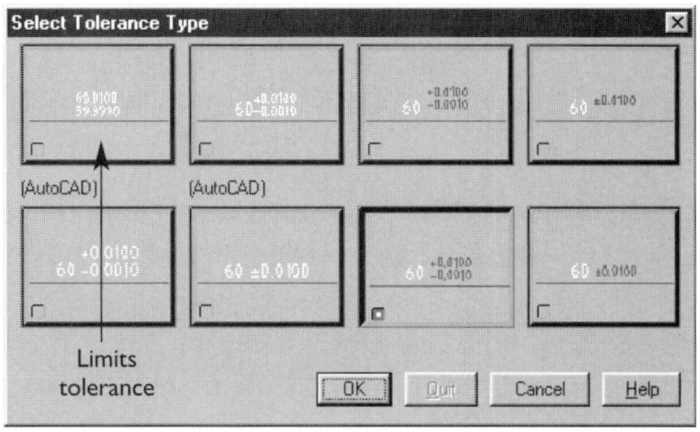

Figure 8.49

Pick: **Limits tolerance** (the upper left tile 60.0100/59.9999)
Pick: **OK**
Pick Primary Precision: **4** (in the tolerance side of the dialog box)
Pick: **OK**

The dimension value changes to show a limit tolerance, as shown in Figure 8.50.

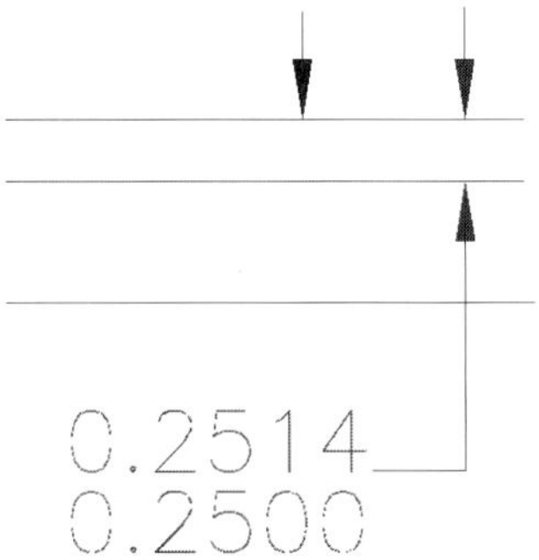

Figure 8.50

Adding a Section View

Part drawings frequently include additional views, such as section and auxiliary views, to describe the part more completely. You will add a section view to show the interior details (such as they are) of the Baseplate and to get practice with creating a section view.

You must have a base view to add a section view; if you don't have one in the drawing, you need to create one. Section views are created perpendicular to the base view, and the cutting plane line is shown in the base view. You can specify the location of the cutting plane by selecting an existing work plane or specifying points on the part. You will use the top view as the base view and the existing work plane to identify the cut through the Baseplate.

You should still be in Drawing mode.

Reference

Engineering Design Communication

Section Views, pp. 146–147 and p. 492

Right-click: **Ortho corresponding to the Top view** in the Browser window
Pick: **New View**
Pick View Type: **Ortho**
Pick: **Section tab** (in the dialog box)

The section choices appear. You can create a full, offset, or broken-out section. You will make choices shown in Figure 8.51.

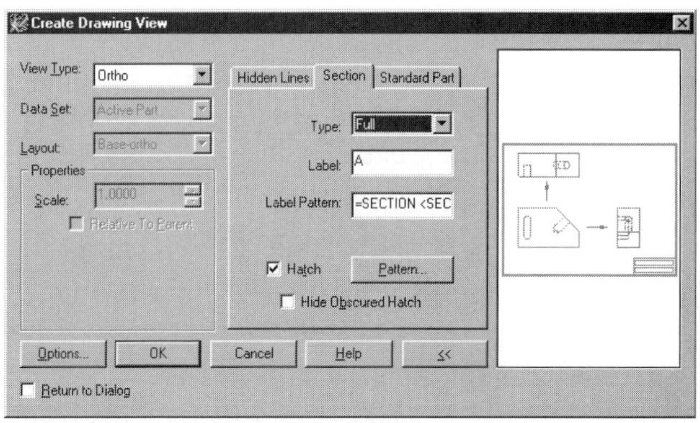

Figure 8.51

Tip

You can easily add an auxiliary view to your drawing by picking New View from the Drawing context menu and selecting Auxiliary as the view type. After you pick OK to exit the dialog box, select a line from the drawing that represents the edge view of the inclined surface for which you are creating the auxiliary view. Then pick the location for the view. It's easy!

Choose Type: **Full**
Choose Label: **A**
Pick: **OK**

You are returned to the drawing view to place the section.

Specify location for orthogonal view: **pick below the front view**
Specify location for orthogonal view: ↵
Enter section through type [Point/Workplane] <Workplane>: ↵
Specify workplane in parent view for depth of section: **pick the work plane that is now visible in the top view**

The section view and cutting line appear. The label you selected, A, is used to identify the cutting plane that corresponds to Section A-A.

Move the view so it fits and reposition the cutting line labels as needed, as shown in Figure 8.52

You have now completed the detail drawing for the Baseplate. You can print it using the same steps you used to print the assembly drawing.

Save your work
Print the drawing
Exit Mechanical Desktop

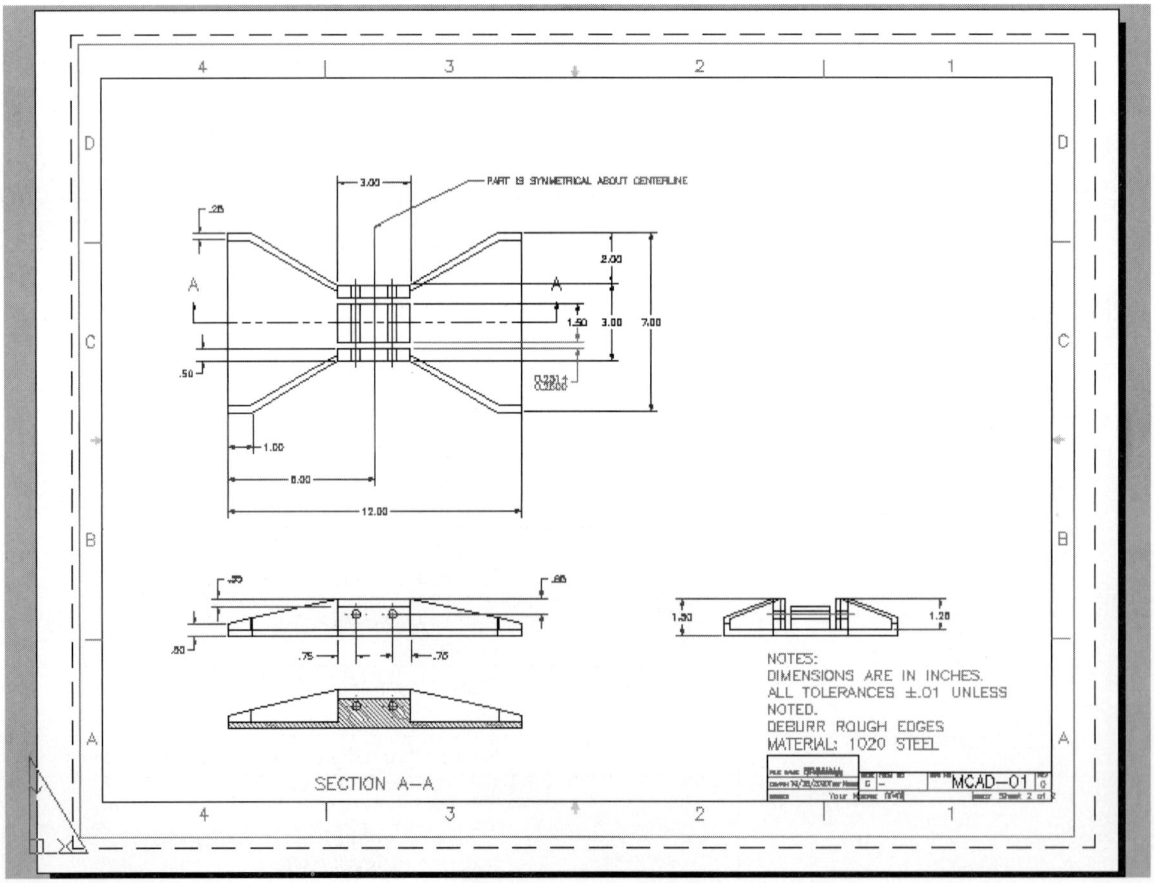

Figure 8.52

Command Summary

Action	Command Name	Description
Annotation	AMANNOTE	Adds, attaches, deletes, and moves annotations (text, leaders, and GD&T symbols) associated with geometry in drawing views
Balloon	AMBALLOON	Creates and organizes balloons, or ball tags, in a drawing
BOM Database	AMBOM	Creates and edits a BOM database
Delete View	AMDELVIEW	Deletes a view in Drawing mode
Add Drawing View	AMDWGVIEW	Creates and adds 2D views to a drawing layout
Layer Control	AMLAYER	Opens the Layer Control dialog box for defining and managing layer definitions
Move Dimension	AMMOVEDIM	Moves dimensions between views in a drawing without losing their association with the geometry
Move View	AMMOVEVIEW	Moves a drawing view within a layout or to a new layout
Leader Note	AMNOTE	Creates a leader or other note symbol and attaches it to a drawing object
Parts List	AMPARTLIST	Creates and adds a parts list to a drawing and controls its properties
Power Edit Dimensions	AMPOWEREDIT	Opens the appropriate dialog box for editing a power object; when a dimension is selected, it opens the Power Dimensioning dialog box
Reference Dimension	AMREFDIM	Creates nonparametric reference dimensions in drawing mode
Import Style	AMSTYLEI	Allows you to import dimension styles from other drawings
Add Leader	AMSYMLEADER	Adds or removes a leader from a drawing sysmbol
Title Block	AMTITLE	Inserts a drawing border and title block; opens and uses the settings in the Drawing borders with Title Block dialog box
Visibility	AMVISIBLE	Controls the visibility of parts, features, symbols, and other entities in parts, assemblies, scenes, and drawings
Edit Attributes	ATTEDIT	Edits block atttributes, such as text in a title block
Update Dimensions	DIM, UPDATE	Updates dimension objects to reflect the current dimension style and units settings; this is a special dimensions command available only at the DIM prompt
Dimension Style	DIMSTYLE	Creates and modifies dimension styles; displays the Dimension Style Manager dialog box
Insert Raster Image	IMAGEATTACH	Adds a raster-type image (such as a TIF or BMP) to a drawing
Plot	PLOT	Plots to a plotting device or file; opens the Plot dialog box
Scale	SCALE	Enlarges or reduces selected drawing objects

Exercises

Create assembly and part drawings for the devices shown in Figures 8.53 through 8.55. (Use your files from the Exercises in Tutorial 7.)

1. Cabinet

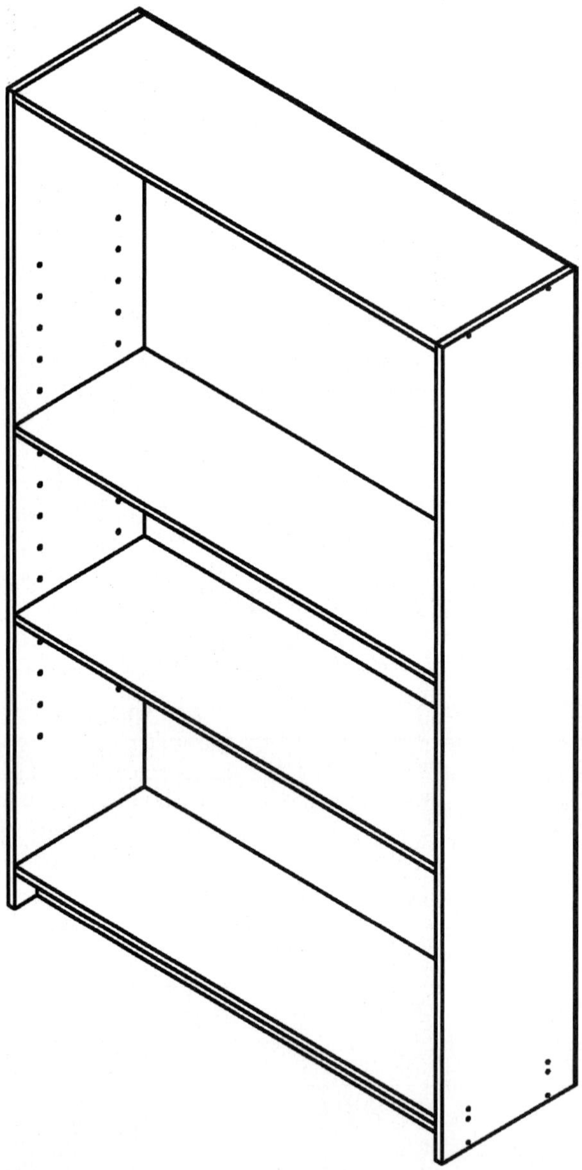

Figure 8.53

2. M6 Hardware Assembly

ITEM NO.	QTY.	PART NO.	DESCRIPTION	MATERIAL
1	1	81006018	M6 X 18 SHCS	18-8 STAINLESS
2	1	81006100	M6 LOCK WASHER	18-8 STAINLESS
3	1	81006000	M6 FLAT WASHER	18-8 STAINLESS

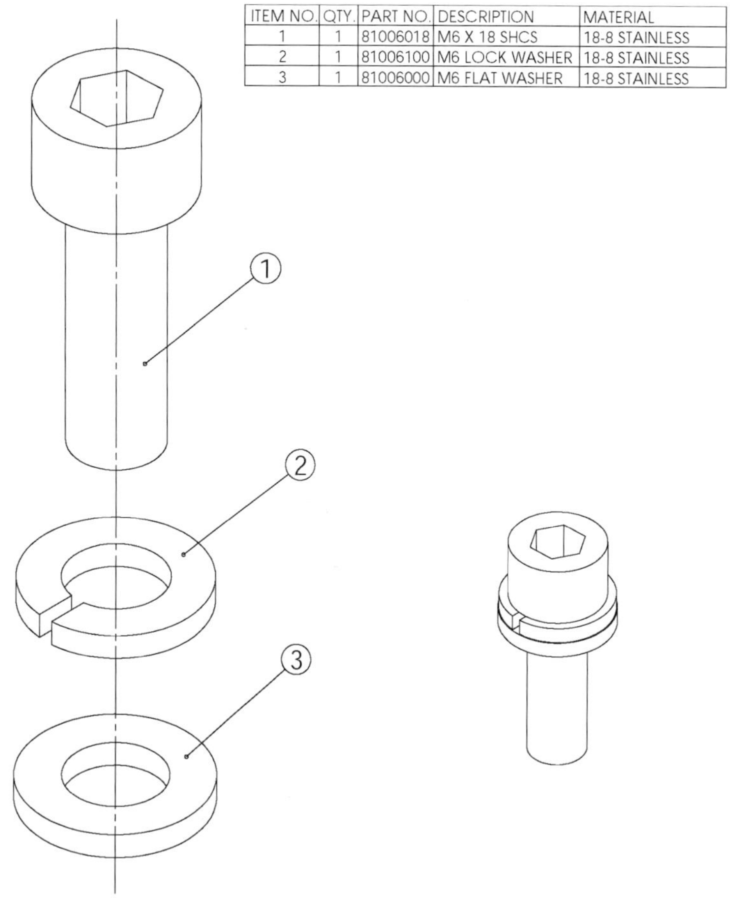

NOTES:
1) ALL DIMENSIONS IN mm

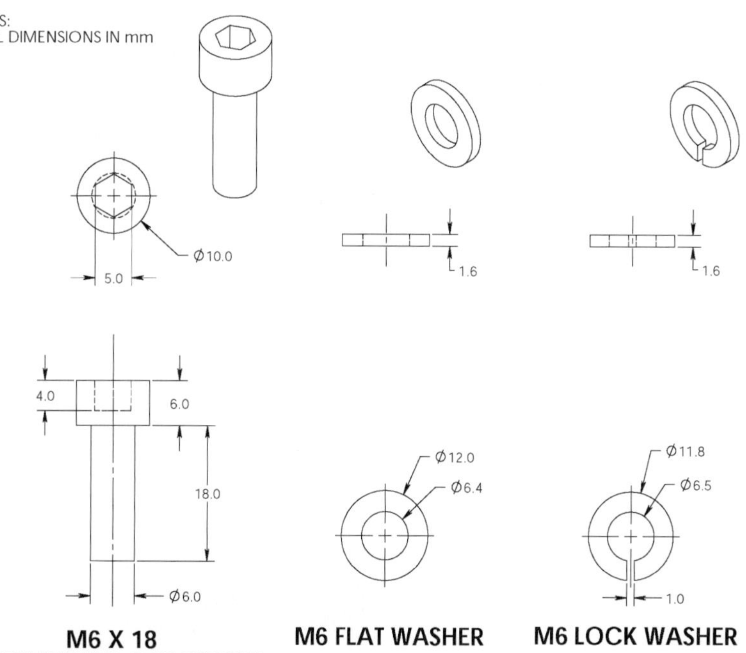

Figure 8.54

3. Adjustment Latch Assembly

ITEM NO.	QTY.	PART NO.	DESCRIPTION	MATERIAL
1	1	60001025	BASE	6061 T-6 AL
2	1	60001023	PIN	18-8 STAINLESS
3	1	60001024	ADJUSTMENT LATCH	6061 T-6 AL
4	4	81006018	M6 X 18mm SHCS	18-8 STAINLESS
5	4	81006100	M6 LOCK WASHER	18-8 STAINLESS
6	4	81006000	M6 WASHER	18-8 STAINLESS

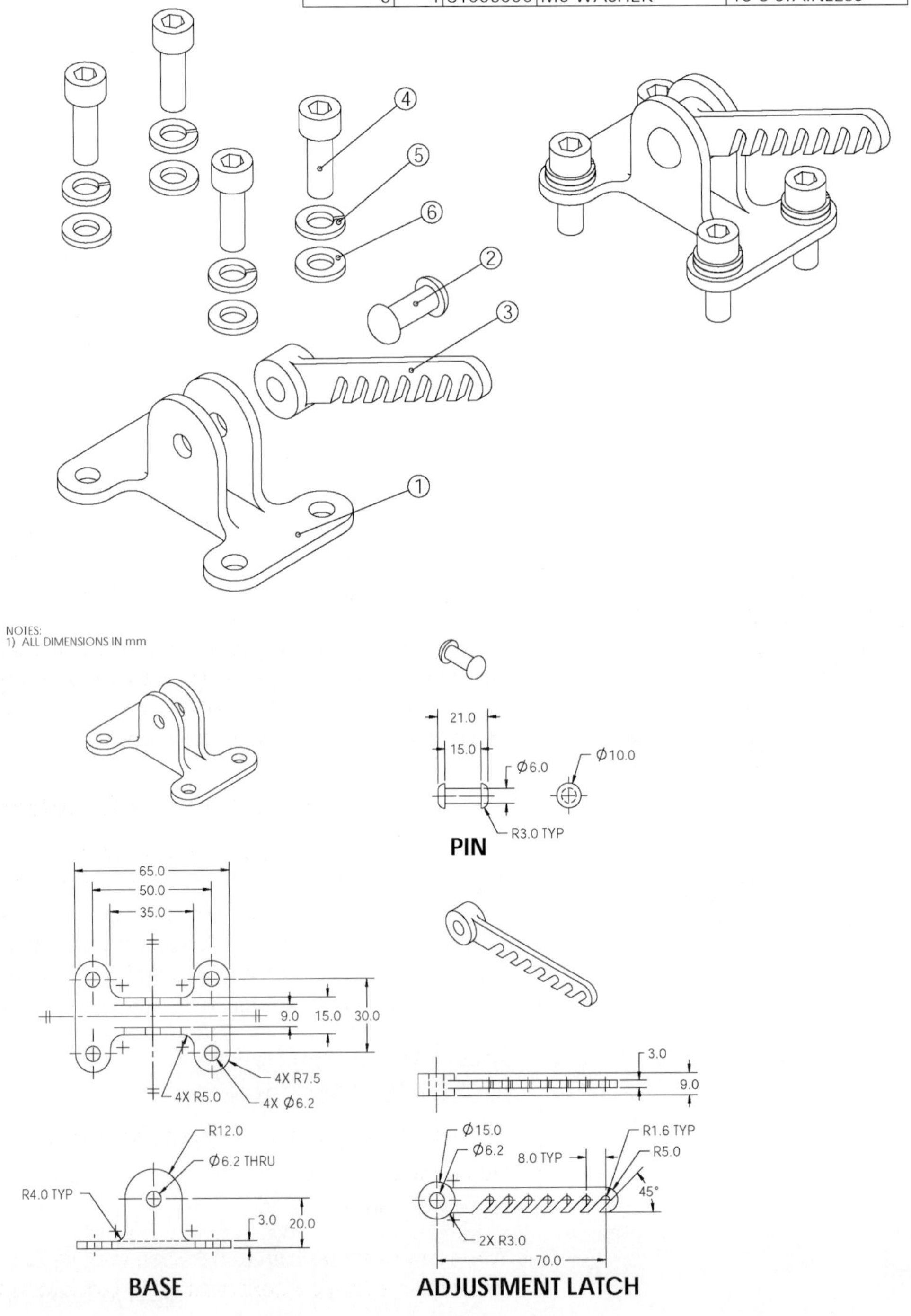

Figure 8.55

Tutorial

9 "Smarter" Parts and Better Assemblies

Objectives

When you have finished this tutorial, you will be able to

- Create a set of default planes to use as a part template.
- Create a swept feature.
- Use equations in dimensioning sketches.
- Model and export a local part.
- Create a bend feature.
- Create a shell feature.
- Create and use subassemblies.

Figure 9.1

Introduction

So far, you have learned the basics of modeling parts and assemblies and creating drawings from them. In this tutorial, you will learn to add even more intelligence to your parts and assemblies. You will create a part template with work features that will make parts easier to assemble and learn to use equations and variables to create smart dimensions that make parts easier to update. In addition, you will learn how to work with subassemblies and how to externalize parts created in an assembly so they can be used in other assemblies. You will learn these skills by modeling the stock cart shown in Figure 9.1.

Using a Part Template

You have already seen how to use a template to start your drawings so that you do not have to redraw standard information such as the title block and border. In a similar way, you can start parts from a template that contains elements you wish to have in all parts. This part template can contain default work planes, view orientations, standard drawing sheets and title blocks, and other items. Using a standard part template not only eliminates the need to add these items to each new part but also can also make it easier to assemble parts by providing some default work planes that you can mate in the assembly.

Starting a New Part from a Template

Mechanical Desktop provides several template files that you can use to start your drawings or as a basis for your own part template. Templates are a special kind of Mechanical Desktop file with the extension *.dwt*. When you start a file from a template, Mechanical Desktop loads a copy of the template into the new *.dwg* file. The template file itself remains unchanged.

You will start with a template for an ANSI C drawing and then customize it.

Launch Mechanical Desktop 5

Pick: **File, New Part File**

Choose: **Template** (on the Create Drawings tab)

Pick: **A** (to expand it)

Pick: **ansi_c.dwt**

You should have a blank drawing on your screen. If you pick the Drawing tab, you should see a border and title block for a C-sized drawing sheet.

Load your user profile, if necessary

Save the file as StartPart.dwg in your MDT-Work directory

Work Planes as Datum Planes

Creating three mutually perpendicular work planes can often be an effective way to start modeling a part. These planes can be used later when you are assembling parts, and they can also save time in part modeling. Think about modeling a cylindrical feature. Without adding a work plane, it can be difficult to align other features tangent to the cylinder. By habitually starting your sketch on a work plane, you have it available later for the creation of other features—although you can always add additional work planes to your parts. Also, many designs are symmetrical in nature. Having a work plane that remains at the center of the object can make it easy to maintain the symmetry of your part when it is updated. You will create a set of three default work planes that will be part of your part template.

Reference

Engineering Design Communication

Datum Planes and Surfaces, pp. 263–265

Pick: **Part, Work Features, Basic 3D Work Planes** (from the menu bar)

Pick origin: **0,0,0** ↵

Computing ...

New part created

Pick: **Zoom All icon**

Pick: **Left Front Isometric view**

The work planes appear in your drawing, as shown in Figure 9.2. (If not, pan the view to center it.)

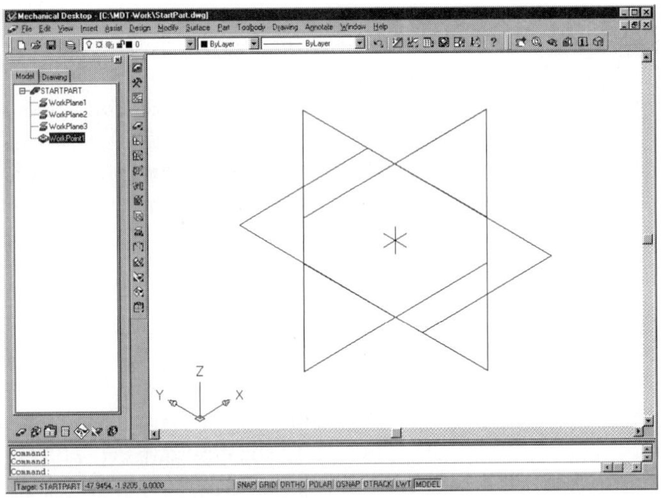

Figure 9.2

Pick: **Top view icon**

The view changes to show Workplane3 from the top, as shown in Figure 9.3.

Next you will change the color of Workplane3 and its name to make it easier to identify.

Right-click: **WorkPlane3** (in the model tree)

Pick: **Rename**

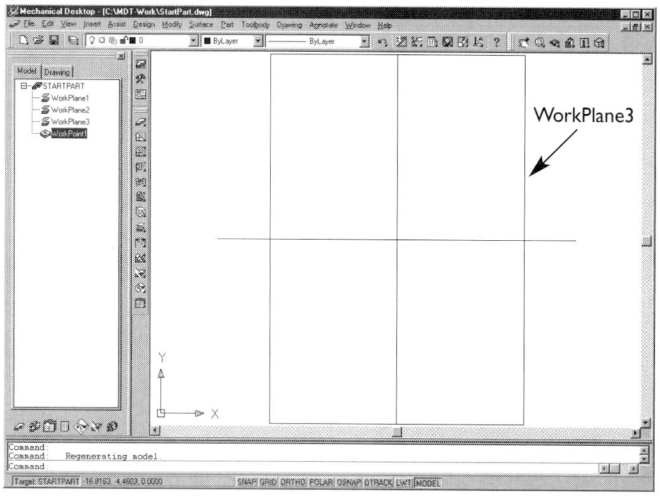

Figure 9.3

Type: **XY-Top-Red** ↵

Right-click: **XY-Top-Red** (in the model tree)

Pick: **Properties, Color**

Pick: **Red**

Pick: **OK**

The name for the work plane is changed and it is drawn with a red line.

Pick: **Right view icon**

The view changes to the right side view. The red plane shows on edge in this view.

Right-click: **WorkPlane1** (in the model tree)

Pick: **Rename**

Type: **YZ-Side-Blue** ↵

Right-click: **YZ-Side-Blue** (in the model tree)

Pick: **Properties, Color**

Pick: **Blue**

Pick: **OK**

The work plane changes to color blue and displays the name YZ-Side-Blue in the model tree.

Pick: **Front view icon**

The view changes to the front view. The red and blue planes show on edge.

Right-click: **WorkPlane2** (in the model tree)

Pick: **Rename**

Type: **XZ-Front-Green** ↵

Right-click: **XZ-Front-Green** (in the model tree)

Pick: **Properties, Color**

Pick: **Green**

Pick: **OK**

The front work plane changes to color green. Using the colors and names associated with the default views can help you orient your model so that parts assemble with the same orientation and default views. When more than one designer is modeling the parts for an assembly, using a standard orientation (where top corresponds to the top view of the assembly) can make it easier to assemble parts.

Pick: **Left Front Isometric view icon**

The view changes again to the isometric view. You should be able to identify the three planes by their colors.

Next you will display the UCS icon so that it is at the origin of the coordinate system. This can also be helpful in understanding the coordinate values as you type them in.

Command: **UCSICON** ↵

Enter an option [ON/OFF/All/Noorigin/ORigin/Properties] <ON>: **OR** ↵

The UCS icon displays at the origin of the coordinate system, which is at the conjunction of the work planes, as shown in Figure 9.4.

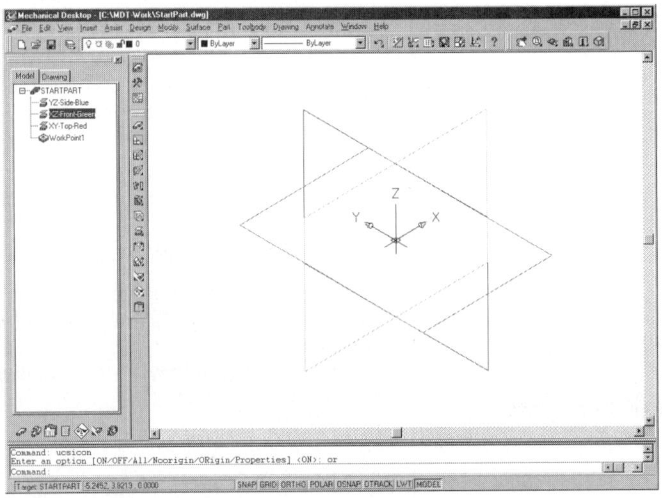

Figure 9.4

Making a Customized Template

Your file now contains the ANSI C drawing elements from the Mechanical Desktop template and the work planes you just created. You will save your file, then save it again as a template file. By saving the template as a drawing file first, you can add the work planes to an assembly by attaching the part file. By saving it as a template, you can use it to start new part files.

Pick: **File, Save**

Pick: **File, Save As**

Choose Files of type: **Mechanical Desktop 5 Drawing Template (.dwt)**

Choose Save in: **C:\MDT-Work directory**

Type: **StartPart** (as the File name)

Pick: **Save**

Type: **Default Datum Planes** (in the Template Description box)

Pick: **OK**

Pick: **File, Close**

Now you will start a new part file using *StartPart.dwt* as the template.

Pick: **File, New Part File**

Choose: **Template** (on the Create Drawings tab)

Pick: **Browse** (to locate StartPart.dwt in your MDT-Work directory)

Pick: **Save As**

Browse to your MDT-Work directory

Pick: **Create New Folder icon**

Replace the name New Folder with Cart

Save your new part as CartRail.dwg in your MDT-Work\Cart directory

You should see an isometric view of the three work planes from the template on your screen.

Sweeping the Side Rail

You will first model the side rail part, which will be formed from a piece of L-shaped angle stock. The side rail is a *swept feature*, which means that it is formed by sweeping a profile along a path. Sweeping is similar to extrusion but allows you to extrude the profile (or cross section) along a curved 2D or 3D path.

Sketching the Path

You will start by sketching the U-shaped path. Then you will sketch the L-shaped profile that will be swept along the path to form the feature. Path sketches can be 2D or 3D. For this feature you will use a 2D path sketch. You will center the path sketch on the green work plane from the template.

Reference

Engineering Design Communication

Swept Shapes, p. 184.

Right-click: in the graphics area

Pick: **New Sketch Plane**

Select work plane, planar face or [worldXy/worldYz/worldZx/Ucs]: **pick the green plane**

Computing ...

Plane = World ZX

Select edge to align X axis or [Flip/Rotate/Origin] <Accept>: ↵ (if the coordinates are aligned as shown in Figure 9.5, if not use the command options to orient them like the figure)

Pick: **Sketch View icon**

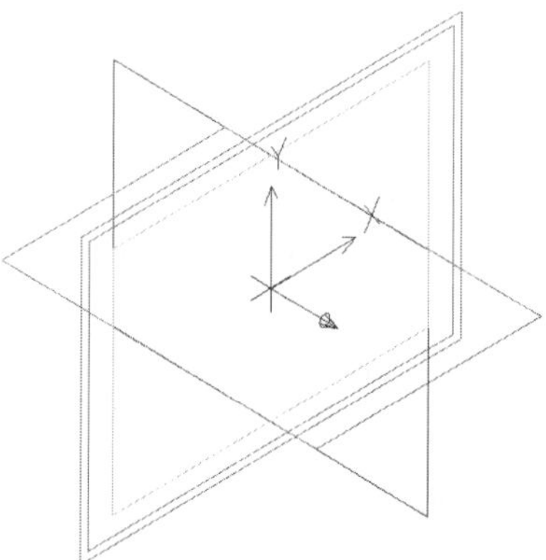

Figure 9.5

Your view is reoriented so you are looking straight onto the sketch plane. Next you will zoom out so that you can create your sketch close to the actual size. This makes it easier to edit the dimension values without the geometry changing drastically.

Command: **Z** ↵

Specify corner of window, enter a scale factor (nX or nXP), or

[All/Center/Dynamic/Extents/Previous/Scale/Window] <real time>: **.1X** ↵

The view doesn't appear to change, but it shows ten times the previous area. On your own, move your mouse around the screen and verify that the coordinate display shown in the lower left of the status bar shows a larger area (about −50 to +50 in the *X*-direction and −40 to +40 in the *Y*-direction). Now you are ready to start drawing the path for the sweep. Refer to Figure 9.6 as you pick points.

Turn Ortho on.

Command: **L** ↵

LINE Specify first point: **pick above and to the left of the work planes** (point A)

Specify next point or [Undo]: **pick a point below point A and above the red plane** (point B)

Specify next point or [Undo]: **pick a point in line to the right of the work planes** (point C)

Specify next point or [Close/Undo]: **pick directly above the last point and in line with the first point** (point D)

Specify next point or [Close/Undo]: ↵

You should have three lines on the screen that resemble a flattened U shape sitting above the red plane and roughly centered on the blue plane, as shown in Figure 9.6.

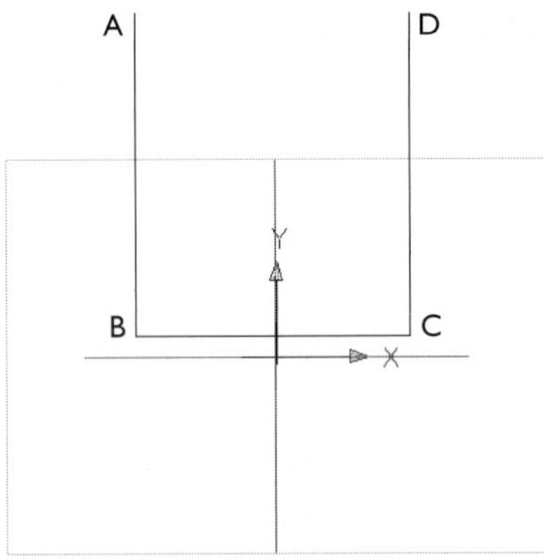

Figure 9.6

Next you will add fillets to the two bottom corners.

Right-click: in the graphics area

Pick: **2D Sketching, Fillet**

Current settings: Mode = TRIM, Radius = .125

Select first object or [Polyline/Radius/Trim]: **R** ↵

Specify fillet radius <0.1250>: **3** ↵

Select first object or [Polyline/Radius/Trim]: **pick the left vertical line**

Select second object: **pick near the left end of the horizontal line**

Right-click: in the graphics area

Pick: **Repeat Fillet**

Current settings: Mode = TRIM, Radius = 3.0000

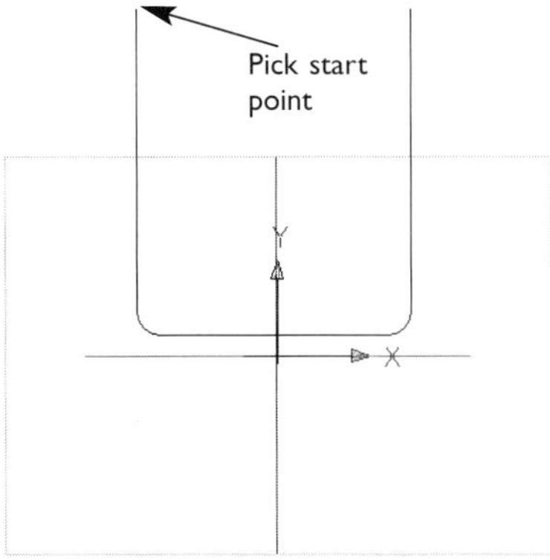

Figure 9.7

Select first object or [Polyline/Radius/Trim]: **pick the right vertical line**

Select second object: **pick near the right end of the horizontal line**

The two fillets should be added to the sketch, as shown in Figure 9.7.

Next you will turn the sketch into a 2D path. When you create a path, you can also choose to create a work plane perpendicular to some location on the path. This can be used later as the sketch plane for the profile that will be swept along the path to create the feature.

Right-click: in the graphics area

Pick: **Sketch Solving, 2D Path**

Select objects: **pick the two vertical lines, the horizontal line, and the two fillets**

Select objects: ↵

Select start point of the path: **pick the upper left endpoint of the path line**

Solved under constrained sketch requiring 4 dimensions or constraints.

Create a profile plane perpendicular to the path? [Yes/No] <Yes>: ↵

Plane = Parametric

Select edge to align X axis or [Flip/Rotate/Origin] <Accept>: ↵ (when you have the axis oriented as shown in Figure 9.8).

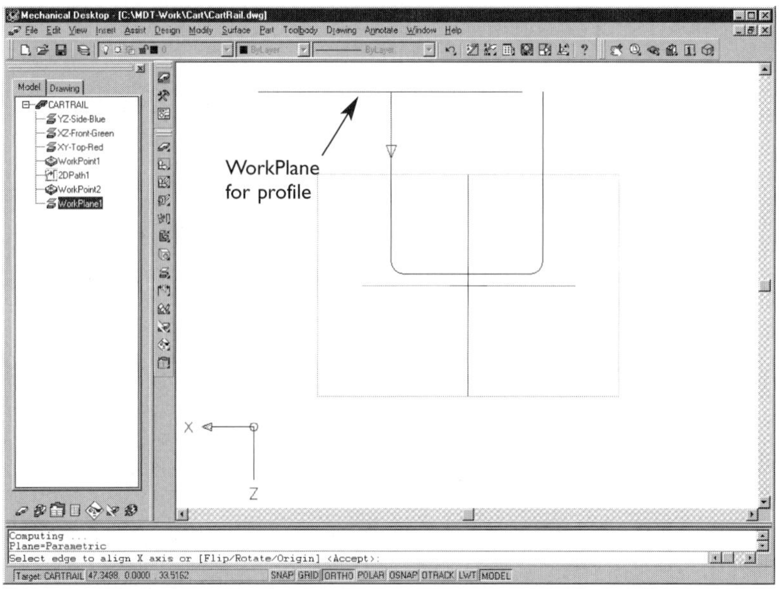

Figure 9.8

Notice that a new work point and work plane have been added to the model tree. Now you are ready to constrain and dimension the path.

Right-click: in the graphics area

Pick: **2D Constraints, Equal Length**

Valid selections: line or spline segment

Select first object: **pick the right vertical line**

Valid selections: line or spline segment

Select second object: **pick the left vertical line**

Solved under constrained sketch requiring 3 dimensions or constraints.

Valid selections: line or spline segment

Select first object: ↵

Enter an option [Hor/Ver/PErp/PAr/Tan/CL/CN/PRoj/
Join/XValue/YValue/Radius/Length/Mir/Fix/eXit]<eXit>: ↵

The right vertical line is adjusted so that it is the same length as the left one.

Next you will delete the Fix constraint that keeps the sketch locked in its position on the coordinate system. If the sketch is fixed on the coordinate system, you cannot dimension it so that it is centered on the work axis.

Right-click: in the graphics area

Pick: **2D Constraints, Delete Constraints**

Select sketch from which to remove the constraint: **pick the sketch**

Select or [Size/All]: **pick on the F in the sketch to delete it**

Solved under constrained sketch requiring 5 dimensions or constraints.

Select or [Size/All]: ↵

Now you are ready to dimension the sketch.

Right-click: in the graphics area

Pick: **Dimensioning, New Dimension**

Select first object: **pick the right vertical line**

Select second object or place dimension: **pick the left vertical line**

Specify dimension placement: **pick below the sketch**

Enter dimension value or [Undo/Hor/Ver/Align/Par/aNgle/Ord/
Diameter/pLace]<value>: **27** ↵ ↵

Making Dimensions Legible

The dimension is added below the sketch, but due to the settings for the dimension style, its value may be barely legible. You will edit the dimension style to change the overall scale for the dimension style. You can choose Mechanical Options from the Assist menu or use the Options icon at the bottom of the Browser window (it looks like a spreadsheet with folder tabs).

Pick: **Options icon**

Pick: **Standards tab**

The Standards tab of the Mechanical Options dialog box appears on your screen, as shown in Figure 9.9. You will use it to modify the dimension style.

Double-click: **Dimension** (from the list at the right of the dialog box)

The Dimensioning dialog box pops up on your screen, as shown in Figure 9.10. You will pick on the Styles button to change the dimension style for AM_ANSI.

Pick: **Styles**

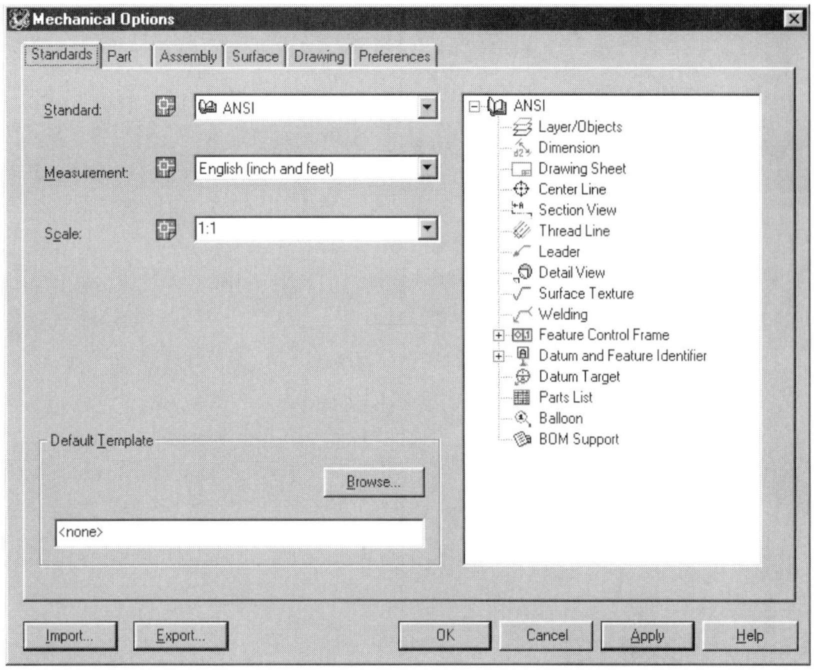

Figure 9.9

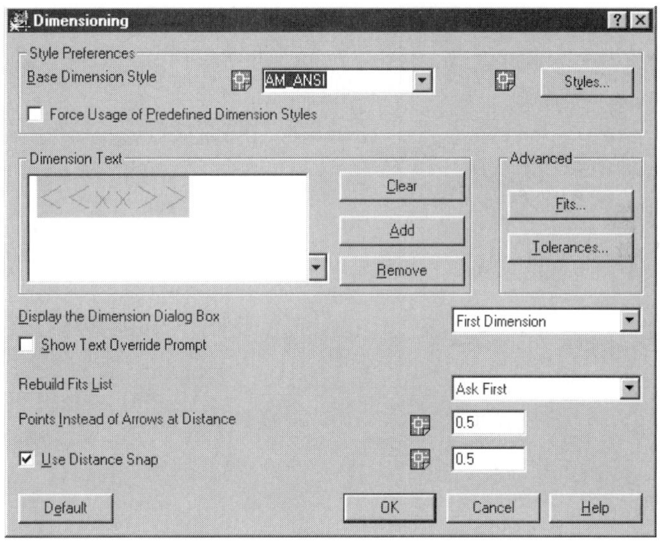

Figure 9.10

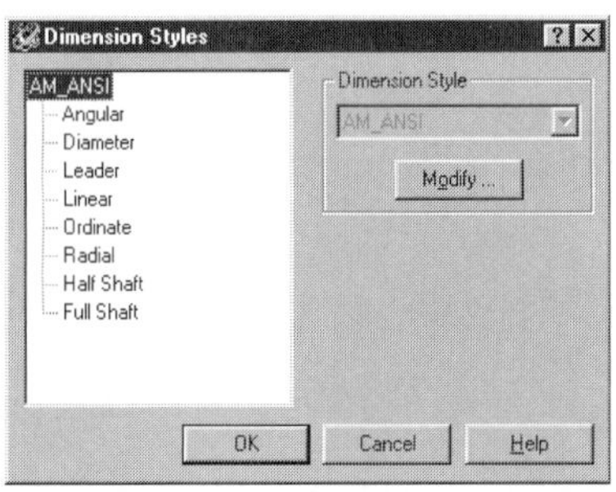

Figure 9.11

The Dimension Styles dialog box now appears, as shown in Figure 9.11. Dimension style AM_ANSI has different child styles for angular, diameter, leader, and other dimensions. You will leave the parent style, AM_ANSI, selected and then

Pick: **Modify**

The typical Modify Dimension Style dialog box appears.

Pick: **Fit tab**, so that it appears uppermost

The Fit tab appears as the top tab on the stack. As shown in Figure 9.12,

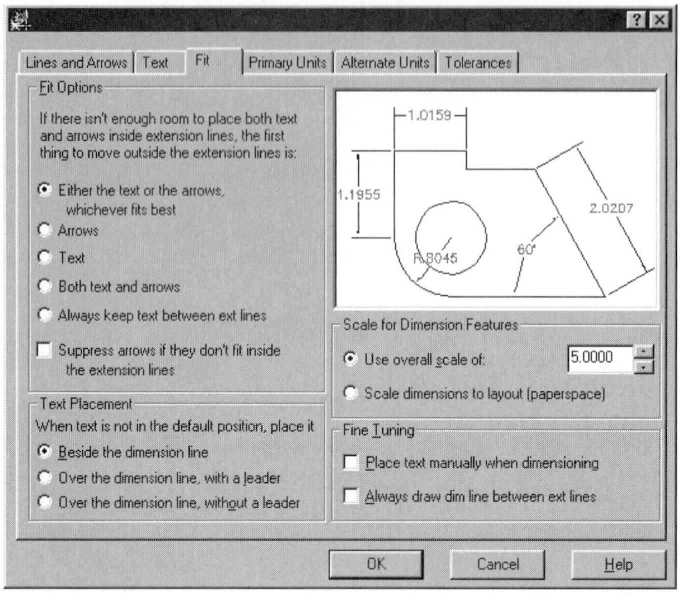

Figure 9.12

Set Use overall scale: **5.0000**

Next you will set the precision for the display of dimensions to two decimal places, as shown in Figure 9.13. For the cart, this will be a reasonable precision and it will make the dimension values easier to read. You can always type dimension values with a higher precision than they are displayed. You can later reset the display to show the higher precision if desired.

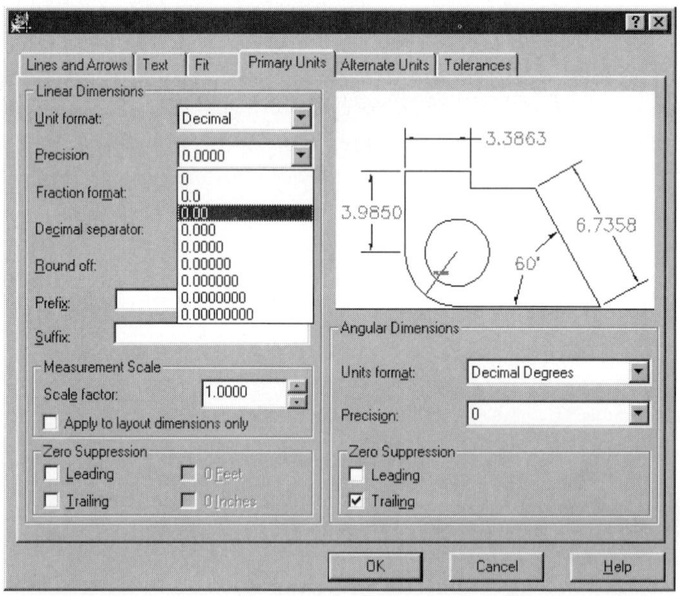

Figure 9.13

Pick: **Primary Units tab**

Set Precision: **0.00**

The dimension values may be hard to read because of their color, depending on what background screen color you are using, black or white. You will change the color to red to display better on a white background.

Pick: **Text tab**

Change the Text color to red, as shown in Figure 9.14

Pick: **OK**

Pick: **OK (to exit the Modify Styles dialog)**

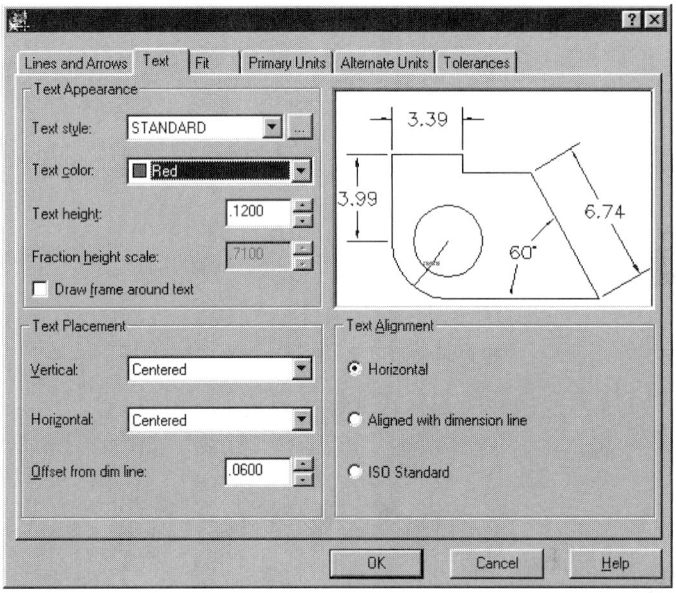

Figure 9.14

You return to the Dimension Styles dialog box, as shown in Figure 9.11. Next you will edit the sub-style for the radial dimensions. Otherwise it will not reflect the changes made to the parent style, AM_ANSI.

Pick: **Radial**

Pick: **Modify**

Repeat the steps you used with the parent style to change the radial child dimension style

Pick: **OK** (three times)

Pick: **Apply** (in the Mechanical Options dialog box)

Pick: **OK**

Next you will add the rest of the dimensions so your sketch resembles Figure 9.15. The distance from the edge view of the red plane is an arbitrary distance; it could be set to 0, but that can make it hard to select the line.

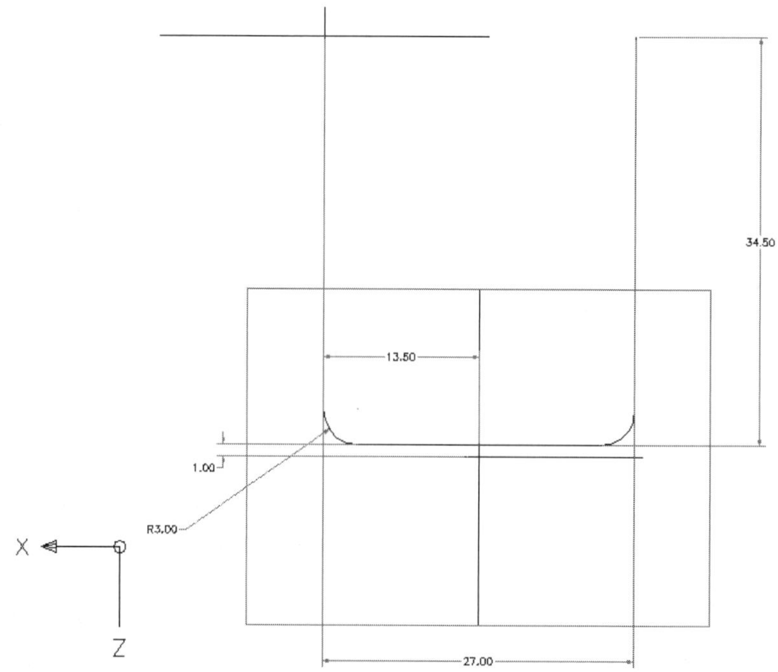

Figure 9.15

Dimension the total height to be 34.5

Dimension the distance from the edge view of the blue work plane to the left side to be 13.5

Dimension the fillet with a radius of 3

Dimension the distance from the red plane to the horizontal edge of the object to be 1

When you have finished adding the dimensions, you should see the message, *Solved fully constrained sketch*. Now you have a fully constrained path sketch that is centered on the blue work plane.

Pick: **Left Front Isometric view**

The view changes to that shown in Figure 9.16, showing the isometric view of the work planes and the sketched path.

Creating the Profile

You will now create the L-shaped profile to sweep along the path. Notice the new work plane in Figure 9.16. You will select the new work plane as the sketch plane for the cross section.

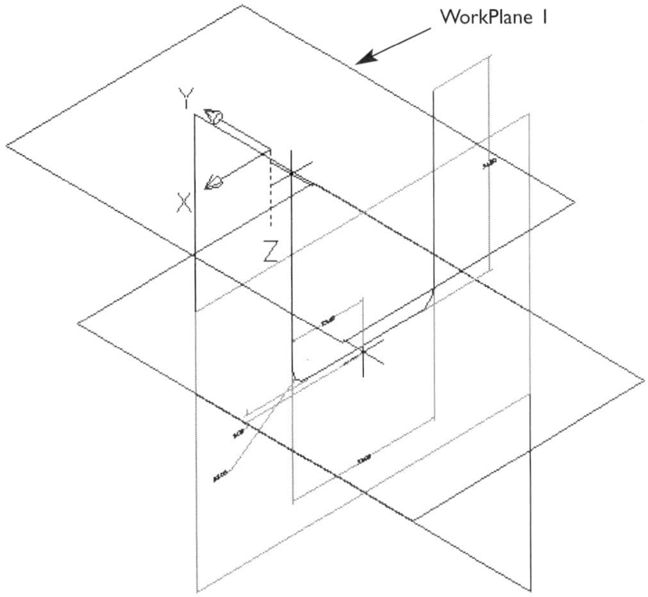

Figure 9.16

Right-click: in the graphics area

Pick: **New Sketch Plane**

Select work plane, planar face or [worldXy/worldYz/worldZx/Ucs]: **pick on the black border lines of Workplane1**

Plane = Parametric

Select edge to align X axis or [Flip/Rotate/Origin] <Accept>: ↵ (to accept the default coordinates, which should be aligned with the WCS in the left corner of your screen)

Pick: **Sketch View icon**

Your view changes so that you are looking straight onto the sketch plane. The 2D path you drew is on edge in this view and appears as a straight line. A work point was added to your sketch at the location you selected when you created the path. This work point is aligned with the endpoint of the 2D path. You will start sketching your cross section from that point, as you want the cross section to sweep directly along the path, not off to the side of it. You will use the Line command to sketch the L-shaped section shown in Figure 9.17 so that it starts from the work point.

Turn OSNAP Endpoint on (for the first pick, then turn it off)

Sketch the shape shown in Figure 9.17, starting from the work point, on your own

Use the Profile command to create a profile from your 6 sketched lines

You will constrain the sketch on your own. As you do, check to see that the Fix point constraint for the sketch is at the work point. If it is not, delete the existing Fix point constraint and add a new one that is at the location for the work point. Use Osnap, Endpoint to find the exact point if necessary.

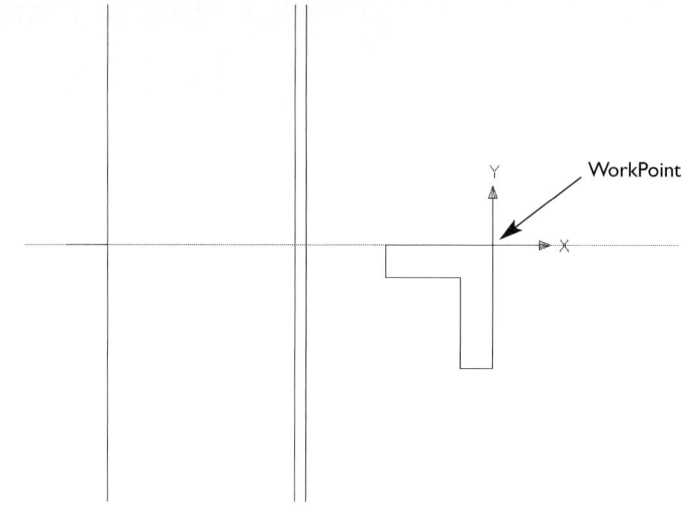

Figure 9.17

Tip

If you have trouble selecting the lines of the sketch, switch to an isometric view temporarily to create the profile.

Set the Fix point constraint at the work point

Constrain the two long outside edges equal to each other

Constrain the two short end segments equal to each other

Dimension the long outside edge to be 2

Dimension the short end segment to be .125

You should now have a fully constrained sketch, as shown in Figure 9.18.

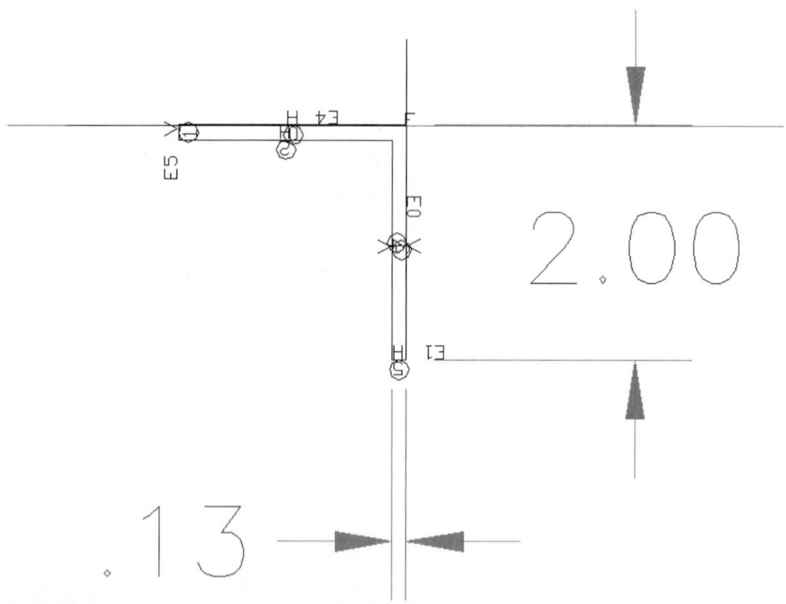

Figure 9.18

Creating the Swept Feature

You are now ready to sweep the profile along the path.

Right-click: in the graphics area

Pick: **Sketched and Work Features, Sweep**

The Sweep dialog box appears on your screen, as shown in Figure 9.19. You can explore the various sweep options and see them illustrated in the

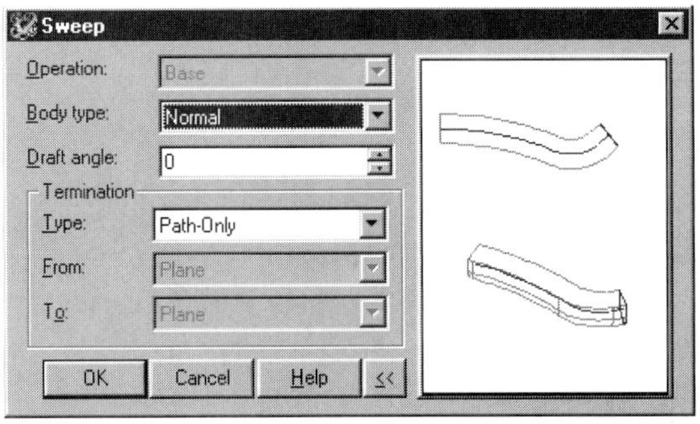

Figure 9.19

image tile. When you are finished, be sure the settings are as shown in the figure.

Pick: **OK**

Pick: **Front Left Isometric view icon**

Right-click: **SweepPathOnly1 feature** (in the model tree)

Pick: **Color**

Pick: **Apply**

Pick: **Green**

Pick: **OK**

Right-click: **SweepPathOnly1 feature**

Pick: **Rename**

Type: **SweptRail** (as the new name for the feature) ↵

Pick: **Flat Shaded icon**

Zoom as needed so that the swept feature and work planes fill your screen

The swept feature for the side rail is created, as shown in Figure 9.20.

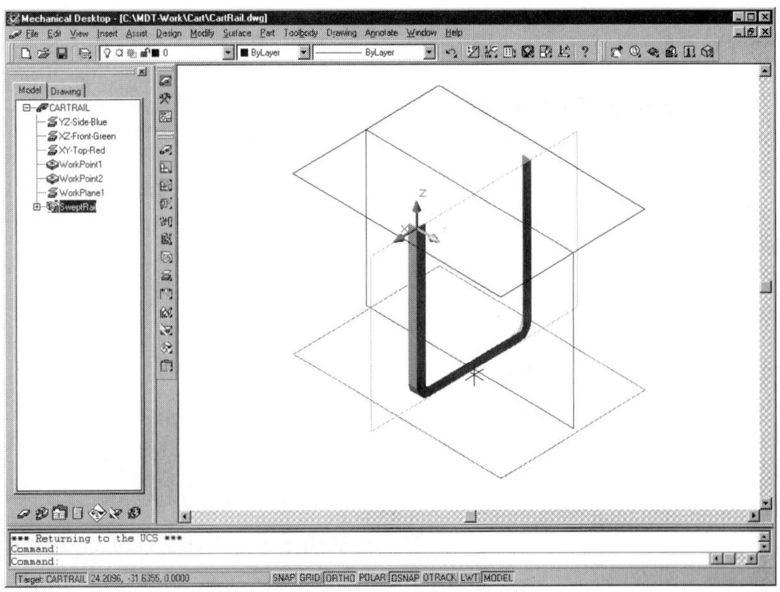

Figure 9.20

Using Equations in Dimensions

Dimensions can be displayed as numbers (representing the current value of the dimension), parameters (which are variable names), or equations. So far you have been seeing dimensions displayed as numbers, but each parametric dimension is also assigned a name (or parameter) when it is created. These names can be used in equations with numeric values and to relate one dimension to another.

You will edit the sketch for the 2D path to use some equations in the dimensions. The equations will relate the width dimensions so that the part will remain centered on the middle work plane if the width dimension is changed.

Reference

Engineering Design Communication

Formulas in Dimensions, pp. 265–268

Pick: **Front view icon**

Right-click: in the graphics area

Pick: **Dimensioning, Dimensions as Parameters**

Right-click: **2Dpath1** (in the model tree)

Pick: **Edit**

Pick: **View, Zoom, Realtime** (and zoom as needed so that you can see all of the sketch and dimensions)

The dimensions for the path sketch appear on your screen. Instead of showing as dimension values, the dimension parameter names appear, as shown in Figure 9.21.

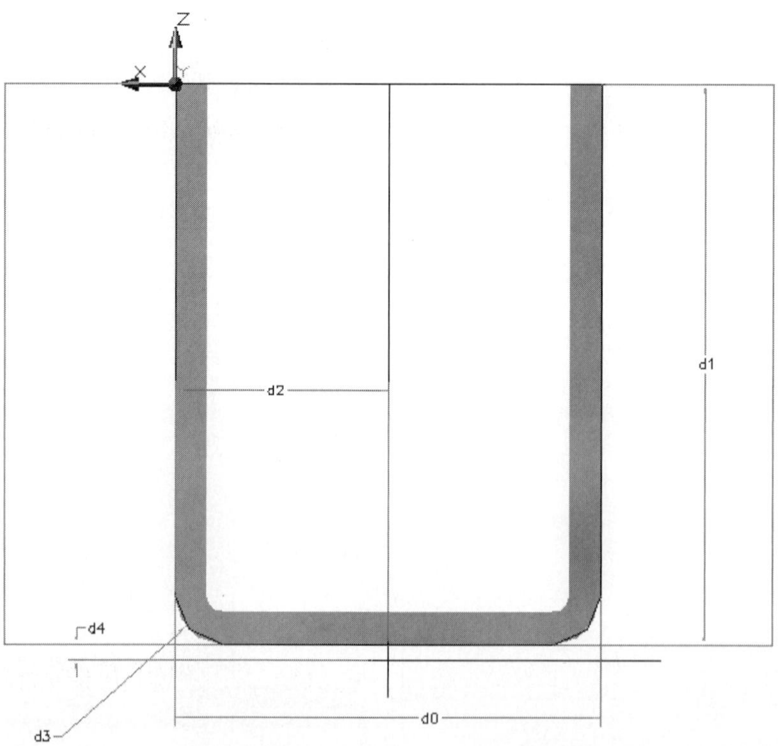

Figure 9.21

Tip

Your dimensions may have different parameter names depending on the order in which you created them. If they do, replace the following parameter names with the correct ones from your drawing.

Next you will change the value for the dimension from the middle blue work plane to the left vertical edge of the path. In the figure, that dimension parameter is named d2. D0 is the dimension parameter for the overall width of the sketch. You will specify that d2 is equal to half of the d0 parameter. This way, however you change the width, the feature will always remain centered on the blue work plane.

Select object: **pick the d2 dimension from the center plane**

Enter dimension value <13.5>: **d2 = d0/2** ↵

Solved fully constrained sketch.

Select object: **pick the d0 dimension for the overall width**

Enter dimension value <27>: **29** ↵

Solved fully constrained sketch.

Select object: ↵

Right-click: in the graphics area

Pick: **Update Part**

The part updates so that it is 29 inches wide, but it still remains centered on the work plane. What would have happened if you did not change the d2 parameter to this equation? (The answer is that the dimension would remain 13.5 from the left side to the blue work plane and the plane would no longer be in the center.)

Right-click: in the graphics area

Pick: **Dimensioning, Dimensions as Equations**

Pick: **to expand the model tree for the SweptRail feature**

Right-click: **2Dpath1** (under the SweptRail feature)

Pick: **Edit**

Now the dimensions display as equations, as shown in Figure 9.22.

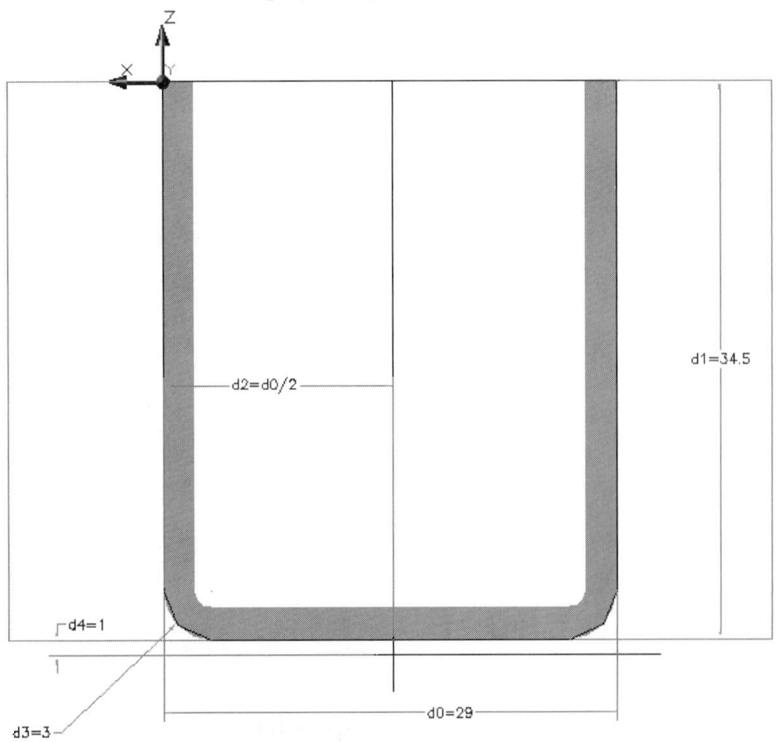

Figure 9.22

Some of the parameters are equal to a value. The d2 parameter is equal to an expression. You can use equations in your dimensioning to create smart parts. The operators that you can use in your equations are as follows:

Operator	Description
^	Exponent
+	Add
-	Subtract
*	Multiply
/	Divide
%	Modulus (remainder)
sqrt	Square root
log	Logarithm
ln	Natural logarithm
floor	Round down to nearest integer
ceil	Round up to nearest integer
sin	Sine
cos	Cosine
tan	Tangent
asin	Arcsin (sin -1)
acos	Arccos (cos-1)
atan	Arctang (tan-1)
sinh	Hyperbolic sine
cosh	Hyperbolic cosine
tanh	Hyperbolic tangent
pi	pi
e	Base number for natural logarithm
exp(x)	ex

Pick: **Front Left Isometric view**

Pick: **File, Save** (to save your CartRail part)

Pick: **File, Close**

You have finished the CartRail part.

Creating the Cart Assembly

Next you will begin the assembly for the cart. Assemblies can be started from template files in the same way as part files are. You will use the ANSI C template as you did for the StartPart file.

Pick: **File, New**

Start from a template and choose the ansi_c template from the list

Pick: **File, Save As**

Choose Save in: **C:\MDT-Work\Cart**

Type File name: **CartAsmb** (as the file name)

Pick: **Save**

Next switch to an isometric view,

Pick: **Front Left Isometric icon**

Now you will begin to add parts to the assembly.

Right-click: in the graphics area
Pick: **Assembly Menu**
Pick: **Catalog**
Pick: **External tab** (if necessary to make it appear on top)
Right-click: **All Devices** (in the Directories area)
Pick: **Add Directory**
Pick: **C:\MDT-Work** (from the list of directories)
Pick: **OK**

The Assembly Catalog dialog box shows the files that are available in your *C:\MDT-Work* directory, as shown in Figure 9.23.

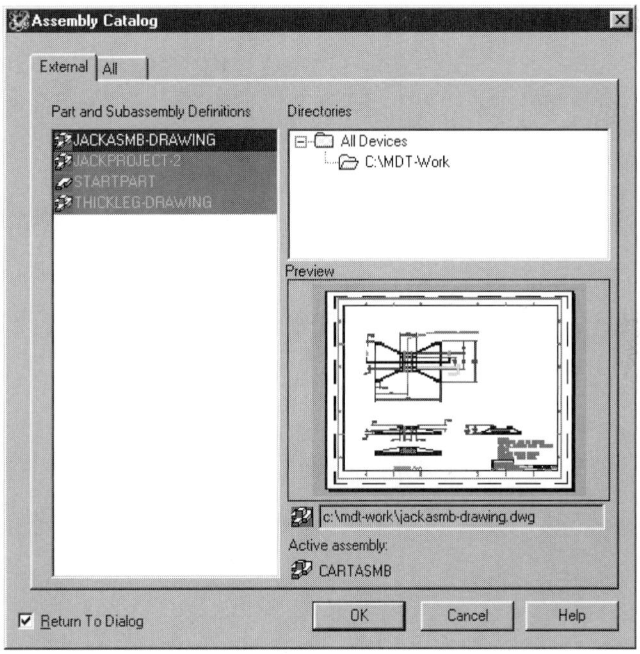

Figure 9.23

Adding a StartPart as the First Part

It is often beneficial to begin assemblies by inserting a blank part with a set of work planes as the first part in the assembly. You can then mate the next part to the blank part. The first, or parent, part in the assembly often cannot be deleted or changed easily. By starting with a blank part for the first part in the assembly, you reduce the likelihood that you will want to delete that part later. Also, the default work planes in the Start-Part can help provide an orientation for assembling other parts.

Double-click: **StartPart**

The colored work planes of the start part appear attached to your cursor in the orthographic view. You will type the location 0,0,0 so that the StartPart planes are centered at the origin.

Specify new insertion point: **0,0,0** ↵

Specify insertion point for another instance or <continue>: ↵

You return to the Assembly Catalog. Next you will add the *MDT-Work\Cart* subdirectory.

Right-click: **in the Directories area**

Pick: **Add Directory**

Pick: **C:\MDT-Work\Cart** (from the list of directories)

Pick: **OK**

Double-click: **CartRail**

The CartRail part is attached to your cursor.

Specify new insertion point: **pick a location to the right of the Start-Part planes**

Specify insertion point for another instance or <continue>: ↵

Pick: **OK** (to exit the Assembly Catalog)

Zoom your drawing so that the parts fill the screen

The Start Part and CartRail appear in an isometric view similar to Figure 9.24.

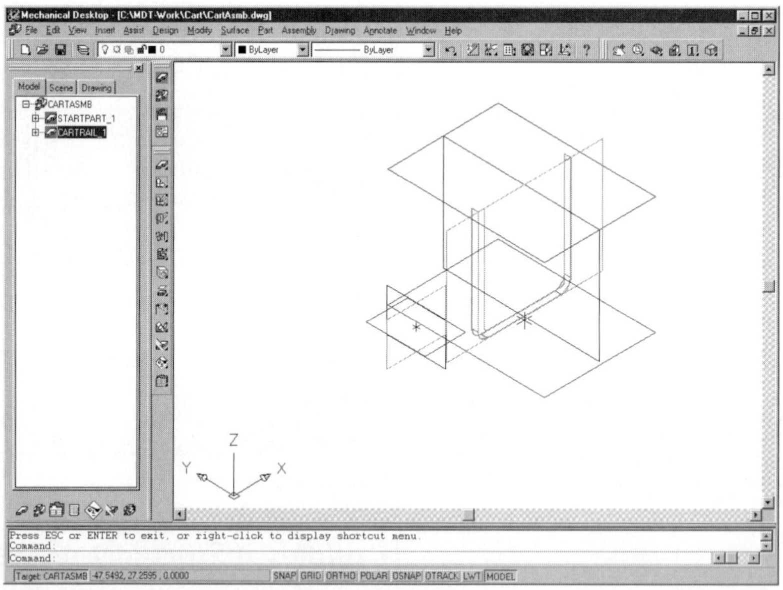

Figure 9.24

Mating CartRail to StartPart

Next you will mate the CartRail to the StartPart using the work planes for each part.

Right-click: in the graphics area

Pick: **3D Constraints, Mate**

Select first set of geometry: **pick on the blue plane of the CartRail**

First set = Plane

Enter an option [Clear/Flip] <accEpt>: **F** ↵ (as needed so that the arrow points in the direction of the Startpart blueplane)

First set = Plane

Enter an option [Clear/Flip] <accEpt>: ↵

Select second set of geometry: **pick the blue plane of the StartPart**

Second set = Plane

Enter an option [Clear/Flip] <accEpt>: ↵ (if the arrows are on each part point toward each other; if not use F ↵ to flip the direction)

Enter offset <.0000>: ↵

The blue work planes are mated, as shown in Figure 9.25.

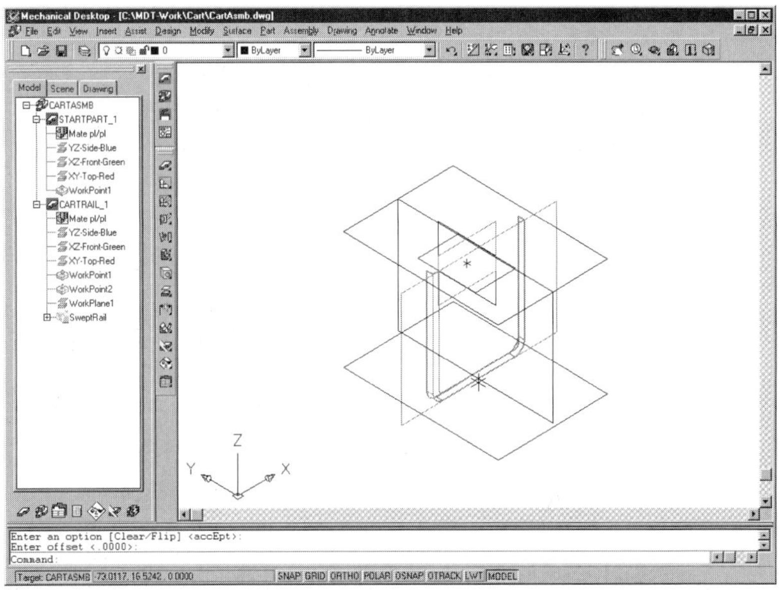

Figure 9.25

Tip

Depending on where you pick, your planes may not look exactly like the figure. Use the 3D Orbiter to confirm that the blue planes are mated; then return to the Left Front Isometric view.

Right-click: in the graphics area

Pick: **Repeat Mate**

Select first set of geometry: **pick on the green plane of the CartRail**

First set = Plane

Enter an option [Clear/Flip] <accEpt>: **F** ↵ (as needed so that the arrow points toward the green StartPart plane)

First set = Plane

Enter an option [Clear/Flip] <accEpt>: ↵

Select second set of geometry: **pick the green plane of the StartPart**

Second set = Plane

Enter an option [Clear/Flip] <accEpt>: ↵ (if the arrows are on each part point toward each other; if not use F ↵ to flip the direction)

Enter offset <.0000>: ↵

The green work planes are mated. On your own,

Mate the red planes together (rotate the view as needed)

Spin the part with the 3D Orbiter to ensure all three planes are mated

Now the CartRail part should be mated to the StartPart planes, as shown in Figure 9.26.

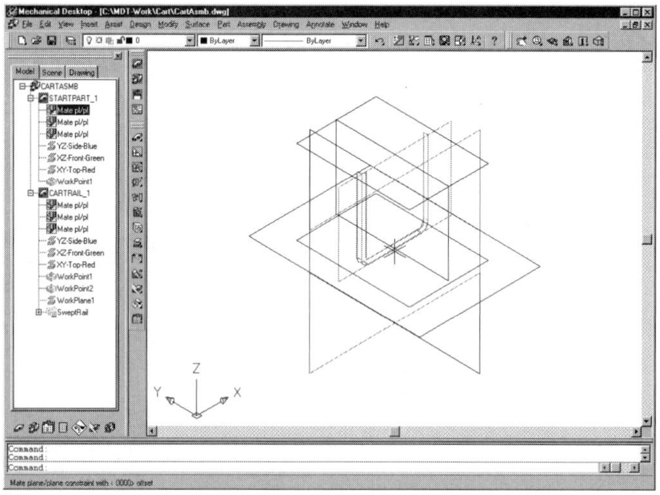

Figure 9.26

Adding Another Instance of the CartRail Part

Now you are ready to insert another instance of the CartRail part that will be used as the back rail for the cart.

Right-click: in the graphics area

Pick: **Catalog**

On your own, insert a second instance of the CartRail in back and to the left of the first instance

You will start by rotating the second instance to get it facing the correct direction. The wider surface of the rail should be facing outward.

Shade the display (if necessary) **to make it easier to see the shape of the rail**

Right-click: in the graphics area

Pick: **3D Manipulator**

Click and drag the end of the X-axis to rotate the rail 180 degrees about the Z-axis until your part is rotated similarly to Figure 9.27

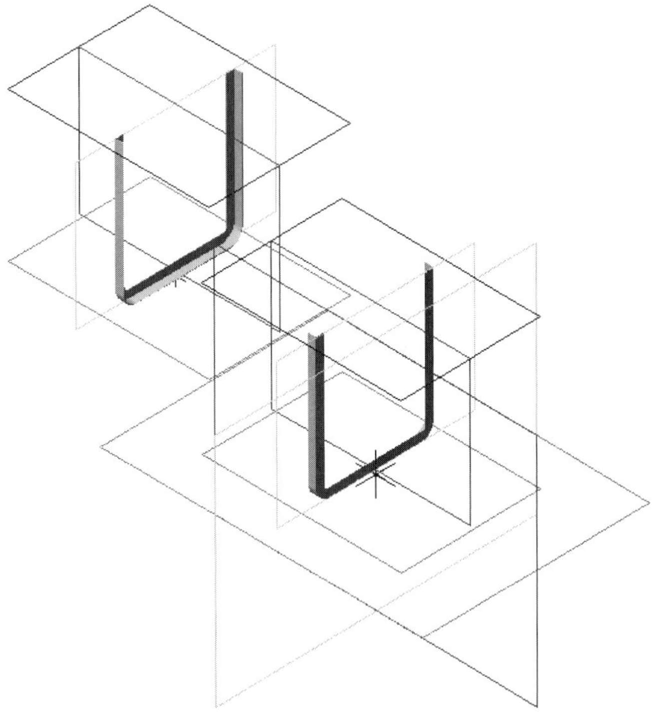

Figure 9.27

Mating the Second Rail

Tip

It can be easier to mate the planes if you adjust your view so you are looking at them on edge.

You will mate the blue and red planes so that the new CartRail instance is aligned with the previous one. Take care to mate the CartRail planes to each other, not the StartPart planes.

Pick: **Front Left Isometric icon**

Zoom the model on your own so that it fills the screen

Right-click: in the graphics area

Pick: **3D Constraints, Mate**

Mate the red and blue planes of the CartRail instances on your own

You will use an offset distance between the two green planes. The green planes should be offset 17.982 so that the rails are spaced appropriately. You can enter the offset as the last step of adding the mate constraint or edit the constraint to add the offset.

On your own, mate the green planes of the CartRail instances with an offset of 17.982

Tip

If your CartRails do not face each other correctly, try a negative number for the offset distance, or flip the direction arrows the other way as needed.

Pick: **Front Left Isometric icon**

Zoom the model on your own so that it fills the screen

Your assembly should look like Figure 9.28.

Next you will activate the CartRail part and change the visibility to hide the work planes for the CartRail. When you change one instance, you are changing the part, so both will change.

Right-click: **CartRail_1** (in the model tree)

Pick: **Activate Part**

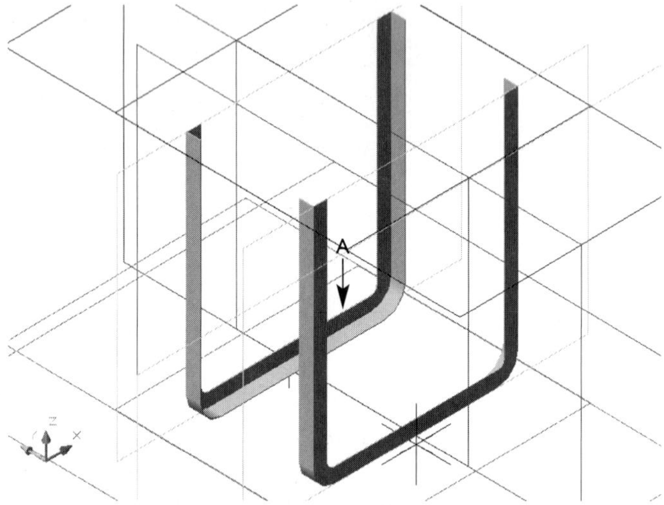

Figure 9.28

Right-click: **XY-TOP-red** (under CartRail_1)

Unselect: **Visible** (so it is unchecked)

The red work plane for the CartRail is no longer displayed on your screen.

On your own, turn off the visibility of the green and blue work planes and WorkPoint1

Turn off the visibility of WorkPlane1 and WorkPoint2 (used for the profile to be swept).

Now reactivate the assembly.

Right-click: in the graphics area

Pick: **Assembly, Activate Assembly**

Enter assembly name to activate or [?] <CARTASMB>: ↵

Modeling a Local Part within the Assembly

Next you will model the braces that will fit between the two side rails. You will model these parts as local parts within the assembly. A local part is one that is stored only in the assembly file. It cannot be used in another drawing until it is exported to a separate part file. You will learn how to export a local part later in this tutorial.

Right-click: in the graphics area

Pick: **Part menu**

Pick: **Part, New Part**

Select an object or enter new part name <PART1>: **CrossBrace** ↵

New part created.

Scroll down in the Desktop Browser. You should see the new part, CrossBrace_1, at the bottom of the model tree. To model this new part, you will first select a sketch plane to use for the base sketch of the CrossBrace. You will draw on the World XY-plane, which corresponds to the top view of the assembly.

Right-click: in the graphics area

Pick: **New Sketch Plane**

Select work plane, planar face or [worldXy/worldYz/worldZx/Ucs]: **X** ↵

Plane = World XY

Select edge to align X-axis or [Flip/Rotate/Origin] <Accept>: ↵

Because the CrossBrace should fit inside the two CartRails, you will copy geometry from the narrow edge of the CartRail to start the sketch. Because the geometry is small,

Zoom so you can clearly see the top face of the rails

Switch to a 3D Wireframe View

Right-click: in the graphics area

Pick: **Sketch Solving, Project Objects to Plane**

The Projection Options dialog box appears on your screen. You will use it to make the following selections.

Choose Projection of: **3D Face**

Choose Projection to: **Sketch Plane**

Pick: **OK**

Select face: **pick on the upper surface identified as A in Figure 9.28**

Enter an option [Next/Accept] <Accept>: ↵ (if the correct surface is selected; if not press N ↵ to move to the next surface until the correct face is highlighted)

Right-click: in the graphics area

Pick: **Repeat Project Objects to Plane**

On your own, repeat the process for the opposite rail

You should have two thin rectangular shapes projected onto the sketch plane, as shown in Figure 9.29.

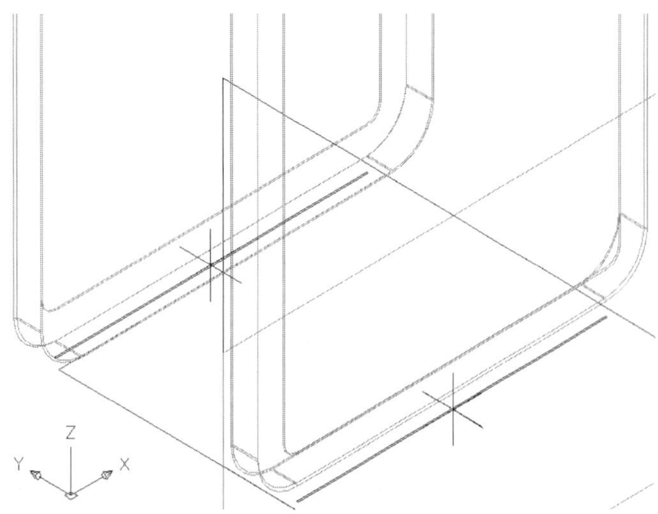

Figure 9.29

You will switch to the sketch view and then turn off the visibility of the other parts in the assembly.

Pick: **Sketch View icon**

Right-click: **CartRail_1** (in the model tree)

Deselect: **Visible** (so it is unchecked)

Repeat this process on your own to hide CartRail_2

The parts are removed from your screen. The resulting sketch looks like Figure 9.30.

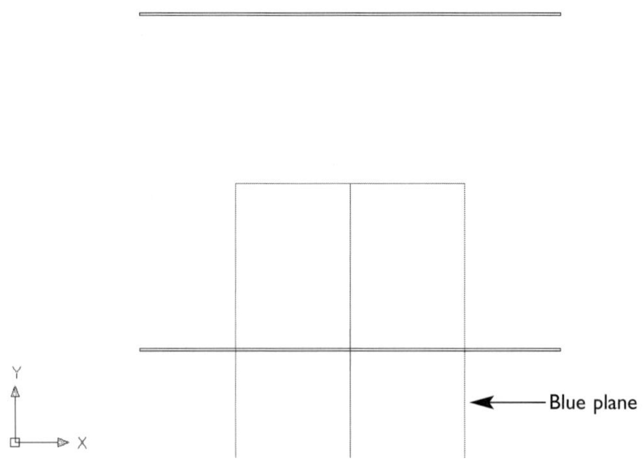

Figure 9.30

Tip
Cross the existing geometry a bit with your lines so you get a good intersection; you will trim the lines later.

On your own, create two new sketch lines, as shown in Figure 9.31

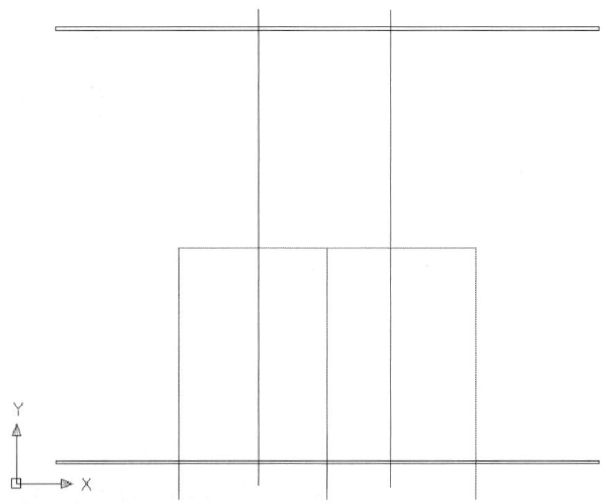

Figure 9.31

Now you will remove the unnecessary lines of the sketch until you have only one rectangle remaining, as shown in Figure 9.32. Be sure not to leave the small end segments of the projected geometry.

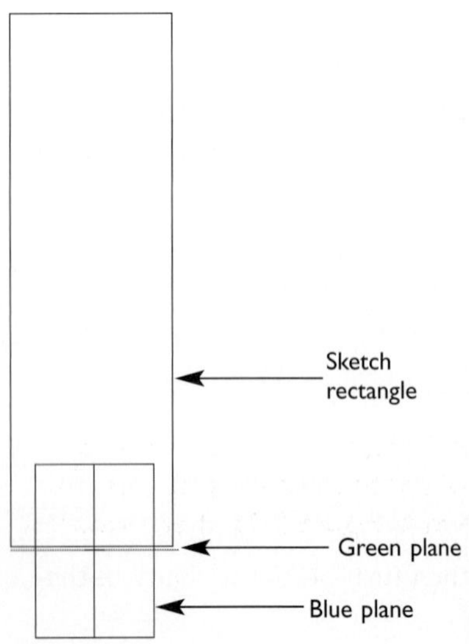

Figure 9.32

Use Trim and Erase on your own until you have only one rectangle, as in Figure 9.32

Right-click: in the graphics area

Pick: **Sketch Solving, Profile**

Select objects for sketch: **use a Crossing window to select all of the sketch lines**

Select objects for sketch: ↵

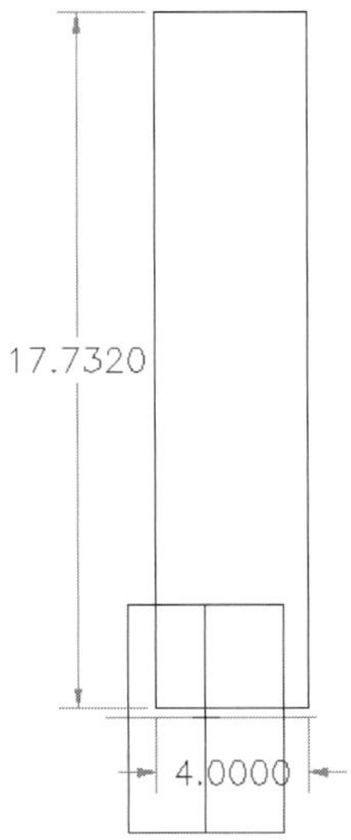

Figure 9.33

Solved under constrained sketch

You will constrain, dimension, and extrude the sketch on your own, as follows.

Add constraints and dimensions to fully constrain the sketch, as shown in Figure 9.33

Change your dimension style on your own so that it is legible

Extrude the completed sketch to a distance of .125

Change the color of the CrossBrace part to magenta for easier viewing

Make the CartRail parts visible

Pick: **Front Left Isometric view icon**

Zoom and shade the drawing

Figure 9.34 shows the new CrossBrace part.

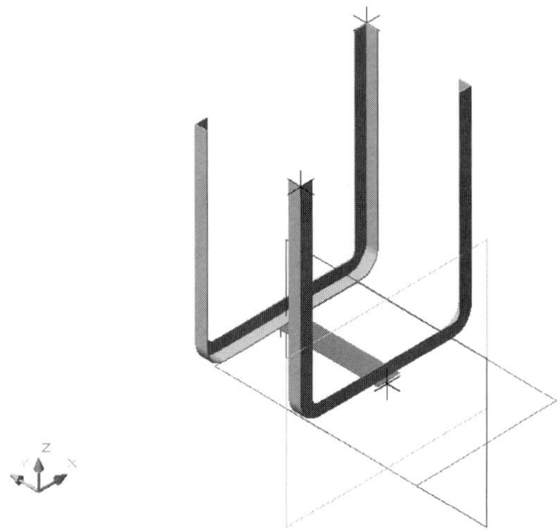

Figure 9.34

Assembling the CrossBrace Part

Next you will add 3D constraints to locate the CrossBrace within the assembly. You will use the 3D Orbiter to rotate the view during the process of adding constraints to make selection easier. You will start the command and then spin the view so that you see the bottom surface of the CrossBrace, as shown in Figure 9.35.

Right-click: in the graphics area

Pick: **Assembly Menu**

Pick: **3D Constraints, Mate**

Select first set of geometry: **pick 3D Orbit icon**

Press [ESC] or [↵] to exit, or right-click to display shortcut-menu: **use the Orbiter controls to spin the model so that you see the bottom surface of the CrossBrace, as shown in Figure 9.35, then press** [ESC]

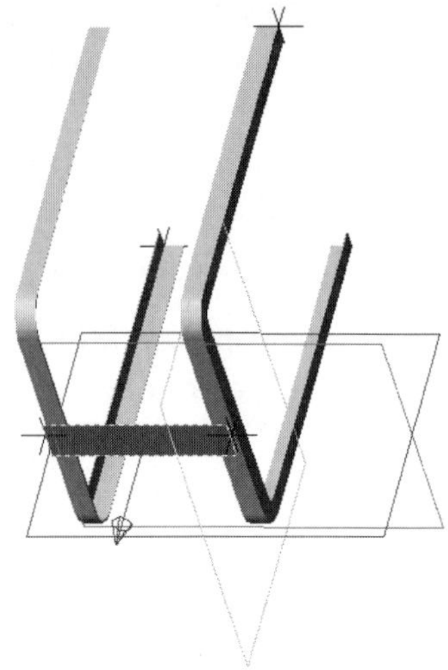

Figure 9.35

Resuming AMMATE command.

Select first set of geometry: **pick the bottom surface of the CrossBrace**

First set = Plane

Enter an option [Clear/Next/Flip/cYcle] <accEpt>: ↵ **if the bottom surface is highlighted; if not use the command options to select it and then press** ↵

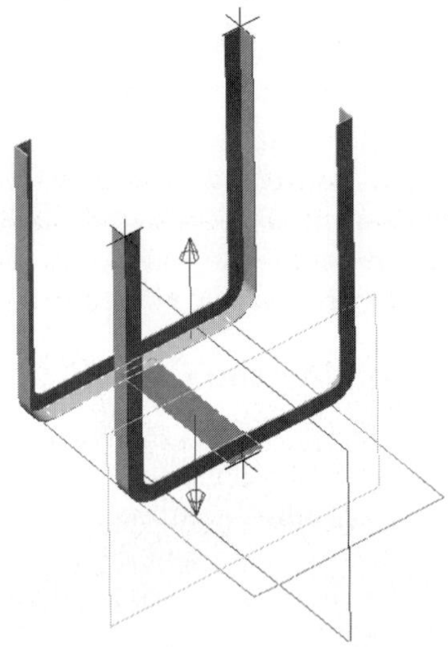

Figure 9.36

Select second set of geometry: **pick 3D Orbit icon**

Press ESC or ↵ to exit, or right-click to display shortcut menu: **spin the model so that you can see the top surface of the horizontal portion of the CartRail, as shown in Figure 9.36, then press** ESC

Regenerating model.

Resuming AMMATE command.

Select second set of geometry: **pick the top surface of the horizontal portion of the CartRail**

Second set = Plane

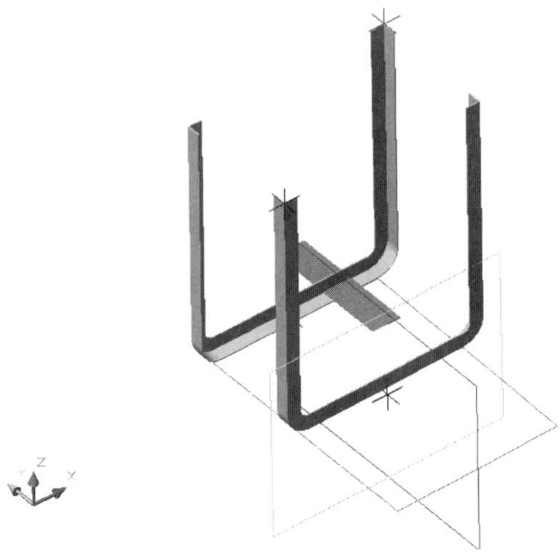

Figure 9.37

Select second set or [Clear/fAce/Point/cYcle] <accEpt>: ↵ **if the correct surface is highlighted, if not use the command options to select it and then press** ↵

Enter offset <.0000>: ↵

The CrossBrace part is now mated to the top side of the horizontal surface of the CartRail, as shown in Figure 9.37.

Next you will mate the narrow surface at one end of the CrossBrace to the inside face of the CartRail.

Right-click: in the graphics area
Pick: **Repeat Mate**

On your own, use the 3D Orbiter and Zoom Window during the mate command to select the surfaces indicated in Figure 9.38. Mate the small end of the CrossBrace to the inner vertical surface of the CartRail.You may find it easier to work in wireframe mode

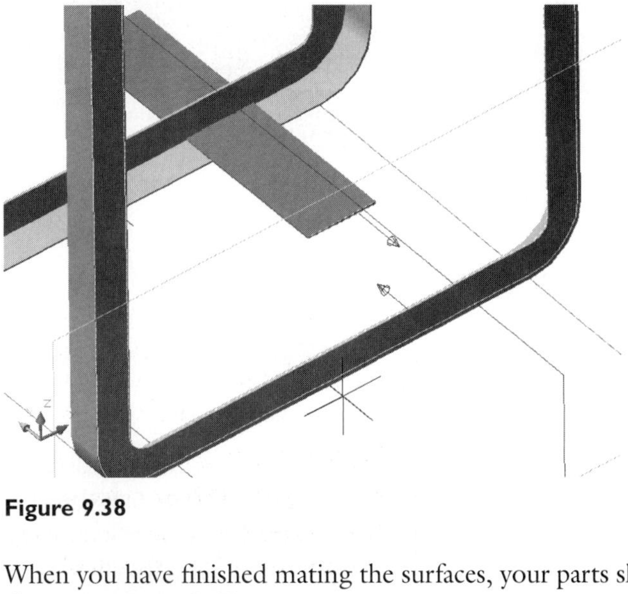

Figure 9.38

When you have finished mating the surfaces, your parts should appear as shown in Figure 9.39.

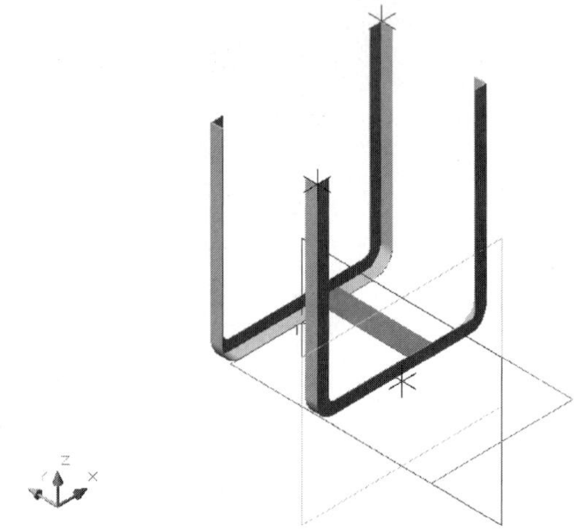

Figure 9.39

Finally, you will use the Mate command again to mate the inside surface of the CrossBrace to the middle plane. This time you will specify an offset distance so that the brace is offset from the middle plane.

You should see the red, blue, and green planes of your start part in the drawing, as shown in Figure 9.39. You will mate the right edge of the CrossBrace to the blue plane. You will activate the assembly first to make sure that you can select the blue plane of the StartPart.

Tip

If you have trouble selecting the blue plane, you may have to activate the StartPart and then reactivate the CartAsmb assembly.

Right-click: in the graphics area

Pick: **3D Constraints, Mate**

Select first set of geometry: **pick the narrow surface labeled A in Figure 9.40**

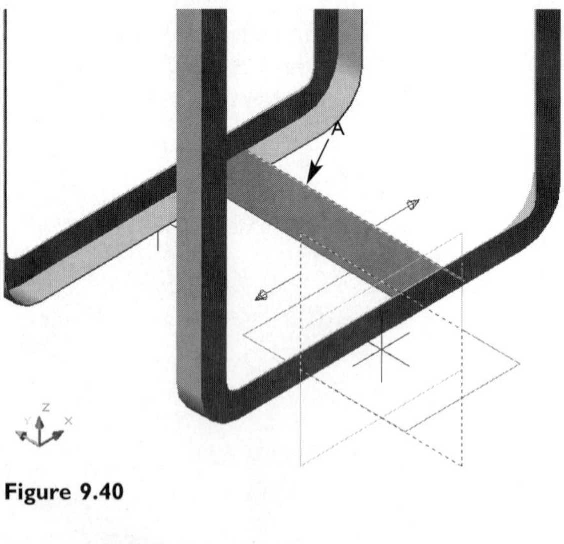

Figure 9.40

First set = Plane

Select first set or [Clear/aXis/fAce/Next/cYcle] <accEpt>: ↵ **if plane A is highlighted, or use the command options to select it and then press ↵**

Select second set of geometry: pick the blue plane of the StartPart

Second set = Plane

Enter an option [Clear/Flip] <accEpt>: ↵ **if the arrows are pointing in opposing directions, as in Figure 9.40, or press F ↵ to flip the direction**

Enter offset <.0000>: **7.5** ↵

The CrossBrace is fully constrained in the assembly, as shown in Figure 9.41.

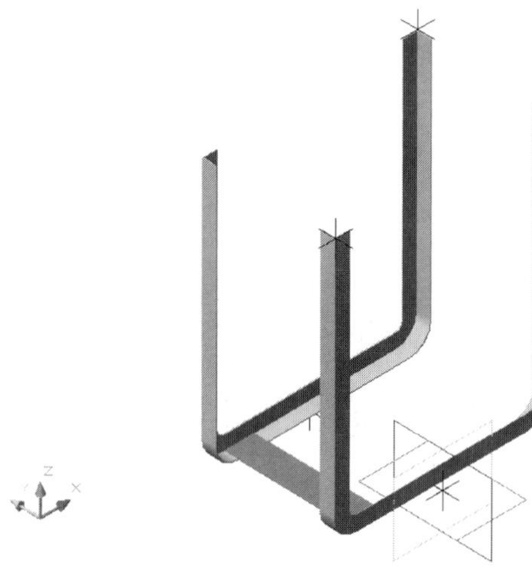

Figure 9.41

Tip

You must have noun/verb selection turned on for the next step.

Creating a New Instance of a Local Part

Now you will add a second instance of the CrossBrace. As it is a local part, you do not need to use the Assembly Catalog.

Pick: **on the CrossBrace part so that its grips appear activated**

Right-click: in the graphics area

Pick: **Instance**

An instance of the CrossBrace part appears attached to your cursor.

Select insertion point: **pick to the right of the assembly**

Select insertion point: ↵

Constrain the new instance as you did the first CrossBrace, as follows.

Mate the new instance to the top of the horizontal surface of the CartRail part

Mate the end of the new instance to the inside edge of the CartRail

Mate the left edge to the blue StartPart plane using an offset of 7.50

When you are finished, your model should look like Figure 9.42.

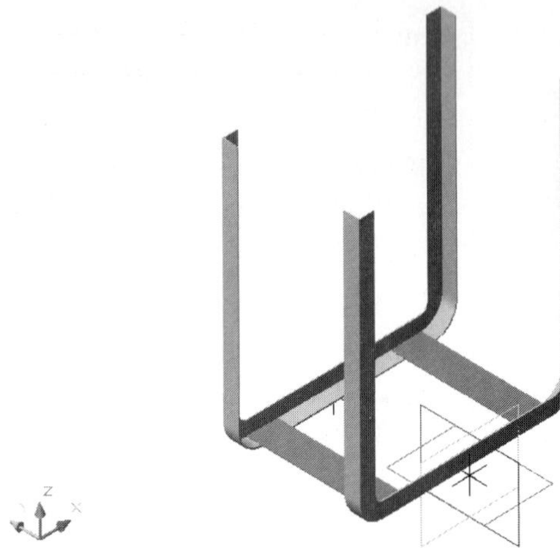

Figure 9.42

Right-click: **CrossBrace_1** (in the model tree)

Pick: **Properties, Color**

Pick: **Red**

Pick: **OK**

Notice that both the original CrossBrace and the instance change color. An instance is like a copy that refers to the original part. It has the same properties as the original. Changing the instanced copy will also change the original part.

Now you are ready to create the tray that will be welded to the cart frame.

Creating the Tray Part

The tray is a rectangular shape with a 1-inch lip all around. To model it, you will sketch a flat pattern with 1-inch tabs on all four sides, then bend the tabs up to form the lip.

On your own, activate the CartAsmb assembly

Right-click: in the graphics area

Pick: **Part Menu**

Pick: **Part, New Part**

Select an object or enter new part name <PART1>: **Tray** ↵

A new local part is created within the assembly. You will sketch it off to the side of the assembly and then mate it to the existing parts. You will sketch two rectangles to form a cross for the tray's basic shape. Refer to Figure 9.43.

Right-click: in the graphics area

Pick: **New Sketch Plane**

Select work plane, planar face or [worldXy/worldYz/worldZx/Ucs]: **X** ↵

Plane = World XY

Select edge to align X-axis or [Flip/Rotate/Origin] <Accept>: ↵

Pick: **Sketch View icon**

Pick: **Pan Realtime and pan your view so that the existing parts are at the right of the screen**

Right-click: in the graphics area

Pick: **2D Sketching, Rectangle**

Specify first corner point or [Chamfer/Elevation/Fillet/Thickness/Width]: **pick point A** (to the left of the assembly)

Specify other corner point or [Dimensions]: **pick point B** (above and to the right of the previous point to form a rectangle nearly the size of the cart)

Right-click: in the graphics area

Pick: **Repeat Rectangle**

Pick: **to create another rectangle that crosses the previous one to form a cross shape, as shown in Figure 9.43**

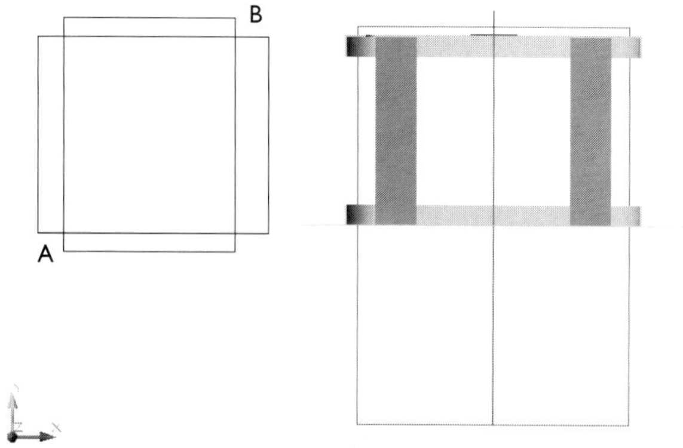

Figure 9.43

On your own, trim the four interior lines so that you have a shape as shown in Figure 9.44

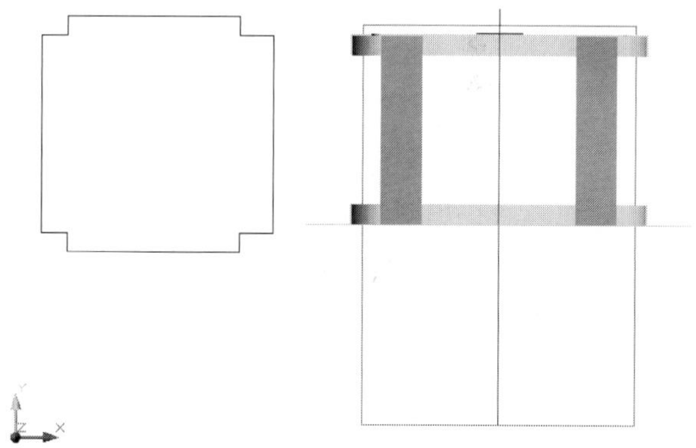

Figure 9.44

Right-click: in the graphics area

Pick: **Sketch Solving, Profile**

Select objects for sketch: **use a crossing box to select all of the lines forming the cross shape**

Select objects: ↵

Right-click: in the graphics area

Pick: **2D Constraints, Equal Length**

On your own, make all of the short line segments equal to the one at the upper left of the part

Right-click: in the graphics area

Pick: **Dimensioning, New Dimension**

On your own, add the three dimensions shown in Figure 9.45.

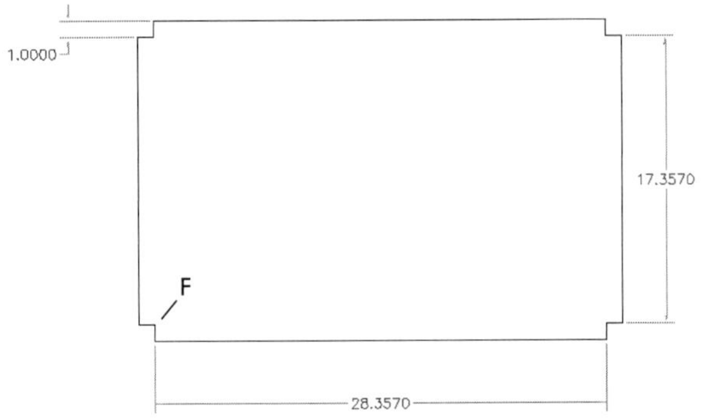

Figure 9.45

The profile should now be fully constrained. (If it is not, try adding a Fix point constraint to the lower left inner corner as shown in Figure 9.45.)

Right-click: in the graphics area

Pick: **Sketched & Work Features, Extrude**

Extrude the profile to a depth of .0625

Change the color of the Tray to magenta

Pick: **Front Left Isometric view icon**

Use Zoom Window on your own to enlarge just the Tray

Figure 9.46

When you have completed these steps, your tray part should look like Figure 9.46.

Next you will create an open profile of a straight line with two endpoints that will be used as the bend line to bend up the side of the tray. First, you will set the sketch plane as the top surface of the tray.

Right-click: in the graphics area

Pick: **New Sketch Plane**

Select work plane, planar face, or [worldXy/worldYz/worldZx/Ucs]: **pick on the top surface of the tray**

Enter an option [Next/Accept] <Accept>: ↵

Plane = Parametric

Select edge to align X axis or [Flip/Rotate/Origin] <Accept>: ↵

You will use Osnap Endpoint to select points for the bend lines.

Right-click: **OSNAP button**

Pick: **Settings**

Confirm that Osnap is on and Endpoint is selected

Pick: **OK**

Right-click: in the graphics area

Pick: **2D Sketching, Line**

Specify first point: **zoom in and target and pick endpoint A in Figure 9.47**

Tip

It may be easier to draw the lines if the view is switched to hidden lines. Make sure to select the endpoints of the top surface.

Specify next point or [Undo]: target and pick **endpoint B in Figure 9.48**

Specify next point or [Undo]: [Enter]

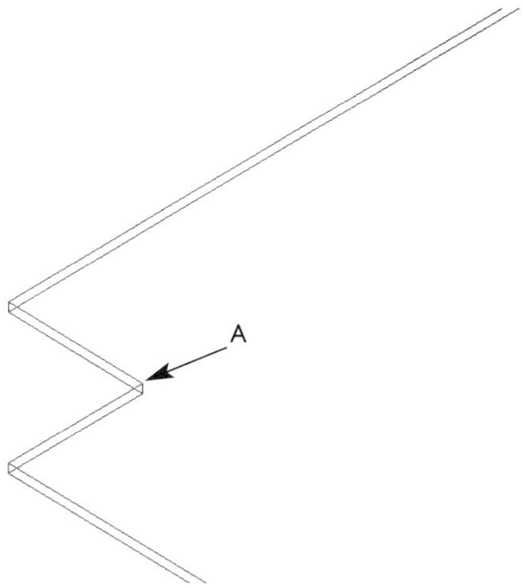

Figure 9.47

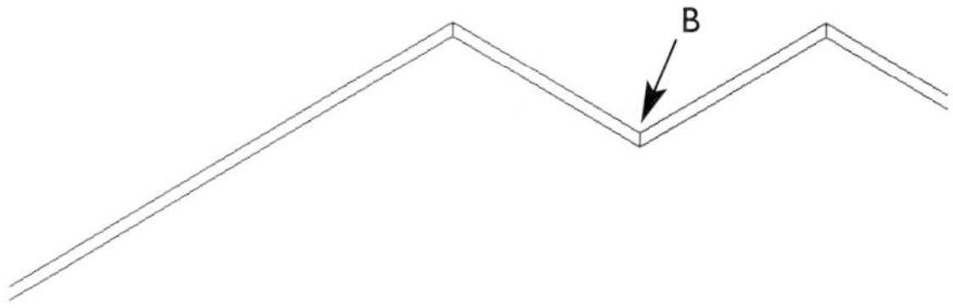

Figure 9.48

Now you should have a line drawn across the upper tab of the tray, as shown in Figure 9.49. You will use it as an open profile to bend up the tab. For most operations, Mechanical Desktop requires the profile to be closed. However, an open profile is used for bend lines, and ribs and for forming thin features.

Figure 9.49

Pick: **Sketch View icon**

Right-click: in the graphics area

Pick: **Sketch Solving, Profile**

Select objects for sketch: **pick the line**

Select objects for sketch: [Enter]

Select part edge to close the profile <open profile>: [Enter]

The OpenProfile is added to the model tree. Next you will bend the part along this line.

Right-click: in the graphics area

Pick: **Sketched & Work Features, Bend**

The Bend dialog box appears on your screen. Make the following selections, as shown in Figure 9.50.

Reference

Mechanical Desktop On-line Help

To learn more about open profiles and how they relate to the neutral axis of a bend, look for Open Profile in the Help Index.

Tip

You can see a preview of the side that will be bent when you are making selections in the Bend dialog box. You may need to Zoom in on the model so it is visible when the dialog box is on screen.

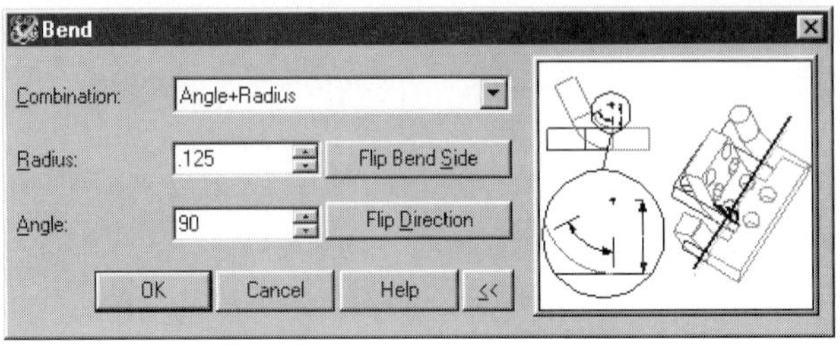

Figure 9.50

Choose Combination: **Angle + Radius**

Radius: **.125**

Angle: **90**

Pick: **OK**

Pick: **Front Left Isometric icon**

Figure 9.51

The tab is bent upward on the part, as shown in Figure 9.51. If the bend goes the wrong direction, right-click on Bend1 in the model tree. Edit the feature by using the Bend dialog box to choose Flip Bend Side; then update the part.

Repeat this process on your own to bend up the three remaining tabs

When you are finished, the part should appear as shown in Figure 9.52.

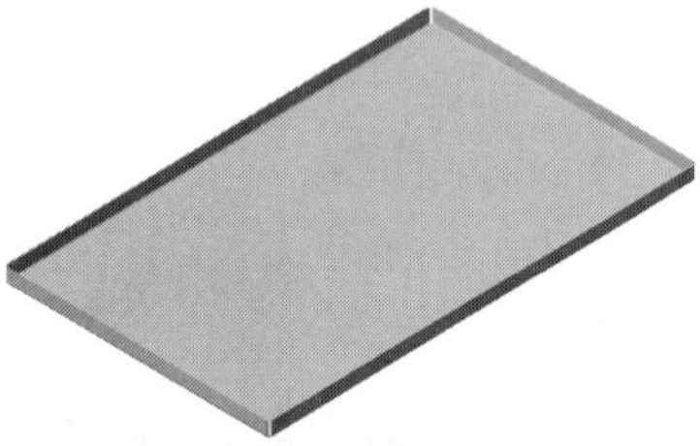

Figure 9.52

On your own, turn off the visibility of StartPart

Save your drawing

Assembling the Tray

To assemble the tray, you will constrain it so it sits inside the corners of the CartRails, with the top edge of the lip flush with the top ends of the CartRails. You will do this on your own, using the Flush and Mate constraints.

Tip

If you have trouble bending the last side, draw the line farther away from the center of the part, out toward the end of the tab. Then create the bend feature. Once you have it working, edit the profile sketch for the bend line and move it to the endpoint.

Use the Flush constraint to match the top surface of the lip of the Tray to the top surface of the CartRail corner

Mate the outside surfaces of the Tray lip to the inside surfaces of the angle forming the CartRail

Tip

When the top edge of the tray lip and the top edge of the CartRail are selected for the Flush constraint, both arrows should be pointing downward.

When you are finished, your assembly should look like Figure 9.53.

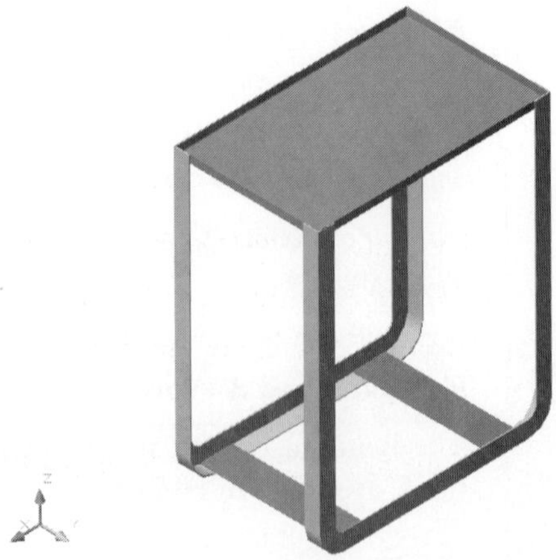

Figure 9.53

Shelled Features

Next you will create another version of the tray part off to the side of the assembly. This time you will make it as an extrusion and then you will remove the inside portion to form a hollow box open on one end.

Pick: **Part, Part, New Part** (from the menu bar)

Select an object or enter new part name <PART1>: **TrayShelled** ↵

Pick: **Sketch View icon**

Your view returns, looking directly onto the XY-plane.

Draw a rectangle off to the side of the cart assembly, as shown in Figure 9.54

Figure 9.54

Pick: **Sketch Solving, Single Profile**

Add the horizontal dimension of 28.357 and the vertical dimension of 17.357 to the sketch, as shown in Figure 9.55

Figure 9.55

Extrude the rectangle to a depth of 1.00

Switch to a Front Left Isometric view

Change the color for the part to cyan

Shade the part

Your new part should look like Figure 9.56.

Figure 9.56

Next you will shell out the inside of the tray.

Right-click: in the graphics area

Pick: **Placed Features, Shell**

The Shell dialog box appears on your screen. You will make the settings shown in Figure 9.57. You can select whether the shell is created to the inside or the outside or the feature, or use a midplane. Excluding faces lets you select which faces will be open to the outside of the part. If you do not exclude any, the hollowed-out portion is entirely inside the part. You can also select to have varying thicknesses instead of a uniform thickness.

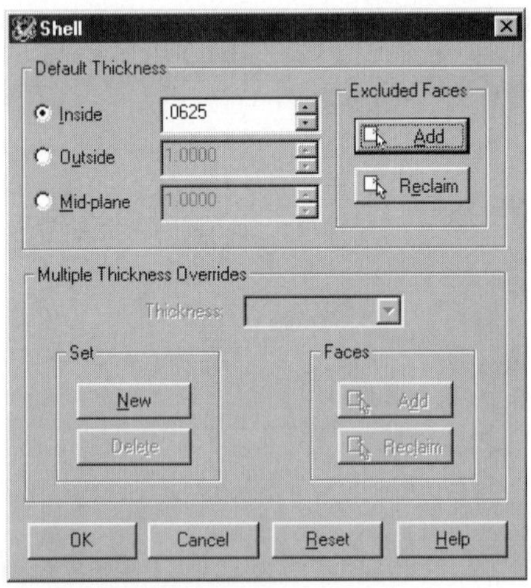

Figure 9.57

Set Default Thickness for Inside: **.0625**

Pick: **Add** (under Excluded Faces)

Select faces to exclude: **pick the top surface of the tray**

Enter an option [Next/Accept] <Accept>: ↵

Figure 9.58

Select faces to exclude: ↵

Pick: **OK**

Your shelled tray should look like Figure 9.58.

Exporting a Local Part

The shelled tray that you modeled as a local part is one that may be useful in other models, but you will not use it in this assembly. To save it as a separate file, so that it will be available outside this assembly,

Right-click: in the graphics area

Pick: **Assembly Menu**

Pick: **Catalog**

Pick: **All tab** (in the Assembly Catalog window)

The right side of this tab of the dialog box reveals the parts in the assembly that are local. To save your local shelled tray part as an external part,

Right-click: **TrayShelled** (in the right-hand list)

Pick: **Externalize**

The New External File dialog box appears so you can select a location for the file and enter a file name.

Choose Save in: **MDT-Work**

Accept TrayShelled.dwg as the file name

Pick: **Save**

The primary definition of the Tray part is now stored in an external file, and the Assembly Catalog confirms that by moving Tray to the left-hand list.

Pick: **OK** (to exit the dialog box)

Right-click: **TrayShelled_1** (in the model tree)

Pick: **Delete**

You see a message stating, *The last instance of the following definition has been deleted and is no longer used in the assembly:TRAYSHELLED. Would you also like to remove this definition?*

Pick: **Yes**

The shelled tray is deleted from your assembly drawing but it is still available as a part drawing on your drive.

On your own, Zoom so that you see the cart assembly.

Adding the Caster Subassembly

You will use parts from the data files to create the subassembly for the casters and then insert them into holes you will make in the CrossBraces. The casters will bolt on.

When designing an assembly, it is often helpful to use CAD models of parts or hardware that already exist. One such source of existing parts can be a vendor's 3D CAD library. Often these existing parts can be downloaded from the vendor's Web page and used in the assembly. Although these parts may not be parametric, the already created 3D geometry can save you design and modeling time. The casters and bolts for the cart are likely to be purchased from a supplier. Files for these parts are included with the data files.

On your own, copy Caster.dwg and Bolt-halfinch.dwg from the MDT-Data directory to the MDT-Work\Cart directory

Right-click: in the graphics area

Pick: **Assembly, New Subassembly**

Enter new subassembly name <SUB1>: **Caster&Bolt** ↵

Right-click: **Caster&Bolt** (in the model tree)

Pick: **Activate Assembly**

The parts of the main assemby appear grayed out. Now you are working on a smaller assembly inside your main assembly. Once you have assembled the new parts together, they will behave as one part at the main assembly level.

Right-click: in the graphics area

Pick: **Catalog**

Insert an instance of the Caster and four instances of the Bolt-halfinch into the lower left of the assembly on your own

Your screen should now look like Figure 9.59.

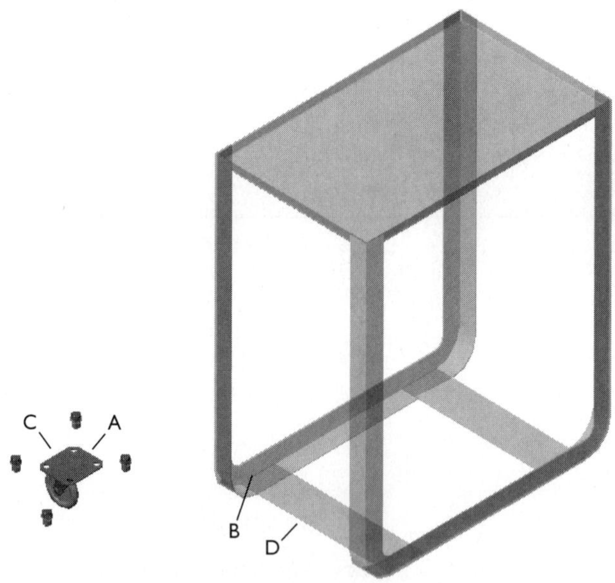

Figure 9.59

Mating the Bolts

Now you are ready to mate the bolts to the caster. This way when you use the subassembly, you will not have to mate the bolts each time individually.

To constrain the bolts to the caster, you will use Insert. Remember that the Insert constraint aligns two circular objects along an axis and a planar face. You will need to Zoom in on the bolts so you can mate the inner surface of the greenish nut to the hole on the underside of the caster, as

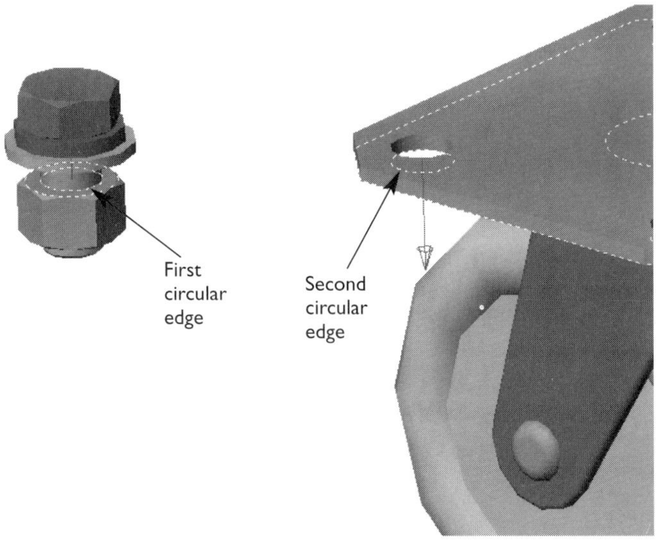

Figure 9.60

shown in Figure 9.60. When you have finished, there should be some space beneath the underside of the flat washer and the top of the caster. This is where the CrossBrace part will fit.

On your own, insert the four bolts into the mounting holes on the caster

You are now ready to constrain the first instance of the Caster&Bolt subassembly.

Right-click: **CartAsmb** (in the model tree)

Pick: **Activate Assembly**

The main level of the assembly is activated. Now the Caster&Bolt subassembly behaves as a single part.

Zoom so that you can see the entire assembly

Pick: **Caster&Bolt_1** (in the model tree)

Notice that the entire subassembly is selected. Next you will mate the caster to the CrossBrace part.

Right-click: in the graphics area

Pick: **3D Constraints, Mate**

Using skills you have already learned, you will mate the Caster to the CrossBrace. Look at Figure 9.61 to see how the caster wheels are oriented. Use the 3D Orbiter as needed to spin the part as you add the following constraints on your own.

Figure 9.61

Mate the top gray surface of the Caster's mounting plate to the bottom red surface of the CrossBrace

Mate the side face of the Caster's mounting plate, surface A in Figure 9.59, to the inner vertical surface of the CartRail, labeled surface B, with a offset distance of 2.0

Mate the face of the Caster, labeled C, to the outer face of the CrossBrace, labeled D

Once you have finished, your assembly should look like Figure 9.61.

Once you have finished mating the Caster&Bolt subassembly to the CrossBrace, you will repeat the steps for the remaining three Casters&Bolt subassemblies. All of them are oriented in the same direction, and all have the same offset from the inner face of the CartRail.

Add three more instances of the Caster&Bolt subassembly into the main assembly (as you did to add the second CrossBrace instance)

Mate the casters to the CrossBraces so your assembly looks like Figure 9.62

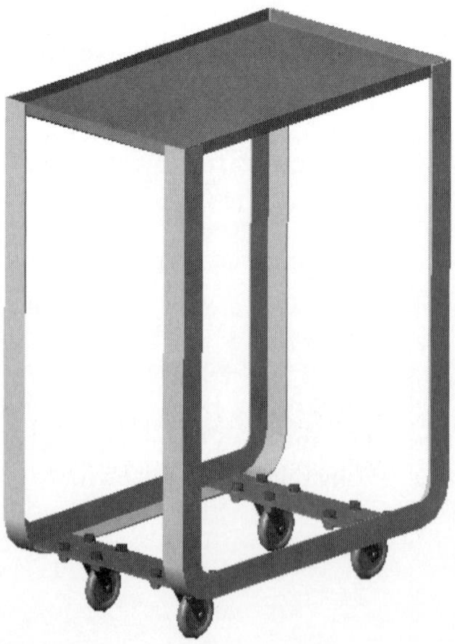

Figure 9.62

Adding the Holes

Now you will edit the CrossBrace part to add the mounting holes that will be used to bolt the Casters to the CrossBraces. You will add all of the holes as a single feature.

Since it is sometimes easier to create a sketch if you hide a part that would make selecting or creating features difficult, you will turn off the Tray's visibility.

Right-click: **Tray_1** (in the model tree)

Deselect: **Visible** (so it is unchecked)

Expand the Caster&Bolt subassembly on your own

Change the visibility for all four of the Bolts so that they do not show

Now activate the CrossBrace part so your new sketch will be a feature of it.

Right-click: **CrossBrace_1** (in the model tree)

Pick: **Activate Part**

Right-click: in the graphics area

Pick: **Part menu**

Pick: **New Sketch Plane**

Select the top face of CrossBrace_1 as the new sketching plane

Pick: **until you have the axes oriented as shown in Figure 9.63**

Right-click: **to accept**

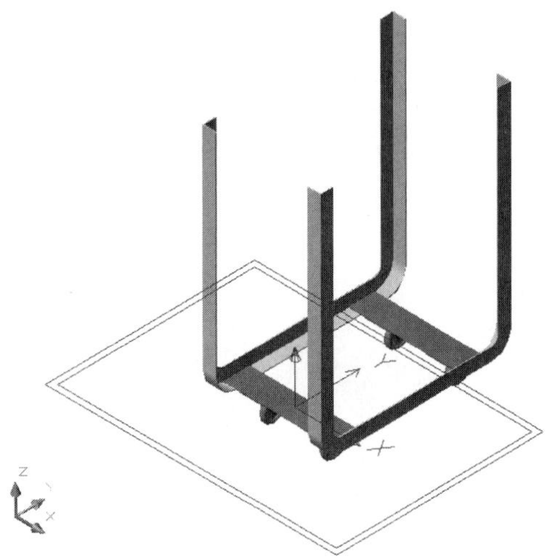

Figure 9.63

Pick: **Sketch View icon** (so you are looking directly onto the sketch plane)

When you do this, your screen should look like Figure 9.64.

Figure 9.64

Since you want the holes on the CrossBrace to be aligned with the holes in the caster, you will now select the outline of the existing eight holes on the caster mounting plate to define the sketch for the hole cut. Switching to wireframe display will make it easier to see the mounting holes.

Pick: **Toggle Shading/Wireframe**

Tip

You may need to Zoom in and use crossing windows to select the hole edges for the profile.

Right-click: in the graphics area

Pick: **Sketch Solving, Copy Edge**

Enter an option [Face/select Edge] <select Edge>: **select all 8 of the hole edges on the two casters mounted to CrossBrace_1** ↵

All eight of the circular edges of the mounting holes on the two casters are projected onto the sketch plane. You will select all eight for a profile. This time you will use an underconstrained profile. You will not dimension the holes; you will let their projected size determine how large they are. Mechanical Desktop will make assumptions about the sketch that will let you work with a profile that is not fully constrained. This practice is not always one you will want to follow because it is best to think carefully about the design intent for your parts and leave nothing in how it updates to chance. However, sometimes when you are modeling, all of the factors influencing the design may not be fully defined. Using an underconstrained profile lets you continue on and create a model for design visualization purposes, and you can always edit the sketch later and define it more fully. With Mechanical Desktop, if the size of the holes in the casters changes, the size of the projected hole in the CrossBrace will not automatically update to reflect this change.

On your own, make a profile out of the eight projected circles

Use Extrude to cut the profile through the CrossBrace

Pick: **Left Front Isometric view icon**

Pick: **Toggle Shading/Wireframe**

Your assembly should look like Figure 9.65.

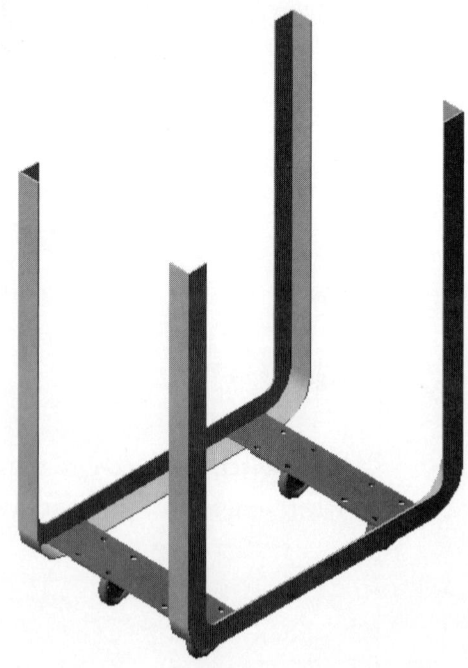

Figure 9.65

Do you notice anything different about CrossBrace_2? It also has eight holes cut through it. Since CrossBrace_1 and CrossBrace_2 are both the same part, anything done to one of the CrossBrace instances will be done to the other.

When you are finished,

Turn on the Visibility for the Tray and the Bolts

Save your work

Your fully assembled Cart should now look like Figure 9.66.

Congratulations! Your cart assembly is now complete. In the next tutorial you will learn how to use design variables and a part table to generate assemblies that you can configure to different sizes.

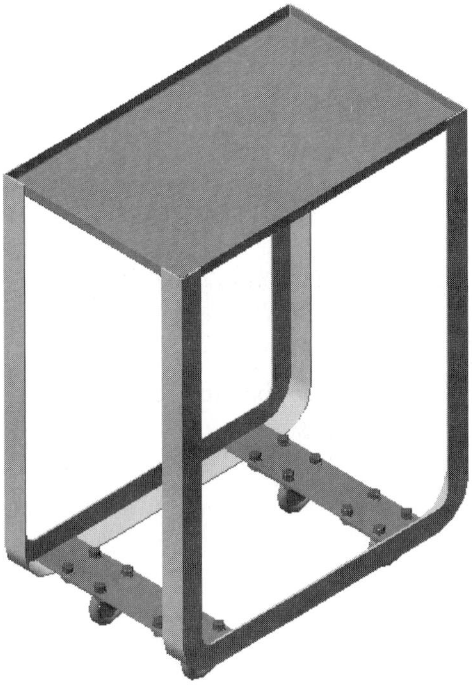

Figure 9.66

Command Summary

Action	Command Name	Description
2D Path	AM2DPATH	Creates a 2D path for a sweep feature from an open or closed profile
Add Set of Workplanes	AMBASICPLANES	Creates three mutually perpendicular work planes and a work point at their intersection
Bend	AMBEND	Bends a portion of a part using an open profile to define the location of the bend
Assembly Catalog	AMCATALOG	Attaches and removes, converts, and organizes local and external parts in an assembly
Change Dimension Display	AMDWGDIMDSP	Changes the display of dimensions to parameters, numbers, or equations
Flush Constraint	AMFLUSH	Creates a parallel assembly constraint between two planes
3D Manipulator	AMMANIPULATOR	Dynamically rotates, copies, or moves parts freely or along 3D axes
New Part, New Part Instance	AMNEW	Creates a new part or part instance
Copy Edge	AMPARTEDGE	Copies an existing edge to create a line or arc that can be used in a new sketch
Project to Plane	AMPROJECT2PLN	Projects 2D objects or edges of planar faces to a work plane, sketch plane, or 3D face so they can be used in new sketch
Shell	AMSHELL	Creates an open or closed shell feature with uniform or multiple wall thicknesses
Sweep	AMSWEEP	Creates a solid feature by sweeping a profile along a 2D or 3D path
Display UCS Icon	UCSICON	Controls the display and location of the icon that indicates the current coordinate system
Set a Zoom Factor	ZOOM, <value>X	Zooms the view to a specified number of magnifications

Exercises

Create the parts and assemblies shown in Figures 9.67 and 9.68.

1. Gyroscope

ITEM NO.	QTY.	DESCRIPTION
1	1	OUTER RING
2	1	INNER RING
3	1	CENTER PIN
4	1	BOTTOM PEG
5	1	TOP PEG
6	1	CENTER WHEEL
7	1	STAND

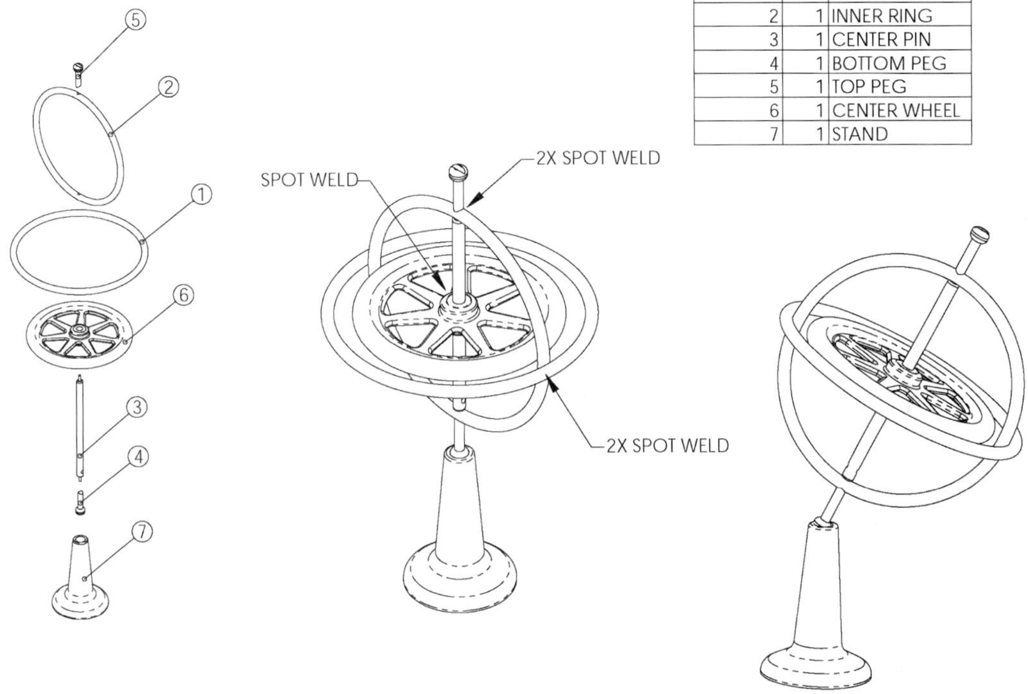

GYROSCOPE ASSEMBLY

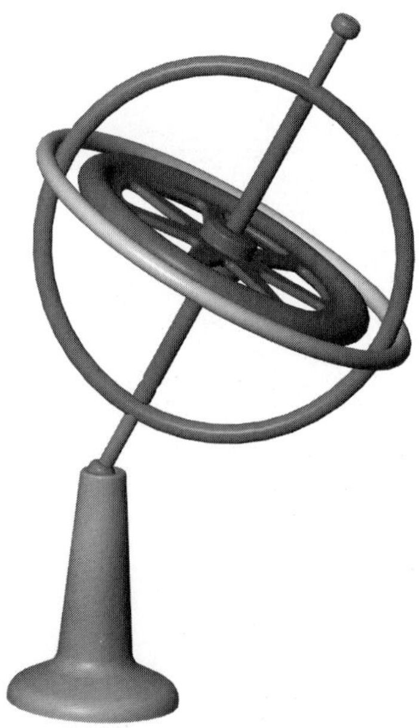

Figure 9.67a

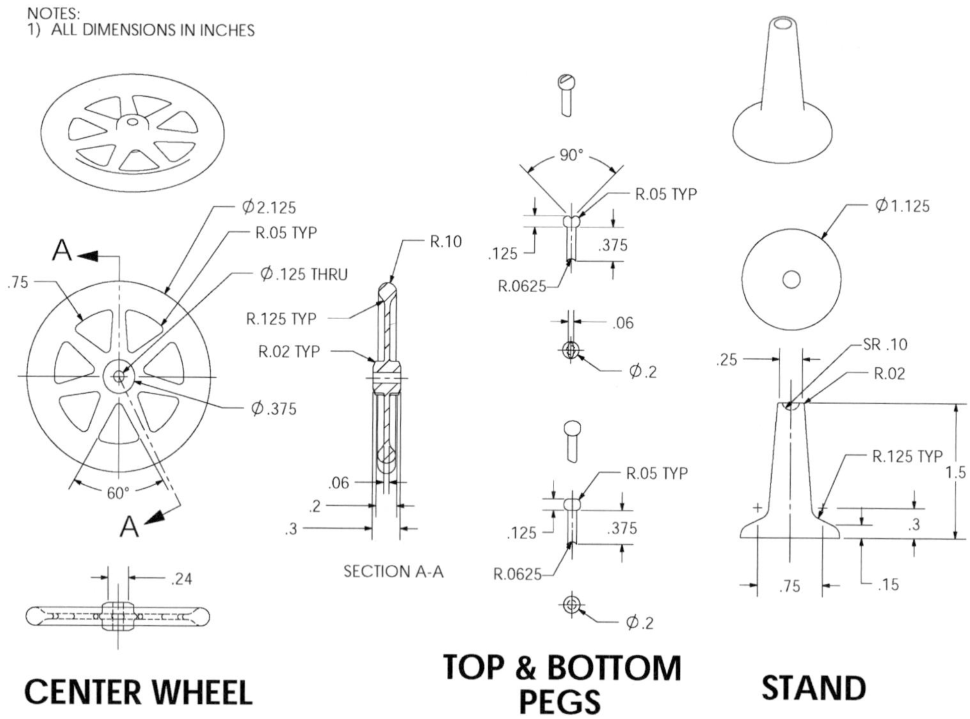
NOTES:
1) ALL DIMENSIONS IN INCHES
Ø2.125
R.05 TYP
Ø.125 THRU
.75
Ø.375
60°
A
A
R.10
R.125 TYP
R.02 TYP
.06
.2
.3
SECTION A-A
.24
CENTER WHEEL
90°
R.05 TYP
.125
.375
R.0625
.06
Ø.2
R.05 TYP
.125
.375
R.0625
Ø.2
TOP & BOTTOM PEGS
Ø1.125
.25
SR .10
R.02
R.125 TYP
1.5
.3
.75
.15
STAND

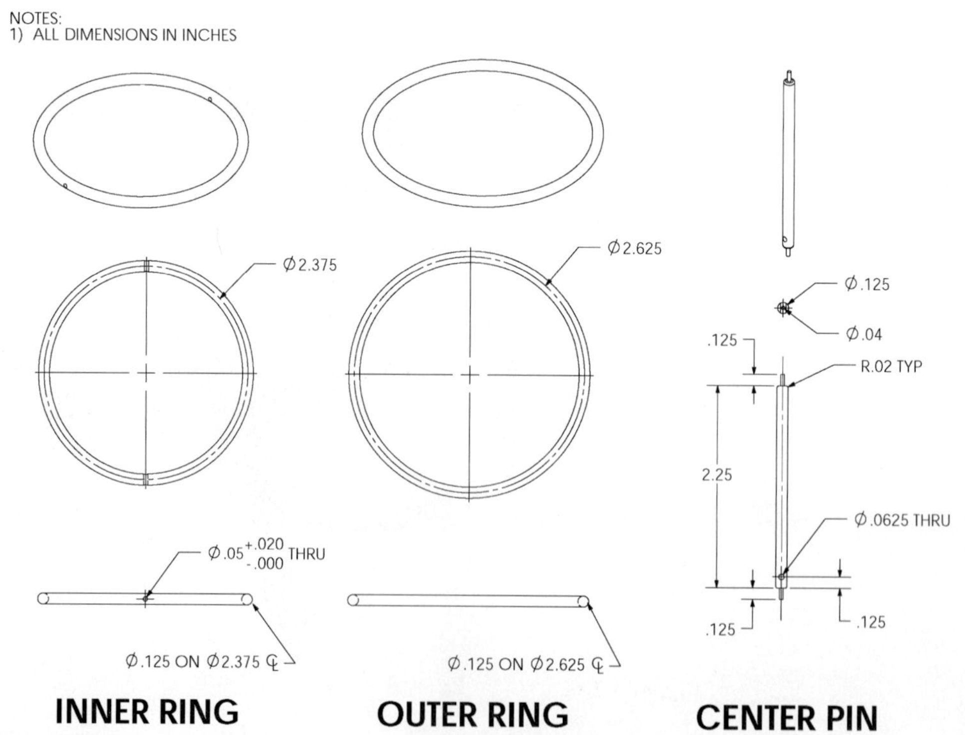
NOTES:
1) ALL DIMENSIONS IN INCHES
Ø2.375
Ø.05 +.020 -.000 THRU
Ø.125 ON Ø2.375 ℄
INNER RING
Ø2.625
Ø.125 ON Ø2.625 ℄
OUTER RING
Ø.125
Ø.04
.125
R.02 TYP
2.25
Ø.0625 THRU
.125
.125
CENTER PIN

Figure 9.67b

2. Compass

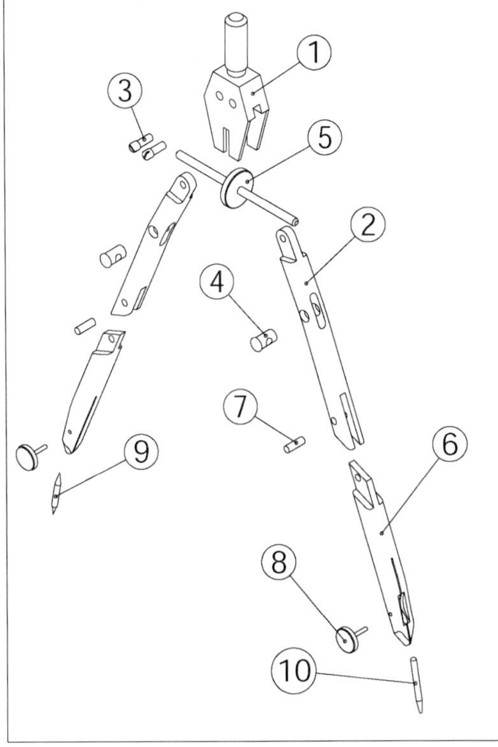

ITEM NO.	QTY.	DESCRIPTION
1	1	TOP
2	2	TOP LEG
3	2	TOP SCREW
4	2	ROLLER GUIDE
5	1	THUMB ROLLER
6	2	BOTTOM LEG
7	2	LEG PEG
8	2	THUMB SCREW
9	1	PIN
10	1	LEAD

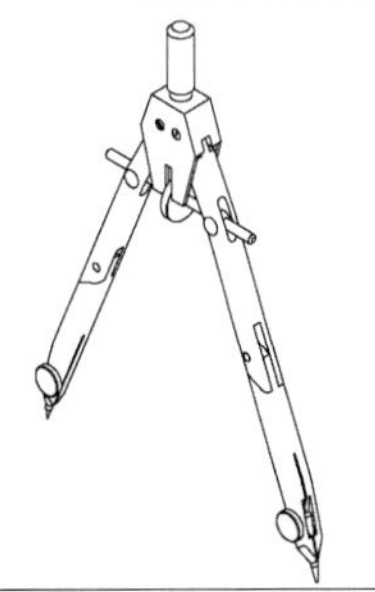

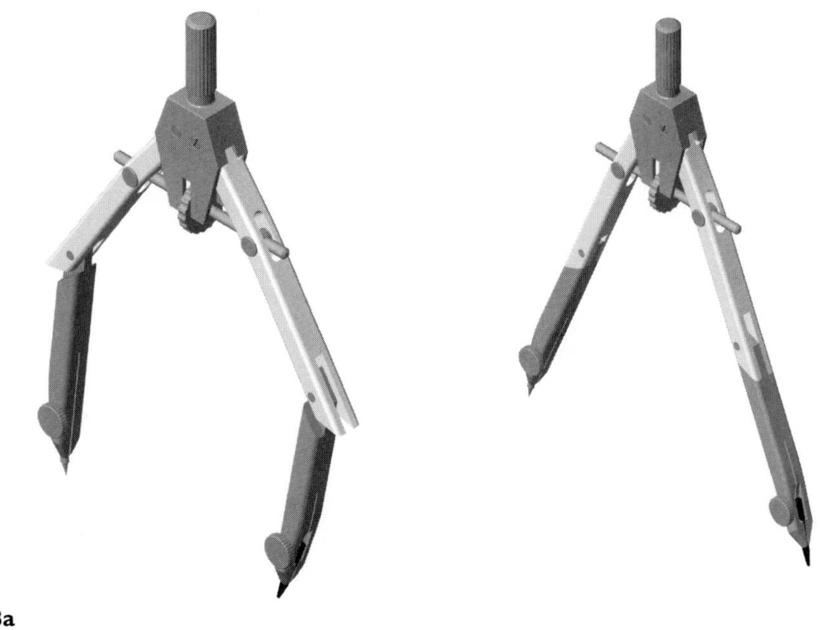

Figure 9.68a

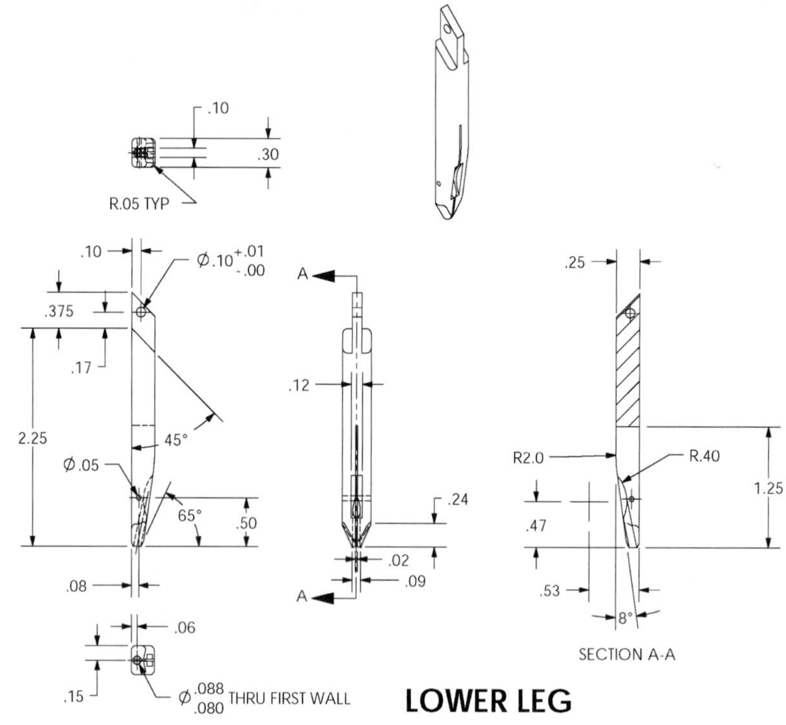

Figure 9.68b

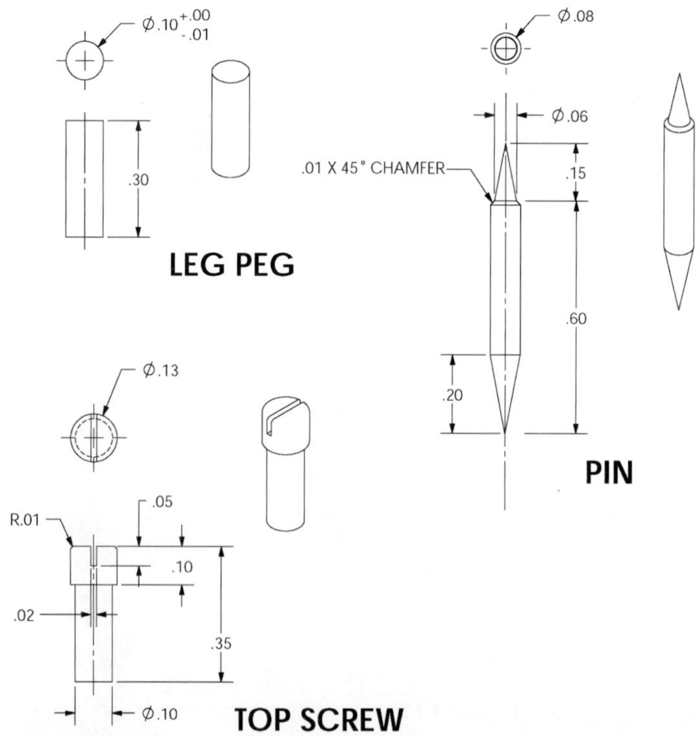

Figure 9.68c

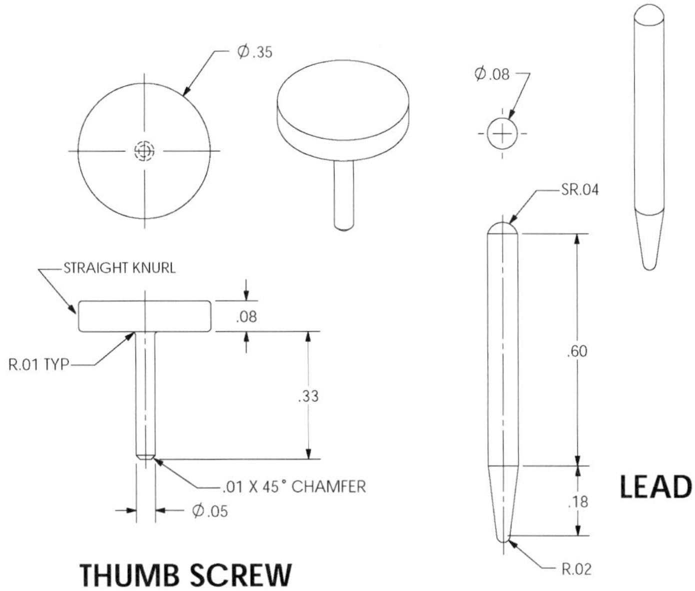

Figure 9.68d

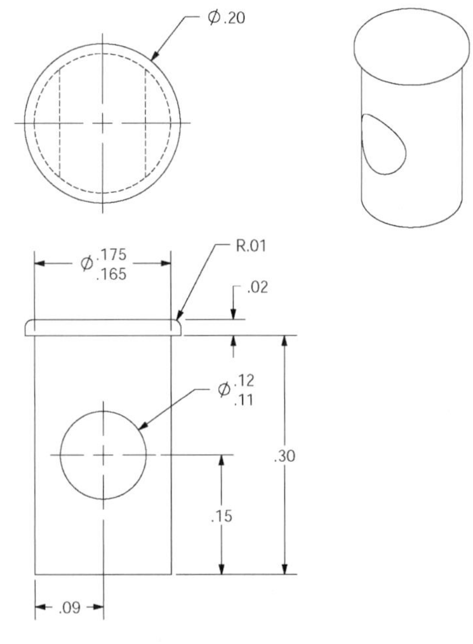

ROLLER GUIDE

Figure 9.68e

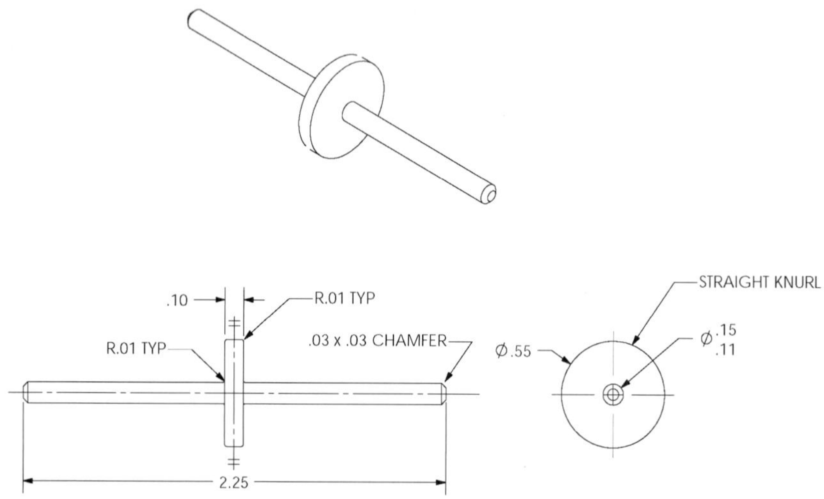

THUMB ROLLER

Figure 9.68f

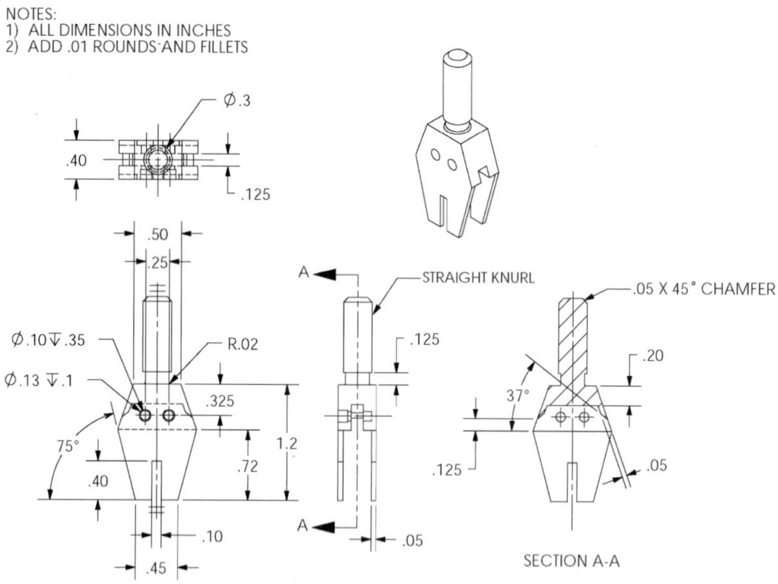

TOP

Figure 9.68g

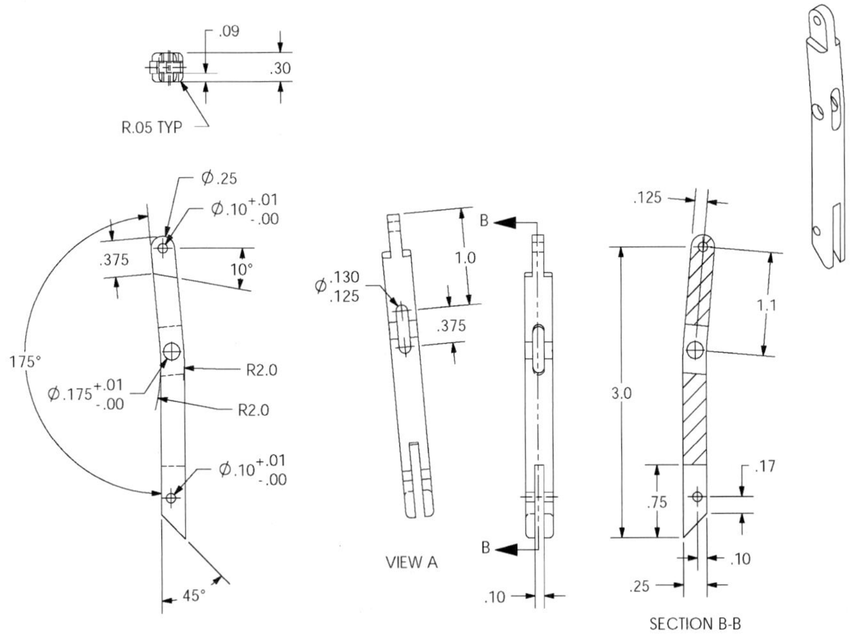

UPPER LEG

Figure 9.68h

Tutorial

10 Getting Results from Your Intelligent Assembly

Objectives

When you have finished this tutorial, you will be able to

- Use a design table to create a family of similar parts.
- Use local and global design variables as parameters.
- Display information about the assembly's mass properties.

Figure 10.1

Reference

Engineering Design Communication

Formulas in Dimensions, pp. 266–286

Introduction

In the previous tutorial, you created the parts and assembly for the cart. In this tutorial, you will use global parameters to create an assembly that you can easily update to create different configurations of the same basic cart shown in Figure 10.1.

Launch Mechanical Desktop 5

Open the file CartAsmb-dta.dwg from the MDT-Data directory

Save the file as CartAsmb-10.dwg to your MDT-Work directory

Load your user profile (if necessary)

The data file for the cart is different from the cart assembly that you created in Tutorial 9. Some of the parameters have been created for you so that you can finish the tutorial in a shorter time. Using what you will learn in this tutorial, you could later make these changes to your cart assembly. Be sure to use the data file provided for the tutorial so that you can follow the steps without problems.

You will start by making the tray a table-driven part.

Table-Driven Parts

A table is a spreadsheet that stores the dimensions for a part and is linked to the part file. Several sets of dimension values can be stored in the table and used to update the part so that a single part can have more than one configuration.

Table-driven parts are perfect to use when modeling objects that are common in more than one size but have the same basic geometry. Most of the size dimensions for a bolt, for example, are derived from its diameter. You could produce a spreadsheet that stored the information for a typical bolt and then save many different configurations of the same bolt at different diameters and lengths so that a single part model could be used to produce any size bolt you would ever need.

Linking to a spreadsheet in this way is a very powerful tool. Not only can a part get its dimensional information from the table, but you can also use the spreadsheet to calculate the size of part features. The part will then update based on the results of these calculations. All of the computational power of the spreadsheet can be used to calculate and optimize the size of the part's features. For this simple cart, you will not do any complex engineering calculations, but you will learn how to use this tool so that you could generate parts based on the results of calculations in the future.

Let's see how it works!

First you will activate the Tray part.

Right-click: **Tray** (in the model tree)

Pick: **Activate Part**

The Tray becomes active for editing. Next you will create some design variables that you can use as you would values in your models.

Creating Active Part Design Variables

Design variables are named parameters that you can use to dimension features or groups of features that are the same size. The nice thing about design variables is that when you change their value, it changes anywhere you used that variable name. This is unlike a typed value, where you would have to edit each one individually. Design variables are similar to the parameter names you used in dimensions in the last tutorial, but they offer even more power in your model. Design variables can be defined only in the active part, or they can be global variables, which can be used for features across multiple parts or to define assembly parameters such as offsets. In addition, design variables can be used in a spreadsheet to control table driven parts. In this tutorial, you will learn to use active part design variables first. Later you will make global variables that you can use across the assembly.

Pick: **Part, Design Variables** (from the menu bar)

The Design Variables dialog box appears on your screen, as shown in Figure 10.2. Make sure that the Active Part tab appears uppermost, as shown in the figure. The Tray part already has one design variable created, named ReliefDia. This is the size for the small relief holes in the tray where it is bent. You will create three more design variables for the tray. (If you had not made the Tray part active before opening the dialog box, you would not see the Tray variable.)

Pick: **New**

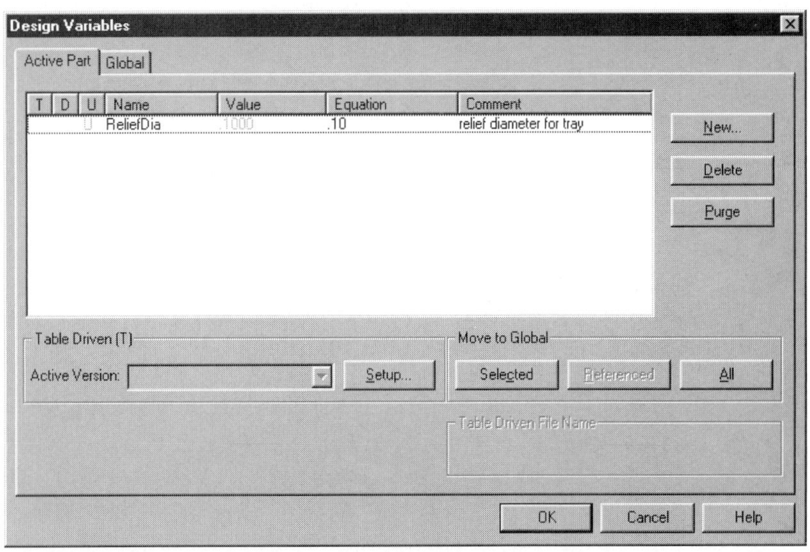

Figure 10.2

The New Part Variable dialog box appears. You will use it to make the settings shown in Figure 10.3. Note that you will enter values to define the design variables, but you could enter an equation in this box.

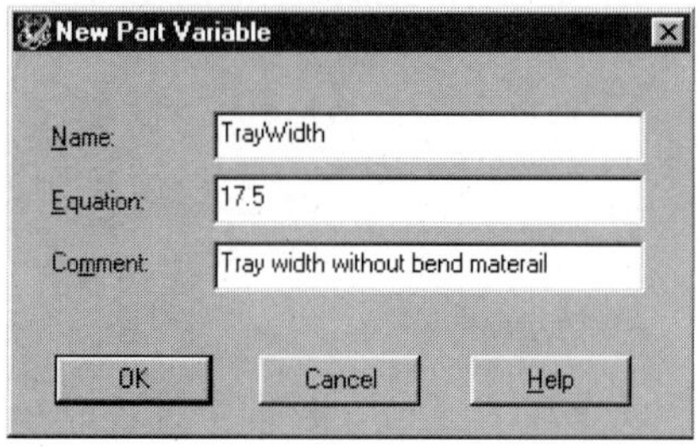

Figure 10.3

Set Name: **TrayWidth**

Set Equation: **17.5**

Set Comment: **Tray width without bend material**

Pick: **OK**

The variable is added to the Design Table dialog box. On your own, add two other new variables, as shown in Figure 10.4.

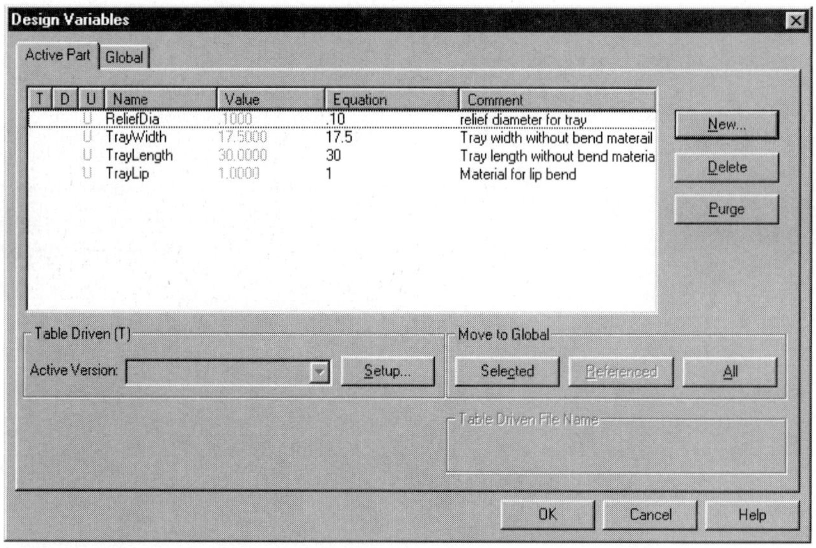

Figure 10.4

Set Name: **TrayLength**

Set Equation: **30**

Set Comment: **Tray length without bend material**

Pick: **OK**

Set Name: **TrayLip**

Set Equation: **1**

Set Comment: **Material for lip bend**

Pick: **OK**

Creating a Table-Driven Part

Next you will set up the part and create a table of values to drive the part. You should still be in the Design Variables dialog box. (If not, restart it.) In the Table Driven portion of the dialog box,

Pick: **Setup**

The Table Driven Setup dialog box shows on your screen. You will use it to create the spreadsheet that is linked to the part to drive the different sizes. You also use it to control whether the spreadsheet information is stored in rows or columns.

Pick Version Names: **Across** (so it is darkened)

To name the spreadsheet, type

Sheet Name: **TRAY**

When you are finished making these selections, the dialog box should look like Figure 10.5.

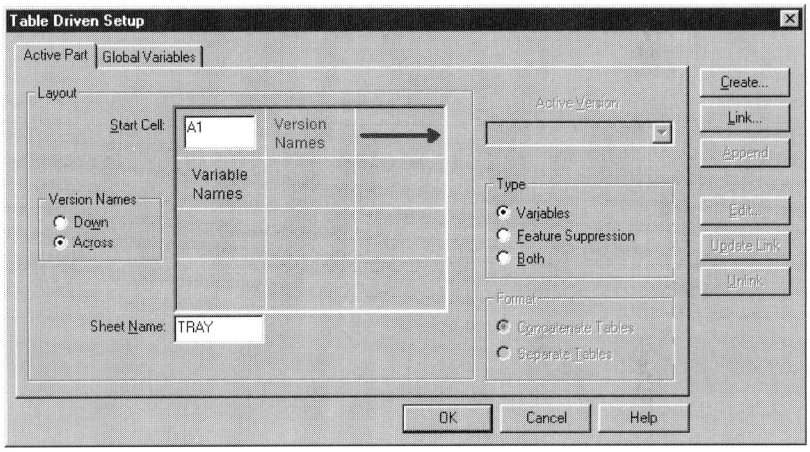

Figure 10.5

Pick: **Create**

The Create Table dialog box appears. You use it to name the spreadsheet file that will be created.

Choose Save in: **C:\MDT-Work**

Type File Name: **Tray-Part.xls**

Pick: **Save**

A spreadsheet is automatically generated from the variables in the design table, as shown in Figure 10.6.

Now you are ready to create the table of different configurations that will be used to update the Tray's dimensions. The column headings will appear in the model tree to identify the configurations available in the table. You will rename the column that is currently named Generic to 17.50 × 30, a more descriptive name for the configuration. Then you will define additional configurations that you will use within the part model later.

Tip

If the spreadsheet does not appear, check to see if you have it on your task bar and pick its name to switch to it. If it still does not appear, make sure that you have Microsoft Excel installed.

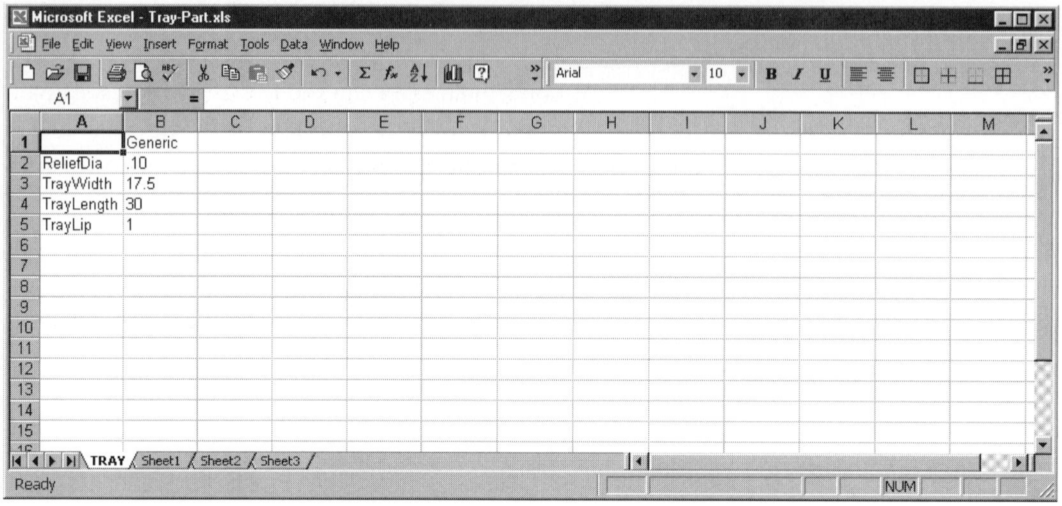

Figure 10.6

Pick: **Generic** (to select the cell in the spreadsheet)

Use the input line above the cells to change the name to 17.50 × 30; then press ⏎

On your own, type the information into the remaining cells so that your sheet looks like that shown in Figure 10.7. Adjust the width of the first column if necessary so that you can see the entire variable name

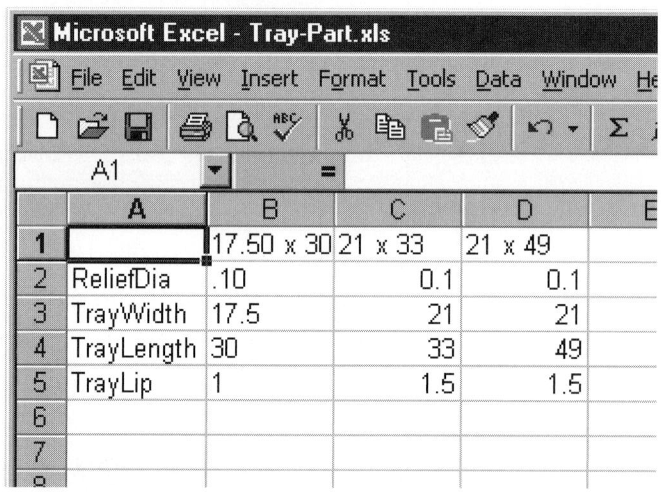

Figure 10.7

Pick: **File, Save** (in Excel to save the spreadsheet (or press CTRL-S))

Pick: **File, Exit** (to close Excel)

The table is complete, but the new data will not be available in Mechanical Desktop until you update the link.

Pick: **Update Link** (from Table Driven Setup dialog box)

Pick: **OK** (to exit the Create Table dialog box)

Pick: **OK** (to exit the Design Variables dialog box)

You should now be able to see the new part configurations listed in your model tree, as shown in Figure 10.8.

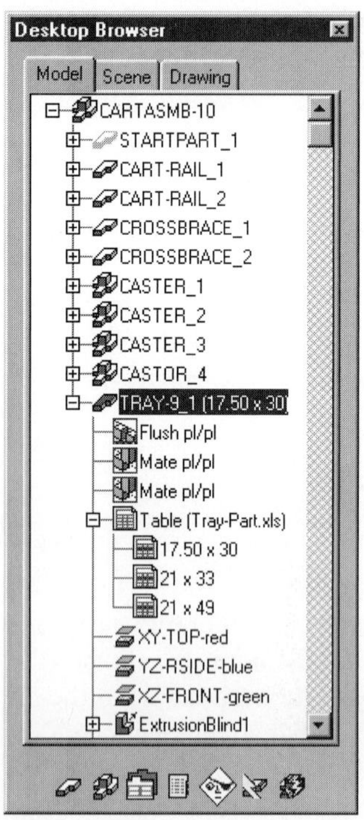

Figure 10.8

Changing the Sketch to Use the Design Variables

Next you will edit the sketch for the Tray part so that it uses the variable names you created to dimension the features. You will expand Extrusion-Blind_1 so you can select the profile.

Pick: **to expand the model tree below ExtrusionBlind1 for the Tray**

Right-pick: **Profile1** (below ExtrusionBlind1)

Pick: **Edit Sketch**

Pick: **Left Back Isometric view**

Pick: **Zoom Window** (to enlarge the view of the dimensions.)

The sketch dimensions for the main extrusion for the tray appear on your screen, as shown in Figure 10.9.

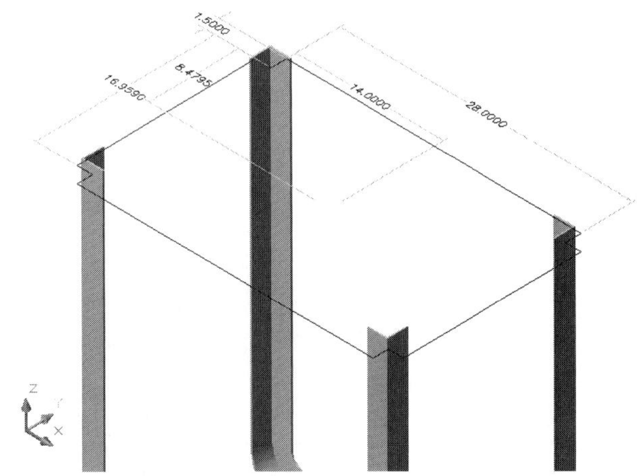

Figure 10.9

You will change the sketch dimensions to the design variable names that you created. First display the dimensions as parameters so that you can see their names.

Right-click: in the graphics area

Pick: **Dimensioning, Dimensions as Parameters**

The display changes to show the sketch dimensions as parameter names, as you see in Figure 10.10.

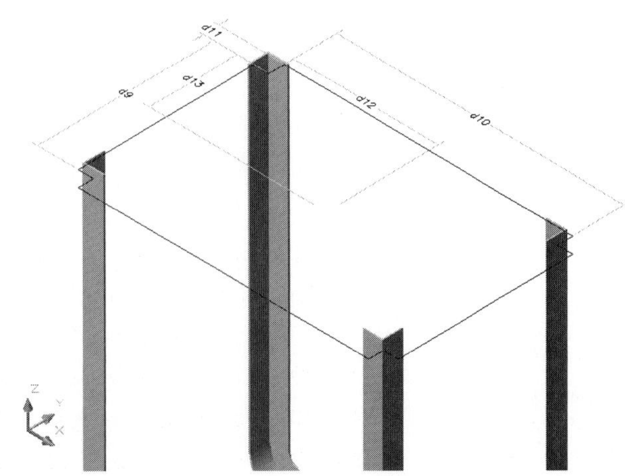

Figure 10.10

Next you will use equations to assign the design variable names you created to certain dimensions.

Right-click: in the graphics area

Pick: **Dimensioning, Edit Dimension**

Select dimension to change: **pick the d9 dimension**

Enter dimension value <16.959>: **d9 = TRAYWIDTH** ↵

Solved fully constrained sketch.

Select dimension to change: **pick the d10 dimension**

Enter dimension value <28>: **d10 = TRAYLENGTH** ↵

Select dimension to change: **pick the d11 dimension**

Enter dimension value <1.5>: **d11 = TRAYLIP** ↵

Solved fully constrained sketch.

Select dimension to change: ↵

Now display the dimensions as equations so that you can verify how this sketch is dimensioned.

Right-click: in the graphics area

Pick: **Dimensioning, Dimensions as Equations**

Figure 10.11 shows the dimensions displayed as equations. Notice that the dimension for the distance from the midplane has already been set to the overall size for the feature divided by 2.

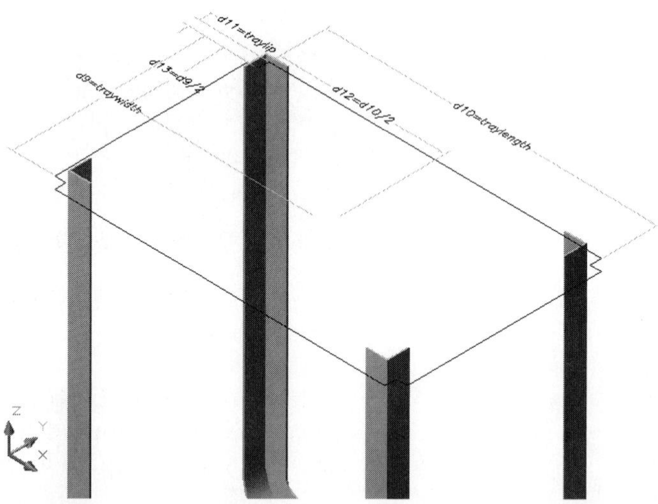

Figure 10.11

Next you will update the part to show the new size.

Right-click: in the graphics area

Pick: **Update Part**

The tray changes size to reflect the 17.50 × 30 inch dimensions for the default tray size.

Displaying a Table-Driven Part Configuration

To change to the display of a different configuration of the part,

Double-click: **21 × 33** (in the model tree)

The tray size updates to 21 × 33 inches plus a slight additional allowance for the bend.

Double-click: **21 × 49** (in the model tree)

The tray updates again to this configuration. Notice that this time the lip also increases to 1.5 inches, as shown in Figure 10.12.

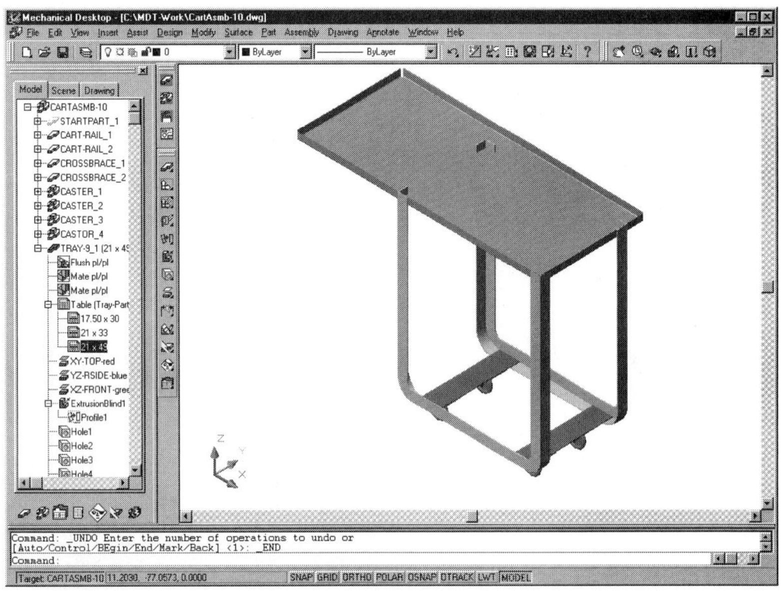

Figure 10.12

The rest of the assembly is still the same size. Next you will make global variables for the width and depth of the cart so that the rest of the cart can update as the tray configuration changes.

Defining Global Design Variables

Global variables are created in the same way as active part variables, but they can be used across all parts in an assembly, as well as in assembly constraints. You will create global design variables for the cart assembly in this section and then see how they can be used to define active part variables as well.

Pick: **Part, Design Variables** (from the menu bar)

Pick: **Global tab**

The Design Variables dialog box appears on your screen with the Global tab uppermost, as shown in Figure 10.13. Next you will create a set of global design variables for the cart.

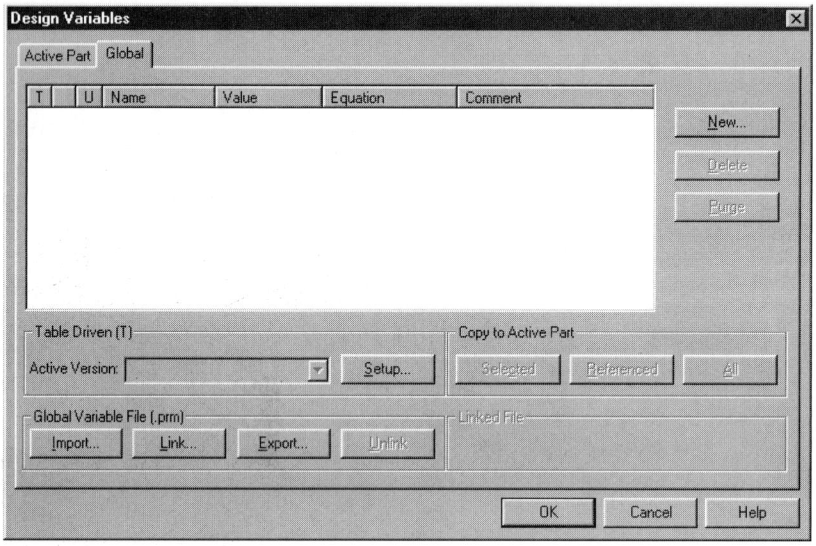

Figure 10.13

Pick: **New**

The New Part Variable dialog box appears. You will use it to make the settings shown in Figure 10.14. The values you will use for these variables correspond to the size of the tray (without the bend material) and the height of the cart without the casters.

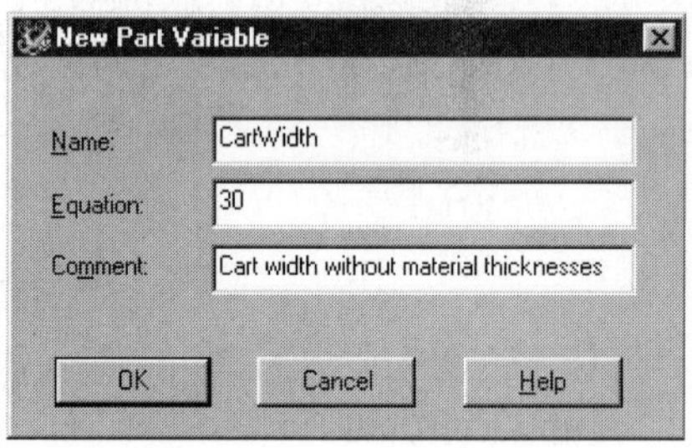

Figure 10.14

Set Name: **CartWidth**

Set Equation: **30**

Set Comment: **Cart width without material thicknesses**

Pick: **OK**

The variable is added to the Design Table dialog box. On your own, add two other new variables, as shown in Figure 10.15.

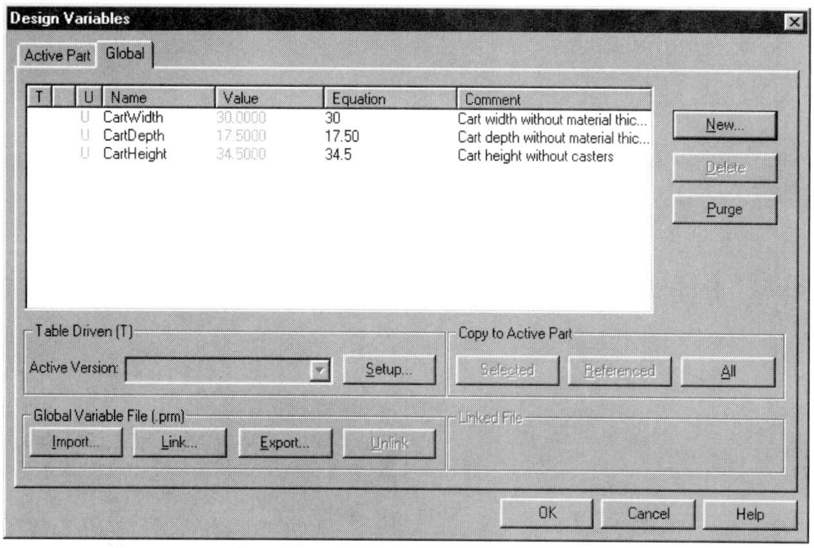

Figure 10.15

Set Name: **CartDepth**

Set Equation: **17.50**

Set Comment: **Cart depth without material thickness**

Pick: **OK**

Set Name: **CartHeight**

Set Equation: **34.5**

Set Comment: **Cart height without casters**

Pick: **OK**

Pick: **OK** (to exit the Design Variables dialog box)

Using Global Variables in a Sketch

Next you will change the sketch for the cart rail so that it uses some of the global variables you just created. Then you will edit some of the assembly offsets and the sketches for other features.

Right-click: **CartRail-1** (in the model tree)

Pick: **Activate Part**

Pick: **to expand the features below CartRail-1**

Right-click: **2Dpath1** (below the SweptRail feature)

Pick: **Edit Sketch**

Pick: **Left Front Isometric view**

The sketch for the sweep path for the rail appears on your screen, as shown in Figure 10.16. Use Zoom Window as needed so the dimensions are visible.

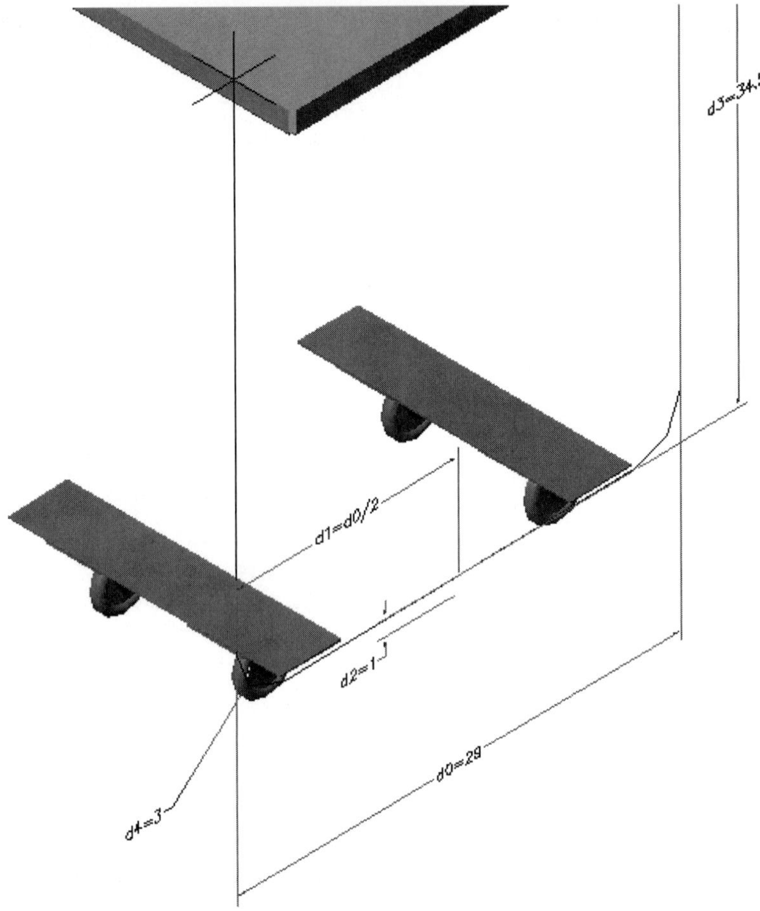

Figure 10.16

On your own, you will use the commands you have learned to edit the dimensions so that they are as follows. Note that the overall width of the CartRail allows .825 inches for the material thicknesses of the rails (.125 × 2), the tray lips (.0625 × 2), and the bend allowance ((.1 + .125) × 2).

Edit the dimension so d0 = CARTWIDTH +.825

Edit the dimension so d3 = CARTHEIGHT

Right-click: in the graphics area

Pick: **Update Part**

The CartRail parts both update to show the new sizes. Since they are both instances of the same part, updating one or the other will update both parts.

Right-click: in the graphics area

Pick: **Assembly Menu**

Pick: **Update Assembly**

The cart is updated, as shown in Figure 10.17.

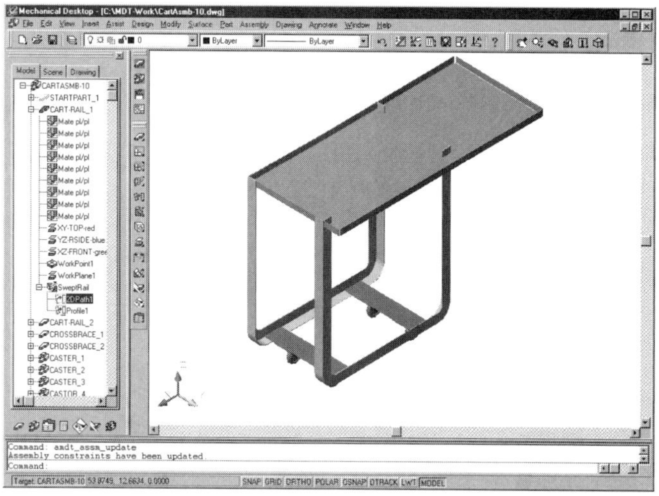

Figure 10.17

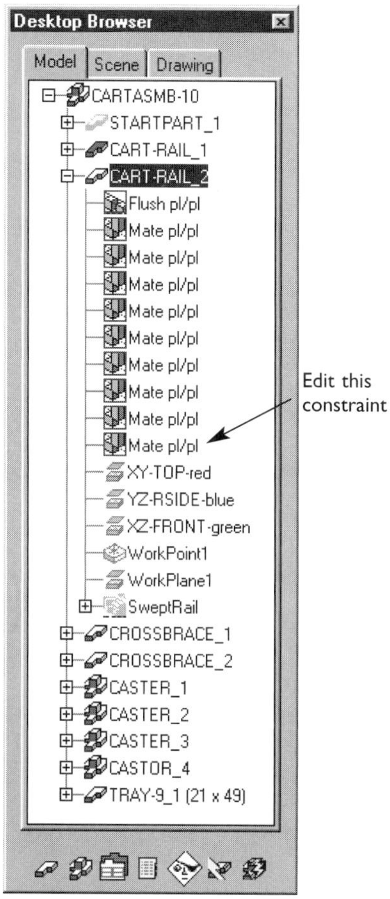

Figure 10.18

Pick: **to expand the features below CartRail_2, as shown in Figure 10.18**

Using Global Variables in Assembly Constraints

The bottom-most Mate constraint in the list for the CartRail is the one that established the distance between the rails when the work planes were mated. The rails were offset 17.9814 inches. You will change the offset for this Mate constraint so that it uses the global design variable for the depth of the cart.

Right-click: **the last listed Mate/pl/pl constraint below CartRail_2**

Pick: **Edit**

The Edit 3D Constraint dialog box appears with the offset value set to 17.9814. (If you see a different offset value, try selecting a different mate constraint until you find the correct one.) You will change the offset, as shown in Figure 10.19.

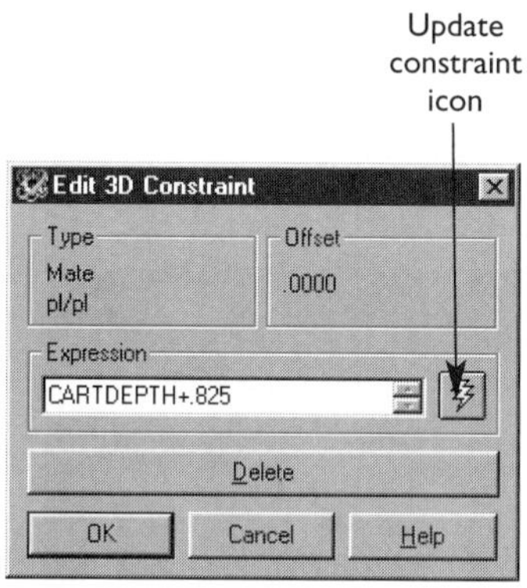

Figure 10.19

Set Expression: **CARTDEPTH +.825**

Pick: **Update Constraint icon** (which is in the dialog box and looks like a lightning bolt)

The offset for CartRail_2 updates to the width of the cart plus an allowance for the material thicknesses.

Pick: **OK**

Next you will change the CrossBrace part so that it uses the CARTDEPTH global variable plus an allowance for the material thickness to get its size.

Right-click: **CrossBrace_1** (in the model tree)

Pick: **Activate Part**

Pick: **to expand the features below CrossBrace_1**

Right-pick: **Profile1** (under ExtrusionBlind1)

Pick: **Edit Sketch**

The Sketch for the CrossBrace appears on your screen. Use the commands you have learned to edit its dimension so that it shows as follows. When you have finished, the sketch should look like Figure 10.20.

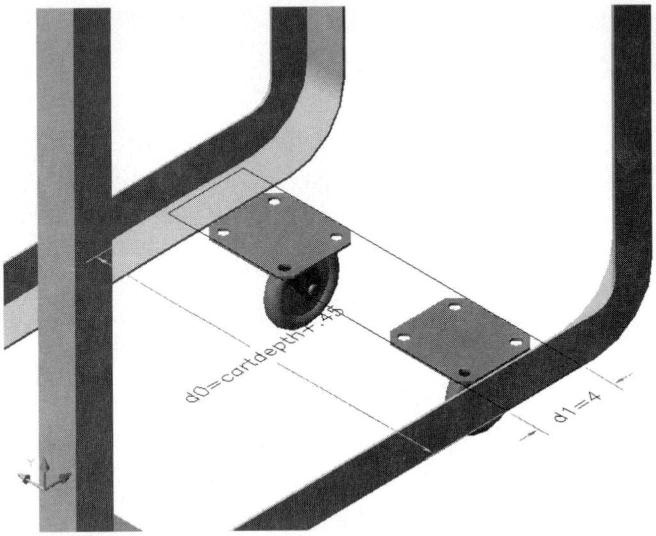

Figure 10.20

Edit dimension so **d0 = CARTDEPTH +.45**

Pick: **Update Part**

The CrossBrace parts update to the new size, but the assembly constraint for the CrossBrace also needs to be changed so it updates with the new size of the CartRails. The Mate constraint that is third from the bottom controls the offset of the inside edge of the brace from the work plane centered on the part. You will edit this constraint to use the new global variable for CartWidth.

Right-click: **the Mate constraint third from the bottom under CrossBrace_1**

Pick: **Edit**

Set Expression: **(CartWidth/2) −7**

Pick: **Update Constraint icon**

The CrossBrace shifts so its inside edge is 7 inches from the end of the cart (or, more precisely, from the end of the tray without the lip).

Repeat these steps to change the same constraint for Cross-Brace_2 on your own

Now the assembly appears correctly mated, as shown in Figure 10.21, and the sizes are controlled by design variables and equations.

Figure 10.21

Creating a Table-Driven Assembly

Next you will use the global variables to make the assembly table driven. The process is very similar to making a table for a table-driven part.

Pick: **Part, Design Variables** (from the menu bar)

The Design Variables dialog box appears on your screen. You will select the global tab, so that it appears uppermost.

Pick: **Global tab**

Next you will use the dialog box to make a spreadsheet to control the assembly.

Pick: **Setup**

The dialog box for Table Driven Setup appears on your screen.

Select Version Names: **Across**

Pick: **Create**

The Save File dialog box appears so that you can name the file and select the directory.

Choose Save in: **C:\MDT-Work**

Type File name: **Cart-Configurations.xls**

Pick: **Save**

The spreadsheet now shows in Excel on your screen. You will use the spreadsheet to change the Generic heading to 17.5 × 30 and to create two new columns, one for a 21 × 33 cart that is 34.5 inches tall and the other for a 21 × 49 cart that is 36 inches tall.

On your own, edit the spreadsheet so that it shows the information in Figure 10.22

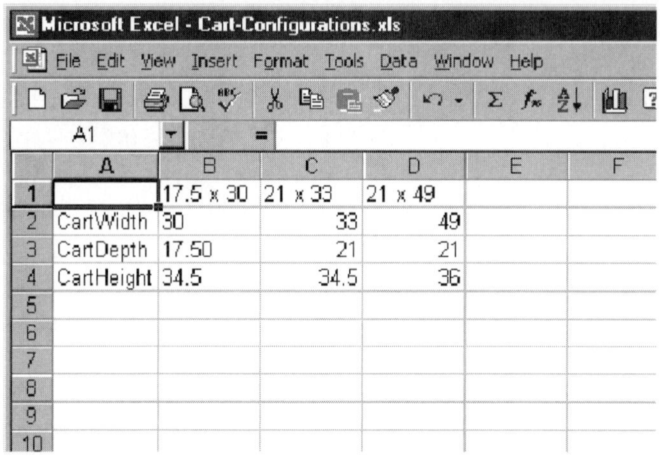

Microsoft Excel - Cart-Configurations.xls

	A	B	C	D	E	F
1		17.5 x 30	21 x 33	21 x 49		
2	CartWidth	30	33	49		
3	CartDepth	17.50	21	21		
4	CartHeight	34.5	34.5	36		
5						
6						
7						
8						
9						
10						

Figure 10.22

Make a heading for each column that labels its depth × width

Pick: **File, Save** (in Excel)

Pick: **File, Exit** (in Excel)

Pick: **Update Link** (to link the new values in the spreadsheet to the assembly file)

Pick: **OK**

Pick: **OK**

Now you should see the global table added to the menu tree. You can double-click on the table configurations to update the assembly.

Double-click: **the 21 × 33 configuration**

The assembly updates, as shown in Figure 10.23. But notice that the tray does not update. It used active part variables for its configuration. Active part variables cannot be converted to global variables when they have been used to control a table-driven part. (They *can* be converted to global variables if they have not been used to set up a table.) But you can edit the table of values for the part variables to include the global variables. This is what you will do in order to make an assembly that will update completely.

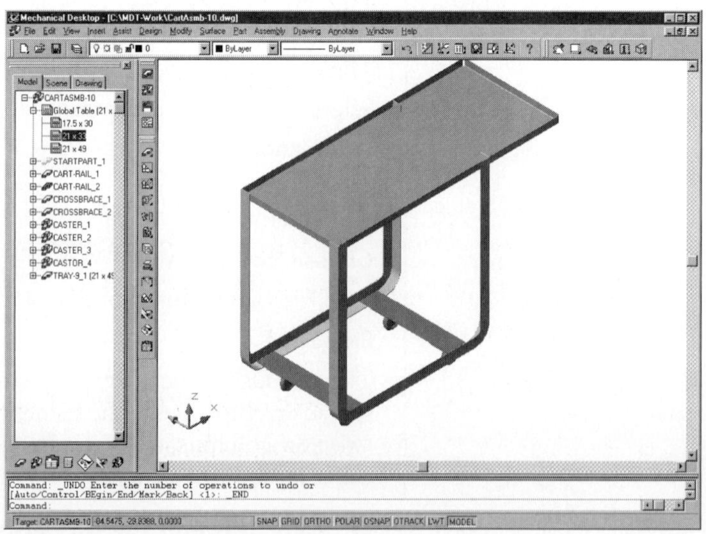

Figure 10.23

Editing a Table of Variables

You can use the context menu for the Tables in the model tree to edit the table of values in the spreadsheet. After you make changes to a table, you can use the same menu to update the link. You can also unlink a table if you no longer want the part to be table-driven.

To edit the spreadsheet for the tray's table,

Pick: **to expand Tray until you can see the variable tables**

Right-click: **Table (Tray-Part.xls)**

Pick: **Edit**

The spreadsheet for the tray reappears on your screen. You will edit it so that the TRAYWIDTH rows contain CARTDEPTH as their value. The TRAYLENGTH rows will contain CARTWIDTH as their values. Figure 10.24 shows the global variables added into the spreadsheet as the values for TrayWidth and TrayLength.

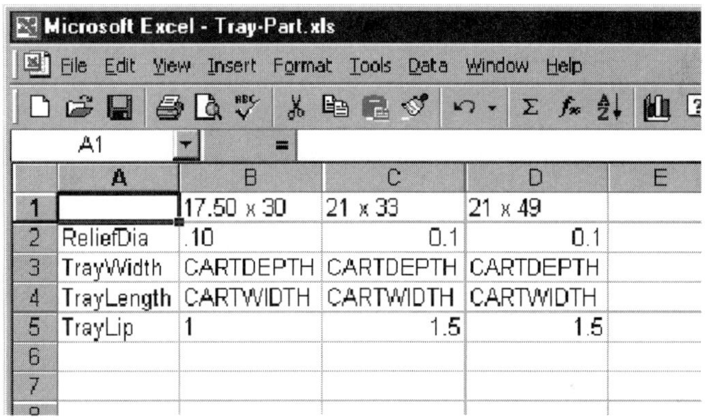

	A	B	C	D	E
1		17.50 x 30	21 x 33	21 x 49	
2	ReliefDia	.10	0.1	0.1	
3	TrayWidth	CARTDEPTH	CARTDEPTH	CARTDEPTH	
4	TrayLength	CARTWIDTH	CARTWIDTH	CARTWIDTH	
5	TrayLip	1	1.5	1.5	
6					
7					

Figure 10.24

Edit the TRAYWIDTH row so it contains CARTDEPTH for all values

Edit the TRAYLENGTH row so it contains CARTWIDTH for all values

When you have finished editing the spreadsheet,

Pick: **File, Save**

Pick: **File, Exit**

Right-click: **Table (Tray-Part.xls)**

Pick: **Update**

The spreadsheet is closed and the global variables are set as the values for the tray features. Next you will select the assembly configurations based on the global table sizes.

Double-click: **21 × 49** (under Global Table)

The assembly updates, as shown in Figure 10.25.

Tip

The Tray Lip is still controlled solely by the part table configurations in this example assembly. You could create a CartLip variable in the global table and use it to define the TrayLip part variable so that all aspects of the tray will update when the global configuration changes.

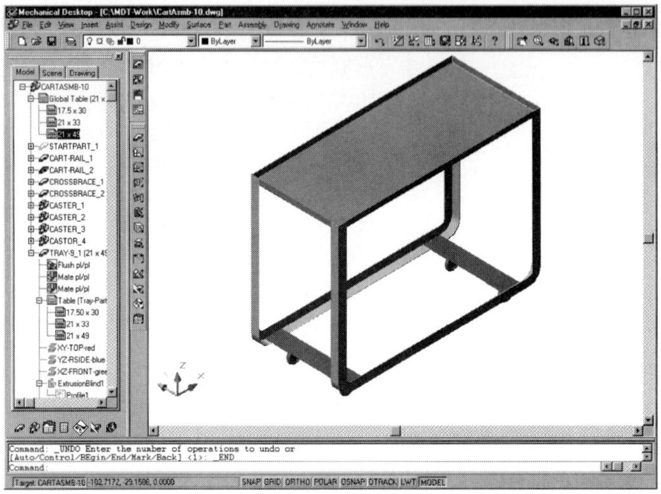

Figure 10.25

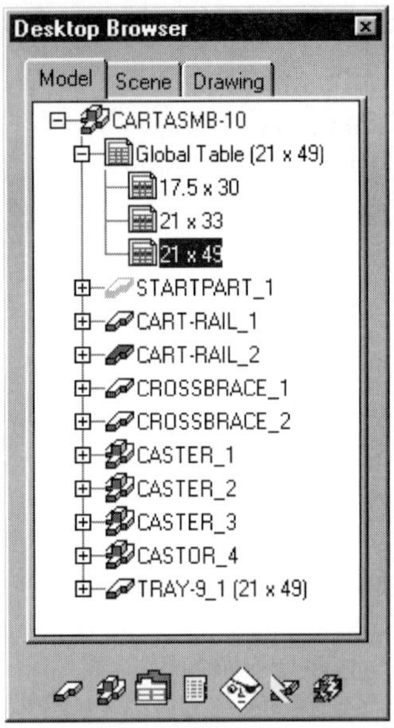

Figure 10.26

Double-click on each of the other table sizes. You should now have three configurations of the assembly, as shown in Figure 10.26. Now you understand how to use design variables and table-driven parts and assemblies to create models that reflect your design intent and update automatically!

Mass Properties

You can calculate the mass properties of single parts, the complete assembly, or any portion of it. You can assign a material to each part that will then be used in the mass property calculations. (This is one reason why you always want to model each individually manufactured piece as a separate part and not join them together as features of a large part. You should assemble separate models to build your device or system just as you would if you were really constructing it.) You will use the cart example to see how mass properties works, but you will exclude the casters to simplify the exercise.

To calculate the mass properties of the cart,

Pick: **Assembly, Analysis, Mass Properties** (from the menu bar)

Enter an option (parts/subassemblies) [Name/Select] <Select>: ↵

Select part and subassembly instances: **pick the two cart rails, the tray, and the two crossbrace parts**

Select part and subassembly instances: ↵

The Mass Properties dialog box appears on your screen.

Pick: **Material**

The Select Material dialog box appears, as shown in Figure 10.27. You will use this to select a material for each of the parts in the list on the right side of the dialog box. Several materials are listed in the drop-down menu. As each material is selected, you can see its properties displayed in the box. (To add more materials, you can edit the *mcad.mat* file stored in the Mechanical Desktop *desktop\support* folder by opening it in NotePad or another text editor.)

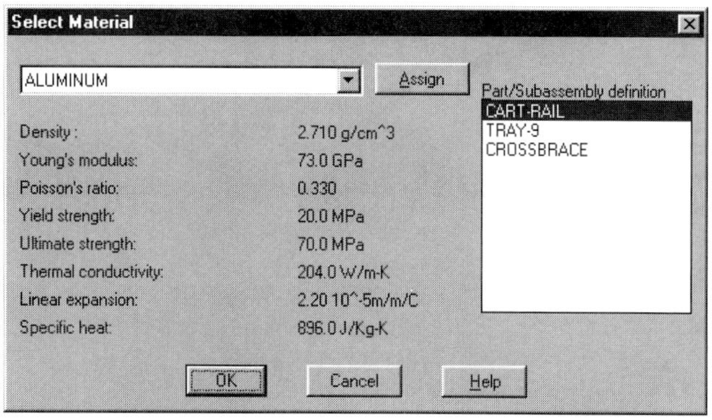

Figure 10.27

You will set materials for the parts as follows:

Pick: **Stainless Steel, Austenitic**

Pick: **CartRail** (so that it appears highlighted)

Pick: **Assign**

Pick: **Aluminum**

Pick: **Tray** (so that it appears highlighted)

Pick: **Assign**

Pick: **Mild_Steel**

Pick: **CrossBrace** (so that it appears highlighted)

Pick: **Assign**

Pick: **OK**

Reference

Engineering Design Communication

Determining Mass Properties, pp. 376–383; Mass Properties industry case study, pp. 414–417

You return to the Mass Properties dialog box. Now that you have materials assigned to the different parts, you will choose the units for the mass properties, as shown in Figure 10.28.

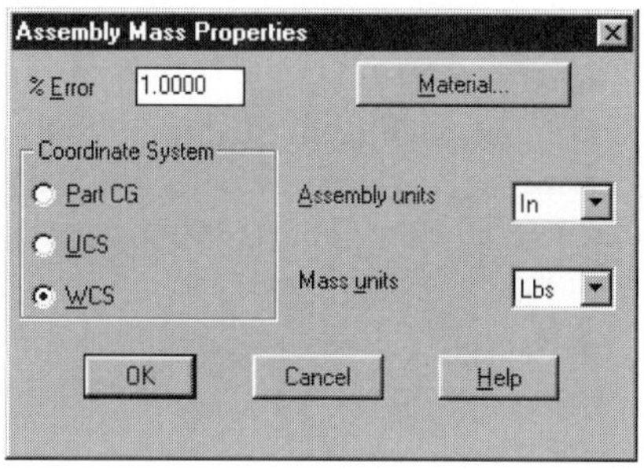

Figure 10.28

Set Assembly Units: **In**

Set Assembly Mass: **Lbs**

Select Coordinate System: **WCS**

Notice the Percent Error setting near the top of the Mass Properties dialog box. The mass properties are calculated using an iterative method. The smaller the percent error you list, the more accurate the results that are reported but the longer the amount of time to perform the calculations. For now you will leave this set at 1% error.

It is also extremely important to set your units correctly to reflect the sizes used in your models in order to get an accurate mass property result.

Finally, the moments of inertia and other values are reported relative to the coordinate system that you have selected. These can be calculated against the part's center of gravity (Part CG), the current user coordinate system (UCS), or the world coordinate system (WCS). Make sure that you understand to which coordinate system the results you are reporting relate.

Pick: **OK**

The mass properties are displayed on your screen, as shown in Figure 10.29. Notice the mass of the cart displayed in pounds. This is without the casters, as they were not selected. If you wanted to save this data, you could pick File to export it to an external file that could be used in another analysis program.

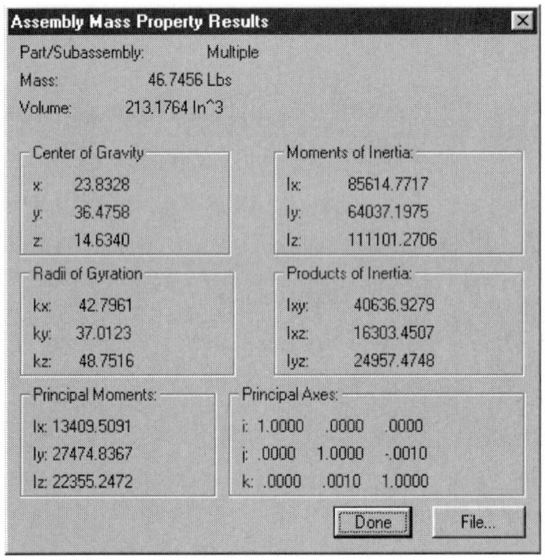

Figure 10.29

Conclusion

Now that you have completed the tutorials in this book, you have learned to model parts and assemblies that reflect your design intent. You have also learned to create legible drawings that include dimensions, tolerances, and even shaded 3D views. In addition, you have practiced making intelligent assemblies that can automatically update for families of similar designs. You have also seen how to make engineering calculations from your models.

Parametric modeling with Mechanical Desktop is a powerful engineering tool. You can explore many avenues of it on your own by applying the basic skills that you have learned and using the Help commands and Web support that are available.

Command Summary

Action	Command Name	Description
Mass Properties	AMASSMPROP	Calculates mass properties information for selected parts and subassemblies in an assembly
Edit Dimension	AMMODDIM	Edits values in existing parametric dimensions in a sketch or in a drawing view
Design Variables	AMVARS	Creates and controls global and active part design variables

Exercises

Create multiple configurations for the assemblies shown in Figures 10.30 through 10.32.

1. Small Part Storage Bin

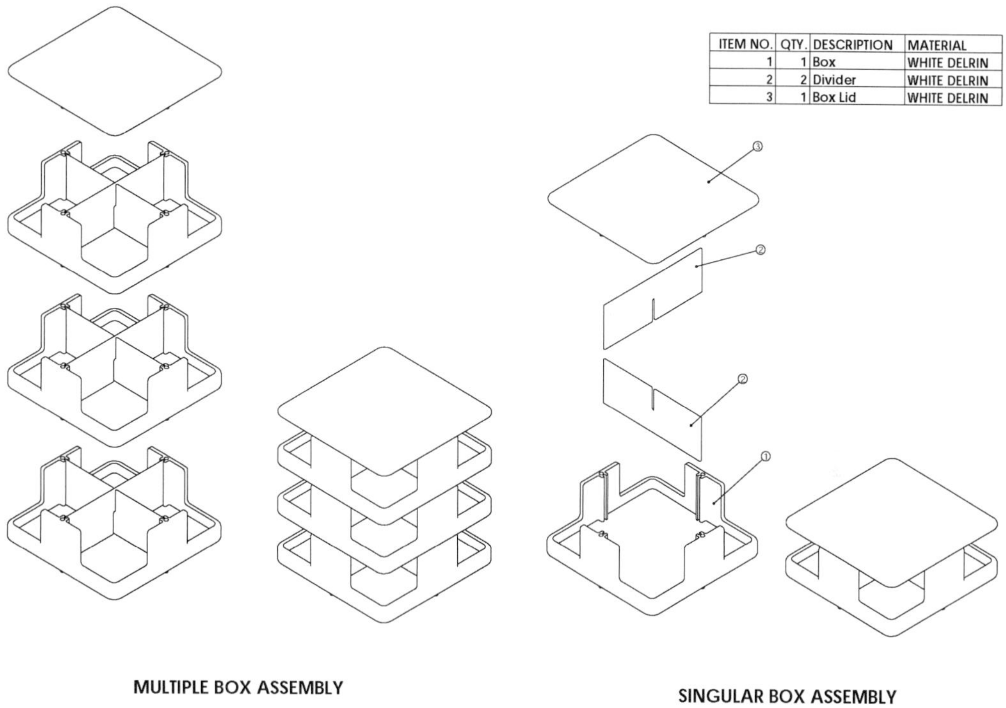

ITEM NO.	QTY.	DESCRIPTION	MATERIAL
1	1	Box	WHITE DELRIN
2	2	Divider	WHITE DELRIN
3	1	Box Lid	WHITE DELRIN

Figure 10.30a

Figure 10.30aa

Figure 10.30ab

NOTES:
1) ALL DIMENSIONS IN INCHES

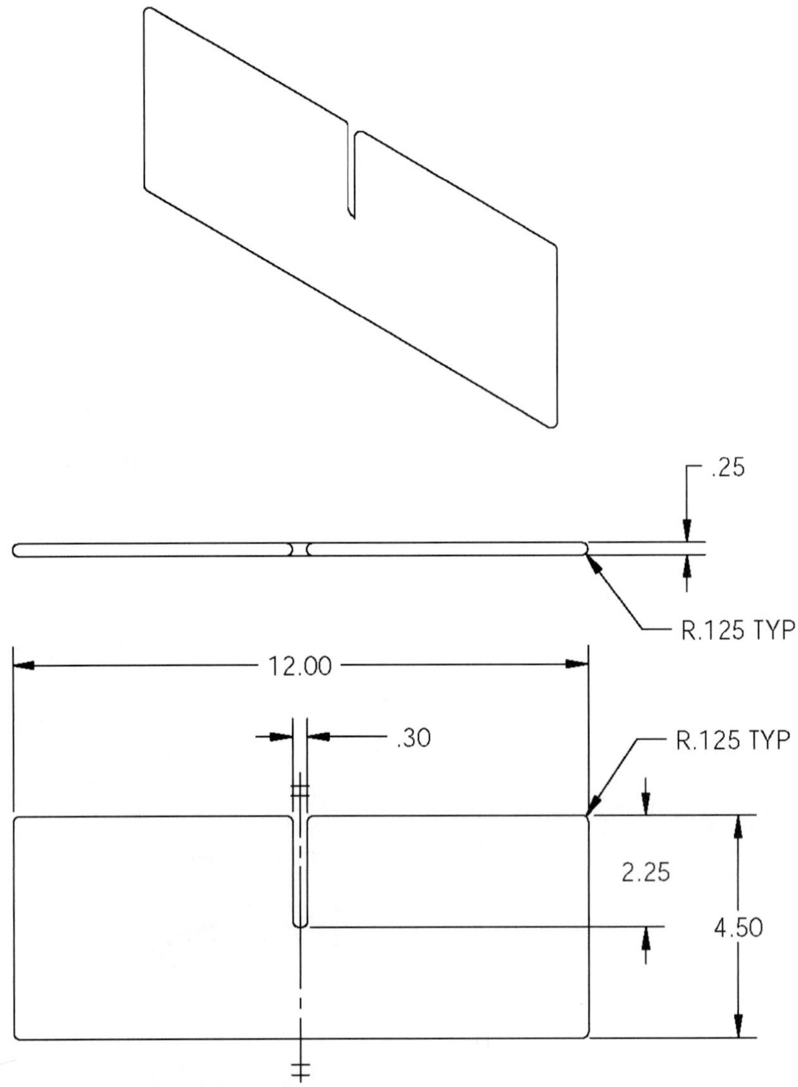

BOX DIVIDER

Figure 10.30b

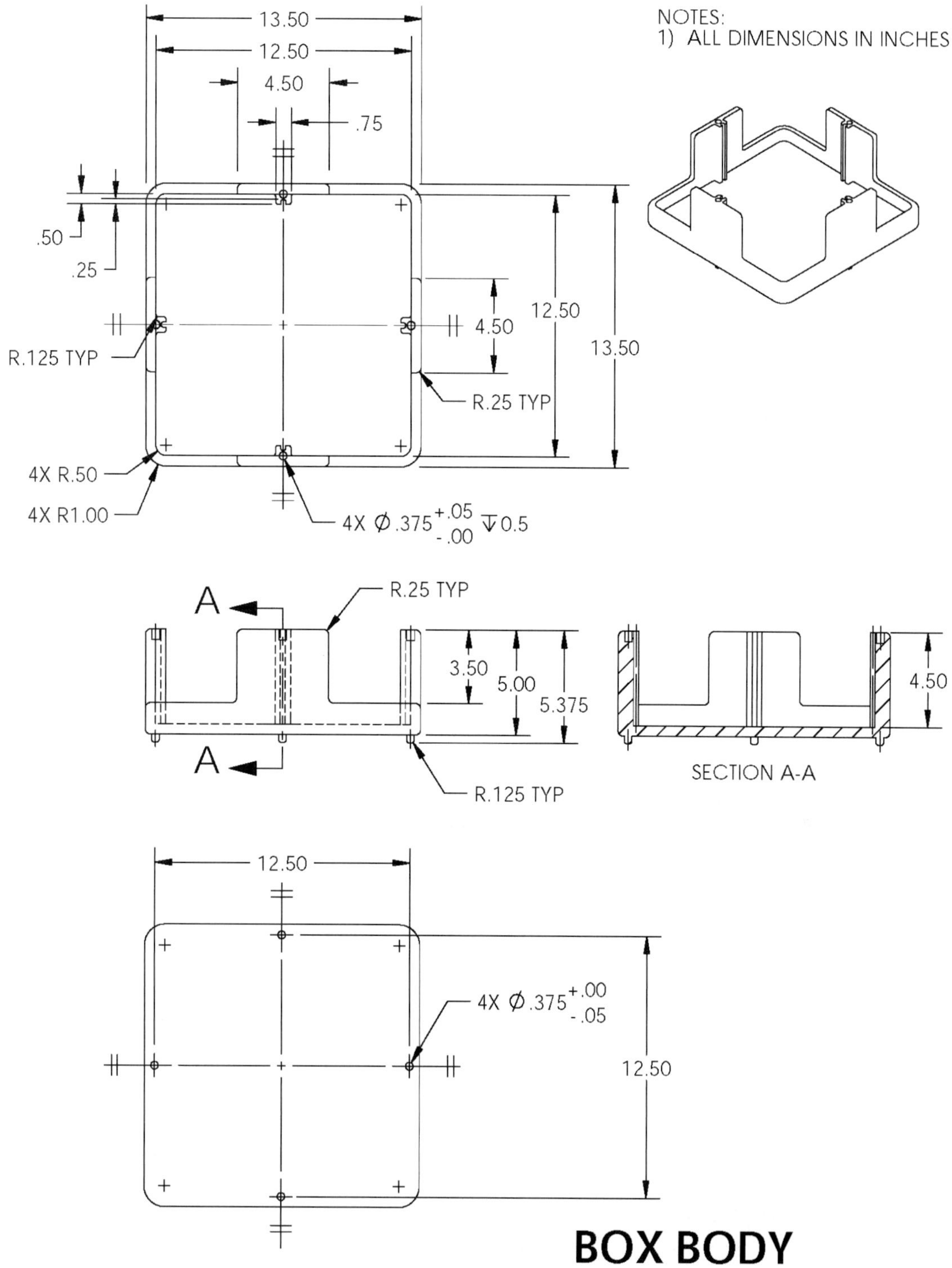

Figure 10.30c

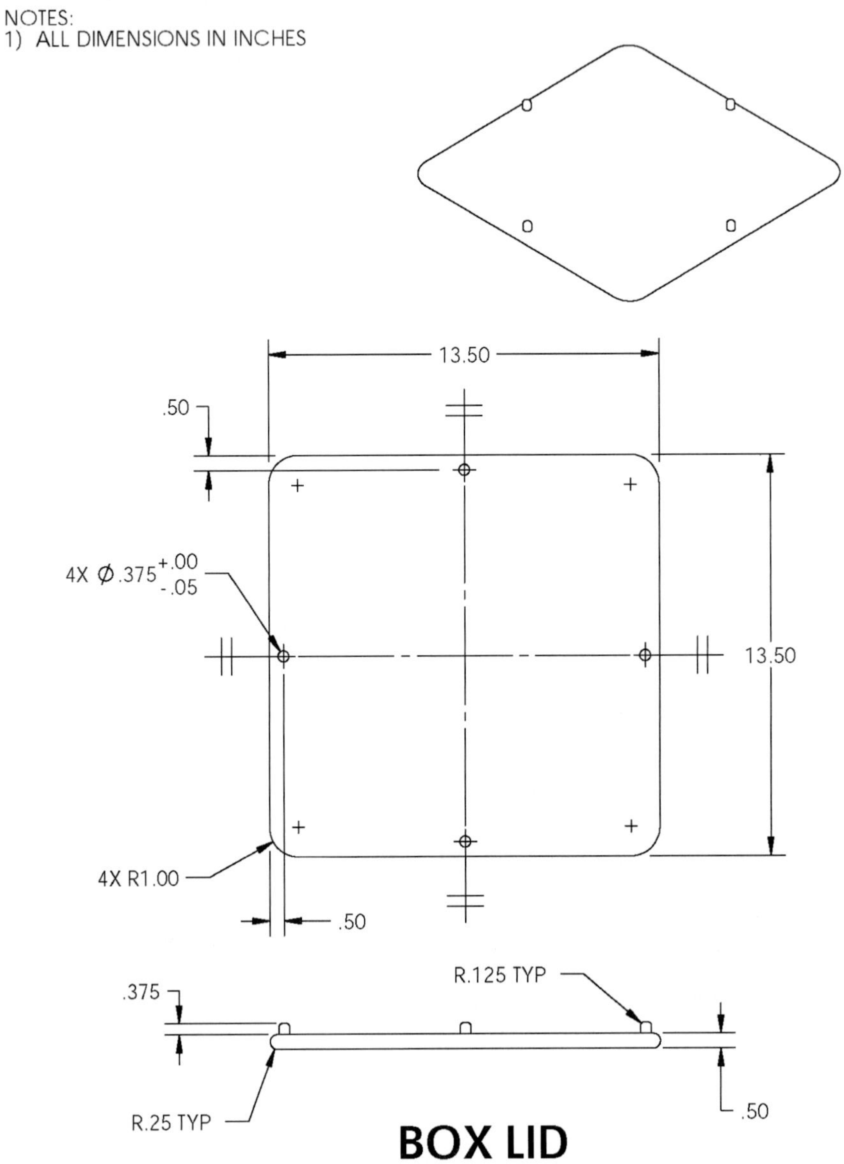

Figure 10.30d

2. **Cabinet** Model parts for the cabinet shown in Figure 10.31, or modify the parts in *MDT-Data\Exercises\CabinetParts* to create multiple configurations for the cabinet assembly. Make one configuration with 12″ deep shelves that are 24″ wide and another configuration with 18″ shelves that are 36″ wide. Make configurations of the overall cabinet that are 4′, 5′ and 6′ tall.

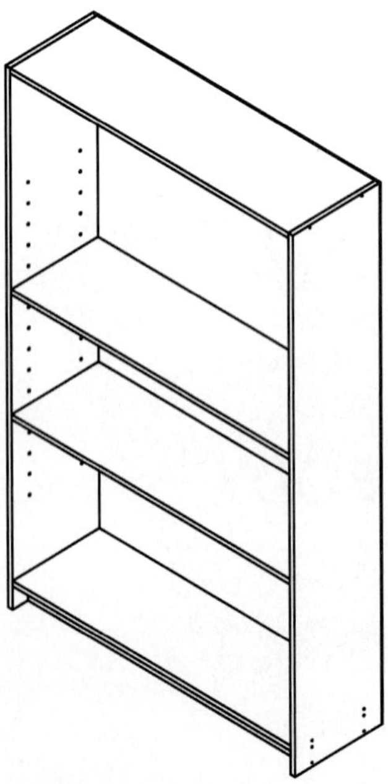

Figure 10.31

3. **Calculator** Use the AMSHELL command to create the case for the calculator; then create the different configurations shown in Figures 10.32a through 10.32d. Alternatively, experiment with your own calculator design, using your own sizes and shapes for the button configurations.

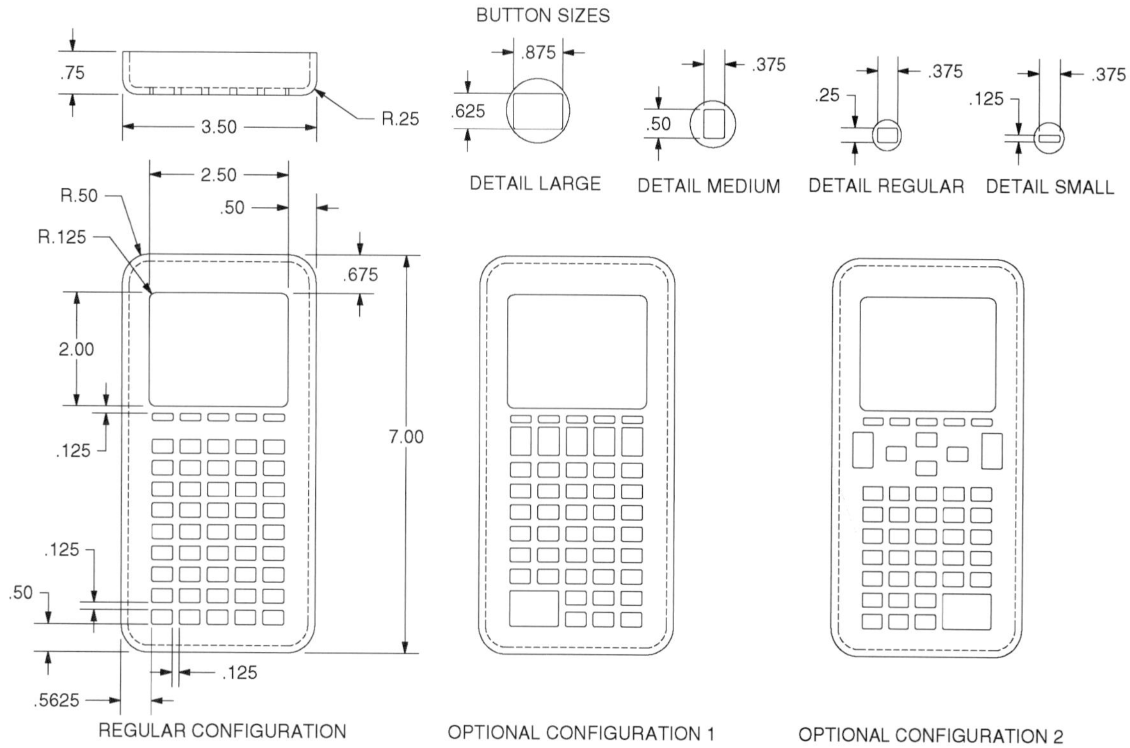

NOTES:
1) ALL DIMENSIONS IN INCHES
2) ALL BUTTON RADII = .03125
3) CALCULATOR BODY THICKNESS = .125

Figure 10.32a

Figure 10.32b

Figure 10.32c

Figure 10.32d

Appendix

A Command Reference

Command Name	Action	Description	Tutorial
—	Hot Grips	A selection method associated with editing commands for quickly modifying existing geometry; must be enabled in the Options dialog box	4
—	Append to Sketch	Adds 2D geometry to an existing profile	6
3DCLIP	3D Adjust Clip Plane	Enables and sets the location of dynamic clipping planes that create section views through a part or assembly	1
3DCORBIT	Continuous 3D orbit	Enables the user to set objects in the 3D view into continuous orbit	1
3DDISTANCE	3D Adjust Distance	Switches to 3D interactive view so that cursor movement simulates the effect of moving the camera closer or farther away	1
3DORBIT	3D Orbiter	Controls the interactive viewing of objects in 3D	1
3DPAN	Pan Realtime	Switches to the interactive 3D view so you can click and drag the view of the object in the graphics window	1
3DSWIVEL	3D Swivel	Switches to 3D interactive view so that cursor movement simulates the effect of turning the camera	1
3DZOOM	Zoom Realtime	Switches to the interactive 3D view so you can click and drag to change the magnification of the view	1
AM2DPATH	2D Path	Creates a 2D path for a sweep feature from an open or closed profile	9
AMACTIVATE	Activate Part	Activates a local or external part in an assembly so it may be edited.	1, 7
AMADDCON	Add Constraints	Adds constraints to a 2D sketch	2, 4
AMADDCON, Fix	Set Fix Point	Option of this command; sets the fixed point on a sketch	4
AMANGLE	Angle Constraint	Creates an angular assembly constraint between parts, subassemblies, or assemblies	7
AMANNOTE	Annotation	Adds, attaches, deletes, and moves annotations (text, leaders, and GD&T symbols) associated with geometry in drawing views	8

Command Name	Action	Description	Tutorial
AMASSMPROP	Mass Properties	Calculates mass properties information for selected parts and subassemblies in an assembly	10
AMBALLOON	Balloon	Creates and organizes balloons, or ball tags, in a drawing	8
AMBASICPLANES	Add Set of Workplanes	Creates three mutually perpendicular work planes and a work point at their intersection	9
AMBEND	Bend	Bends a portion of a part using an open profile to define the location of the bend	9
AMBOM	BOM Database	Creates and edits a BOM database	8
AMBROWSER	Toggle Desktop Browser	Controls the display of the Desktop Browser	1
AMCATALOG	Assembly Catalog	Attaches and removes, converts, and organizes local and external parts in an assembly	7, 9
AMCHAMFER	3D Chamfer	Creates a beveled edge (placed feature) on 3D objects	6
AMCONSTRLINES	Construction Line	Adds a construction line to the sketch that can be used for dimensioning but does not add to the 3D geometry	4
AMDELCON	Delete Constraint	Removes constraints from a 2D sketch	4
AMDELETE	Delete part	Deletes a part, part instance, or scene	7
AMDELFEAT	Delete Feature	Deletes a feature and all of its dependent features	3
AMDELTWEAKS	Delete Tweaks	Removes tweaks in a scene so part moves back to originalpositions	7
AMDELVIEW	Delete View	Deletes a view in Drawing mode	8
AMDWGDIMDSP	Change Dimension Display	Changes the display of dimensions to parameters, numbers, or equations	9
AMDWGVIEW	Add Drawing View	Creates and adds 2D views to a drawing layout	1, 3, 8
AMEDITCONST	Edit Assembly Constraints	Changes offsets in or deletes assembly constraints	7
AMEDITFEAT	Edit Feature	Edits a feature by changing the operation and settings used to create it	1, 2, 5
AMEDITVIEW	Edit View	Edits scale, visibility, and other aspects of drawing views	3
AMEXTRUDE	Extrude Feature	Creates an extruded solid feature from a 2D profile	2, 4
AMEXTRUDE, C	Extrude, Cut	Extrudes a feature and removes it from the model (equivalent to Boolean subtraction)	4
AMEXTRUDE, I	Extrude, Intersection	Extrudes a feature and retains the portions of the feature and model that intersect (equivalent to Boolean intersection)	4

Command Name	Action	Description	Tutorial
AMEXTRUDE, J	Extrude, Join	Extrudes a feature and adds it to the model (equivalent to Boolean union)	2
AMFILLET	3D Fillet	Creates a fillet or round (placed feature) between faces of a 3D part	2, 5
AMFLUSH	Flush Constraint	Creates a parallel assembly constraint between two planes	9
AMHELP	Help	Opens online help: Mechanical Help Topics	1
AMHOLE	Hole	Creates parametric hole features; opens the Hole dialog box	2, 4
AMINSERT	Insert Constraint	Constrains two circular edges or faces in an assembly so that they share the same axis and the faces are coplanar	7
AMINTERFERE	Interference	Analyzes selected parts and reports any interference between them	7
AMLAYER	Layer Control	Opens the Layer Control dialog box for defining and managing layer definitions	3, 8
AMMANIPULATOR	3D Manipulator	Dynamically rotates, copies, or moves parts freely or along 3D axes	7, 9
AMMATE	Mate Constraint	Creates a mate constraint between points, planes, axes, or nonplanar faces	7
AMMODDIM	Edit Dimension	Edits values in existing parametric dimensions in a sketch or in a drawing view	10
AMMODE	Change active mode l between Drawing and Mode	Controls whether Model Mode (Part, Assembly, Scene) or Drawing Mode (Layout) is active	1, 3
AMMOVEDIM	Move Dimension	Moves dimensions between views in a drawing without losing their association with the geometry	3, 8
AMMOVEVIEW	Move View	Moves a drawing view within a layout or to a new layout	3, 8
AMNEW	New Part, New Part Instance	Creates a new part or part instance	2, 9
AMNOTE	Leader Note	Creates a leader or other note symbol and attaches it to a drawing object	8
AMOPTIONS	Adjust Constraint Size	Manages part, assembly, surface, scene, drawing, annotation, and desktop preferences; can set size of constraint display	2, 4
AMOPTIONS	Mechanical Options	Manages settings, standards, and preferences to be used in parts, assemblies, and drawings	6
AMPARDIM	New Dimension	Adds a parametric dimension to a sketch	2
AMPARDIM	New Dimension	Creates parametric dimensions to constrain sketches and define features in Model mode	2, 4

Command Name	Action	Description	Tutorial
AMPARTEDGE	Copy Edge	Copies an existing edge to create a line or arc that can be used in a new sketch	9
AMPARTLIST	Parts List	Creates and adds a parts list to a drawing and control its properties	8
AMPARTPROP	Mass Properties of a Part	Lists the mass properties of the currently active part	7
AMPATTERN	Pattern Feature	Creates a rectangular, polar, or axial pattern of features from an existing feature	5
AMPOWEREDIT	Power Edit Dimensions	Opens the appropriate dialog box for editing a power object; when a dimension is selected, it opens the Power Dimensioning dialog box	3, 8
AMPROFILE	Make a Profile	Solves 2D geometry to create a constrained sketch.	2, 4
AMPROFILE	Single Profile	Solves a single 2D entity to create a constrained sketch	2, 4
AMPROJECT2PLN	Project to Plane	Projects 2D objects or edges of planar faces to a work plane, sketch plane, or 3D face so they can be used in new sketch	2, 4, 9
AMREFDIM	Reference Dimension	Creates nonparametric reference dimensions in drawing mode	8
AMRENAME	Rename	Renames parts, instances, scenes, and drawing views in the Browser Window	2
AMRENAME	Rename (in Browser)	Renames an element in the Browser window	2, 5
AMREORDFEAT	Reorder Feature	Changes the order of a feature in the model tree hierarchy	5
AMREPLAY	Replay Part	Replays the steps used to create the features of a part	5
AMREVOLVE	Revolve	Creates a solid feature by revolving a profile about an axis	6
AMROTCENTER	New Rotation Center	Calculates a center of rotation at the 3D center of all visible objects	1
AMRSOLVESK	Re-solve Sketch	Displays the number of constraints needed to fully constrain a sketch, or appends new geometry to a sketch	2, 4
AMSELROT	Set Rotation Center	Lets the user set a point about which the model will rotate	1
AMSHAFT3D	Generate Shaft	Creates rotationally symmetrical shaft parts and inner and outer shaft contours	7
AMSHELL	Shell	Creates an open or closed shell feature with uniform or multiple wall thicknesses	9
AMSHOWCON	Show Constraints	Displays 2D constraint symbols on the selected sketch	2, 4

Command Name	Action	Description	Tutorial
AMSKPLN	New Sketch Plane	Sets the sketch plane location and the orientation of its coordinate axes	2
AMSTYLEI	Import Style	Allows you to import dimension styles from other drawings	8
AMSUPPRESSFEAT	Suppress Features	Suppresses a feature or specifies a group of features by type to suppress	5
AMSWEEP	Sweep	Creates a solid feature by sweeping a profile along a 2D or 3D path	9
AMSYMLEADER	Add Leader	Adds or removes a leader from a drawing sysmbol	8
AMTITLE	Title Block	Inserts a drawing border and title block; opens and uses the settings in the Drawing borders with Title Block dialog box	8
AMTWEAK	Tweak	Positions a part or subassembly in an exploded scene.	7
AMUPDATE	Update	Recalculates the model to show the effect of changes made to features, constraints, or dimensions on parts, assemblies, scenes or drawing views	1, 3, 5
AMVARS	Design Variables	Creates and controls global and active part design variables	10
AMVIEW	Change the 3D View	Controls the view in Model mode	1
AMVIEW, Sketch	Sketch View	Controls the view in Model mode; the Sketch option sets the view to a plan view of the current sketch	2
AMVISIBLE	Visibility	Controls the visibility of parts, features, symbols, and other entities in parts, assemblies, scenes, and drawings	4, 8
AMWORKAXIS	Work Axis	Creates a work axis from a sketched location or from an existing circular edge	6
AMWORKPLN	Create Work Plane	Creates a work plane at a specified location	4, 5
AMXFACTOR	Set Explode Factor	Controls the explosion factor for a scene	1, 7
ARC	Draw an Arc	Creates 2D arc geometry	6
ATTEDIT	Edit Attributes	Edits block atttributes, such as text in a title block	8
CIRCLE	Draw Circle	Creates 2D circle geometry	5
DIM, UPDATE	Update Dimensions	Updates dimension objects to reflect the current dimension style and units settings; this is a special dimensions command available only at the DIM prompt	3, 8
DIMSTYLE	Dimension Style	Creates and modifies dimension styles; displays the Dimension Style Manager dialog box	3, 8

Command Name	Action	Description	Tutorial
ERASE	Erase	Removes objects from the sketch or drawing	4
EXTEND	Extend	Extends the length of existing 2D lines, arcs, and polylines to a selected boundary edge	4
EXTEND	Extend	Extends existing 2D geometry to a specified boundary edge	6
FILLET	Fillet	Rounds or fillets the edges of 2D objects; can be used to trim lines when radius is set to 0	4
GRID	Grid	Controls a grid of dots on screen that can be used as a drawing aid	4
IMAGEATTACH	Insert Raster Image	Adds a raster-type image (such as a TIF or BMP) to a drawing	8
ISOLINES	Set View of Curved Surfaces	System variable that controls the number of tesselation lines used to represent a curved surface in wireframe display	6
LINE	Line	Draws line segments	4
LIST	List	Displays information such as length, layer, and location about any object	2
LTSCALE	Scale linetype	Sets the linetype scale factor so line patterns can be adjusted for different size drawings	3
MIRROR	Mirror	Creates a mirror-image copy of 2D objects	4
MOVE	Move	Moves selected geometry to a new location on the coordinate system	7
MVSETUP	Add Title Block	Sets up the specifications of a drawing, including the insertion of a title block	3
OFFSET	Offset	Creates a new 2D object (parallel line, curve, or conventric circle) at a specified distance from an existing object or through a specified point	4
OPEN	Open a File	Opens an existing *.dwg* file	1
ORTHO	Ortho	Constrains cursor movement to horizontal and vertical on the current UCS (or with the Grid rotation angle)	2
OSNAP	Object Snap	Sets object snap modes	4
PLINE	Polyline	Creates a 2D polyline (of lines and arcs) that may also have a width and linetype pattern	2
PLOT	Plot	Plots to a plotting device or file; opens the Plot dialog box	3, 8
POLYGON	Create Polygon	Creates a regular (equilateral) polyline object; object can be inscribed or circumscribed about the circle defined by the radius	6

Command Name	Action	Description	Tutorial
PROPERTIES	Change Color	Changes the color of a 3D object	2
RECTANGLE	Rectangle	Creates a rectangular polyline 2D object.	2, 4
REDO	Redo	Reverses the last Undo	4
REGEN	Regenerate View	Regenerates the display of the model to reflect current settings and the model data	6
RENDER	Render	Creates a photorealistic image of the 3D model	1
SAVEAS	Save As	Saves the current file under a new file name and makes the new file name current	1
SCALE	Scale	Enlarges or reduces selected drawing objects	3, 8
SHADEMODE	Toggle Shading/Wireframe	Controls whether an object appears shaded or wireframe in the current view	1
SNAP	Snap	Controls the increment to which the cursor will jump when sketching commands are active	4
STYLE	Text Style	Creates and modifies named styles and sets the current style; opens the Text Style dialog box	3
TEXT (or DTEXT)	Add Text	Adds single line of text; text is displayed on screen as it is typed and formatting options are presented at the command line	3
TRIM	Trim	Trims objects at a cutting edge defined by other objects	4
UCSICON	Display UCS Icon	Controls the display and location of the icon that indicates the current coordinate system	9
UCSMAN	Named UCS	Displays and modifies defined user coordinate systems, and activates named and orthographic UCSs	2
UNDO	Undo	Reverses the effect of the last command (or set of commands)	4, 5
VIEWRES	Set screen resolution	Sets the resolution at which objects are displayed on screen; can be used to change the number of line segments used to display arcs and circles	3
ZOOM	Zoom	Controls the magnification at which objects are viewed on screen	1
ZOOM, <value>X	Set a Zoom Factor	Zooms the view to a specified number of magnifications	6, 9
ZOOM, A	Zoom, All	Zooms to the maximum magnification possible so that all of the objects are visible	2
ZOOM, E	Zoom, Extents	Zooms to the maximum magnification possible so that the extents of the drawing (visible or not) fit on screen	6

Appendix

B Glossary

angle constraint — Constrains two planes, two vectors, or a combination of a plane and a vector on different parts by controlling the angle between them.

balloon number — A circled number identifying each part shown in an assembly drawing; also called a *ball tag*.

base feature — The first feature created in a parametric model and the one from which other features are defined.

base grip — The grip used as the base point for hot grip commands; the base grip is specified by picking it to fill it with color.

circumscribed — Drawn around the outside of a base circle.

clipping plane — An invisible plane used to slice through an object (or set of objects) to define the portion to be displayed. The portion of the object on either side of the clipping plane can be displayed, or two clipping planes can be used to display a slice of an object.

close box — A box in the upper right corner of a window that, when clicked, closes the window.

command line area — The lines of text at the bottom of the screen that indicate the status of commands and prompt for user input.

constraint — A condition that determines the geometric and size properties of a sketch, or the relationship between parts in an assembly.

construction line — A line used to constrain and dimension a sketch, but which does not become part of the 3D model.

context menus — Menus that pop-up when you right-click with the mouse.

context sensitive — Describes software that presents options and menus relevant to the current command or the current location of the cursor.

crossing window (implied Crossing) — A means of selecting objects by forming a selection rectangle from right to left; all objects fully enclosed in or crossed by the rectangle are selected.

cursor — A mark that shows the location of the pointing device in the graphics window of the screen; used to draw, select, or pick icons, menu items, or objects. The appearance of the cursor may change, depending on the command or option selected.

cutting edges — Drawing objects that define where another object should be cut off.

default — The value that the software will use unless you specify otherwise; appears in angle brackets < > after a prompt.

degrees of freedom (DOF) — The number of axes along which a part can move.

design variables — Named parameters that can be used to dimensions features or groups of features that are the same size; can be defined for a single part or for all parts in an assembly.

dimension style — A group of dimension features saved as a set.

dimension variables	Features of dimensions that can be altered by the user; you control the features by setting the variables using the Dimension Style dialog box.
dock	To attach a toolbar to the edge of a window so it does not overlap the window's contents.
draft angle	The taper on a molded part that makes it possible to easily remove the part from the mold.
external part	A part used in an assembly whose definition is stored in a separate file.
extrusion	Creating a three-dimensional object by defining a closed two-dimensional shape and a length.
feature	Any definable aspect of an object—a hole, a chamfer, etc. Also the basic unit of a parametric solid model.
file extension	The part of a file name that is composed of a period, followed by one to three characters, and that helps you to identify the file.
floating	Refers to a toolbar or window that is located on top of other windows anywhere on screen.
flush constraint	Constrains two items on different parts so that they are parallel and face the same direction; the items may be offset from one another by a specified distance.
flyout	A sub-toolbar that becomes visible when its representative icon (marked with a small triangle) on the main toolbar is chosen.
font	A set of characters with a particular shape.
freeze	The practice of making a layer invisible and excluding it from regeneration and plotting.
graphics area	The central part of the screen, which is the main workspace used to create and display parts, assemblies, scenes, and drawings.
grid	Regularly spaced dots covering an area of the graphics window to aid in drawing.
highlight	A change of color around a particular command or object, indicating that it has been selected and is ready to be executed or worked on in some way.
hot grips	A method for editing an already drawn object that lets you select the object(s) and activate certain editing commands directly from the mouse or pointing device.
image tile	An active picture which displays dialog box choices as graphic images rather than words.
inscribed	Drawn inside a base circle.
insert constraint	Constrains selected circular edges or faces on different parts so they are aligned on the same axis, and so that their faces lie in the same plane or are separated by an offset distance.
layer	A method of separating drawing objects so that they can be viewed individually or stacked like transparent acetates, allowing all layers to show. Used to set color and linetype properties for groups of objects.
layout	The workspace in Drawing mode, similar to a single piece of paper, on which drawing views are located and stored.
local part	A part used in an assembly whose definition is stored in the assembly file.

mate constraint — Constrains points, axes, planes, or non-planar faces on different parts so that the selected items are parallel and facing each other; the items may touch or be offset by a specified distance.

maximize button — A button in the top right of an on-screen window that enlarges the active window so that it fills the entire screen.

menu bar — The strip across the top of the screen showing the names of the pull-down menus, such as "File", "Edit", etc.

minimize button — A button in the top right of an on-screen window that reduces the active window to a title button at the bottom of the screen.

mirror line — A line that defines the angle and distance at which a reversed image of a selected object will be created.

model space — The mode used for creating objects in Mechanical Desktop's 3D database.

noun/verb selection — An option that allows you to select an object first, then select the command to be used on it.

object — An element of a model or drawing; lines, arcs, circles, text, and other elements that you create, select, and edit using Mechanical Desktop's commands and menu options.

offset distance — The distance from an existing object at which a new object will be created.

options — The choices associated with a particular command or instruction.

paper space icon — A triangular icon that appears in lieu of the UCS icon when paper space is active.

paper space — A mode that allows you to arrange, annotate, and plot different views of a model in a single drawing, as you would arrange views on a piece of paper.

parametric dimension — A dimension used to define model geometry; changing the dimension value will change the size of the related feature.

parametric — A modeling method that uses parameters to define the size and geometry of features and to create relationships between features and parts. Changing a size or relationship will result in the model being updated to the new appearance.

parent-child relationship — A description of the dependency created between features; the way in which one feature is related to another feature through the constraints and dimensions applied to it.

parts list — A listing of the parts in an assembly drawing, including information such as the item number, description, quantity, material, and part number.

placed feature — A feature, such as a hole or chamfer, created not from a sketch but by specifying values for predefined parameters.

polar coordinates — A means of specifying the location of a point by providing a distance and an angle from another specified point, using the input format @DISTANCE<ANGLE.

polygon — A closed shape with sides of equal length.

polyline — A connected series of lines and arcs that behave as a single object.

profile — A two-dimensional, constrained sketch used to create a three-dimensional solid object.

prototype drawing — A drawing saved with certain settings that can be used repeatedly as the basis for starting new drawings; also called template drawing.

reference dimension A dimension added to a drawing that was not used to create the model geometry; changing the dimension value will not change the size of the related feature.

relative coordinates The location of a point in terms of the *X* and *Y* distance, or distance and angle, from a previously specified point; @X, Y.

rendered A view in which lighting effects, color, and shape are applied to an object in a single two-dimensional view.

resolution Refers to how sharp and clear an image appears to be and how much detail can be seen; the higher the resolution, the better the quality of the image. This is determined by the number of colors and pixels the computer monitor can display.

revolution Creating a three-dimensional object by revolving a two-dimensional shape about an axis.

selection set All of the objects chosen to be affected by a particular command.

selection window (implied windowing) A means of selecting objects by forming a selection rectangle from left to right; only objects fully enclosed in the rectangle are selected.

sketch plane A temporary, infinite plane with *X* and *Y* axes that defines the plane on which sketched objects will be located. A sketch plane can correspond to a planar surface on a feature, or to a predefined work plane.

status bar The rectangular area at bottom of screen that displays time and some command and file operation information.

sweeping Extrusion along a curved 2D or 3D path.

table A spreadsheet that stores the dimensions for a part and that is linked to that part file.

template drawing A drawing with the extension .dwt having certain settings that can be used repeatedly as the basis for starting new drawings. Any drawing can be renamed and used as a template.

tessellation lines Lines displayed on a curved surface in a wireframe view to help you visualize it.

through point The point through which the offset object is to be drawn.

title bar The rectangular area at the top of the screen which displays the application and document name and path.

toggle A button that cycles through different states, such as on and off, as it is clicked.

tolerance The amount that a manufactured part can vary from the specifications laid out in the plans and still be acceptable.

tool tip The name of the icon that appears when you hold the cursor over the icon.

toolbar A strip containing icons for certain commands.

transparent command A command that can be executed while another command is operating.

UCS icon Indicates the current coordinate system in use and the direction in which the coordinates are being viewed in 3D drawings.

update To regenerate the active part, assembly, or drawing using any new dimension values or constraints.

User Coordinate System (UCS) A set of *x*, *y*, and *z* coordinates whose origin, rotation, and tilt are defined by the user. You can create and name any number of User Coordinate Systems.

window border The outside edge of a window; can be dragged to resize the window.

work axis a parametrically defined axis used to aid in the construction of other objects.

work plane an infinite plane used to aid in the construction of other objects; can be parametrically defined or fixed on the world Coordinate System.

work point a parametrically defined point used to aid in the construction of other objects.

World Coordinate System (WCS) AutoCAD's system for defining three-dimensional model geometry using *x*, *y*, and *z* coordinate values; the default orientation is a horizontal *X* axis with positive values to the right, a vertical *Y* axis with positive values above the *X* axis, and a *Z* axis that is perpendicular to the screen and has positive values in front of the screen.

INDEX

A

Action buttons
  fit option, 24
Active part
  action description, 57, 260
  creating design variables, 359–364
Add constraints
  action description, 91, 166
Add drawing view
  action description, 57, 301
Adding geometry, 200
Adding multiple views, 98–99
Adding new feature, 179–181
Adding stock parts, 248–250
Add leader
  action description, 301
Add multiple drawing views
  action description, 131
Add Set of Workplanes
  action description, 346–350
Add text
  action description, 131
Add title block
  action description, 131
Adjust constraint size
  action description, 91, 166
Adjusting views, 115
Align, 213
Alignment, 234–236
Alternate units tab
  New Dimensions Style, 112
AMBOM command, 273
Angle, 213
Angle constraint, 226–227
  action description, 260
Angular dimensions, 111
Annotation, 21
  action description, 301
  associating, 294
  with leaders, 291–292
  tab, 292
Append sketch
  action description, 216
Apply to Linetype area, 192
Arc
  command, 200
Arc
  drawing
    action description, 216
Arrayed features, 182
Arrow size, 107
Arrow styles, 107
Arrow tab
  New Dimension Style, 107
Assembly
  adding, 343–344
  advanced skills, 305–357
  catalog, 221–222
    action description, 260, 346–350
  constraints, 226–230
    global variables, 368–370
  creating, 219–262
  Cross Brace part, 331–334
  drawings, 252
    creating layout, 263–264
    exploded, 269
  global parameters
    create, 358–381
  parts
    localizing, 258–260
  preparing for, 187
  printing, 280
  starting new, 220–221
  tray, 340
  viewing, 250
  working drawing, 263–304
Attachment tab, 292–293
Attdia, 15
AutoCAD drawing commands, 63
Auxiliary view
  add, 300

B

Backing out
  commands, 19–20
Backup
  commands, 19–20
Balloon
  action description, 301
  numbers, 276–279
    position, 279
Balloon Properties
  for ANSI dialog box, 278

Ball tags, 276–279
Based modeling software
  constraint, 59
Base feature, 62, 171
  extruding, 149–150
  selecting, 62–63
Base grip, 144
Base point
  setting, 68
Basic
  displaying tolerances, 113
Bend
  action description, 346–350
  dialog box, 340
Bidirectional associativity, 120
Bill of materials (BOM), 273–274
  database
    action description, 301
    creating, 273–274
Bolt
  creating, 211
  inserting, 248
BOM, 273–274
  database
    action description, 301
    creating, 273–274
Boolean operation union, 80
Borders, 264, 268–269
Browser mode, 37–45
Browser window, 16
  setting properties, 72–73
Button
  dialog box, 23

C

Cart assembly
  creating, 322–323
Caster subassembly
  adding, 343–344
Center block
  sketching
    with rectangle, 161–163
Centermark, 107–108
  size, 107–108
Center option, 212
Chamfer
  adding, 215
  3D
    action description, 216
Change color
  action description, 91
Change dimension display
  action description, 346–350
Character tab, 295
Choose
  defined, 1
Circle
  drawing
    action description, 188
Circular edges, 244
Circumscribed polygon, 212
Click
  defined, 1
Clipping, 50
Clipping plane, 50
Close box, 16
Cmddia, 15
Collinear constraint
  adding, 78–79
Color, 14
  change
    action description, 91
  dialog box, 72
  dimension lines, 106
  dimension values, 314
  extensions lines, 107
  layers, 103
Command line options, 213
Command prompts, 2
Complex shapes, 155
Configuration
  accepting changes, 14
  changing, 10–15
Constraint
  action description
    adding, 91, 166
    adjust size, 91, 161
    angle, 260
    delete, 166
    edit assembly, 260
    flush, 346–350
    insert, 260
    show, 91, 161
  adding, 66–67
  angle, 226–227
  assembly, 226–230
    global variables, 368–370
  based modeling software, 59
  collinear
    adding, 78–79
  2D, 65–66
  3D, 331
  deleting, 66
  dimensions, 67–68
  editing, 244–246
  flush, 226
  geometric
    list and description, 66
  insert, 226, 232–239
  mate, 226, 227, 229, 234–236, 260
  part, 239–241
  pins, 244
  second leg projected geometry, 84

show, 65–66
size, 145, 192
sketch, 65
Construction lines, 153–154
action description, 166
Contents tab, 27
Context sensitive
defined, 28–29
menus, 1
transparent help, 28–29
Controlling view
curved surfaces, 198–199
Coordinates
working with, 136
Copy edge
action description, 346–350
Create Drawings tab, 60
Create Drawings View dialog box, 271
Create polygon
action description, 216
Create work plane
action description, 166
Crossing
multiple objects, 243
Crossing to select, 143–145
Current dimension style
setting, 114–115
Curved surfaces
controlling view, 198–199
Cut
action description, 166

D

3D adjust clip plane, 51
action description, 57
3D adjust distance, 49
action description, 57
Data files
checking for, 219
tutorial installation, 7
Datum planes
work planes, 306–308
Datum planes and surfaces, 62
3D chamfer
action description, 216
2D constraints, 65–66
3D constraints
add, 331
3D continuous orbit, 48–49
Default configuration
for Mechanical Desktop Power Pack, 10
Default extrusion direction
changing, 214–215
Degrees of freedom (DOF), 233–234
Delete constraint
action description, 166
Delete feature
action description, 131
Delete part
action description, 260
Delete tweaks
action description, 260
Delete view
action description, 301
Design variables
action description, 376
dialog box, 359
Desktop browser, 36–37
Developing drawing
views, 284
Deviation
displaying tolerances, 113
3D fillet
action description, 91, 188
Dialog box, 22
settings, 15
Dimension
action description
display, 346–350
edit, 376
move, 131, 301
new, 91, 166
power edit, 131, 301
reference, 301
style, 131, 301
update, 131, 301
adding, 68–69
reference, 288–290
angular, 111
zero suppression, 112
appearances, 106
color
lines, 106
values, 314
command options, 154
constraints, 67–68
edit, 21
editing
changing value, 120–121
equations, 320–322
extraneous
hide, 287–288
hot grips, 286–290
lines
color, 106
setting, 106
making legible, 312–316
modifying, 89–90
moving, 101
nonparametric, 289
options, 213
parameters, 320–322
parametric, 283, 288–290

Dimension *(continued)*
part, 280–282
drawing, 280–282
placement, 69
primary units, 110
scaling factor, 110
style, 105
creating names, 105–106
current setting, 114–115
technique, 283–288
Dimension Style dialog box, 106, 313
Display tab, 10
Display UCS icon
action description, 346–350
3D Manipulator, 230–232
action description, 260, 346–350
handles, 240
Docked, 16
desktop browser and toolbars, 36–37
Docked toolbar, 20
Documentation drawings, 96
Documents
working with, 30
DOF, 233–234
3D Orbiter, 46, 229, 331
action description, 57
Double-click
defined, 1
feature, 81
2D path
action description, 346–350
sketch, 309
Draft angle option, 70
Drafting tab, 13–14
Drag
defined, 1
Drawing
action description
adding view, 57, 301
add multiple views, 131
arc, 216
circle, 188
adding notes to, 290–291
assembly, 252
creating layout, 263–264
exploded, 269
AutoCAD commands, 63
developing views, 284
different views, 282–283
dimensioned part, 280–282
documentation, 96
exploded assembly, 269
improving appearance, 101–102
Mechanical Desktop, 18–26
open and save, 17–18
mode
create engineering drawing, 96–131
moving views, 101, 294
parts, 281
pictorial views, 116
printing, 117–118
scenes, 269–270
tab, 39–41, 60
working
assemblies, 263–304
models, 263–304
Drawing As dialog box
Save, 30
Drawing View dialog box, 271
Edit, 116
2D sketching commands
using, 63–65
3D swivel, 49
action description, 57
3D view
action description, 57

E

Edge option, 212
Edit assembly constraints
action description, 260
Edit attributes
action description, 301
Edit dimensions, 21
action description, 376
Edit Drawing View dialog box, 116
Edit feature
action description, 57, 188
Edit view
action description, 131
Endpoint
object snap mode, 140–141
Entering command options, 18–19
Entering Mechanical Desktop commands, 19
Entering points, 17–18
Equal distance, 215
Equations
in dimensions, 320–322
Equilateral polyline, 212
Erase, 25
action description, 166
command, 142
ESC button
keyboard, 19–20
Exiting
Mechanical Desktop, 30–31
Exploded assembly drawing, 269
Explode factor, 250–251
Exporting
local part, 343
Extend
action description, 166, 216
command, 138–139
and fillet, 159
Extensions lines, 107

Extents
action description, 216
External, 221
adding part, 223–226
Extraneous dimensions
hide, 287–288
Extrude
action description, 91, 166
base feature, 149–150
command, 70
cut
remove material, 156
profile, 59
Extrusion dialog box, 80, 214

F

Feature
action description
delete, 131
edit, 57, 188
pattern, 188
reorder, 188
suppress, 188
adding, 179–181
arrayed, 182
base, 62, 171
extruding, 149–150
selecting, 62–63
based modeling, 37–38
software, 59
creating
pattern, 181–183
profile, 77
second leg, 81
from sketch, 69
swept, 318–319
double-clicking, 81
editing, 173–176, 206
intersecting, 155
Mechanical Power Pack, 248
modifying, 89
placed, 86–89
recording, 178–179
renaming, 81, 171
shelled, 340–342
sketched, 69
suppress, 183–186
and unsuppress, 173
work, 150
Filedia, 15
File Management, 41–42
File Output Configuration dialog box, 55
Files
backing up, 42
card, 9
Fillet, 87–88
action description, 166
to trim and extend, 159–161
Fit tab, 24
New Dimension Style, 109–110
Fixed point
specifying, 67
Flat shaded, 53–54
Flip button, 214
Flip direction, 71
Floating toolbar, 20
Flush constraint, 226
action description, 346–350
Flyouts, 47
Font
changing, 104–105
Formulas
in dimensions, 320
for regular polygons, 212

G

General notes, 292–296
Generate shaft
action description, 260
Geometric constraints
list and description, 66
Geometry
existing sketch, 200
Global design variables
defining, 365–366
Global variables
in assembly constraints, 368–370
in sketch, 367–368
Gouraud shaded, 53–54
Graphics area, 16
Graphics formats, 55
Grid, 134–135
action description, 166
and snap toggles, 135–136

H

Helps, 26–29
action description, 57
Hidden lines, 283
Hidden option, 53
Holes, 86–87
action description, 91, 166
adding, 163–164, 346–349
Hot grips, 142
action description, 166
dimensions, 286–290
menu, 144–145

I

Implied crossing, 143
Implies windowing, 144
Import style
action description, 301
Index cards, 8
Index tab, 27

Inscribed polygon, 212
Insert constraint, 226, 232–239
  action description, 260
Inserting
  bolt, 248
Insert raster image
  action description, 301
Instance
  adding, 326
  creating
    local part, 334–336
Interference
  action description, 260
  checking for, 254, 256–257
Intersection
  action description, 166
  feature, 155
Invisible plane, 50
Isolines, 198–199
Isometric views
  create, 270–271
  exploded, 269
  scales, 99

J

Jack parts
  assembling, 221
Join
  action description, 91
  operation
    to extrude, 80

L

Launch Mechanical Desktop, 33–34
Layers, 103
  changing color, 103
  control
    action description, 131, 301
Layout, 98
Layout2
  creating, 122–127
Leader
  adding, 301
  with annotations, 291–292
  note
    action description, 301
Leader Settings dialog box, 292
Limits
  displaying tolerances, 113
Line
  action description, 166
  command, 136
  tab
    New Dimension Style, 107
Linear Dimension
  zero suppression, 111
Linetype
  interpreting, 103
  scale setting, 102
List
  action description, 91
  command, 79
Local part, 221
  exporting, 343
  modeling
    within assembly, 328–331
L-shaped profile, 309

M

Make a profile
  action description, 91, 166
Mass properties, 257–258, 373–376
  action description, 260, 376
Mate
  bolts, 344–346
  cart rail to start part, 324–326
  constraint, 226, 227, 229, 234–236
    action description, 260
  second rail, 326–328
Material
  removing, 156–158
Maximize button, 16
MCADPP, 10
Measurement scale, 111
Mechanical Desktop
  configuration
    for tutorials, 7–9
  drawing, 18–26
    open and save, 17–18
  launching, 33–34
  and Microsoft Windows, 15
  online help, 26
  shortcuts, 4–6
  workspace, 35–36
Mechanical Desktop Today window, 33
Mechanical options
  action description, 166, 216
Microsoft Windows
  and Mechanical Desktop, 15
Minimize button, 16, 17
Mirror
  action description, 166
  command, 141
  line, 141, 203
Model
  analyzing, 254
  environment
    setting up, 190–192
  feature-based, 37–38
  first decisions, 62

investigating, 171
local part, 328
within assembly, 328–331
modes
parametric, 2
parametric, 32
making changes, 42–44
parts, 59
within assembly, 328–331
redefining, 173–176
replaying, 172–173
software
constraint, 59
feature-based, 59
parametric, 59
space, 98
tab, 37–38, 42–43
tree, 171
working drawing, 263–304
Model Accuracy, 199
Modify Dimension Style dialog box, 313
Modifying parts, 170–189
Mouse techniques, 1
Move
action description, 260
dimension, 131, 301
view, 131, 301
reposition part, 241
Multiline Text Editor dialog box, 295
Multiple views
adding, 98–99

N

Named UCS
action description, 91
Named views, 49–50
Neutral axis, 339
New dimension
action description, 91, 166
New Dimension Style dialog box, 106
New part
action description, 91
file
starting, 60–62
instance, 346–350
New rotation center
action description, 57
New sketch plane
action description, 91
selecting, 73
New View selection, 39
Nonparametric dimensions, 289
Nonparametric work planes, 150
Notes
add to drawing, 290–291
general, 292–296
Noun/verb selection, 334
Numbers
balloon, 276–279

O

Objects
activating grips, 143
defined, 25
snaps, 140–141
action description, 166
Object Snap Endpoint, 200–203
Object Snap Overrides, 203–205
Offset
action description, 166
command, 137–138
Open a file
action description, 57
Open and save tab, 11–12
Open profiles, 339
Operators
equations, 322
Options dialog box, 8
Ortho
action description, 91
Orthographic base view, 270–273
Orthographic UCS tab, 74
Orthographic view
scales, 99
Over-and under-constrained sketches, 65

P

Page Set Up dialog box, 265
Pan real-time, 45
action description, 57
Paper size
setting up, 264–269
Paper space icon, 98
Paper space mode, 97–98
Parameters
control the geometry, 211
in dimensions, 320–322
Parametric assemblies, 221
Parametric dimensions, 283, 288–290
Parametric mode, 37–45
Parametric modeling, 32
making changes, 42–44
modes, 2
software, 59
Parametric work plane
creating, 150
Parent-child relationships, 170–171
Parts
activating and editing, 254–255
active
action description, 57, 260

Parts *(continued)*
creating design variables, 359–364
add, 248
external, 223–226
stock, 248–250
advanced skills, 305–357
assembly
localizing, 258–260
constraining, 239–241
creating, 59–95
instance, 334–336
list, 274–276
pin, 196–198
revolving, 190–218
deleting, 260
dimensioning, 280–282
drawing, 281
editing, 190–218
exporting, 343
list
action description, 301
adding, 273
creating, 274–276
rename, 276
local, 221, 328–331, 343
modeling, 59
within assembly, 328–331
organizing, 221
remodeling, 134–169
replay, 188
reposition, 241
revolving
creating, 190–218
standard, 248
starting new
from existing one, 205–211
from template, 305
table-driven, 358
configuration, 365
creating, 361–362
template, 305
tray
creating, 336–339
undo, 177–178
updating, 174–175
Path
sketching, 309–312
Pattern feature
action description, 188
creating, 181–183
Pen assignments, 119
Pick
defined, 1
Pictorial drawing views, 116
Pin
constraining, 244
create, 196–198
Placed features
adding, 86–89
Placement, 213
Plot
action description, 131, 301
style table, 119
Plot Device Tab, 117
Plotter configuration area
printer selected, 118
Plotter Configuration Editor, 124, 266
Plotting tab, 12
Point
defined, 1
Polygon
circumscribed, 212
command, 212–213
create
action description, 216
formulas for regular, 212
inscribed, 212
Polyline, 63–64
action description, 91
Power dimension editing
action description, 131, 301
changing value, 120–122
Power pack, 248–250
Preset views, 45
Preview, 340
Primary units
for dimensions, 110
tab, 23–24
New Dimensions Style, 110–111
Printing, 117–118
assemblies, 280
Profiles, 59
constraining, 65–69
creating, 65, 316–318
action description, 91, 166
feature, 77
second feature, 77
editing, 175–177
extruded, 59
L-shaped, 309
open, 339
projected geometry, 83–85
revolve, 194–195
revolved, 59
revolving, 194–195
sweeping, 59
tab, 9, 14–15
user, 9
Projected geometry
using profile, 83–85
Projecting sketch geometry, 82
Project objects to plane
action description, 91, 166
command, 82
Project to plane
action description, 346–350
Pull-down menus, 21

R

Radio buttons
  fit option, 24
Real-time Zoom, 44
Recognizing symmetry, 150
Recording
  features, 178–179
Rectangle
  action description, 91, 166
  sketching center block, 161–163
Redefining model, 173–176
Redo
  action description, 166
  command, 138
Reference dimension
  action description, 301
  adding, 288–290
Regenerate view
  action description, 216
Regular polyline, 212
Relative coordinates, 136–137
Remodeling
  parts, 134–169
Removing objects
  from selection set, 26
Rename (in browser)
  action description, 91, 188
Renaming features, 171
Render
  action description, 57
Rendered view
  printing, 54–56
Rendering options and algorithms, 54
Reorder feature
  action description, 188
Repeating commands, 25
Replaying model, 172–173
Replay part
  action description, 188
Reposition part
  move, 241
Re-solve sketch, 67
  action description, 91, 166
Revolution dialog box, 195
Revolve
  action description, 216
  parts
    creating, 190–218
  profile, 59, 194–195
Right-click
  defined, 1
Rotate, 230–232
  center
    setting, 47–48
  commands
    reposition part, 241
  view, 331
Route dialog box, 232

S

Save as, 41–42
  action description, 57
Save Drawing As dialog box, 30
Scale
  action description, 131, 301
  area, 99
    Primary Units tab, 111
  linetype, 102
    action description, 131
Scenes
  drawing views, 269–270
  tab, 38–39
Screen resolution, 3
Search tab, 27–28
Search window, 28
Second leg features
  creating, 81
Section view
  adding, 299–300
Select
  defined, 1
    set, 25
  objects, 25
  tab, 14
Select File dialog box, 34
Set a zoom factor
  action description, 216, 346–350
Set explode factor
  action description, 57, 260
Set fix point
  action description, 166
Set rotation center
  action description, 57
Set screen resolution
  action description, 131
Set view of curved surfaces
  action description, 216
Shaded image
  adding, 279–280
SHADEMODE, 151
Shading modes, 52
Sheet size
  setting up, 264–269
Shell
  action description, 346–350
  features, 340–342
Shell dialog box, 342
Shortcuts, 4–6
  Mechanical Desktop, 4–6
Show constraints
  action description, 91, 166
Side rail
  sweeping, 309
Simple part creation, 59–95
Single profile
  action description, 91, 166

Sketch
  assemblies, 252
  center block
    with rectangle, 161–163
  changing
    vs. design variables, 363–364
  constraints, 65
  editing, 200–203
  features, 69
  front view, 152
  path, 309–312
  plane, 73
  re-solving, 67
  Stopper Part, 192–193
  view
    action description, 91
Sketch Setting area, 192
Snap, 135
  action description, 166
Solid Model Accuracy, 199
Solid models
  first decisions, 62
Standard parts, 248
Standard sheet size, 265
Start End Radius, 200
Start part
  adding, 323–324
Stopper Part
  sketching, 192–193
Storing electronic files, 41
Stretch command, 144
Style, 107–108
Suppress features
  action description, 188
Suppressing features, 173
Surfaces, 306
Sweep
  action description, 346–350
  feature
    creating, 318–319
  profile, 59
  shapes, 309
  side rail, 309
Symbols
  keyboard, 2, 19–20
Symmetry
  displaying tolerances, 113
  recognizing, 81
System tab, 12–13

T

Table-driven assembly
  creating, 370–372
Table-driven part configuration
  displaying, 365
Table-driven parts, 358
  creating, 361–362
Table Driven Setup dialog box, 361
Table of variables
  editing, 372–373
Template
  creating file, 129–130
  making customized, 308–309
  parts, 305
Terminating options, 70
Tessellation lines, 198–199
Text
  adding
    title block, 128
  style
    action description, 131
    changing, 104–105
  tab
    New Dimension Style, 108–109
Text Editor dialog box, 295
Three-D adjust clip plane
  action description, 51, 57
Three-D adjust distance, 49
  action description, 57
Three-D chamfer
  action description, 216
Three-D constraints
  add, 331
Three-D continuous orbit, 48–49
Three-D fillet
  action description, 91, 188
Three-D Manipulator, 230–232
  action description, 260, 346–350
  handles, 240
Three-D Orbiter, 46, 229, 331
  action description, 57
Three-D swivel, 49
  action description, 57
Three-D view
  action description, 57
Title bar, 16
Title block, 268–269
  action description, 301
  adding, 264–269
  adding text, 128
Toggle desktop browser
  action description, 57
Toggle shading, 52
  action description, 57
Tolerance
  specifying, 296–299
  tab
    New Dimensions Style, 113–114
Toolbars, 36
  flyouts, 47
  icons, 20–21
Transformation, 44
Transparent command
  defined, 29

Tray
  assembling, 340
  parts
    creating, 336–339
Trim, 139–140
  action description, 166
  and fillet, 159
Tweak
  action description, 260
  command, 251–252
Two-D constraints, 65–66
Two-D path
  action description, 346–350
  sketch, 309
Two-D sketching commands
  using, 63–65
Typographical conventions, 2

U

UCS, 73–76
  icon, 75
    action description, 346–350
  named, 91
  orthographic tab, 74
Undo
  action description, 166
  commands, 138
    action description, 188
  parts, 177–178
Update
  action description, 57, 188
    dimensions, 131, 301
    layout, 131
User Coordinate System (UCS), 73–76
  icon, 75
    action description, 346–350
  named, 91
  orthographic tab, 74
User-defined set, 73
User profiles, 9
U-shaped path, 309

V

Verb selection, 334
View
  action description
    add drawing, 57, 301
    add multiple drawing, 131
    curved surfaces, 216
    3D, 57
    delete, 301
    edit, 131
    regenerate, 216
    relocating, 131, 301
    sketch, 91
  adding multiple, 98–99
  adding section, 299–300
  adjusting, 115
  assembly, 250
  auxiliary
    add, 300
  controlling, 44
    curved surfaces, 198–199
  drawing
    developing, 284
    different, 282–283
    moving, 101, 294
    scenes, 269–270
  editing, 116
  front
    sketching, 152
  isometric
    create, 270–271
    exploded, 269
    scales, 99
  multiple
    adding, 98–99
  named, 49–50
  orthographic
    base, 270–273
    scales, 99
  pictorial drawing, 116
  preset, 45
  preview, 340
  rendered
    printing, 54–56
  rotate, 331
Visibility
  action description, 166, 301

W

Window border, 16
Windows configuration
  Mechanical Desktop, 3
Wireframe, 52
  action description, 57
Work axis, 150
  action description, 216
  adding, 207–209
Work features, 150
Working directories
  create and prepare, 7–9
  setting, 6–7
Working drawing
  models, 263–304
Work plane, 150
  adding
    for assembly, 164–165
  datum planes, 306–308
Work Plane Feature dialog box, 150–151
Work points, 150

Z

Z-direction, 214
Zero suppression
for angular dimensions, 112
for Linear Dimension, 111
Zoom
action description, 57, 91, 216
real-time, 57
button, 47
command, 2, 44
window, 154